PROGRAMMABLE CONTROLLER HANDBOOK

ROBERT E. WILHELM, JR.

HAYDEN BOOK COMPANY
a division of Hayden Publishing Company, Inc.
Hasbrouck Heights, New Jersey

Acquisitions Editor: PRIJONO HARDJOWIROGO
Production Editor: ALBERTA BODDY
Text Design: JOHN M-RÖBLIN
Compositor: McFARLAND GRAPHICS
Printer: HAMILTON PRINTING COMPANY
Cover Design: PETER TYRAS
Cover Photo: HERBERT W. FRANKE/PETER ARNOLD INC.
Illustrated by: GREGG D. JORDON
             FRANK J. MASLEY
             GENE S. SKINNER

Library of Congress Cataloging in Publication Data

Wilhelm, Robert E.
    Programmable controller handbook.

    Bibliography
    Includes index.
    1. Programmable controllers.   I. Title.
TJ223.P76W55   1984   001.64   84-12844
ISBN 0-8104-6311-3

Due to the variety of uses for the information contained in this text, the user of, and those responsible for applying this information, must satisfy themselves as to the acceptability of each application and use of equipment. In no event will the author, publisher, and/or any programmable controller manufacturer be responsible or liable for indirect or consequential damages resulting from the use or application of this information.

The illustrations, charts, and examples detailed in this text are intended solely to illustrate the principles of programmable controllers and some of the methods used to apply them. Because of the many variables and requirements associated with any particular installation, the author, publisher, and/or any programmable controller manufacturer cannot assume responsibility or liability for actual use based on the illustrative uses and applications cited in this text.

No patent liability is assumed with respect to use of information, circuits, equipment, or software described in this text.

2	3	4	5	6	7	8	9	PRINTING
85	86	87	88	89	90	91	92	YEAR

*In dedication to
my parents*

PREFACE

My decision to write a text on programmable controllers was partly based on comments made by students in my PC classes at a local college. They often asked where they could obtain additional PC information that was more generic than the manufacturers' literature passed out in class. In addition, I found that many students wanted to become knowledgeable on the topic without subjecting themselves to a possibly overzealous salesperson or applications engineer. Through my teaching experiences, I was also finding out that only several of the manufacturers actually offered information that could be used to educate someone wishing to learn more about PCs. In fact, I found that the material presented in class as well as the slides and examples used were based on only a few of the more than one hundred PC systems available.

In November 1981, I decided to author a PC text provided that a word-processor program could be obtained for my home computer and that the figures for the text could be developed with a computer-aided design system. When I found that a commercial word processor was not yet available for my home computer system, a colleague at work, Jim Smith, came through with a magazine article that described a word-processor program written in BASIC. I adapted that program for my system, and the first draft was generated with this program. Several commercial programs became available prior to the first revision, and one was selected to finish the text. A spelling correction program was added for the last revision to find all those "keyboard errors" that seemed to find their way into the text.

Several computer-aided design systems (CAD) were possible candidates for use in developing the figures. The system used for this text was an INTERGRAPH system, employing a Digital Equipment Corporation VAX 780 processor. Writing *Programmable Controller Handbook* has been more demanding than I ever anticipated. I hope that you find the material presentation simple to read and understand as well as informative.

ACKNOWLEDGMENTS

I would like to thank the following individuals who contributed to this text by providing me with the latest vendor information as it was generated or for their assistance whenever I requested it.

Michael F. Andrayo, General Electric
Debra D. Bann, Westinghouse Electric Company
William Barrow, DuPont
Douglas C. Brown, General Electric
Gene Bruni, United Electric Supply
Edward Casey, DuPont
Dennis Christensen, United Electric Supply
Lee J. Cline, Tecot Electric Company
Robert O. Deitz, Miller & Deitz
Thomas A. Delia, Gould Incorporated, Modicon Division
Tricia Flynn, Westinghouse Electric Company
Bruce C. Freer, Allen–Bradley
Pete Fundinger, Eagle Signal
Phillip M. Gaffney, Allen–Bradley
Steve Haleowski, Trola Industries
G. Douglas Haring, DuPont
Atlas Harmon, DuPont
Fred U. Henderson, Jr., Gould Incorporated
Ronald M. Hilmer, Square D Company
Max Hitchens, HEI Corporation
Robert D. Irwin, Comptrol Incorporated
Gregg Jordan, DuPont
Walter R. Keithly, Allen–Bradley
Peter L. King, DuPont
Lance Lauletta, Rumsey Electric Company
Bernard LuBow, Gould Incorporated, Modicon Division
Victor J. Maggioli, DuPont
Joseph F. Marullo, Industrial Data Terminals
Frank J. Masley, DuPont
Kent E. Matson, Trola Industries
Harry C. Miller, Trola Industries

Donald Myers, Cusick Electrical Sales
F. Joe Poulouin, Rumsey Electric Company
R. Lawrence Roby, General Electric
Ernest P. Ross, DuPont
Keith W. Routson, Trola Industries
Gerald L. Safranski, DuPont
William H. Schweizer, Eaton Corporation, Cutler–Hammer
Gene Skinner, DuPont
James T. Smith, Jr., DuPont
Robert M. Swartwout, Square D Company
Richard C. Veihl, Microswitch
William F. Wagner, DuPont
Robert A. Whitehouse, General Electric
Larry Zgrabik, Allen–Bradley

Special Thanks

Work on the illustrations was initiated by Frank Masley in April 1982, based on sketches that I had generated while writing the first draft. When this task became more time-consuming than originally anticipated, Gene Skinner and Gregg Jordan provided much-needed assistance. I have the utmost admiration and respect for these three assistants, as they contributed greatly to the development of this book. Even when the figure files were accidentally corrupted by hardware failures, they were not discouraged from getting the work done on time. I am deeply indebted to Gregg for his long hours of work during the final months, when it seemed that every figure he corrected was returned with additional comments and corrections. I was determined from the outset of this project that the figures must enhance the textual material, and I believe that Frank, Gene, and Gregg definitely helped me achieve that goal.

Everything that I wrote was proofread by Harry C. Miller. His comments and suggestions regarding the presentation of material are greatly appreciated. His years of PC experience, as well as his support, definitely facilitated the work required to produce this text.

A final thanks goes to Victor Maggioli for whom I worked during my first years at the DuPont Engineering Department. It was his expert guidance that laid the groundwork upon which I built my PC knowledge and experience. I also owe thanks to Fred U. Henderson, Jr. of Gould Incorporated, Modicon Division, and to Larry Zgrabik of Allen-Bradley for their time and assistance.

Introduction

The programmable controller, or simply PC, has revolutionized the industrial controls market since its introduction in the early 1970s. The wide acceptance of PCs and the rapid growth of the PC industry further support the idea that the PC might be called the "industrial revolution of the '70s." As with many rapid growth industries, there is often insufficient general technical reference material available to those who desire to learn and apply the most recent advances in engineering and technology. This text is an attempt to provide a single source reference of information and technical data on the subject of programmable controllers.

The first chapters discuss the birth of the PC in the automotive industry, and how PCs have evolved to their present-day form. Typical PC applications are discussed, as well as many questions often asked by those unfamiliar with the application of a PC. Chapters 3 and 4 examine the input/output structures commonly found in use with PCs, and provide detailed information and examples as to the proper design of an input/output system for a PC. For those interested in knowing more about the internal workings of a PC processor, Chapter 5 "removes the industrial covers" and looks at the "nuts and bolts" found inside. Chapter 6, Power Systems and Supplies, examines the proper methods of providing ac power to a PC system, including various input/output power system designs for a typical PC system, and the operation of various dc power supplies used to power a PC system. This chapter concludes with a discussion of the more common battery systems employed to maintain a processor's solid-state memory in the event of system power interruption.

A major portion of the text, Chapters 7 to 13, investigates the various programming instructions commonly understood by most PC processors.

Chapter 7 begins with a review of relay ladder schematic diagrams as a basis on which to build an understanding of fundamental PC programming. The workhorse instructions of a PC, the contact and coil instructions, are introduced and compared to their equivalent relay ladder diagram functions. These discussions are then brought together with an example problem which converts a simple relay ladder schematic to a relay ladder language program compatible with a PC.

Once the reader has a thorough understanding of basic PC programming, Chapter 8 begins to enhance this background by describing many of the special relay-type functions available, including latches and one-shot contact instructions. Also included are those that begin to show some of the PC's computer heritage, including such instructions as jumps and subroutines. Timer and counter functions are examined next, followed by a detailed discussion on the math capabilities provided with many PC systems. Rounding out the discussions on mathematics are examples of the various number systems commonly employed with PC systems.

Chapters 11, 12, and 13 examine the computerlike capabilities of a PC with a thorough discussion of the many bit and data manipulation instructions. Many PCs provide the ability to handle and manipulate data, and/or other process or machine information, without the use of a computer. Operating parameters for a batch recipe process can be stored for later recall, and reports can be generated to list the immediate past operating history of the controlled system and/or hardware. Finally to round out the discussion, special programming languages such as Boolean and BASIC are examined, as well as many of the specialty instructions available on selected PC models to perform special industrial control routines.

Chapter 14 views the many programming devices commonly available for PC use. These peripheral devices range from simple hand-held calculator-size miniloaders to large 8-inch CRT-based programming terminals. A large portion of the chapter is devoted to proper programming procedures required to program a typical PC. PC communications with other PCs and computers via special PC-based local area networks (LANs) are examined in Chapter 15. For those applications where a PC must function without significant downtimes or failures, the use of various commonly used redundancy schemes is discussed. Chapter 15 concludes with a quick discussion of various interfaces, such as RS-232C and IEEE 488, commonly employed with PC hardware.

Shortly after the birth of the PC, users began to develop special-purpose accessories for PC use. These accessories include program documentation packages, color terminal graphic systems, simulation hardware, and special-purpose displays and input/output devices. Chapter 16 examines a cross section of typical add-on accessories and peripheral devices available for PC use.

The proper selection and purchase of a PC to meet a particular application will depend on many factors. Chapter 17 examines some of the more basic considerations which should be addressed when selecting a PC. Once a PC has been selected, Chapter 18 provides the PC user with guidelines regarding the assembly, installation, operation, and maintenance of a typical PC system.

The remaining chapters of the text provide supplemental information for the PC user. Chapter 19 details several simple programs for common PC applications. Even though these programs may seem simple and straightforward, they do provide additional examples of PC flexibility. Since many readers may not be totally familiar with the "lingo" used by more experienced PC personnel, Chapter 20 comprises a glossary of terms. This glossary contains many expressions encountered during PC implementation and use, and includes definitions and descriptions to help clarify each term. Closing out the formal text is Chapter 21 with a survey of PC models available from various manufacturers. This survey was compiled to provide the user with a "broad brush" sampling of the many PC models available, as well as their individual system characteristics.

Most of the text's chapters reference one or more specific PC models as part of the many examples found in this text. While the specific PC model used as part of an illustration was selected at random, the use of an actual PC and its instruction set and hardware should provide the reader with a more accurate understanding of real-world PC applications or programs. The reader should not infer that the particular model selected for an example implies that that model is the best for the application being discussed. Likewise, the reader should understand that the methods discussed may not represent the only solution, or necessarily the best solution, to a particular application or problem.

This text has been written assuming that the reader has a basic knowledge of electrical engineering and industrial control system design, as well as an elementary acquaintance with simple computers. Where the reader's knowledge is insufficient, the glossary should help enhance the understanding. The information presented in this text has been compiled from the author's experience in applying PCs to industrial applications. It has been further supplemented with information and assistance from many of the various PC manufacturers. The reader is strongly urged to contact the various manufacturers listed in the Appendix for additional and more specific information, and to continue his or her education by attending some of the excellent PC schools offered by many of the manufacturers, distributers, and educational institutions. The Appendix also lists several organizations and magazines which can provide additional information on the subject of PCs and their applications.

Contents

PART I

History and Applications

1

What Is a Programmable Controller?

The programmable controller has been described as the industrial revolution of the seventies. It has, in the short span of time since its introduction, provided industrial control capabilities never dreamed possible in prior years. Industrial control systems incorporating a programmable controller, or PC as it is commonly called, are now able to operate machines and processes with an efficiency and accuracy never before achievable with conventional industrial relay-based control systems. While the relay control devices used in industrial control systems will never be totally obsolete, the advent of the PC has changed the thinking of engineers and designers of industrial control systems.

Industrial control of machinery and processes prior to the birth of the PC was performed using specially designed industrial control relays, similar to those depicted in Fig. 1.1. These special-purpose relays are great-great-great-grandchildren of an electromechanical device developed in the early 1800s.

The electromechanical relay was invented in 1836 by Samuel F. B. Morse as a means to increase the station-to-station distance of his newly developed telegraph system. He required a device that could sense minute telegraphic-signal currents and "relay" these signals in an increased current mode to another "relaying" station 15 or 20 miles distant, as illustrated in Fig. 1.2. The device Morse invented would later become the foundation for a family of electromechanical devices which today are used in practically every segment of our society.

The relay is responsible for many of humanity's greatest achievements, as well as for some of its most perplexing problems. The false tripping of a Canadian substation relay initiated the East Coast Blackout of 1967. A single relay signaled

3

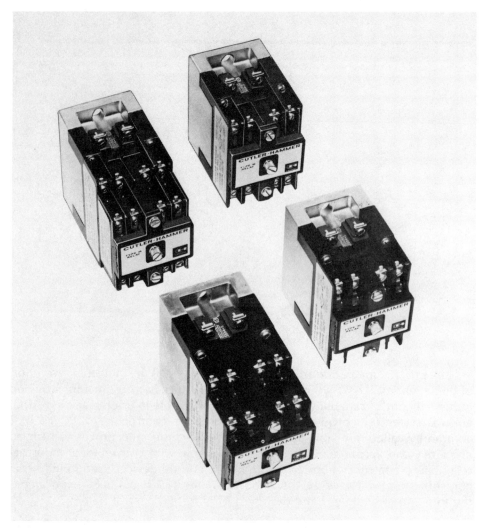

Fig. 1.1.   Typical Industrial Control Relay

the GO for Apollo 11 to make history by carrying Neil Armstrong to man's first steps on earth's only natural satellite. As our technology has increased, the development of the relay has kept pace. Today relays are found in the home as well as throughout industry. The relay of today is much more reliable than were its ancestors. However, there can be problems associated with the use of numerous relays. Most control relays are mechanical devices subject to wear and fatigue. The contacts of a relay can arc and eventually weld together. Enclosures housing the hundreds of relays required for a particular application can be large and heavy. Large relay cabinets are noisy and generate a great deal of heat when in full operation, and relay-controlled systems must be hand wired, making the installation, or even a simple change, both time-consuming and expensive to perform.

THE BIRTH OF THE PROGRAMMABLE CONTROLLER

With advances in technology, especially in the field of electronics, solid-state replacements for the relay cabinet were investigated. As transistors and simple integrated circuits became more cost effective to use, companies such as General Electric, Westinghouse, and Allen-Bradley developed solid-state control systems. These systems increased the reliability of a control system immensely, and decreased the cost of an installation. However, these systems, still available and sold today, had several characteristics which made them unsuitable as replacements for the conventional relay control system.

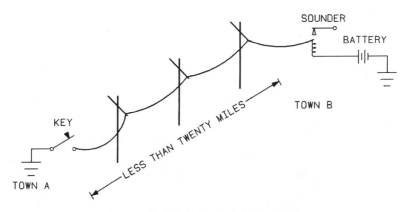

SIMPLE TELEGRAPH CIRCUIT

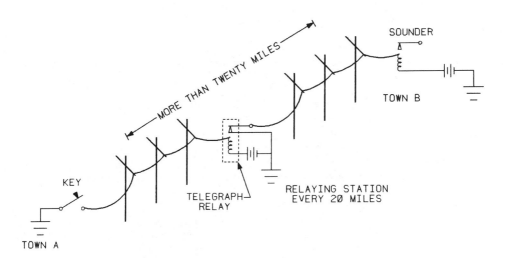

SIMPLE TELEGRAPH CIRCUIT WITH RELAY

Fig. 1.2. Morse Telegraph System

A large user of industrially designed relays, called control relays, is the automotive industry. The control relay is used to operate and control the many machines that assist plant operators in the manufacture of an automobile. Looking to reduce relay control system costs during each model year changeover, the Hydromatic Division of General Motors in 1968 prepared a specification detailing a "programmable logic controller." This specification reflected the sentiments of many control relay users associated not only with the automotive industry, but with practically every other manufacturing industry as well. The specification generated interest from such companies as Reliance Electric, Struthers-Dunn, Modicon, Digital Equipment Corporation, and Information Instruments. The results of these companies working with General Motors spawned the specialized-purpose computer we now call a *programmable controller* (PC) in late 1969 and early 1970.

Entries in the PC market were primitive at first. Information Instruments, Incorporated (often referred to as Triple I) produced the PDQ II. Triple I was later purchased by Allen-Bradley and became the foundation for one of the largest PC manufacturers in the world. Modicon Corporation was founded in a small facility in Bedford, Mass., later to become another one of the world's largest PC manufacturers. The Modicon 084 (pronounced "oh-eighty-four") became the first of a series of Modicon eighty-four series controllers. Reliance Electric produced the Automate 33 with a unique instruction set providing a great deal of flexibility. Since the General Motors specification required the new device to be programmable, Digital Equipment Corporation applied its well-established computer technology to the application and developed the PDP-14 programmable controller based on their successful PDP-8 series computer. Not to be outdone by its competitors, Struthers-Dunn introduced the VIP programmable controller. The PC caught on, and in 1970 a new industry was born. Soon other well-known electrical controls manufacturers began either producing PCs of their own design or repackaging a system manufactured by a competitor.

Even though the early PCs were primitive by today's standards, they did meet the basic requirements of General Motors' specifications. These requirements can be summed up in ten primary categories as follows:

1. The control hardware and/or device must be easily and quickly programmed and reprogrammed at the user's facility with a minimum interruption of service.
2. All system components must be capable of operation in industrial plants without special support equipment, hardware, or environments.
3. The system must be easily maintained and repaired. Status indicators and plug-in modularity should be designed into the system to facilitate repairs and troubleshooting with minimum downtime.
4. The control hardware must occupy less plant space than the relay control system it replaces since space costs money. In addition it shall consume less power to operate than present relay control systems.
5. The programmable logic controller must be capable of communication with central data-collection systems for the purpose of system status and operation monitoring.
6. The system shall be capable of accepting 120-volt ac signals from standard existing control system push buttons and limit switches.

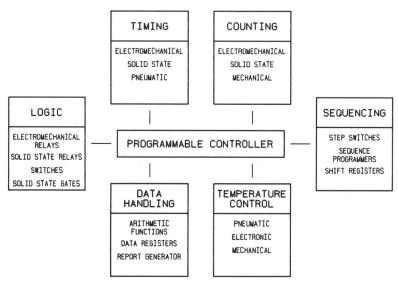

Fig. 1.3. Functions of a Programmable Controller

7. Output signals from the logic controller must be capable of driving motor starter and solenoid valve loads operating at 120 volts ac. Each output shall be designed to switch and continuously operate a load device of 2-ampere rating.

8. The control hardware must be expandable from its minimum configuration to its maximum configuration with minimum system alteration and downtime.

9. The unit is to be competitive in purchase and installation cost with relay and solid-state logic systems currently in use.

10. The memory structure employed in the programmable logic controller shall be expandable to a minimum of 4000 memory words or elements.

Today every PC manufactured not only meets the original criteria established by the automotive industry, it exceeds those simple requirements many times over. The PC is actually a special-purpose computer designed to provide a more flexible and reliable alternative to an industrially designed relay-based control system. As Fig. 1.3 illustrates, the PC of today is a total control system in a small package capable of assuming a variety of control system functions. Ironically once the PC began infiltrating the industrial controls market, it established itself as an excellent computer replacement device for many applications.

Many control system specialists had experienced poor performance from computer-based control systems which could provide the flexibility of a PC, but not the environmental ruggedness. The environmental and industrial ruggedness of the PC solved many of their concerns. While PCs were once thought of as special-purpose control devices, they are now regarded as the standard solution to most industrial control applications.

This high degree of acceptance has resulted in a formal definition for a PC by the National Electrical Manufacturers Association (NEMA). NEMA standard ICS3-1978, Part ICS3-304, defines a PC as follows:

A digitally operating electronic apparatus which uses a programmable memory for the internal storage of instructions for implementing specific functions such as logic, sequencing, timing, counting, and arithmetic to control, through digital or analog input/output modules, various types of machines or processes. A digital computer which is used to perform the functions of a programmable controller is considered to be within this scope. Excluded are drum and similar mechanical type sequencing controllers.

COMPUTERS VERSUS PROGRAMMABLE CONTROLLERS

The major difference between a PC and a commercial mainframe computer or a minicomputer lies in the programming language used and the industrial ruggedness of a PC's design. As Fig. 1.4 illustrates, there are many methods available to describe and perform a control function, such as operating a motor starter or contactor. Relay ladder schematics are the general industry standard. Solid-state logic and computer programming techniques can be used to provide the same information in an alternate "language" of symbols and commands, but only the *relay ladder language* of the PC emulates best the relay ladder schematic. As will be explored in later chapters, a computer is designed to be user friendly to programmers by accepting and implementing instructions entered in a word and/or broken sentence format similar to the spoken word. The PC is used by a specialized clientele accustomed to conveying information by an elaborate convention of symbols, the relay ladder schematic symbols. These same symbols and conventions are utilized to form a specialized high-level language for the PC called *relay ladder language symbology*. Both a computer and a PC can usually perform identical control functions. However, PC relay ladder language symbology, coupled with specialized PC design criteria, usually makes PCs the more cost-effective solution for an industrial-control application.

Several PCs are available that offer other programming languages in addition to the standard relay ladder format. One of the most common of these languages is a programming format based on Boolean equations and algebra. The Boolean format expresses in algebraic equation format the identical control requirements as the relay ladder format. The use and implementation of the Boolean format will be discussed in depth in a later chapter. Some PC manufacturers have developed specialized languages in order that their systems can better meet more specific applications. Since there are many manufacturers who describe their product offerings as a "programmable controller," including manufacturers of programmable temperature and analog loop controllers, this text will limit itself to those products that meet the NEMA definition of a PC *and* offer at least one relay ladder formatted programming language. Inclusion of devices called "programmable controllers" that are not programmable in a relay ladder format would greatly extend the overall extent and scope of this text. Readers are alerted that these products do exist, and that they may wish to consider also a portion of them for their particular needs, in addition to the capabilities of an industrially designed PC.

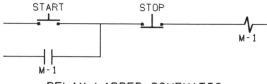

RELAY LADDER SCHEMATIC

IF THE STOP BUTTON IS DEPRESSED, TURN THE STARTER OFF. OTHERWISE,
IF THE STARTER BUTTON IS DEPRESSED, TURN THE STARTER ON. OTHERWISE,
LEAVE THE STARTER IN ITS PRESENT STATE.

ENGLISH DESCRIPTION

```
PBSTART:    EQU         *           ORA     B
PBSTOP:     EQU         *           MOV     B,A
STARTER:    EQU         *           LDA     PBSTOP
            PUSH        PSW         CMA
            PUSH        B           ANA     B
            LDA         PBSTART     STA     STARTER
            MOV         B,A         POP     B
            LDA         STARTER     POP     PSW
```

MACHINE LANGUAGE PROGRAM

```
10 INPUT A 'PUSHBUTTON START'
20 INPUT B 'PUSHBUTTON STOP'
30 IF A=1 AND B=1 THEN C=1
40 IF B=0 THEN C=0
50 OUTPUT C
60 GO TO 10
```

COMPUTER PROGRAM

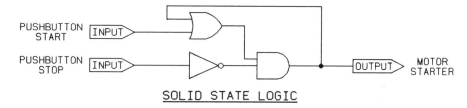

SOLID STATE LOGIC

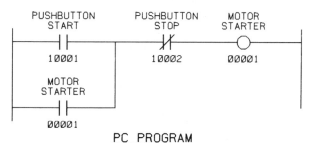

PC PROGRAM

Fig. 1.4. Motor Starter Programs

Thus far the PC has been defined as a special-purpose computer utilizing a unique programming language. However, it has also been described as being designed for a harsh industrial environment. PCs operate without special climate controls in applications as severe as 140°F and at humidity levels reaching 95 percent. They generally do not require special installation considerations above standard good electrical practices found in most industrial applications. Most important, they are designed to withstand moderately rough handling. The PC affords these major advantages over a computer by its basic architecture. Each component of a PC system is independently designed for optimum performance under the worst industrial conditions. To underscore the PC's ruggedness in design and operation, the STS-7 mission of the Space Shuttle carried a PC in the cargo bay as part of an experiment. It was the function of the PC to operate various aspects of the experiment while in orbit. Not only did the STS-7 mark many firsts and successes for the U.S. space effort, the experiment controlled by the PC performed flawlessly.

All hardware associated with a PC falls into one of two functional areas. The actual intelligence of the PC is derived from electronic computer-based hardware, which comprises the processor, or CPU, portion of the system as represented in Fig. 1.5. The processor section of a PC includes a power supply, a microprocessor or special-purpose electronic circuitry, as well as a computer-type memory for the storage of programming instructions and system data. All activity of the PC system is handled by the processor. The processor is responsible for the analysis of incoming as well as previously stored data, and for responding to that information according to a detailed control plan stored within the unit by the user. Every processor is designed to perform a wide variety of functions. Most systems offer the standard relay, latch, timing, counting, and simple mathematical functions of addition, subtraction, multiplication, and division as part of their instruction capabilities. Also found in many medium- to large-sized systems are data and bit manipulation instructions, including shift, rotate, compare, and sense functions; binary to binary-coded decimal (BCD) and binary-coded decimal to binary conversions; analog control functions, including proportional, integral, derivative control (PID); report generation and graphics; computer, printer, and intelligent terminal interfaces; subroutines; jumps, skips, sequencers, and master control relay instructions; as well as a complete complement of self-diagnostic instructions and trouble-shooting aids. All of these functions give the processor the flexibility required to meet a diverse listing of applications and industrial environments.

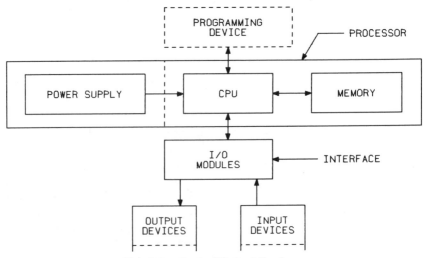

Fig. 1.5. Basic PC Architecture

Since the processor must receive data from the industrial environment as well as control hardware in the same environment, the second major hardware grouping of a PC is the *interface* or *input/output* (I/O) *structure.* The I/O structure is designed to be modular for ease of replacement and simplicity of future expansion. Signals from the harsh industrial world are terminated at a PC's I/O system. Fig. 1.6 illustrates the simplified operation of a typical PC. Within the hardware of the I/O structure, incoming field device signals rich in electromagnetic noise and interferences are scoured and reduced to electronic clean-room condition prior to use by the processor. These signals are transferred to the processor for use during the interpretation and solving of the user's relay ladder logic program. Conversely the commands of the processor, which are based on the solution of the user's program, are boosted from logic voltage levels by the I/O system to drive power-hungry industrial controls.

Typically an I/O system is capable of interfacing the processor's logic level signals with industrial control voltages from 5 to 240 volts ac and/or dc, as well as of handling low-level analog signals and computer communications. The use of *intelligent I/O* permits the user to interface complex signals, such as high-speed pulses or stepper motor drive and axis positioning functions, to a PC with a minimum of processor programming. Special interface systems and hardware enable multiple PCs to communicate at will among themselves via *data communications networks* and/or *control networks.* The interface portion of the PC allows the processor to reach out thousands of feet to touch the control devices of an industrial complex. The processor and its I/O system together form an industrial control team ready to adapt to the most demanding industrial control requirements, applications, and environments.

THE PC INDUSTRY

Acceptance of the PC has had many effects on other industrial markets as well. Electrical distributors have included various PC systems as part of their offerings. This type of "in stock" service has given the PC an off-the-shelf availability similar to the relay components it replaces. Many electrical suppliers have sent one or more of their staff persons through extensive PC manufacturer training classes and application seminars so that they can meet the diversified needs of their customers more effectively. This training gives a customer the benefit of having a person located within minutes from his plant who is knowledgeable in the latest PC developments as well as the customer's applications.

Many colleges, universities, and technical schools offer classes dealing with the selection, installation, operation, and maintenance of a PC-based control system. The PC industry has a magazine, *Programmable Controls,* dedicated to the manufacturers and users of PCs and PC-related equipment. Every PC manufacturer offers training classes lasting from a day to several weeks on such areas as PC basics, advanced PC programming, PC applications, and PC system troubleshooting and maintenance. Articles abound in practically all industrial and electrical trade magazines dealing with some aspect or application of a PC.

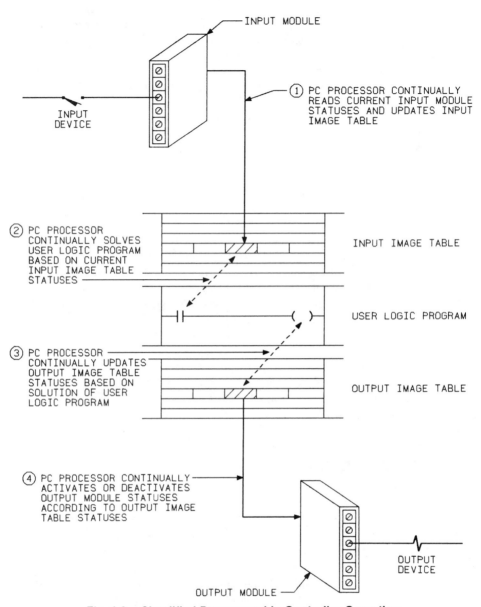

Fig. 1.6. Simplified Programmable Controller Operation

The PC industry organizes a trade show dedicated to the newest offerings of the manufacturers, as well as participating at many society and industry trade shows.

Numerous companies have been established to develop and implement PC-based control systems. Panel shops that once specialized in the design and fabrication of electrical, pneumatic, and electropneumatic control panels, now offer control system designs based on the PC of the customer's choice. A typical control system house, as they are now referred to, can provide an analysis of a customer's needs and desires. The details are developed into a system specification for customer approval. Once approval is received, a complete drawing package is developed consisting of all necessary mechanical, electrical, and other related drawings. The PC program is developed, simulated for customer approval, and documented, while necessary control enclosures of PC hardware are fabricated and electrically tested. Once the equipment has been shipped from the control systems house, assistance can be provided to the customer during installation and startup of the new system. Later the systems house is available for consultation regarding possible system modifications and expansions as required by the customer.

Many diversified companies have discovered the need for specialized hardware to support the PC industry. Computers have been developed specifically for the documentation of a PC program. These specialized computers look into the memory of a processor and produce written printouts in user-understandable symbols and mnemonic descriptions of the processor's programming. Numerous processor and user program statistics are available, detailing such areas as memory configuration and utilization, instruction cross references, and other related data.

Specialized simulation computers are offered as peripheral devices to test and simulate a PC program prior to actual plant installation. Operation of simulation systems permits a PC to operate as if it were actually installed in the facility. The PC initiates an operation, and the simulation computer graphically responds to show the real-world actions taking place, as well as signaling the PC to current I/O statuses. The computer-based simulator is menu programmed to emulate the mechanical and electrical hardware that will be controlled by the PC. Any omissions or errors in the PC program can be spotted easily and corrected during simulation instead of during startup of the actual plant system. To test further the user's logic, abnormal system operations and even component failures can be simulated to test the response of the PC logic.

Numerous operator display devices have been designed and developed for PC use. Color-graphic intelligent cathode-ray tubes (CRTs) are available to replace control panels of push buttons, indicators, dials, and meters. An operator can now visualize the operation of the facility and effect changes with the touch of a finger. The color CRT not only indicates the current statuses of a system graphically, but together with the PC it tells the operator any possible control options, as well as alerting him or her to improper requests. Highly sophisticated systems may even make minor decisions or control changes for the operator.

Special-purpose message display units are available for use with PC-based control systems. These alphanumeric display devices allow user flexibility to program them with messages pertaining to system statuses and conditions. The PC transmits a binary or ASCII code to the display, which is converted to a visual message to alert the operating staff of potential problems. The intelligent display allows written communication between a PC and its operator when the cost of an intelligent color CRT is not justified.

The future of the PC is unlimited. Its uses and applications are only limited by the limitations of the user. The future development of the PC will be mostly a result of user demands. There will always be a quest for improved operating efficiency and product quality. The desire to produce more at better economics will guarantee the future success of the PC and its supporting industries.

PCs are doubling their capabilities in the same amount of time it takes them to halve their size. As Fig. 1.7 outlines, the PC industry has continually evolved, pushing beyond its present capabilities to meet the demanding requirements of

PC HISTORY

● 1968 DESIGN OF PC'S DEVELOPED FOR GENERAL MOTORS CORPORATION TO ELIMINATE COSTLY SCRAPPING OF ASSEMBLY-LINE RELAYS DURING MODEL CHANGEOVERS.

● 1969 FIRST PC'S MANUFACTURED FOR AUTOMOTIVE INDUSTRY AS ELECTRONIC EQUIVALENTS OF RELAYS.

● 1971 FIRST APPLICATION OF PC'S OUTSIDE THE AUTOMOTIVE INDUSTRY.

● 1972 ADDITION OF TIMING & COUNTING PC INSTRUCTIONS.

● 1973 INTRODUCTION OF 'SMART' PC'S FOR ARITHMETIC OPERATIONS, PRINTER CONTROL, DATA MOVE, MATRIX OPERATIONS.

● 1974 INTRODUCTION OF CRT PROGRAMMING TERMINALS.

● 1975 INTRODUCTION OF ANALOG PID (PROPORTIONAL, INTEGRAL, DERIVATIVE) CONTROL WHICH MADE POSSIBLE THE ACCESSING OF THERMOCOUPLES, PRESSURE SENSORS, ETC.

● 1976 FIRST USE OF PC'S IN HIERARCHICAL CONFIGURATIONS AS PART OF AN INTEGRATED MANUFACTURING SYSTEM.

● 1977 INTRODUCTION OF VERY SMALL PC'S BASED ON MICROPROCESSOR TECHNOLOGY.

● 1978 PC'S GAIN WIDE ACCEPTANCE. SALES APPROACH $80 MILLION.

● 1979 INTEGRATION OF PLANT OPERATION THROUGH A PC COMMUNICATION SYSTEM.

● 1980 INTRODUCTION OF INTELLIGENT INPUT AND OUTPUT MODULES TO PROVIDE HIGH SPEED, ACCURATE CONTROL IN POSITIONING APPLICATIONS.

● 1981 INTRODUCTION OF 'COMMUNICATION NETWORKS' PERMITTING PC'S TO TALK WITH ANY INTELLIGENT DEVICE SUCH AS COMPUTERS, CODE READERS, ETC.

● 1982 INTRODUCTION OF MINI & MICRO PC'S

● 1983 INTRODUCTION OF 'CONTROL NETWORKS' PERMITTING MULTIPLE PC'S TO ACCESS EACH OTHER'S I/O IN A USER TRANSPARENT FASHION.

Fig. 1.7. History of the Programmable Controller

its users. PCs are continually becoming less mystical to install and operate. Developments will continue with current systems as well as the introduction of more sophisticated models designed to suit more customized applications. Computers and PCs will share data communication networks and transfer data among themselves with increased simplicity and speed. Advances in computer technology and hardware will spill over into the PC marketplace to add even more capabilities never dreamed possible several years ago.

The PC is, and will continue to be, a sound investment with extremely fast and profitable returns. It was born of computer technology and space-age electronics. It is literally "the control of the future and beyond."

2

The Application
of a Programmable
Controller

The PC has become widely accepted because of its ability to be adapted to a majority of applications. However, the selection of the proper PC for a particular purpose, the proper installation of a chosen model, and the proper use of the PC in an application are largely dependent on the control system engineer and/or designer being familiar with PCs and their peripherals. In order to be adaptable to a wide variety of applications and uses, the PC is inherently highly flexible and diversified in its design. While this flexibility and diversification can make the PC a sort of "one key that fits all locks" type of control device, the actual selection, implementation, and utilization of the PC can leave those unfamiliar with the device a bit bewildered.

The selection of the control devices to be used in a particular application will never be a simple cut-and-dried group of decisions. Today's technology has produced a host of electronically based systems and components from which to choose. During the '60s and '70s one had to decide whether a system would remain mechanical, electromechanical, an in-between hybrid, or whether it would be equipped with the latest electronic wizardry. With the flood of electronic systems offered for practically every industrial application, the choice of the correct one for the application now becomes paramount. The application of a PC-based control system is not only dependent on the answers to such standard questions as: How much will it cost? Does it fill the requirements of the application? Will the operators and maintenance personnel accept it? Will it be expandable in the future? Is it easily maintained? The application of the PC is also dependent on answers obtained from such questions as: Is it the best technology available for the money being spent? Is the proposed system the best configuration for the particular hardware involved? Will the PC be efficient and production-effective? The questions that must be answered, as well as asked, are diverse and far-reaching in nature. The final answer to any one can dynamically change the final design of a system from its original conception.

The industries implementing a PC-based control system are extremely diverse, as indicated by the partial listing in Fig. 2.1; and the uses that a PC finds are equally diverse, as partially indicated in Fig. 2.2. It is interesting to note that even though the PC was originally perceived as an industrial control relay replacement device, many applications take the PC into nonindustrial fields. One of the larger well-known casinos in Las Vegas uses a PC to monitor and operate the drives and lifts of its moving theatrical stages. The PC is programmed to raise and lower as well as to move and/or rotate stage platforms according to the needs of the current production. The PC is solely responsible for seeing that the proper conditions are met for a stage platform change, and that that change is implemented in the fastest time possible, with greatest safety to the performers on as well as near the stage floors being moved.

The operations to which a PC can be applied are limitless. It is often stated that the only limits to the application of a PC are the limits of imagination of the designer or engineer. While this statement is very often true, it should not imply that the PC is a solution to every industrial and commercial control problem. The PC is a definite-purpose machine, with defined limits to its abilities. For example, while a PC may have the capability to perform proportional, integral, derivative (PID) type control, there are other special-purpose process controllers designed to fill the needs of the larger systems more economically. The highly accurate machining of materials and components can be done under the direction of a PC. However, many computer-numerical control systems (CNC) are designed to be more cost-effective, faster, and efficient. The proper application of a PC requires that the final result be the most cost-effective method of acquiring the desired outcome.

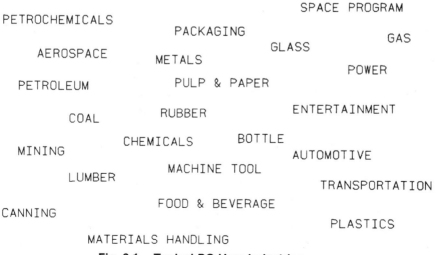

Fig. 2.1. Typical PC User Industries

```
                                          BOTTLING
   SAW MILL CONTROL        BLENDING
                                        BATCH PROCESSING
           DRILLING
                         BORING
 CANNING
           HEAT TREATMENT              CONVEYOR CONTROL

     PIPELINE                          MATERIALS HANDLING
                      WEIGHING
   PELLETIZING                            BURNER CONTROL

       ORE HANDLING      WATER TREATMENT

   FILLING                             PALLETIZING

     WASTE TREATMENT      COAL HANDLING
```

Fig. 2.2. Typical PC Applications

Determining the most economical all-around solution for a control system necessitates that the engineer and/or designer be familiar with many types of control devices and hardware configurations. While this text is intended to address the capabilities and use of the PC, first-hand experience will be the ultimate teacher. This text cannot teach the reader everything about the application of a PC. Knowledge of PCs is obtained by a bootstrapping process. Use of this text will give readers a broad base of information about PCs from which they can proceed to learn more about a particular facet of the subject.

The proper application of a PC not only requires that the end result be cost-effective, but the final control system must meet the needs of its users. The needs of the application can only be defined by a detailed analysis of the overall system. This often means that detailed examinations must be performed on every facet of a machine or process operation. Again, since every application is different, there is no clear, concise routine to evaluating the needs for all applications. General areas of concern will include such topics as maximum system throughput, minimum operator interface, most efficient use of energy and materials during the production process, floor space requirements, ease of maintenance, maximum system utilization, and of course minimum installation and repair time. Often it will be necessary to examine similar systems and installations in order to evaluate whether one system will meet the needs of a particular application better than an alternate system.

Equally important to the economics and usefulness of a PC-based control system application is whether or not the plant personnel will accept the new system. This will depend largely on whether it is simple to operate. Many control system designers do not "human engineer" a facility to the people that must operate it. For example, a lack of sufficient operating status indicators often leaves plant operators or maintenance mechanics agitated at an installation. Their continual complaints can soon convince fellow workers that the system is indeed poor in design and/or implementation.

Often overlooked at the time of system design and installation is the proper training of those who will have direct contact with the controls. Many plant sites and engineering departments postpone operator and maintenance training until after the system has been installed and is operating, since it is often felt that it will be easier to train people on actual operational hardware. Keeping those who will ultimately maintain and operate a future system informed throughout installation will make them feel part of the new system. Through proper training, the reasons a system is being installed in a particular manner can be understood better by those who must operate it. If forced to learn the "why fors" of a system during an emergency maintenance or rush operation environment, people's attitudes and feelings may shadow their perspective. All operators and maintenance personnel have times when they must be away from their jobs due to physicals, performance reviews, safety meetings, and so on. Extending an employee's lunch period an hour or two once a week for several weeks to attend in-house training will usually not affect production signifi-cantly. The operator learns about the new system as it is being installed. The company does not lose a skilled operator totally during the training period, and when the new system is operational, those who will interface with it are knowledgeable with the hardware and its operation and maintenance.

A final major consideration in the application of a PC involves the future growth of the system. The PC is modularily designed and therefore expandable to meet future needs. Often only a single control system has to be replaced or upgraded. However, many times numerous identical machines may be involved during a control system installation. The application of a PC should always consider future benefits and expansions as well as current needs. A control system implemented for one system may be easily duplicated for hardware, installation, and startup costs on duplicate equipment which may be slated for future replacement. The possibilities of adding additional manufacturing hardware should always be kept in mind.

A sales statement made by practically every PC salesperson is: "If there's nothing else that programmable controllers do, they do continue to sell more of themselves!" A great many PC-based control systems are installed with their memory and I/O hardware capabilities being utilized only partially at the outset. Once the systems are in operation, plant operators, production personnel, and maintenance technicians devise all sorts of additional things for the PC to do. There are systems which, by nature of the application, can be installed and forgotten, but it seems that the vast majority of systems are refined and modified to the physical limitations of the PC installation.

SELECTION OF A PC SYSTEM

Once the use of a PC has been decided upon for a particular application, the sizing and selection of the PC for the control system is a rather straightforward operation. Each PC usually offers a diverse assortment of possible configurations and capabilities by its design. The sifting out of the proper system configuration, followed by selection of the make and model to be used, begins with an examination of the I/O requirements.

Three factors must be considered in order to properly define the I/O requirements for a proposed PC system. The first is the location of the I/O devices with respect to the proposed location of the PC processor. Many systems offer remote as well as local I/O configurations. While local I/O has the advantage of keeping system hardware together, remote I/O may be advantageous when the wiring costs of the installation are examined.

Once the selection of remote versus local I/O has been made, the types of I/O interface modules to be implemented must be chosen. In most applications more than one type of I/O module may be required. Standard operator controls, such as push buttons, selector switches, and indicator lamps, will commonly use 12, 24, or 120 volts ac, or possibly 12–48 volts dc. In addition many control devices, such as limit switches, solenoids, and motor starters, will use similar voltages for operation. Solid-state devices like light-emitting diodes (LEDs), digital displays, and solid-state alarm systems often require transistor-transistor-logic (TTL) level interfaces. If instrumentation is required to measure flows, temperatures, pressures, weights, or other analog quantities, the use of analog-to-digital I/O interfaces may be necessary. Likewise the control of a process control valve or temperature controller may require the selection of a suitable digital-to-analog interface. Interfaces are also available for a host of specialty control applications, including stepper motors, high-speed pulse counters, servomotors, tachometers, and thermocouples, to list just a few.

In determining the total I/O requirements of a system, the quantity of each type of interface module needed must be tallied. This tallying is done according to voltage level and operation of the field sensors and controls needed for the application. The number of interface modules required is tallied according to function (input or output) as well as voltage and operation. The overall size of the PC can then be partially determined from these data.

Actual selection of the PC required for an application is determined by three major items in addition to the final installed system costs. The quantity of I/O field devices and the amount of processor memory for the control system's logic program are the two most important items of the three. The instruction set of the processor is the third. However, the instructions available in a particular PC are usually a function of its size and intended market. Since a good estimate now exists on the total I/O requirements for the system as well as on the amount of user memory required, the choice of a micro, mini, small, medium, or large PC can be determined based on these criteria.

Whatever the PC finally selected, it must have the capability to handle the I/O requirements of the system. It is always good practice to consider that at least 20 percent additional I/O may be required above and beyond the initially calculated quantity. The I/O requirements, including the 20 percent future growth, will determine the PC installation as requiring a *micro, mini, small, medium, large,* or *super* unit. Generally, small-sized PCs handle from 128 to 256 total I/O points. Medium-sized systems range from 256 to 1024 discrete I/O points, while any system handling more than 1024 I/O devices is considered a large PC system. Latest additions to the PC market include micro PCs (less than 32 I/O points), mini PCs (from 32 to 128 I/O points), and super PCs (more than 4000 I/O points per processor). Note that these limitations are based on the total number of discrete input and output points available, and do not include analog I/O points as part of the classification. Detailed discussions of I/O system operation, selection, and configuration are presented in later chapters.

Once the I/O requirements of an application and the general size of the PC system are defined, the amount of processor memory required for the application needs to be determined. Each manufacturer offers guidelines recommending the amount of *user* memory needed for a given application and I/O count. These recommendations range from allowing 8 to 16 words of user memory per installed I/O point of the system. For example, a manufacturer recommending 10 words of user memory per installed I/O point would suggest that the user memory be at least 3770 words for a system composed of 230 input points and 147 output points. In conjunction with the amount of user memory required, additional amounts of processor *data* memory may need to be included for calculation and data manipulation, or message-intense programs. The type of memory, either random access memory (RAM), programmable read-only memory (PROM), or core memory, will also be a consideration. A discussion of the types of memories available and their operation and use follows in later chapters.

Many PC systems offer various levels of relay ladder programming. All PCs offer basic relay, timing, counting, and latch instructions of some form. The offering of extended instruction sets and specialty programming languages varies between makes and models. As a rule of thumb, the larger the PC model, the more sophisticated the instruction set offered. It should be noted that a reduction in memory requirements is often possible with the use of higher powered extended instruction sets since they offer the ability to perform complex operations with minimal memory requirements.

The final items to be selected in the application of a PC are the peripheral devices. Usually a peripheral device is used to enhance the overall operation of a PC-based control system. The first and most necessary of the peripherals is the programming terminal. While this device can be rented or leased from many manufacturers, its purchase is usually always recommended. A programming terminal allows the user to program and troubleshoot the user software programmed in the processor. A tape cassette loader is highly recommended for the storing and reloading of user programs. Additional peripherals include printers and data terminals for documentation purposes, telephone modems for remote communications between one or more PCs and a programming terminal, as well as computer and communication network interfaces. A later chapter is dedicated to the types and functioning of typical programming terminals. Additional peripherals are discussed in greater depth in a later chapter.

PC CONTROL SCHEMES

One area that must be touched upon when considering the application of a PC is how the overall control scheme will be implemented. The majority of applications are either *continuous* or *batch* in nature. There are two ways to determine whether an operation is continuous or batch. However, there will still be a few operations that could be classified as either. The first method of determining whether an operation is batch or continuous examines whether the product is made in batches or lots. Processes such as the manufacture of various paint colors are usually batch processes. A run or "campaign" is produced of a specified color. Once that particular color has been produced in the desired quantity, a new color is chosen and the system is configured for the new color production. The manufacturing facility uses the same hardware to produce each color, but the actual color is dependent on the amounts of raw products used and the manner in which they are assembled and blended.

There are facilities that produce the exact same product every operating hour. This would seem to indicate that the operation is continuous in nature. In fact, there are paint facilities that produce the same shade of paint, such as white or black, day after day. The determination of whether a process is continuous or batch in nature then rests on a second test. This test involves the examination of the operating characteristics of the facility. Continuous processes produce the same product day after day with the exception that they come to a complete stop for routine maintenance or monthly clean-out operations.

Batch operations produce the end product in short-time intervals lasting from minutes to days. The interval production of an end product may be affected by a need to hold or cure the product during its production, or due to a production inability to handle large, continuous amounts of the product being manufactured. The paint process mentioned earlier might appear to be continuous since the same color is manufactured during the life of the facility. If the paint is produced in discrete repetitive quantities, the operation would be batch in nature. A manufacturing facility capable of continuously blending and packaging paint would most likely involve a continuous control scheme. It should be noted that the determination of whether or not an operation is continuous lies in the production capabilities of the manufacturing equipment, and not the actual production schedule. If the manufacturing hardware is not capable of continuously processing the end product due to some process or hardware limitation, the operation will usually be classified as a batch operation. Lack of product flow due to insufficient orders and demand does not indicate that the system could not handle 100 percent capacity production schedules.

Batch and continuous process control schemes can be *centralized* or *distributed* in layout. Centralized control layouts utilize a single large control system or processor to control many diverse manufacturing processes and operations. Each individual step in the manufacturing process is handled by a central control system PC. The distributed control layout differs from the centralized layout in that each step is handled by a dedicated control system or PC. In the paint manufacturing example cited earlier, the filling, packaging, inspection, and shipping of the individual paint containers could be handled by a centralized control system or by individual control systems dedicated to specific functions.

The final distinction that can be made of a control scheme is with regard to the interactions between various control systems involved in a facility. The simplest interaction is none at all. Many control schemes can be designed as *stand-alone* systems. They await the arrival of a product, process it, discharge it to the next manufacturing function, then await the arrival of another unit. Dedicated control systems are totally independent and could be removed from the overall control scheme if it were not for the manufacturing functions they perform.

Control systems which rely on adjacent control systems upstream in the manufacturing process are termed *peer-to-peer* control schemes. Each control system relies on surrounding systems for functional operation. Peer-to-peer schemes might have numerous identical control systems, but the operation of each control system depends on the operation of either a neighbor system or one or more other plant control systems.

Any control scheme, either stand-alone or peer-to-peer, that is directed by an overall master stand-alone control system is termed a *hierarchical* control scheme. None of the slave control systems can operate without the direction of a central intelligence. The central intelligence directs the manufacture and flow of the product throughout the manufacturing facility, as well as monitoring the operations along the way.

PC RELIABILITY

The PC is an electronic device composed of technology's latest offerings. Like any electronic device, it is prone to possible hardware and/or electronic malfunction and even failure. PCs are not fail-safe by themselves, nor is any other electronic device or system of devices currently used in the controls industry. However, the PC, like any other electronic piece of hardware, can achieve a very high degree of safety and reliability if its inherent safety pitfalls are considered at the time of overall system design. "Fail-safeness" is achieved by careful system design, hardware selection, and control strategy, as well as by a practice of thoroughly training all management, operations, and maintenance personnel on all aspects of the system, its design, and its operation.

Anyone applying a PC for the first time, or even an experienced PC system designer or engineer applying a PC to a critical system, may have concerns relating to the possible failure modes of a PC. There are a handful of common PC failure modes that should be considered since for each of these possible failure modes, the PC manufacturer's design of the unit will determine how the failure is detected and dealt with. Probably from a user standpoint, it is the PC's self-detection of potential problems that is of paramount importance. Once a problem is recognized, it is usually quite simple to deal with.

The most critical of any PC-related failure is the stalling of the processor. This type of failure can be categorized by the PC electronics failing to interpret and carry out the user programming. Since the PC scans the user logic, it is quite easy to detect this form of malfunction. Every PC begins its scan by examining its input devices. The PC processor then interprets and executes each and every user instruction in the same order as it was originally placed in the PC's memory by the user. Upon completing the solution of all user instructions, the

PC updates all output devices which it controls. Having finished the output update, the unit performs any housekeeping that may be necessary prior to repeating the sequence again.

It is usually simple to detect when a PC deviates from its standard scan in any manner. Most PCs incorporate a *watchdog timer* to monitor the scan process of the system. The watchdog timer is usually a separate timing circuit or circuits (often a simple RC network and one-shot) which must be set and reset by the processor within a predetermined time period. One or more watchdog timer circuits are automatically set and reset by the processor during the normal operation of the PC. For example, a pair of watchdog circuits (two may be used for redundancy) may be set at the start of the solution of the user programmed instruction sequence, and later reset when this operation is complete. A second pair of watchdog timers may be set at the beginning of the output device update, and reset when the input update is complete. Should any of the watchdog timer circuits actually time out, either due to watchdog timer hardware failure or to PC set/reset failure, the timed-out circuit immediately halts the operation of the PC. The watchdog timers are designed to detect the failure of a set or reset pulse to occur, as well as a failure due to a continuous set or reset pulse being present. The watchdog timer provides a quick way to detect processor scan failures as well as even the most subtle system hiccups.

Another related PC processor failure is one where the microprocessor forming the heart of the PC fails to interpret or execute the user programmed logic instructions correctly. This type of PC processor failure is handled by simply programming the processor to execute canned instructions during the solution of the user instruction sequence. The user logic solving portion of a PC processor is often required to solve one or more user transparent logic routines already stored in a section of processor PROM. The results of this solution are compared with the correct answer(s) also stored with the test routine or routines in PROM. Should the PC processor return an answer different from the one prestored, an immediate shutdown of the system occurs.

One area of possible PC processor concern is the integrity of the instructions stored by the user in the processor's memory over time. Solid-state memory failures could cause an instruction to change in meaning or function. These types of system malfunctions are usually handled by the addition of mathematical check sums, parity, and other error detection schemes, transparently programmed by the programming terminal in conjunction with each user instruction at the time of user program entry or editing. During the processor's scan of the user instructions, each instruction is first checked to ensure that it has not been changed prior to the instruction being carried out by the processor. Any changes are cause for an immediate shutdown of the complete PC system.

In addition to a PC verifying the accuracy of an instruction prior to execution, some systems actually read and relocate a small section of the user instruction sequence to another portion of memory. The section of processor memory vacated is then checked for proper operation through a predetermined set of error-checking routines. If the vacated memory segment is found to be in total operating order, the previously stored user instructions are returned and verified prior to moving on to another area of user memory for checking.

The user memory and logic solving portions of a PC are not the only areas subject to possible failure. The input and output statuses must be handled properly within the processor. Both input and output information is stored in an intermediate memory storage area prior to use by the processor and/or I/O hardware. The transference of current I/O statuses is subject to error conditions.

One such failure involves the processor using input statuses that have been stored in the input status storage area but have not been properly updated with the latest real-world conditions. Conversely the output hardware could be updated with statuses stored in the output status storage area, which have not been updated with the results of the latest processor user logic solutions. Both of these failures are the result of the I/O system not properly verifying and/or updating the latest system states and statuses.

Most PCs rely on a system of electronics to transfer the latest input module statuses into processor memory while transferring the latest output states from processor memory to the output modules. Failure of this circuitry, often called the *input-output processor* (IOP) circuitry, will result in an update failure similar to that just described. These types of failures can be detected by the processor and/or the IOP circuitry as each sends error-check information back and forth. Should either the processor or the IOP circuitry detect that the self-checking statuses have not changed according to a predetermined arrangement, a system fault can be flagged and the proper system shutdown sequences initiated.

An often considered failure relates to the IOP circuitry incorrectly addressing the correct input or output module or processor input or output status storage area. This type of failure results from either of two sources. The first is a failure of the addressing encoder/decoder electronics making up the IOP circuit. The second is a failure of incorrect addressing of an I/O module by the installer during system assembly. Either of these failures is detectable through the use of proper error detection routines and procedures. These procedures include multiple transmissions of data and addresses between the I/O modules and the IOP circuitry, coupled with voting processes to ensure accuracy. Again error detection schemes included as part of the transmission protocol can also uncover possible failures. An overall PC I/O system design that includes electronic circuitry in each I/O module as well as in the processor to identify the "fingerprint" and overall health of an installed module can eliminate many system addressing errors.

A related failure often occurs with the I/O module electronics. Any module could fail to recognize its assigned address. Again, proper I/O module diagnostics, combined with data and address transmission diagnostics from the processor IOP circuitry to the I/O modules, can detect most of these failures. As later chapters will explore, many PC manufacturers practically eliminate these forms of errors by using a fixed I/O system addressing structure where each I/O module is addressed in a hard-wired fashion back to the IOP circuitry, rather than using a user configurable I/O addressing structure.

A final area of frequent concern is the actual failure of an I/O module. For example, the electrical isolation portion of an I/O module could fail in such a manner that even though an input device is still sending correct signals to the input module, the module's isolation circuitry is sending a continuous ON or OFF signal back to the processor. Likewise the IOP circuitry could be sending correct processor statuses to an output module, but the isolation circuitry of the output module could continuously turn the output point either ON or OFF. While these failures are PC-related, the best correction lies in using redundant I/O points on multiple modules to ensure reliability.

For example, an input signal could be brought into a PC through three separate input points, each located on a different input module. The status of these inputs can be compared in a voting process, with the majority ruling the actual state that is used. Anytime the three points do not agree, an alarm can be sounded so that immediate attention can be called to the mismatch to effect a correction. Likewise multiple output points can be connected to ensure that an output device is either activated or deactivated on processor command. One possible wiring configuration is to wire two output points on different output modules in series to the output device, ensuring that the device can be deactivated if one of the output points fails ON. To ensure that a particular output device is turned ON, two output points, possibly on different output modules, could be wired in parallel to the output device. Should an output device require positive OFF/ON control, the two schemes could be combined to provide the necessary fail-safe operation.

While every PC incorporates diagnostics to locate internal problems and malfunctions, every PC manufacturer addresses the possibility of component failure during the actual manufacturing of the system. A typical PC manufacturer will require that all electronic components (resistors, capacitors, integrated circuits, transistors, etc.) undergo continuous quality control inspections. Incoming components, which have been purchased according to stringent specifications, are tested in test chambers that operate these components at specified loads and ratings during temperature cycling between -55 and $+150°$ C. This type of testing forces marginal units to fail early and provides an indication of those units that have internal weaknesses which might lead to early failure in actual operation.

Assembled circuit cards and subassemblies are also subjected to the same battery of tests. Additional tests may include optical/visual inspections to detect other imperfections, again the object being to detect any marginal components or assemblies and internal weaknesses. At this stage of production the test chambers and hardware are usually designed to pinpoint the source of a problem for quick repair, the repaired assembly being reentered into the testing procedure to ensure its proper operation before being passed to the next production stage.

Finally, the finished processors, I/O modules, power supplies, and associated accessories are individually tested. Temperatures, humidity, and line voltages are varied with the unit operating under various loads to ensure that the PC is ready for shipment. Many manufacturers even select units randomly from their distributors for return to their manufacturing facilities. These units are retested to ensure quality operation, and in some cases they are actually destroyed to ensure that product reliability is being maintained.

While no electronic device is absolutely free of possible malfunction, the majority of malfunctions can be detected and dealt with before serious problems have time to develop. Most manufacturers realize the investment made by a PC user in PC control hardware and in physical hardware controlled by the PC. Failures of a single PC component can quickly turn a multimillion dollar production facility into red-ink operation. PC manufacturers recognize the possibility of failures similar to those previously outlined, and allow for error detection routines and internal operation redundancy to reduce the possibility that a malfunction will occur and possibly cause damage. Every manufacturer will generally discuss the methods used to combat the PC failures noted above. Of course PC manufacturers will not take responsibility for the actual use of their equipment, but most of them go to great lengths to assist the user in the proper design, fabrication, installation, operation, and maintenance of any system relying on their equipment.

The PC and the light bulb share one thing in common. They both must come from the box, in working order, barring any shipping damage or rough handling, and provide the user with the expected service and operation life that was advertised. Any manufacturer allowing quality control, delivery dates, or product support to slip quickly sees the competition increasing sales and profits. Nothing is harder to overcome than a poor reputation in the marketplace. Addressing the fail-safe qualities of a PC as well as the quality control during manufacturing goes hand in hand with offering a quality product.

The remaining portions of this text address the design and hardware associated with the typical PC. The reader's ultimate selection of a PC to meet the requirements of a particular application must not be solely based on typical PC capabilities, as will be discussed in the following chapters. The product being purchased must be evaluated for its quality of construction and degree of reliability as well. The ability to perform a particular operation, or provide a particular instruction, is worthless if the PC will not operate.

PART II

Hardware

3

Input/Output
Structures

The input/output (I/O) system comprises the interface section of a PC's architecture. In the majority of PC systems, the I/O system will usually be the largest single component, with the exception of the process or mechanical hardware being controlled. The I/O structure is the interface between the user's field I/O devices and the control logic programmed in the PC memory. While the actual components that comprise an I/O system will vary, all I/O systems require a power source, a rack for mounting of the I/O interface modules, the individual I/O modules themselves, and the necessary electronics to communicate with the processor. As will be discussed in the section on I/O system design, all I/O systems are arranged for either local or remote operation.

The best way to understand the functional operation of a PC's I/O structure is to examine each of various components with respect to its design requirements and functional use. Since it is the I/O interface module that forms the heart of any I/O system, it will be examined first. There are two basic types of I/O interface modules utilized for I/O systems. The *input module* is designed to receive signals from the user's field input devices and to condition and isolate those signals for use by the processor. An *output module* conditions and isolates a signal from the processor for use in activating or deactivating a user output load device.

Throughout the discussion of I/O systems, the term "I/O module" will be used to refer to the electronics assembly that provides the interface between the user's field devices and the control logic of the processor. Various manufacturers use the equivalent terms "I/O module," "I/O board," or "I/O card" to refer to the electronics assembly that interfaces the user's field devices with the processor, and the designer should not be confused by the

31

interchangeability of these terms. The terms "I/O module," "input module," and "output module" are used in this text as generic names for the I/O interface electronics.

INPUT MODULES

Examination of a block diagram for an input module, Fig. 3.1, reveals that there are six major subdivisions of electronic circuitry. The input module receives signals from the user's input devices through wired connections to

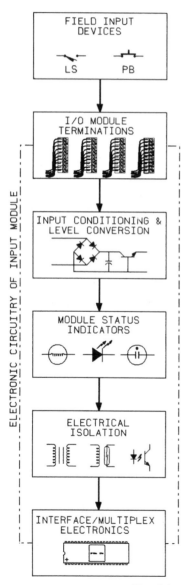

FIELD INPUT DEVICES
PROVIDES THE "SIGHTS AND SOUNDS"
OF THE CONTROL SYSTEM TO THE PROCESSOR

I/O MODULE TERMINATIONS
THE INTERCONNECTION MEANS
BETWEEN THE REAL WORLD DEVICES
AND THE PC CONTROL SYSTEM

CONDITIONING AND LEVEL CONVERSION
CONVERTS THE FIELD SIGNALS TO
LOW-VOLTAGE LEVELS AND FORMATS
USABLE BY THE PC PROCESSOR

MODULE STATUS INDICATORS
PROVIDES A VISUAL INDICATION
AS TO THE FUNCTIONAL STATUS
OF THE INPUT MODULE POINTS

ELECTRICAL ISOLATION
ELECTRICALLY ISOLATES THE
FIELD SIGNALS FROM THE
PROCESSOR SIGNALS

INTERFACE/MULTIPLEX ELECTRONICS
ELECTRONIC CIRCUITRY THAT PROVIDES
HIGH SPEED "PICTURES" OF INPUT
DEVICE STATUSES TO THE PROCESSOR

Fig. 3.1.   Input Module Block Diagram

terminals on the module itself. These signals from the field devices can be of many voltage levels and signal forms. Once received on the module's printed circuit board, they are passed to a conditioning block for the necessary level conversions and signal conditioning required by the processor. In order to provide full electrical isolation between the power levels of the field devices and the low-signal levels of the processor, the conditioned signal is fed through an electrical isolation block. Once electrically isolated from the environment, the signal is multiplexed with other similar signals generated on the same module or on adjacent modules for transfer to the processor I/O communications section.

The input module terminals provide a means to attach the various field signal levels to the electronics of the module itself. In addition one or more signal level references may be brought to additional input module terminals for internal use by the conditioning electronics. Various types of input field wiring terminations are used, some being fixed to the module, others being removable by some sort of plug-in arrangement. Many manufacturers place the wiring terminations on the module itself, while others mount them as part of the support structure for the module.

The electronics of the signal conversion block of an input module will greatly vary depending on the manufacturer, the system architecture, and the type and class of signal conversion required for the particular module in question. The type of conversion can be classified as either singular or complex. The devices listed in Table 3.1 are representative of field devices that are commonly connected to the input modules of a PC. Any device listed as analog will usually require a complex conversion to be performed by the input module. Those devices listed as discrete generally require a single input point on an input module.

Singular or *discrete* conversion, as it is often called, pertains to one and only one input point per conversion circuit. Discrete input conversions are the simplest since they usually comprise signals of an ON/OFF nature. Complex input conversions, often called *analog* conversions, require the conversion and scaling of a continuously variable input signal. Instead of transmitting the ON/OFF status of the input device to the processor, as in the case of a simple conversion, an input module that performs a complex conversion will transmit

Table 3.1. Typical Input Devices

DISCRETE	ANALOG
LIMIT SWITCH	PRESSURE TRANSDUCER
PUSHBUTTON	TEMPERATURE TRANSDUCER
THUMBWHEEL SWITCH	LOAD CELL
PRESSURE SWITCH	FLOW SENSOR
PHOTOCELL	VIBRATION TRANSDUCER
RELAY CONTACTS	CURRENT TRANSDUCER
SELECTOR SWITCH	VOLTAGE TRANSDUCER
KEYBOARD CONTACTS	VACUUM TRANSDUCER
MOTOR CONTROLLERS	STRAIN GAGE
	VIBRATION TRANSDUCERS
	FORCE TRANSDUCERS

the input device's status as data to the processor in the form of a group of binary signals. The input modules listed in Table 3.2 are representative of those that would require the input module to perform either a singular or a complex conversion before transmission to the processor.

The signals received by an input module from the field devices will be either ON/OFF ac, ON/OFF dc, pulse trains, or continuously variable dc levels. Generally, the ON/OFF ac or dc signals are the simplest to convert to ON/OFF processor level dc signals. The ON/OFF dc input signal is easiest to convert since, in most cases, a simple resistor network can be employed. For the ON/OFF ac input signal, a bridge rectifier is used in conjunction with a resistor network to convert the ac input signal to direct current prior to level adjustment. Fig. 3.2 illustrates the more common forms of input conversion circuits found in PC I/O modules.

An additional part of the ac or dc conversion circuit is the addition of resistor/capacitor/inductor networks for the removal of electric noise spikes and the reduction of false input triggering due to field device contact bounce. Input modules without noise and contact bounce filtering are called *fast response* modules. The longer the filter time of the module, the less susceptible the input point will be to electric noise and contact bounce.

In the case of either a pulse train or a varying dc input signal, electronic counter circuits or analog-to-digital converter circuits must be employed for signal conversion. When the incoming signal is pulsed, a counter on the input module counts up the number of pulses per unit of time and sends the data to the processor as binary information. Variable dc input signals are compared to a reference level either supplied to the input module by the user or generated by the module, and the results of the comparison are sent as data to the processor. The variable dc signal supplied to the module may be either a current signal or a voltage signal, depending upon the application. Both of these conversion schemes are simply referred to as analog conversions.

Table 3.2. Types of Input Modules

AC TYPES	SPECIAL TYPES
12 VOLT AC	TTL SOURCE
24-48 VOLT AC	TTL SINK
120 VOLT AC/DC	5-30 VOLT SELECTABLE
220/240 VOLT AC	5 VOLT DC ENCODER/COUNTER
120 VOLT AC - ISOLATED	12-24 VOLT DC ENCODER/COUNTER
	THERMOCOUPLE
	ASCII INPUT
	GRAY ENCODER
DC TYPES	HIGH SPEED PULSE
12-60 VOLT DC	
12-24 VOLT DC FAST RESPONSE	
24-48 VOLT DC	ANALOG TYPES
12-24 VOLT DC SOURCE	1 TO 5 VOLT DC
12-24 VOLT DC SINK	0 TO 10 VOLT DC
48 VOLT DC SOURCE	-10 TO +10 VOLT DC
48 VOLT DC SINK	4 TO 20 MILLIAMPERES

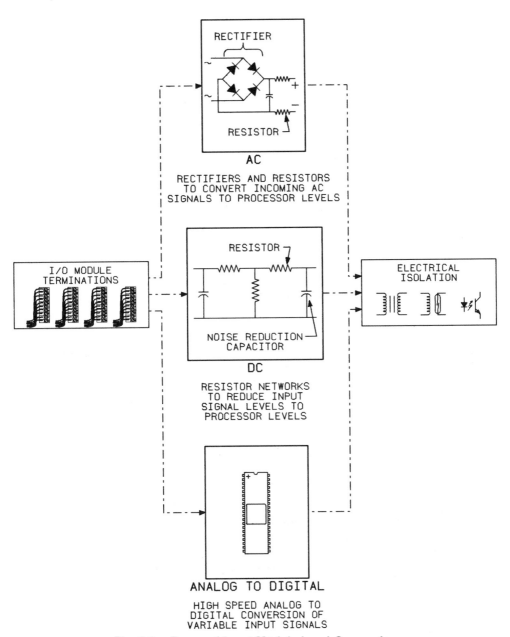

Fig. 3.2. Forms of Input Module Level Conversion

Once the incoming signal has been processed by the conversion electronics, it is passed to the isolation circuitry of the input module. The purpose of the isolation block is to separate electrically and mechanically the converted field input signals from those of the processor. This total isolation of field and processor electrical circuits and signals ensures that any electrical/electronic noise or voltage spikes that are not entirely removed during the signal conversion process do not cause electrical damage to the expensive processor electronics.

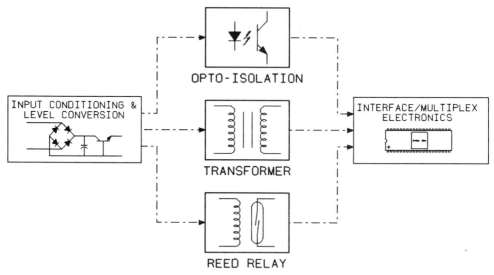

Fig. 3.3. Input Module Isolation Techniques

Three forms of isolation are generally employed, as indicated in Fig. 3.3. Most common is the use of an optoisolator. This device contains an encapsulated LED (Light Emitting Diode), which is powered by the input signal. A photosensitive transistor encapsulated in the same package acts like a light-actuated switch. The transistor is incorporated as part of the processor electronic signal circuit, and it is the use of light to transmit information and the mechanical separation of the LED and transistor that provide the needed isolation. Before the use of optoisolators, transformers were commonly employed. In lieu of the transformer, a reed relay was often used. The major drawback of the reed relay is its limited mechanical life when used for cyclic operation. Both transformer and reed relay forms of isolation are still employed by various manufacturers.

The indicator block of an input module (Fig. 3.4) assists the user in troubleshooting the module itself, and in checking the integrity of the field wiring. This diagnostic aid is in the form of indicator lamps located on the module for user reference. Many input modules have status indicators for

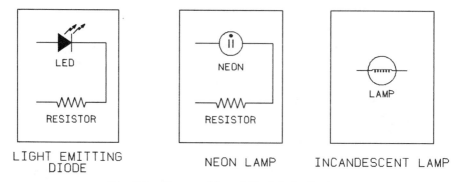

Fig. 3.4. Types of Input Module Indicators

determining whether the module is active, or that it is functioning properly. In addition, there are often indicators that provide status as to incorrect module operation or the current mode of operation. Where ON/OFF-type signals are supplied to an input module, a status light is often provided to verify the actual ON/OFF status of the input signal. The most common indicator devices used are LEDs and neon or incandescent lamps. The LED or the neon lamp is usually preferred due to its long operating life.

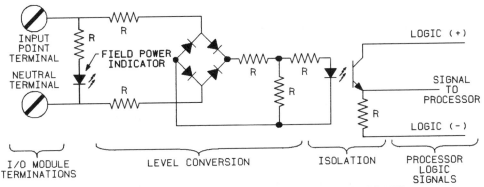

STATUS INDICATOR POWERED FROM FIELD POWER

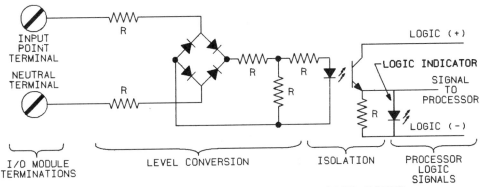

STATUS INDICATOR POWERED FROM LOGIC POWER

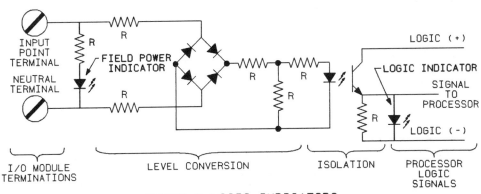

POWER & LOGIC INDICATORS

Fig. 3.5. Location of Input Point Status Indicators

The placement of the indicating device in the electric circuit of the input module is also of importance. Referring to Fig. 3.5, locating a status indicator in the module's field wiring circuitry prior to the isolation block provides a status indication of the field device only. When the indicator is placed as part of the electronics located after the isolation block, it will not only provide indication of the field device, but also that the signal is being properly converted and passed to the interface block on the module. Many of the newer input module designs provide both the logic and the field power indicators for user diagnostic aid. Modules offering both indicators provide the best user indication as to the proper functioning of the input point.

The final input module circuit block consists of the interface/multiplex electronics (Fig. 3.6). This circuit block is responsible for gathering all incoming conditioned and isolated field device signals, and transmitting them to the processor as requested. Operation of this block often requires various processor signals, such as system clock pulses, reset and enable signals, or input module address signals.

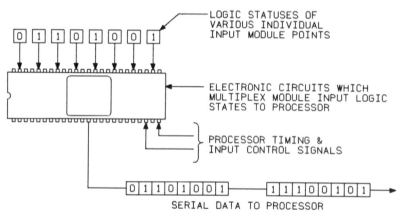

Fig. 3.6. Input Module Interface/Multiplex Electronics

OUTPUT MODULES

Now that the input module and its six subdivisions have been thoroughly investigated, an examination of the output module is in order. As the second type of I/O interface module, the output module is the PC component that permits the processor to communicate with its environment.

Examination of the block diagram for an output module (Fig. 3.7) reveals that there are seven major subdivisions of electronic circuitry. The output module contains an interface/multiplex block similar in function to that of the input module to receive signals from the processor. These signals are periodic in transmission from the processor and therefore must be "remembered" in some form of memory or latch block until the next processor update. The latch block continuously updates the isolation block, which provides full electrical and mechanical isolation between the processor signal levels and the field device

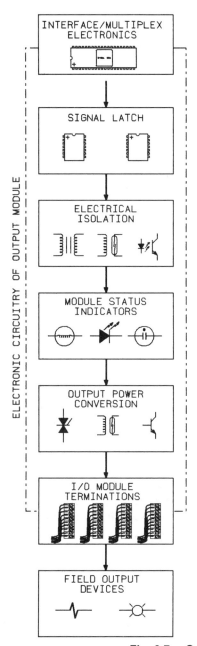

INTERFACE/MULTIPLEX ELECTRONICS

ELECTRONIC CIRCUITRY THAT PROVIDES HIGH SPEED "PICTURES" OF THE PROCESSOR'S DECISIONS AND ACTIONS

SIGNAL LATCH

ELECTRONIC CIRCUITRY TO 'GRAB AND HOLD' THE 'PICTURES" FROM THE INTERFACE/MULTIPLEX ELECTRONICS BLOCK

ELECTRICAL ISOLATION

ELECTRICALLY ISOLATES THE FIELD SIGNALS FROM THE PROCESSOR SIGNALS

MODULE STATUS INDICATORS

PROVIDES A VISUAL INDICATION AS TO THE FUNCTIONAL STATUS AND FUSE STATUS OF THE OUTPUT MODULE POINTS

OUTPUT POWER CONVERSION

TRANSFORMS THE LOW POWER LOGIC SIGNALS OF THE PROCESSOR TO A HIGHER POWER LEVEL CAPABLE OF OPERATING VARIOUS OUTPUT DEVICES

I/O MODULE TERMINATIONS

THE INTERCONNECTION MEANS BETWEEN THE REAL WORLD DEVICES AND THE PC CONTROL SYSTEM

FIELD OUTPUT DEVICES

ELECTROMECHANICAL DEVICES WHICH ARE THE "MUSCLES & SPEECH" OF THE PC CONTROL SYSTEM

Fig. 3.7. Output Module Block Diagram

power levels. The isolated processor signal is then applied to the output conversion/power block, where the power conditions or levels required to operate the field devices are switched to match the status of the incoming processor signal. Like the input module, the output module also contains a status block for monitoring the various operating and signal statuses, as well as terminals for the connection of the actual field output devices to the module's electronics.

The interface/multiplex block of the output module (Fig. 3.8) must gather the signals coming from the processor, decode them as to proper destination and condition, and pass them on to the proper output point destinations. This block requires numerous signals from the processor, including clock pulses, reset and enable signals, and addressing data, to sort out the processor signal information properly and direct it to the correct output destination. The interface/multiplex block will also generate reply and module status signals for communication and use by the processor.

The latch block of an output module receives the signals from the interface/ multiplex block and retains them until the next update is provided. The circuitry of the latch block usually consists of a series of electronic latches that store the current status of a single output or a group of outputs for later use by the isolation and output conversion/power blocks which follow. Communications between the processor and the interface/multiplex block ensure that the latch block is updated with current output statuses within a specified time period, or the latch block is cleared and all outputs are switched off.

The electrical isolation block of the output module is identical in design and function to its counterpart in the input module, as represented in Fig. 3.9. Electrical and mechanical signal isolation is provided by an optoisolator, a transformer, or a reed relay. The manufacturer's selection and implementation of the isolation circuitry used in the output module is usually more critical than the selection and implementation of the isolation circuitry of the input module. This is due in part to the specific character of the two types of modules. Input modules process low-current switch contact signals that generally do not produce much electric noise outside of that generated by the possible bouncing of the field switch contacts as they open or close. On the other hand, the output module must be able to control the power and current levels of the field devices connected to it. In many cases the output module contains a power transistor or triac as part of its output conversion/power block. The combination of power transistors or triacs and field device current levels reaching 1 or 2 amperes at initial turn-on can create sufficient electric and magnetic noise spikes to disrupt the sensitive electronic circuits of the

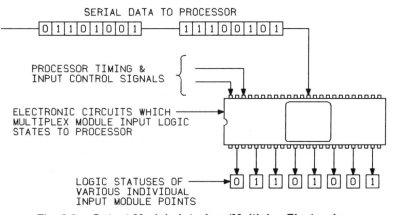

Fig. 3.8. Output Module Interface/Multiplex Electronics

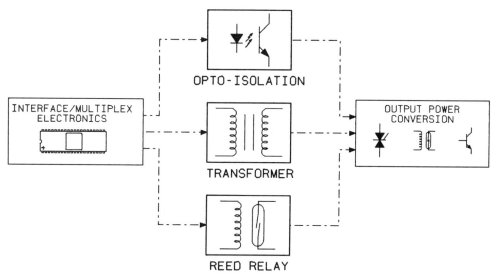

Fig. 3.9. Output Module Isolation Techniques

interface/multiplex block or even the processor. Proper selection and implementation of the isolator block within an output module will ensure that the electric and magnetic noise created by the control action of the module will have no effect on the operation of the PC or any of its associated parts.

The output conversion/power block transforms the signal or signals from the isolation block into field signal levels with sufficient current capacity to drive the output field device. The output conversion/power block requires that the user provide the module with the power source that will be used to power the output device. In addition, the output block may contain any necessary circuit protection devices to protect the output module, field wiring, and field devices from accidental short circuits or overloads.

The output block will provide either singular or complex output signal conversion as is required for the particular application. Singular power conversion will apply to only one output device or circuit per conversion. Table 3.3 lists typical output devices that would normally require a singular-type

Table 3.3. Typical Output Devices

<u>DISCRETE</u>

SOLENOID VALVE
MOTOR STARTER
PANEL INDICATOR
POWER CONTACTOR
LED DISPLAY
RELAY COIL
ALARM UNITS
HORNS
BELLS
WHISTLES

<u>ANALOG</u>

FLOW VALVE
AC DRIVE
ANALOG METER
TEMPERATURE CONTROLLER
FLOW CONTROLLER
DC DRIVE

conversion of an ON/OFF nature. In conjunction with the singular conversion, the output power block will usually employ a power-switching device to control the field device. Typical power switches include transistors, triacs, and power reed relays, as indicated in Fig. 3.10.

Connected in series with the power switch is the circuit protector. In most cases a fuse is employed as the circuit protector, the fuse being sized and rated

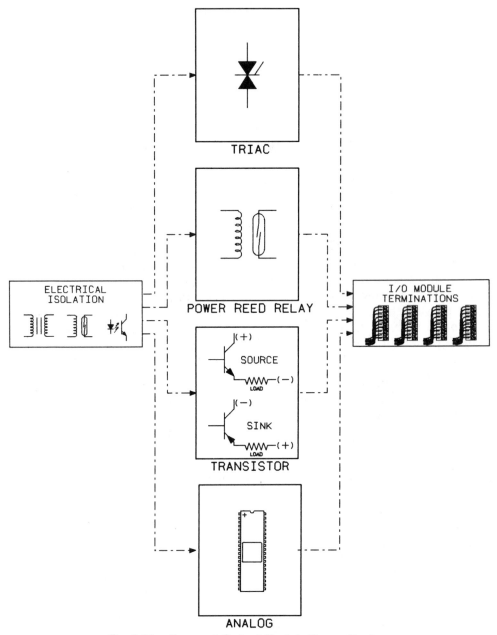

Fig. 3.10. Forms of Output Module Power Devices

to protect the module's power switching device and output module electronics. Depending on the PC manufacturer, the fusing arrangement may vary. The majority of manufacturers provide one fuse for each individual output point. Others choose to gang two or more output points together under the protection of a single fuse. The best protection for the solid-state switching components of an output module is provided where the manufacturer supplies a separate fuse for each output point of the module.

The user should be alerted to the fact that the fuse or fuses provided on an output module are specified by the manufacturer to protect the solid-state switching circuits of the output point. The characteristic curve of an output module fuse is selected to fall below the characteristic curve of the module's solid-state switching device it protects so that should a short circuit occur, the fuse can interrupt the flow of fault current before it exceeds the limits of the switching device. The fuses supplied as part of an output module are short-circuit protection fuses, selected by the manufacturer for the protection of the solid-state switching devices on the module. They *must* always be replaced with an *exact* duplicate that matches the PC manufacturer's specifications. If the user requires that the field output device circuit also be protected from overloads below the maximum rating of the output module points, additional fusing may be needed. Where optional overload fusing is supplied by the user in addition to the normal output module fusing, consultation with the PC manufacturer is recommended to ensure proper overload fuse selection and placement.

Complex output conversions are employed where the processor sends binary-coded data to the output module for use by the conversion/power block. These data may by converted to RS-232C signals for printer or computer communication, or they might be converted to an analog current or voltage by a digital-to-analog converter included as part of the conversion/power block electronics. Table 3.4 lists various output module types generally available.

The output signals controlled by the conversion/power block are provided to the user through terminals located as part of the output module assembly. The terminals may be either fixed to the module itself or designed in such a manner

Table 3.4. Types of Output Modules

AC TYPES	SPECIAL TYPES
12 VOLT AC	TTL SOURCE
24-48 VOLT AC	TTL SINK
120 VOLT AC/DC	5-30 VOLT SELECTABLE
220/240 VOLT AC	NORMALLY OPEN CONTACT
120 VOLT AC - ISOLATED	NORMALLY CLOSED CONTACT
	ASCII OUTPUT
	SERVO MOTOR
DC TYPES	STEPPER MOTOR
12-60 VOLT DC	
12-24 VOLT DC FAST RESPONSE	
24-48 VOLT DC	ANALOG TYPES
12-24 VOLT DC SOURCE	1 TO 5 VOLT DC
12-24 VOLT DC SINK	0 TO 10 VOLT DC
48 VOLT DC SOURCE	-10 TO +10 VOLT DC
48 VOLT DC SINK	4 TO 20 MILLIAMPERES

as to permit removal of the module without disturbing the wiring to the unit. The terminal section of the output module provides access to all output signals controlled by the module, as well as a termination point for any voltage or signal references required for operation of a field device.

An indicator block is provided as part of the output module to assist the user in the troubleshooting of the module, and in checking the integrity of the field wiring. Fig. 3.11 illustrates the more commonly found indicator types and their function. The output module may also have an indicator or indicators to signal the failure of circuit protection devices. Neon lamps and LED indicators are usually employed by the various manufacturers due to their long operating lives.

The electrical circuit location of the indicator with respect to the overall electric circuitry of the output module is important in providing a complete understanding of the true information being provided by the indicator, as shown in Fig. 3.12. When the indicator is connected as part of the latch block, it provides information on the signal status between the processor and the output module only. Modules that place the indicator as part of the conversion/power block indicate the status of the output signal to the field device, which indirectly provides an indication that the electronics of the module are operating correctly. Many of the newer output module designs provide both logic and field power indicators for user diagnostic aid. Modules offering both indicators provide the best user indication as to the proper functioning of the output point.

MOUNTING THE I/O MODULE

Closer examination of the electronics and hardware of an I/O system reveals that there is more than just I/O modules. The I/O module must be mounted in some form of holder and must be powered from a power source. I/O modules are mounted in holders called *I/O racks, I/O housings, I/O tracks, I/O chassis,* or *I/O bases.* While the various manufacturers use all of these terms interchangeably, there are several mounting methods in common use which

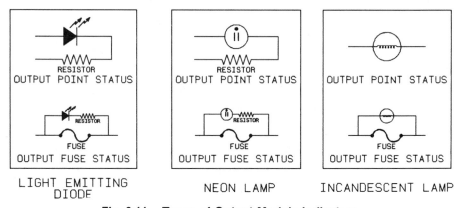

Fig. 3.11. Types of Output Module Indicators

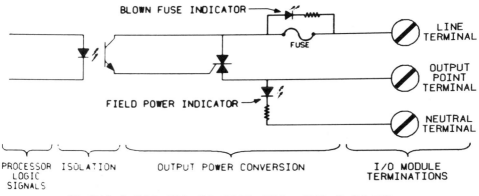

STATUS INDICATOR POWERED FROM FIELD POWER

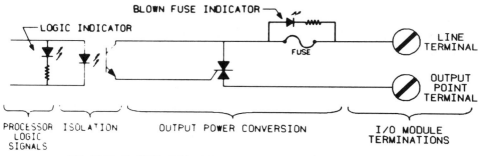

STATUS INDICATOR POWERED FROM LOGIC POWER

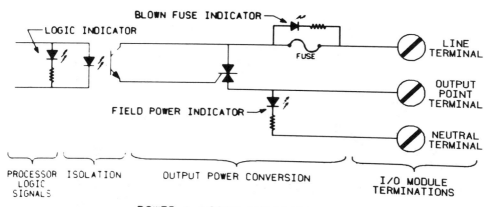

POWER & LOGIC INDICATORS

Fig. 3.12. Location of Output Status Indicators

should be understood. Many manufacturers offer several styles of I/O hardware assemblies, each style based on a different mounting design.

In order to familiarize the reader with the components and terminology of the various I/O structures available, the figures will cite various Industrial Solid State Controls, Incorporated (ISSC) PC hardware. ISSC manufactures several models of PCs and offers an extensive array of I/O modules and systems. In addition to ISSC's PC line, it offered the first public PC communications network. Called *COPNET*, this universal network permits diverse types of control hardware and competitive PCs to intercommunicate on a common communications line with little user programming.

The term *I/O housing* generally refers to an assembly that permits the I/O modules to be mounted in a vertical arrangement, one module beneath another. Multiple housings can be assembled together side by side to provide an I/O assembly of vertical columns and horizontal rows of I/O modules. Fig. 3.13 shows the ISSC B321 I/O housing containing various I/O modules.

In most PC systems, several I/O housings will be required to complete a *channel* of I/O capacity. Practically all PC I/O systems are subdivided into channels, with various criteria being used by a manufacturer in determining the final size of an I/O channel. Some of the more important guidelines used in sizing an I/O channel include the current capability of the power supply that will be used, the physical size of the complete I/O channel, and the format of the communications that will occur between the processor and the individual I/O modules in the channel. The I/O housing will contain multi-pin interconnected receptacles for the I/O modules to plug into, as well as a fastening method to secure the module.

A form of I/O mounting that is similar to the I/O housing approach is the I/O base or I/O track system often used by some manufacturers. This mounting method involves mounting the I/O modules side by side horizontally on a mounting assembly commonly referred to as the mounting base or track. A

Fig. 3.13. ISSC B231 I/O Housing with Various I/O Modules

Fig. 3.14. ISSC IPC-90 System with I/O Base and Simulator

group of module bases or tracks are then connected together vertically to form a complete I/O channel. ISSC offers the I/O base method of I/O module mounting with its IPC 90 series of PC. An IPC 90 I/O base with I/O modules is shown in Fig. 3.14.

A benefit found with using base-type mounting systems as compared with the housing type is the flexibility of arrangement provided by the short interconnection cables of the base mount system. These cables, required to interconnect the various I/O bases or tracks, permit the user to separate the physical distance and arrangement of the bases by several feet to accommodate the installation of other control system parts in the I/O enclosure. Each base or track assembly will contain the necessary number of interconnected multi-pin receptacles for the unit's I/O module capacity as well as a mechanical means to secure each module.

The I/O chassis system of I/O module mounting involves the use of an electronics-type card cage for the assembly of I/O hardware. The chassis consists of a metal frame in the form of an open box into which individual I/O modules can be inserted. The ISSC model 621-7000 I/O chassis is shown in Fig. 3.15. The bottom of the "box" consists of a circuit board referred to as a motherboard, which carries the necessary circuitry and sockets for the I/O

Fig. 3.15. ISSC 621-7000 I/O Chassis with Various I/O Modules

modules to plug into. The ends of the box are usually solid metal plates that support the motherboard and the top guide rail supports. They are also designed to support the finished assembly in an enclosure. The guide rail supports are located along the top and bottom edges of the box and provide support to the individual I/O module guide rails running between the supports and the motherboard.

The chassis mounting system is a nonmodular mounting system, and an I/O chassis will usually be designed to mount the full capacity of I/O modules available for an I/O channel. For I/O systems where the user may not require the full capacity of the I/O channel, some manufacturers provide a smaller I/O chassis, designated half-size and quarter-size chassis. The half-size chassis would permit mounting of half the I/O channel's capacity of I/O modules, and the quarter-size chassis would mount one-quarter of the I/O channel's I/O module capacity.

The term I/O rack is used interchangeably with I/O housing, I/O chassis, and I/O base by many PC users and manufacturers. I/O rack will be used in this text to indicate generically the hardware used to support a group of I/O modules. Where a specific system of mounting is referred to, the proper name of that mounting system will be used to avoid confusion.

I/O CHANNELS AND NODES

As previously noted, most PC I/O systems have their I/O modules grouped into channels. The arrangement of I/O modules into groups or channels is usually determined at the time of PC design and cannot be altered by the user. For the majority of PCs the number of I/O points contained or addressed by a single PC channel will usually be between 128 and 256. A channel may be composed of strictly input points, strictly output points, or a mixture of input and output points. A couple of quick tests, which can often be used to distinguish one channel of I/O points from another, is as follows: First, groups of I/O points in a channel may be numbered to reference the I/O rack in which they are mounted. Second, a group of modules may be fed from a common remote or auxiliary PC power supply. Those modules, which share a common power supply, may be part of the same I/O channel. Finally, a group of I/O modules may be attached to the processor using common I/O communications cabling and electronics. While these tests may not be foolproof, they can provide an indication of the I/O channel structure for a specific PC in question.

To provide more flexibility, several PC manufacturers permit a channel of I/O points to be further subdivided into *nodes*. An I/O node is simply a smaller portion of an I/O channel. To illustrate the difference, consider a PC that can address a maximum of 1024 I/O points. The PC processor has been designed to address four channels of I/O points, each channel containing a maximum of 256 I/O points. To increase the flexibility of the PC, the manufacturer permits any channel to be further subdivided into eight nodes, each node addressing a maximum of 32 I/O points. Any combination of I/O channels and nodes can be configured, from four channels of 256 I/O points per channel to 32 nodes of 32 I/O points per node. It should be noted that there may be restrictions on the distance a node may extend from its main channel, on the I/O-to-processor communications hardware, or in how it is applied. These restrictions, if any, can be obtained from the manufacturer.

I/O MODULE POWER SUPPLIES

All I/O systems require some form of low-voltage power source for the operation of the individual I/O module electronics. This source of power can be supplied either by the processor power supply or from an auxiliary or remote power supply mounted as part of the I/O system. The power supply will be designed to convert either 240 or 120 volts ac at 50 or 60 hertz to the proper dc voltage levels and current capacities required by the processor and/or the I/O system. The chapter dealing with power supplies and power sources provides a more in-depth examination of the power sources used with various I/O systems and configurations. The ISSC model 320-C2 remote I/O power supply is shown in Fig. 3.16. The 320-C2 power supply powers ISSC's B231 I/O housings. In contrast, Fig. 3.17 illustrates the more compact ISSC model 621-9931 I/O chassis power supply. Fig. 3.18 illustrates an ISSC model 621 I/O chassis full of assorted I/O modules powered by the 621-9931 power supply.

I/O POINTS

During the examination of input and output modules various subdivisions of each type were considered. In examining the functional operation of each module type, only one I/O point was implied per I/O module. Many I/O structures permit more than one I/O point to be contained on a module. The availability of more than one I/O point per module often decreases the cost of manufacturing the module since some of the module's electronics can be modified to access the additional I/O points instead of requiring additional duplicated electronics for each additional point. A cost savings may also result from the denser packaging possible with multiple I/O points per module.

Fig. 3.16.   ISSC 320-C2 Remote Power Supply

Fig. 3.17.   ISSC 621-9931 I/O Chassis Power Supply

Fig. 3.18.   ISSC 621 I/O System

The most common I/O module point capacities are one, two, four, eight, or sixteen I/O points per I/O module. These point capacities are for standard I/O modules. However, in the case of specialized modules any number of points per module can be found in addition to those listed above.

Most manufacturers offer I/O modules as strictly input modules or strictly output modules. There are I/O systems available which contain both input and output points on the same I/O module, as well as the more traditional dedicated input or output modules. All I/O systems generally have the same number of input points on input modules as they have output points on output modules for I/O modules of similar functions and voltage levels.

ISOLATED AND NONISOLATED I/O MODULES

I/O modules can be designed for either *isolated* or *nonisolated* field power source connection. In the isolated configuration each input or output circuit is electrically separate from the other circuits on the module. Isolated input and output modules enable the designer to wire field devices of the same voltage level to the same module, even though the field devices are powered from different sources (that is, various I/O devices fed from panelboards connected to different phases of the incoming power distribution system). Fig. 3.19 and 3.20 show typical isolated input and output module schematics, respectively, for the ISSC chassis I/O structure.

I/O modules that have one voltage rail of the supply system internally wired to more than one module's I/O point are considered to be nonisolated or common-connected. Nonisolated I/O modules require that all I/O points on a module which have been connected together via a common or neutral rail receive power from the same source. Often a manufacturer will subdivide a module with four, eight, or sixteen I/O points into subgroups of two, four, or eight I/O points per group, respectively, in order to provide greater flexibility for the user. Due to a limited number of wiring terminals available on some I/O modules, I/O points may be forfeited on some systems in order to provide isolated capability. Figs. 3.21 and 3.22 show typical nonisolated input and output module schematics, respectively, for the ISSC chassis I/O structure.

I/O MODULE POWER RATINGS

All I/O modules have definite power ratings, which must be observed for proper I/O system operation. While the power ratings for input modules are usually specified as a function of the load current that the module draws from the field source, it is the power rating of the output card that is frequently overlooked. All output module points are rated for a maximum current per point. This maximum current is determined by the manufacturer at the time of module design and is usually derived from the heat dissipation capacity of the module. In addition an output module may have a maximum total module power capacity which must not be violated. The output module's total power dissipation is again determined at the time of design and is directly related to the total heat dissipation qualities of the module when installed with other modules in an I/O rack.

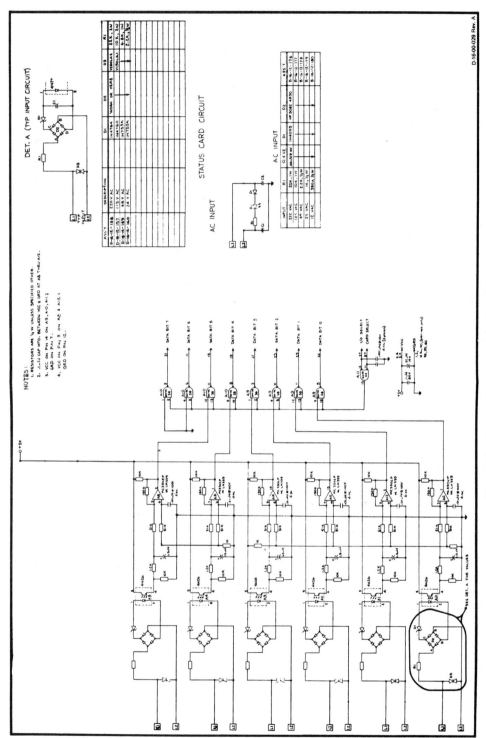

Fig. 3.19. Input Converter—Isolated Inputs

53

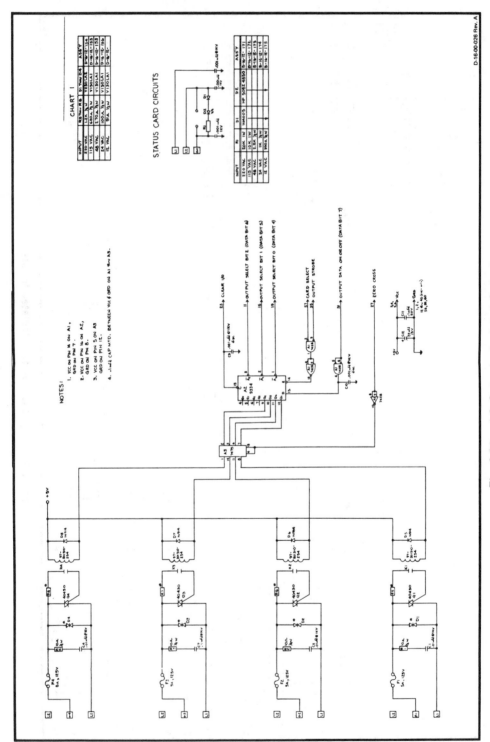

Fig. 3.20. AC Output—Isolated Outputs

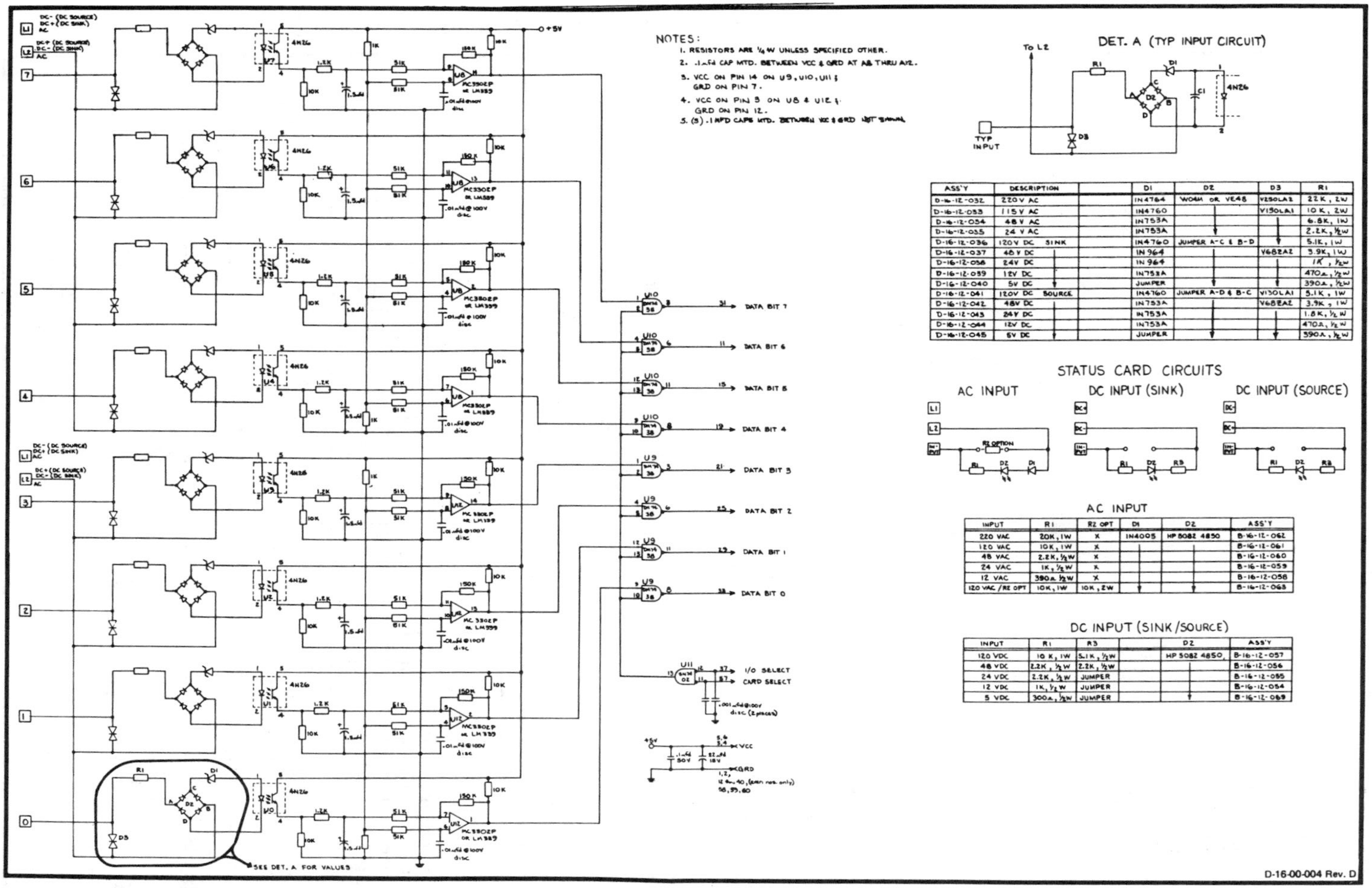

ASS'Y	DESCRIPTION		D1	D2	D3	R1
D-16-12-032	220V AC		IN4764	WO4M OR VE48	V250LA2	22K, 2W
D-16-12-033	115V AC		IN4760	↓	V130LA1	10K, 2W
D-16-12-034	48V AC		IN753A	↓	↓	6.8K, 1W
D-16-12-035	24 V AC		IN753A	↓	↓	2.2K, ½W
D-16-12-036	120V DC SINK		IN4760	JUMPER A-C & B-D	↓	5.1K, 1W
D-16-12-037	48V DC ↓		IN964	↓	V682A2	3.9K, 1W
D-16-12-038	24V DC ↓		IN964	↓	↓	1K, ½W
D-16-12-039	12V DC ↓		IN753A	↓	↓	470Ω, ½W
D-16-12-040	5V DC ↓		JUMPER	↓	↓	390Ω, ½W
D-16-12-041	120V DC SOURCE		IN4760	JUMPER A-D & B-C	V130LA1	5.1K, 1W
D-16-12-042	48V DC ↓		IN753A	↓	V682A2	3.9K, 1W
D-16-12-043	24V DC ↓		IN753A	↓	↓	1.8K, ½W
D-16-12-044	12V DC ↓		IN753A	↓	↓	470Ω, ½W
D-16-12-045	5V DC ↓		JUMPER	↓	↓	390Ω, ½W

AC INPUT

INPUT	R1	R2 OPT	D1	D2	ASS'Y
220 VAC	20K, 1W	X	IN4005	HP 5082 4850	B-16-12-062
120 VAC	10K, 1W	X	↓	↓	B-16-12-061
48 VAC	2.2K, ½W	X	↓	↓	B-16-12-060
24 VAC	1K, ½W	X	↓	↓	B-16-12-059
12 VAC	390Ω ½W	X	↓	↓	B-16-12-058
120 VAC / R2 OPT	10K, 1W	10K, 2W	↓	↓	B-16-12-063

DC INPUT (SINK/SOURCE)

INPUT	R1	R3		D2	ASS'Y
120 VDC	10 K, 1W	5.1K, ½W		HP 5082 4850,	B-16-12-057
48 VDC	2.2K, ½W	2.2K, ½W		↓	B-16-12-056
24 VDC	2.2K, ½W	JUMPER		↓	B-16-12-055
12 VDC	1K, ½W	JUMPER		↓	B-16-12-054
5 VDC	300Ω, ½W	JUMPER		↓	B-16-12-069

Fig. 3.21. Input Converter

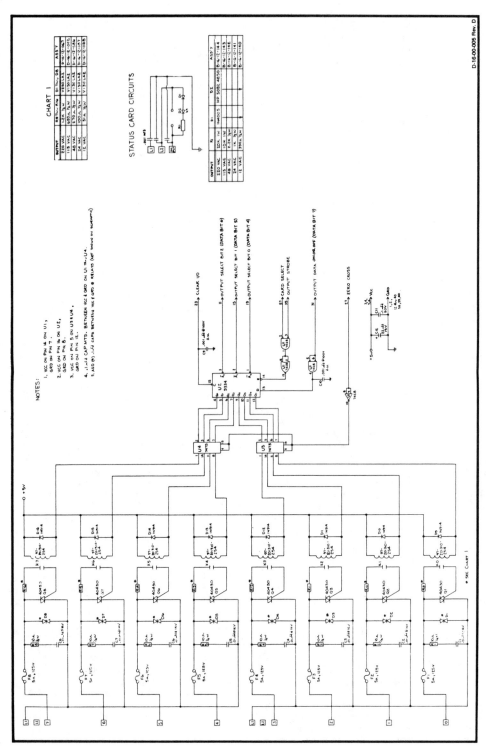

Fig. 3.22. AC Output

In conjunction with the maximum power ratings are minimum power ratings which must be observed for proper module operation. An input module will specify a minimum current, which must be supplied by the field input device if the module is to detect the input signal. Output modules usually carry a specification listing the minimum current that the field output device must draw for the module to function correctly. This minimum output current is of prime importance for ac triac-type output modules if the module is to turn OFF completely. Often operator console indicators are found to glow faintly when the indicator does not draw sufficient triac OFF-state load currents. This glow can be corrected by the addition of a properly sized resistor and/or capacitor connected in parallel with the load device. Chapter 4 provides detailed information and examples relating to the proper calculation, selection, and connection of a load resistor/capacitor. Note that the use of a resistor alone requires that the resistor dissipate the full power developed by its load current. This often requires the use of resistors of 15 watts or more. A bank of 15-watt resistors, all ON, can generate quite a lot of heat, which must be removed from an enclosure if the melting of insulation and the overheating of electronics are to be avoided. The use of a resistor/capacitor combination can eliminate most of this problem.

GENERAL I/O MODULE SPECIFICATIONS

In addition to the power specifications for an I/O module, specifications are usually quoted concerning the general operation of the module. The number of I/O points for which the module is designed, along with the voltage range at which it will operate, are of prime importance. Some I/O modules require special installation requirements in the I/O rack, such as the number of module positions occupied by the module or the installation of a special power source or additional cooling. All special considerations for the module will be noted in the specifications.

Often an I/O system must be installed near heating devices or in an outdoor watertight enclosure. In these types of installations the temperature and humidity specifications are especially important. Special heating and/or cooling apparatus may be required along with humidity control to ensure proper system operation.

Most I/O modules employ 1500-volt isolation devices between the field power supply and the controller logic power system. The isolation capability of the module should be stated for the user.

Many I/O structures are designed to allow the user the option of keying and interlocking the I/O modules in the I/O rack. A plastic key is provided with the module for the installer to insert in the I/O rack at the time of assembly, which only allows a particular catalog number I/O module to be inserted in the keyed I/O rack slot. I/O modules may also be interlocked electrically or mechanically as an option to permit the user the flexibility to interlock all modules in an I/O rack so that removal or tampering of any one module will initiate an alarm or shut down the I/O rack and/or processor.

A schematic must be provided from which the installer wires. This schematic need not include the module's electronics, but should include a diagram of the input conversion or output power block for each I/O point located on the module. Included as part of this diagram are all terminal designations and circuit busing for the module. The wire size range and type of terminal system used on the module for the user's field wiring must be indicated.

Input modules should specify the input impedance for each input point, while output modules should note the OFF-state leakage and the ON-state voltage drop for each output point. All I/O modules have a certain amount of delay from the time a field input changes to the time it is recognized by the processor, or from the time the processor changes an output state to the time when the output module actually changes the field device's state. This delay time is specified as the response time of the module, and represents the maximum switching time or frequency of the module. Many modules carry a dual response specification, which includes an OFF-to-ON time as well as an ON-to-OFF response time.

All I/O modules that incorporate circuit protection should specify the type of protection, and in the case of fusing, should note the manufacturer and catalog number of the fuse used in the module.

I/O modules that incorporate special electronics, such as counters, digital-to-analog converters, or analog-to-digital converters, usually specify such parameters as resolution, accuracy, linearity, and common-mode rejection. It should also be stated how and when the module is to be recalibrated.

4

Input/Output
Configurations

The proper design and implementation of an I/O structure for a PC system is one of the most critical jobs the designer or engineer must perform. The design of the I/O structure involves gathering information not only on the various PCs available to meet the application, but also on the actual application itself. All I/O system designs will be dictated in part by the application, but the majority of the decisions that must be made will depend on the designer or engineer thoroughly understanding the operation or process, so that the proper questions can be asked to obtain the correct answers. Many hardware decisions can have an impact on the processor's software, and as such can make the programmer's job either easier or more difficult.

Even though the processor is usually the single most expensive component in a PC system, the cost of the I/O hardware can easily run several to many times the processor's cost, depending on the system size. Proper selection of the I/O hardware and its implementation begin with the designer or engineer gathering data on the control requirements for the system. This information will usually be compiled from numerous sources, and is best formulated in written form. Data on the operating requirements and human operator interfaces to the control hardware will require input from operating and production personnel. Operators and maintenance technicians will be knowledgeable in diagnostic aids to speed downtime troubleshooting. Installation and startup of any system will require coordination between the construction personnel installing the control system and mechanical hardware, and the plant operators and maintenance personnel who will eventually run and maintain the system. In any control system there will be new hardware components to consider as replacements for older ones, as well as improvements in existing technology.

Before design of the I/O structure and the overall control system can commence, all of the areas noted above must be considered. The results of the above considerations should produce a document detailing the general design and operational requirements for the new control system and hardware.

In final form, the document should contain all nonconfidential operating requirements for the system, such as operating duty, environment, hazardous areas or chemicals involved, and operator interfaces. A thorough description of the mechanical hardware or process should follow, detailing all hardware operation, and where and how the PC system will interface and operate that hardware. If prepackaged units are to be purchased, details of their operation should be requested as part of the vendor's initial quotation in order that the PC designer can consider the interface between the PC controls and the control system that may come as part of the prepackaged unit. Any special control requirements and hardware must be defined and described. All safety devices and considerations must be examined and included where necessary. For an existing plant or operation, the manufacturers of any existing control system hardware should be indicated for compatibility and consideration when similar components are selected for the new system. Finally a list of those manu-facturers that have equipment suitable for the new application should be given for consideration and evaluation.

Once prepared and agreed upon, this document, often called a *system specification*, becomes the basis for the design of the new control system. It will be used as the building block for all control system hardware and specifications. This document will enable the designer or engineer to lay out the initial system and to begin the proper assignment of control system field devices and hardware. Once the overall system has been designed, consultation with the various PC manufacturers' applications engineers can commence.

Design of a PC system begins with the selection of remote and/or local I/O hardware for the application. Once the I/O system configuration is decided upon, selection of the individual I/O components can begin. Selection of the I/O system components will often require consultation with the various PC manufacturers' applications engineers in order to select the best possible I/O configuration for the application at hand. Once the system configurations are complete for all models of PCs under consideration, cost comparisons can be performed. Selection of the final system design and the model of PC(s) to be used will then be reduced to an evaluation of system costs and hardware capabilities.

LOCAL AND REMOTE I/O SYSTEM SELECTION

PC I/O hardware is designated as *local* when the I/O modules are installed in the immediate vicinity of the main processor. This distance is commonly determined by the interconnection cable length between the processor and the I/O hardware. Due to voltage drop considerations and other criteria, this cable length is usually a maximum of 75 to 100 cable feet. When it is desirable to locate the I/O structure some distance from the processor, the I/O hardware will be configured for *remote* operation. Fig. 4.1 illustrates the differences between local and remote I/O hardware configurations. While it is possible to configure a remote I/O system for local operation, use of a remote configuration will permit the I/O hardware to be located up to a maximum of 15,000 to 20,000 cable feet from the processor. Note that the processor of Fig. 4.1 has the I/O hardware configured for both local as well as remote operation, a configuration commonly found in industry.

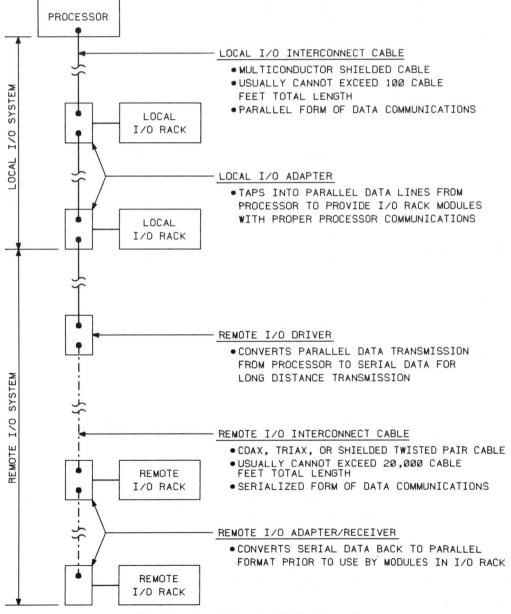

Fig. 4.1. Local/Remote I/O Configuration

There are several considerations that must be examined when the choice is made between a local and/or a remote I/O configuration. The user must first determine the quantity and function of the I/O devices. Then the overall locations of these devices must be known to determine whether they can be arranged conveniently into function groupings. A *function grouping* is a gathering of I/O devices that can be grouped together on a machine or within a plant or process because of their close proximity to each other, or because of their control function similarity. Once the various I/O groupings, if any, have been decided upon, selection of either a remote or a local I/O configuration can begin. In some instances a combination of both local and remote I/O hardware will provide the most cost-effective solution for the overall I/O design. The user should also keep in mind that some PC systems permit one or more I/O channels to be subdivided into nodes for additional flexibility.

LOCAL I/O SYSTEMS

Local I/O configurations are usually best suited where only one or two function groups of I/O devices exist and it is desirable to have the field I/O devices located within a short distance from the processor. Fig. 4.2 details the I/O system components of a local I/O installation. Usually local I/O systems are installed in single enclosures with the processor, and the PC system serves only a single application or function. When performing service or maintenance on local I/O systems, a total control system shutdown is usually required since both the processor and the I/O structure may receive power from the same common power supply.

REMOTE I/O SYSTEMS

If the user has a control requirement where the function groups are distributed over a large portion of the site, or where the control system will be utilized to control hardware in a multiple application or function environment, it is often more cost effective to use a remote I/O hardware configuration connected to a single processor. Remote configurations of I/O hardware can be located several feet to several thousand feet from the processor, and include the hardware detailed in Fig. 4.3. When a PC user applies the remote I/O concept to I/O devices located in various plant locations, the user must install the PC I/O hardware in multiple enclosures referred to as *remote I/O enclosures.* This type of installation practice affords the user the ability to distribute the I/O hardware according to the needs of a particular application or system without the need for multiple processors. Remote I/O may often be employed when it is desirable to shut down a portion of a process or control system for modification, troubleshooting, or maintenance while the remaining equipment operated from other remote I/O locations continues to function normally.

MULTIPLE PROCESSOR SYSTEMS

As an alternate to the user employing a central processor with remote I/O, several PC manufacturers offer a *control network* that allows for instant, user-

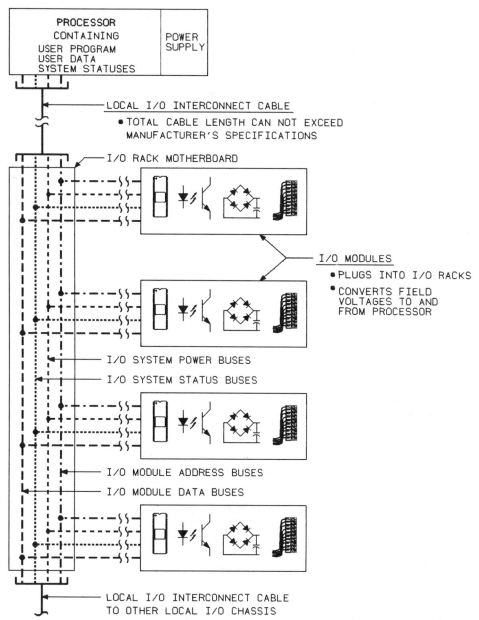

Fig. 4.2. Local I/O Configuration Components

transparent intercommunication of data and system statuses between a select groups of PCs. The user may want to consider the use of several smaller-sized PCs coupled together by a control network in lieu of a single noncritical processor with a large remote I/O structure for any application where loss of the PC processor for short periods of time leads to large areas of a facility being inoperative. Each of the smaller PCs might be configured with a local I/O arrangement, and each PC would then transmit needed system statuses and conditions to a requesting PC via the control network. Besides allowing the user many of the benefits of a remote I/O system controlled by a single large processor, an overall system cost savings may be realized for certain applications.

When large amounts of noncontrol-type information must be passed between two or more PCs, numerous manufacturers have developed industrial *local area networks* (LANs) for use with PCs and other industrial control hardware. These networks permit large amounts of data to be communicated between multiple PCs and other computer-based systems at high rates of speed.

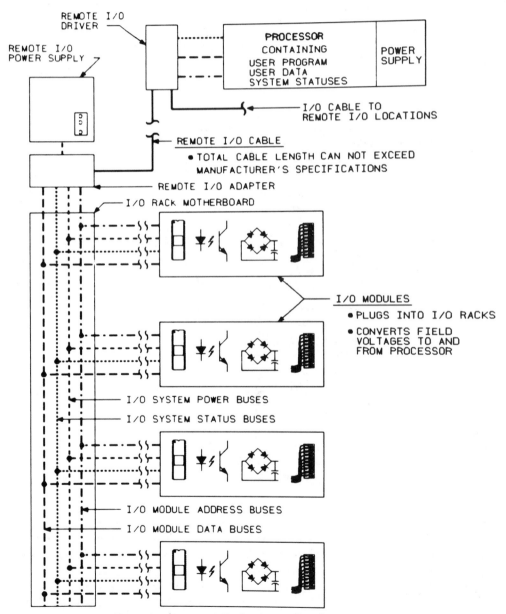

Fig. 4.3. Remote I/O Configuration Components

While these local networks, often referred to as *PC data highways,* cannot usually provide the same split-second control accuracy of the control network, they do provide for the transmission of large quantities of data between network devices, something a simple control network is not capable of. Chapter 15 deals in greater depth with PC-to-PC communication networks and should be consulted for additional information.

I/O CHANNEL DESIGNATION

Having decided on the functional groupings to be employed in the application, the control system designer must now assign I/O channels for each of the various functional groups. As a reminder from the previous chapter, an I/O channel is a group of input or output points or a mix of input and output points that is serviced as a unit by the processor. Before assigning the I/O channels, the designer must become acquainted with the I/O channel configurations for each of the various PC models being considered. Most smaller PCs on the market usually encompass a single channel, while larger systems may be able to handle more than 60 separate I/O channels. Each channel will be allotted a specified number of input points and a specified number of output points, the output points usually equaling the input point specification. As an alternate possibility there are I/O systems that do not specifically dedicate a fixed number of input and output points per channel, but rather a maximum number of I/O points per channel, and leave the mix ratio of input and output points to the user to determine as need dictates. In either arrangement it is the total number of input points and the total number of output points per channel times the number of channels that determines the total I/O capacity of the PC system. The number of I/O points per channel and the number of I/O channels are determined by the manufacturer during PC design and cannot be changed by the user. Some PC models offer the user the ability to designate which I/O points will be active within a particular I/O channel, while others allow total freedom in the assignment of the individual I/O point addresses as well as the final utilization of each I/O channel.

Knowing the I/O capacity per I/O channel, the number of I/O channels available on a particular PC, and the number of I/O points per function group, the designer can now assign I/O channels to meet application needs. It may be necessary to assign multiple I/O channels for a single function group due to the fact that there is not enough I/O point capacity in a single channel to handle the quantity of I/O points required by the functional group assignment. Before finalizing the channel assignments, it might be advantageous to modify the functional group assignments in order to utilize the channel capacities better. Often there are I/O devices in heavily loaded channels that might be relocated to lesser loaded channels in order to free I/O space in the full channel. Careful consideration paid to I/O channel capacity, I/O system hardware utilization, and function groupings at this time may result in decreased hardware costs and a less complex system for later programming and troubleshooting.

It would also be advisable to include space for spare I/O points and future growth in the I/O channels at this time. Most vendors recommend allotting 10 percent spare input points and 10 percent spare output points per I/O housing. This recommendation can be adjusted downward for duplicate systems and upward for prototype systems.

While the next step in PC I/O design may appear to be the simplest to perform, it can be the most time-consuming. Before any PC system can be evaluated financially, the hardware necessary for its operation must be specified. This step also requires that the designer understand the I/O hardware available from the various vendors under consideration, along with the various ways in which it might be electrically and mechanically configured. Perhaps it might be best at this time to consult the various vendors' applications engineers for assistance since there may be several attractive I/O hardware configurations from which to choose.

I/O DESIGN EXAMPLE

To better understand the steps necessary in selecting the proper components for a manufacturer's I/O system, a hypothetical control system will be used as an example. The Struthers-Dunn Director 4001 PC will be used as the PC under consideration for this example. The Director 4001 PC provides 2 K, 4 K, or 6 K of RAM; local or remote I/O capacity to 384 discrete I/O and 32 digital/analog data I/O; plus such options as PID control, EPROM, data handling, and computer interfaces. The control application for which the Director 4001 is being configured involves one function group of local I/O devices within several feet of the processor and a second function group located 2200 cable feet from the processor. Fig. 4.4 has been prepared as a reference of the types and quantities of field I/O devices that are to be connected to the processor at each functional group.

The Director 4001 will address up to 12 channels of discrete I/O points (eight channels on the discrete channel bus and an additional four as discrete channels on the data channel bus) and four channels of data I/O. These I/O channels may be located as local and/or remote channels with respect to the processor. A discrete I/O channel addresses 32 I/O points per channel, giving the system a total capacity of 384 discrete I/O points for the 12 channels. The four data channels will each address eight 16-bit data I/O modules, for a total system capacity of 32 16-bit data I/O points. The remote I/O channels can be located a maximum of 4000 cable feet from the processor for a wired remote I/O system, or 3300 cable feet from the processor for the fiber-optic remote I/O system. In addition to the standard discrete I/O structure, a high-density discrete I/O structure is also available. This high-density structure allows two channels' worth of I/O hardware to be combined into a single I/O rack. Fig. 4.5 diagrams the I/O capacity for the Director 4001 PC.

The first design step in configuring the Director 4001's I/O structure for the example application is the sorting out and tallying up of each specific type of I/O point per functional group. The various I/O points are sorted out by function (discrete versus data), voltage level (120 volts ac; 24, 12, or 5 volts dc), and type (input or output). At this time it is often advantageous to list the

LOCAL FUNCTION GROUP DEVICE QUANTITY/VOLTAGE	I/O DEVICE	REMOTE FUNCTION GROUP DEVICE QUANTITY/VOLTAGE
10/120 VAC	FVNR MOTOR	
3/120 VAC	FVR REVERSING MOTOR	
26/120 VAC	PUSHBUTTON	14/12 VDC
16/120 VAC	INDICATOR	23/12 VDC
	LIMIT SWITCH	17/120 VAC
	SINGLE SOLENOID	16/120 VAC
	DUAL SOLENOID	18/120 VAC
	ALARM HORN	6/ CONTACT TYPE
2/24 VDC	7 6 9 3 THUMBWHEEL SWITCHES (4 DECADE)	6/24 VDC
4/5 VDC	LED DISPLAY (4 DECADE)	15/5 VDC

Fig. 4.4. Example Problem I/O Requirements

particular use a group of I/O points will have, as noted in Table 4.1. Once the I/O is sorted out by type, function, and voltage, specific I/O modules can be selected from the manufacturer's total listing. Table 4.2 lists a portion of the various types of modules that are available for use with a Director 4001 system. Since this PC has a high-density option for the discrete I/O structure, there is the choice of using either four-point or eight-point style modules. In order to get a

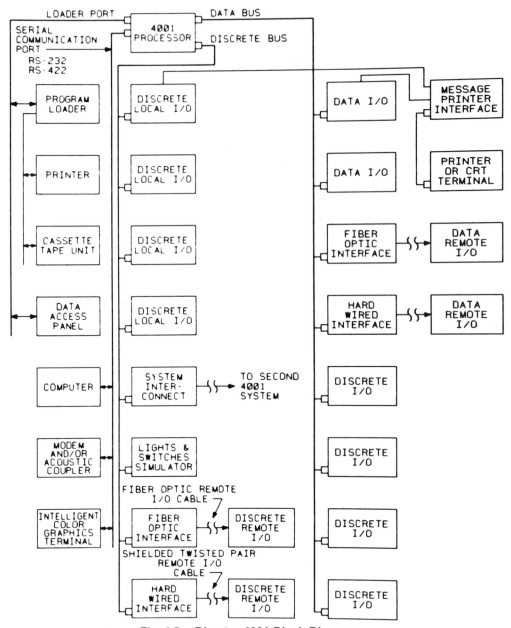

Fig. 4.5. Director 4001 Block Diagram

complete evaluation of this I/O system, both styles of I/O modules will be considered part of this example.

From the I/O tallies for each function group the number of I/O channels per function group can be assigned. The remote function group will require two discrete I/O channels, consisting of 14 12-volt dc input points, 23 12-volt dc output points, and 6 reed relay points. This totals 43 discrete I/O points for the remote function group, which will be divided between the two discrete remote I/O channels. The 15 LED displays and six thumbwheel switches will require three remote data I/O channels.

Examination of the local function group's discrete requirements indicates that 82 120-volt ac input points will be required, and 66 120-volt ac output points will be used. In order to accommodate this amount of discrete I/O points, five channels of discrete local I/O structures will be specified. The two thumbwheel switches and four LED displays will require a single local data channel for operation.

Table 4.1. Example Problem I/O Module Utilization

QUANTITY OF REMOTE FUNCTION GROUP I/O POINTS	I/O MODULE TYPE	QUANITY OF LOCAL FUNCTION GROUP I/O POINTS
	DISCRETE INPUT 120 VAC	10 - FVNR STARTER AUX 6 - FVR STARTER AUX 23 - PHOTOEYES 17 - LIMIT SWITCHES 26 - PUSHBUTTONS 82 - TOTAL
	DISCRETE OUTPUT 120 VAC	10 - FVNR STARTER COILS 6 - FVR STARTER COILS 16 - INDICATORS 16 - SINGLE SOLENOID VALVES 18 - DUAL SOLENOID VALVES 66 - TOTAL
14 - PUSHBUTTONS	DISCRETE INPUT 12 VDC	
23 - INDICATORS	DISCRETE OUTPUT 12 VDC	
6 - ALARMS	DISCRETE OUTPUT REED RELAY	
6 - THUMBWHEEL SWITCHES (4 DECADE)	DATA INPUT 24 VDC	2 - THUMBWHEEL SWITCHES (4 DECADE)
15 - LED DISPLAYS (4 DECADE)	DATA OUTPUT 5 VDC	4 - LED DISPLAYS (4 DECADE)

Table 4.2.   Types of Director 4001 I/O Modules

DISCRETE INPUT MODULES

LOW DENSITY	TYPE	HIGH DENSITY
75M01	5VAC/VDC	75B25
75M02	12VAC/VDC	75B26
75M03	24VAC/VDC	75B27
75M04	48VAC/VDC	75B28
75M05	120VAC/VDC	75B24
75M06	220VAC/VDC	75B29

DISCRETE OUTPUT MODULES

LOW DENSITY	TYPE	HIGH DENSITY
75M07	0-48 VDC SINKING	75B31
75M20	0-48 VDC SOURCING	75B49
75M08	0-125 VDC SINKING	75B32
75M09	0-120 VAC	75B30
75M10	0-240 VAC	75B33

REED RELAY MODULES

75M39 FORM 'C' REED CONTACTS
75M11 LATCHED FORM 'A' CONTACTS

DATA INPUT MODULES

75M15 5VDC
75M19 5-30VDC

DATA OUTPUT MODULES

75M16 5-30VDC

Once the I/O channels have been assigned for each of the function groups, actual selection of the I/O components can begin. For this application the high-density I/O structure cannot be used for the remote I/O system since there is no reed relay contact type of output module. In order to handle the discrete I/O requirements for the remote system, six 12-volt dc output modules, four 12-volt dc input modules, and three reed relay contact modules will be required. This quantity of discrete remote I/O modules will provide two spare 12-volt dc input points, one spare 12-volt dc output point, and no spare contact points.

Installation of the remote I/O modules will require two remote I/O tracks to house the 13 I/O modules, leaving three empty slots for future expansion. Each I/O track will be assigned a unique channel number. Selection of the remote I/O track will involve picking between a wired remote I/O communications link or a fiber-optic remote I/O communications link from the remote I/O tracks to the processor. In addition to the proper I/O track being selected for the communications link, the matching remote I/O interface unit must be selected for connection to the processor.

Selection of the I/O modules for the remote data I/O will require 6 24-volt dc input modules and 15 5-volt dc output modules. These 21 data modules will be housed in three remote I/O tracks, leaving three empty slots for future data I/O expansion. As noted earlier, there is no high-density data I/O structure available for this system. Each data I/O track will be assigned a unique data I/O channel number for system addressing. The selection of either a wired or a fiber-optic communications link between the remote I/O and the processor needs to be made, along with the proper remote I/O track and interface units.

There are two configurations for the discrete local I/O system. The first configuration will use the standard-density I/O system, the second a combination of both the high-density and the low-density systems. With the high number of discrete I/O points required for this example application, two I/O channels will be dedicated for the input points, an additional two discrete channels for the output points, along with a single discrete channel for the remaining discrete inputs and outputs. The first two discrete local I/O channels will be allocated for 120-volt ac input modules. Since there are two I/O density systems available on the Director 4001, these two channels can be two separate 32-point low-density systems, or a single 64-point high-density system comprised of both channels in the same structure. Because the two I/O channels only accommodate 64 of the required 82 input points, the additional input points will be assigned to a third discrete local I/O channel. The third discrete local I/O channel will house five 120-volt ac input modules, finishing out the input requirements and providing two input points as spares for future use.

Discrete local I/O channels 4 and 5 will house 64 of the required 66 120-volt ac output points. As was the option for channels 1 and 2, channels 4 and 5 can be composed of two low-density 32-point I/O tracks or a single 64-point dual-channel high-density I/O track. Channel 3 will be assigned a single 120-volt ac output module to complete the 66 output points required. This single output card will provide two spare output points for future use.

The local data channel will consist of two 24-volt dc input modules for the thumbwheel switches and four 5-volt dc output modules for the LED displays. The local I/O channels will provide two spare discrete and two spare data module spaces for future use. All local channels will be assigned a unique I/O channel number for I/O addressing purposes.

Fig. 4.6 shows the hardware arrangement for the low-density discrete I/O structure, and Fig. 4.7 shows the same I/O structure but with both low- and high-density I/O hardware. For both discrete I/O structures either the wired or the fiber-optic communications link between the remote I/O track and the

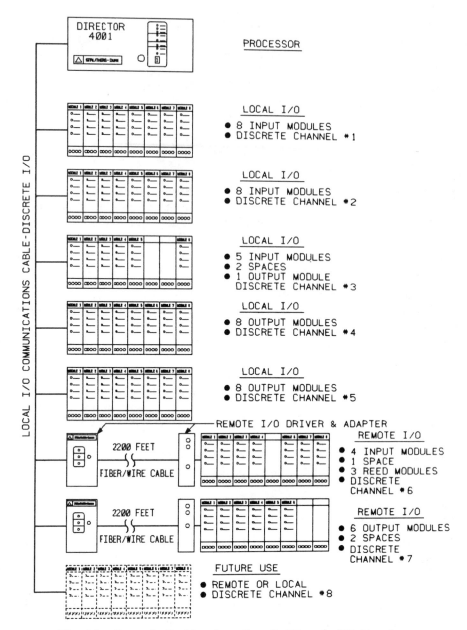

Fig. 4.6. Example Problem Low Density Discrete I/O Structure

processor interface can be used. Fig. 4.8 is a diagram of the data I/O structure for the example application. As the diagrams indicate, this example requires all of the data I/O channels to be used, as well as seven of the eight available discrete channels. None of the four discrete I/O channels which are part of the data bus need to be used; they can be designated as available for future use. The unused discrete channel is available for future remote or local I/O use.

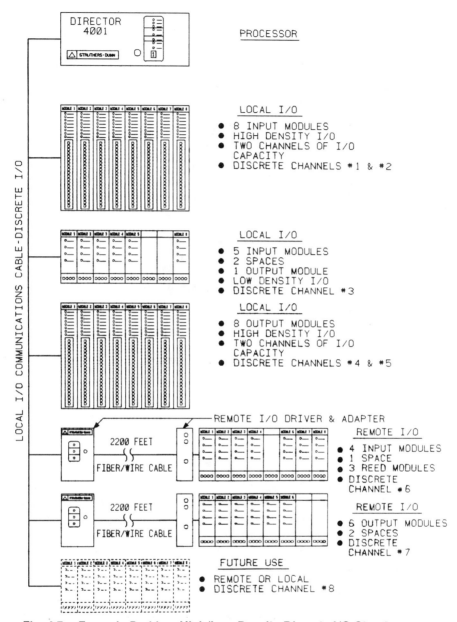

Fig. 4.7.   Example Problem High/Low Density Discrete I/O Structure

For our example application the design of the various I/O structures for the Struthers-Dunn Director 4001 is complete. Each of the I/O structures formulated in the example now needs to be priced with the assistance of the manufacturer's applications engineer. Assistance of a qualified applications engineer should be requested whenever a PC system is being designed. There will always be numerous ways as well as newer and more efficient methods that

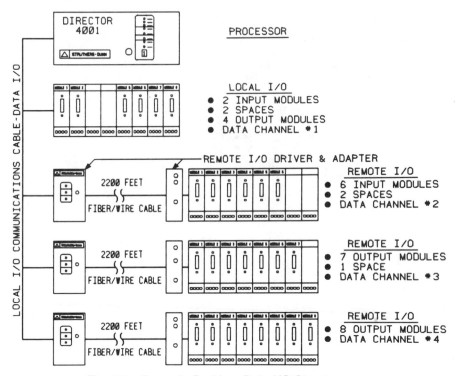

Fig. 4.8. Example Problem Data I/O Structure

a PC can be applied to a given problem. Even the most experienced control systems designers and engineers will find design consultation with an applications engineer well worth the time spent reviewing the current state-of-the-art procedures in PC applications.

In addition to comparing the costs of various PC systems and configurations, cost comparisons must also be included for PC system support hardware. All PC manufacturers recommend that the user mount the PC hardware in some sort of enclosure. These enclosures will vary in cost depending on the design requirements of the PC hardware contained within the enclosure. The enclosure containing the PC hardware must be sized for proper installation of the PC equipment and support hardware, as well as for ample service and installation access of the contents. Each vendor publishes installation data pertinent to his systems, which must be observed. Since no two manufacturers' systems are the same, and the installation of hardware will vary with each application and system, the costs of purchasing, fabricating, and field installing a PC enclosure should also be considered when evaluating various PC systems. For any business to survive it must be cost competitive, and often the difference between two manufacturers' systems will be very small for the PC hardware, but much larger for the PC enclosures and the installation costs to put a system into operation.

I/O SCHEMATICS

Once the cost evaluations are complete and a PC system has been chosen for the intended application, actual design of the I/O system must be completed. The individual field I/O devices must be assigned to dedicated I/O points, and the interconnection wiring diagrams between the field devices and the I/O modules must be completed. This type of information is easily transmitted through the use of input and output schematic diagrams.

A properly prepared input or output schematic will indicate the function of every I/O device connected to the PC system. The schematic will show the wiring between the I/O device and its I/O module. Any junction boxes or terminal boxes through which the field wiring runs must be shown along with the terminal numbers to which the wiring connects. The schematic should show all power sources, their level, and their origin, as well as any internal I/O point busing done inside the I/O module or field I/O devices. A complete functional description should be provided to indicate concisely the function and the operation of the field I/O device. The use of terms such as "forward, reverse, in, out, up, down" should be avoided where possible, and replaced with such specific terms as "north, south, east, west, retracted, or extended."

Fig. 4.9 represents a typical output schematic for a PC system. Noted on this schematic are references to recommended methods of relaying information about the system wiring. Fig. 4.10 is a typical input schematic, again showing preferred methods to document the system wiring. All I/O points including spares should be documented in a manner similar to the methods shown.

PC INSTALLATION PRACTICES

All local, state, and national wiring codes and practices should be observed during the design and installation of a PC system. Drawings should include detailed bill of materials listings for all PC hardware as well as any PC or control system support hardware. In addition to the schematic diagrams, mechanical drawings must be prepared to show the arrangement and mounting details for all PC hardware and control system support hardware contained within a PC enclosure. Block diagrams are also helpful to give quick reference information as to overall system component layout, power distribution, and so on.

Fig. 4.11 shows a suggested mounting arrangement for PC hardware within an enclosure. The National Electrical Manufacturers Association (NEMA) has defined various enclosure standards based on the type of environment in which the enclosure will be installed and the degree of protection required by the electrical hardware mounted inside. Fig. 4.12 lists the various NEMA enclosure classifications and the requirements needed to meet each particular classification. For most solid-state control devices a NEMA 12 enclosure is recommended. This type of enclosure is for general-purpose areas and is designed to be dust tight.

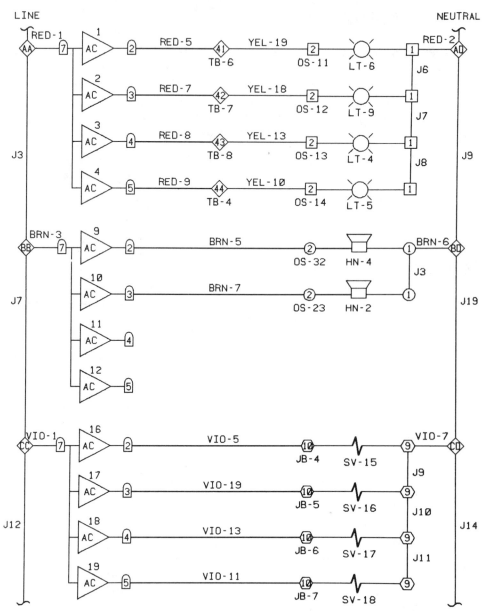

Fig. 4.9. Typical Programmable Controller Output Wiring Schematic

One area of a PC's installation that cannot be overlooked is the heat dissipation calculations for the processor enclosure and for each remote I/O enclosure. Every PC will dissipate heat from its power supplies and processor. This heat accumulates in the enclosure as it is dissipated from the heat sinks and cooling fans of the PC system itself. It must be further dissipated from the enclosure into the surrounding air. In addition, any relays, contractors, transformers, or other related control system electrical hardware will dissipate

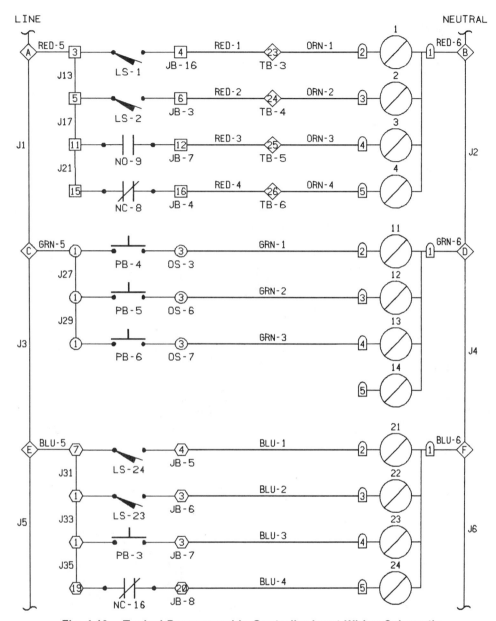

Fig. 4.10. Typical Programmable Controller Input Wiring Schematic

additional heat into the enclosure which must then be transferred to the exterior environment.

Every PC manufacturer specifies an acceptable operating range for his PC equipment. This specification generally relates to the maximum permissible temperature of the air that surrounds each individual PC component. When the PC hardware is installed in an enclosure that is essentially airtight, the temperature of the air inside the enclosure will rise. This elevated air

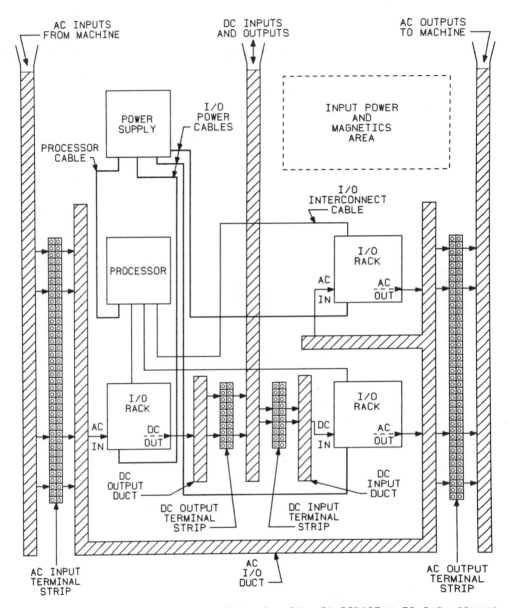

TERMINAL STRIPS OPTIONAL; FIELD WIRING CAN BE MADE DIRECTLY TO I/O MODULES

Fig. 4.11. Suggested PC Enclosure Layout

temperature heats the walls of the enclosure to a temperature within 10 degrees or so of that of the interior air. As long as the enclosure walls are in an environment where they are warmer than the surrounding air space, heat will be dissipated. A natural heat flow path is then developed from the PC hardware to the enclosure air space, from the enclosure air space to the enclosure walls, and from the enclosure walls to the outside air space. Since this heat path is not 100 percent efficient, temperature differences will exist with the PC hardware

NEMA TYPE 1
For indoor use, to provide protection against contact with the enclosed equipment.

NEMA TYPE 2
For indoor use, to provide protection against limited amounts of falling water and dirt.

NEMA TYPE 3
For outdoor use, to provide protection against windblown dust, rain, sleet, and external ice formation.

NEMA TYPE 3R
For outdoor use, to provide protection against falling rain, sleet, and ice formation.

NEMA TYPE 3S
For outdoor use, to provide protection against windblown dust, rain, and sleet, and for operation of external mechanisms when ice covered.

NEMA TYPE 4
For indoor use, to provide protection against windblown dust and rain, splashing water, and hose directed water.

NEMA TYPE 4X
For indoor or outdoor use, to provide protection against corrosion, windblown dust and rain, splashing water, and hose directed water.

NEMA TYPE 5
For indoor use, to provide protection against dust and falling dirt.

NEMA TYPE 6
For indoor or outdoor use, to provide protection against the entry of water during occasional temporary submersion at a limited depth.

NEMA TYPE 6P
For indoor or outdoor use, to provide protection against the entry of water during prolonged submersion at a limited depth.

NEMA TYPE 7
For indoor use in locations classified as Class 1, Groups A,B,C,or D, as defined in the National Electrical Code.

NEMA TYPE 8
For indoor or outdoor use in locations classified as Class 2, Groups E, F, or G, as defined by the National Electrical Code.

NEMA TYPE 10
Constructed to meet the applicable requirements of the Mine Safety and Health Administration.

NEMA TYPE 11
For indoor use, to provide protection to enclosed equipment against the corrosive effects of liquids and gases, by means of oil immersion.

NEMA TYPE 12
For indoor use, to provide protection against dust, falling dirt, and dripping non-corrosive liquids.

NEMA TYPE 12K
With knockouts, intended for indoor use to provide protection against dust, falling dirt, and dripping noncorrosive liquids other than at knockouts.

NEMA TYPE 13
For indoor use, to provide protection against dust, spraying water, oil, or noncorrosive coolants.

Fig. 4.12. NEMA Enclosure Classifications

being the warmest and the enclosure exterior air space being the coolest. A problem develops when the heat generated inside the enclosure is greater than the flow rate of heat from the enclosure. The interior air space of the enclosure can heat to a point where the PC manufacturer's temperature specifications are exceeded.

A manufacturer of enclosures, Hoffman Engineering Company of Anoka, Minn., provides information relating to the proper method to calculate heat rise and dissipation from its enclosures. Fig. 4.13 illustrates the steps necessary to perform a typical heat rise calculation. Based on tests of its enclosures, Hoffman has developed the temperature rise graph shown in Fig. 4.14 for its gasketed NEMA 12 enclosures. Note that the calculation should only consider those surfaces exposed to an external air space. Enclosure bottoms resting on a floor and backs mounted against a wall (within 6 inches) are not included.

Should the temperature rise calculation indicate that the enclosure air space will exceed the manufacturer's recommendations, a fan may be required for forced-air ventilation. Fig. 4.15 illustrates the steps necessary to calculate how many cubic feet of air per minute would be required to remove excess heat adequately from an enclosure. This calculation is also based on Hoffman Engineering studies of its enclosures equipped with fans. It should be further noted that any air introduced to an enclosure will require filtering to ensure that contaminants do not enter the enclosure. Often a filtered drop from an air-conditioning duct can replace a fan, or temperature controls can cycle a fan ON and OFF as required to maintain proper enclosure temperatures.

Many manufacturers provide detailed information on the proper grounding methods to use in an enclosure. The general recommendations are for all ground connections to be made with star washers between the grounding wire lug and the metal equipment or enclosure surface. The minimum ground wire

```
ENCLOSURE CONTENTS:
    ● PROCESSOR POWER SUPPLY
    ● ANALOG POWER SUPPLY
    ● CONTACTORS
    ● EMERGENCY STOP RELAYS
    ● TRANSFORMER
TOTAL HEAT DISSIPATION = 656 WATTS
```

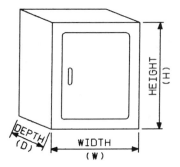

CALCULATION OF TEMPERATURE RISE:

```
    1) CALCULATE TOTAL AREA IN SQUARE FEET OF ALL EXPOSED
       ENCLOSURE SIDES

            FRONT: H X W          BACK AND BOTTOM ARE
            SIDES: 2 X H X D      NOT EXPOSED, THEREFORE
              TOP: W X D          THEY DO NOT CONTRIBUTE
                                  TO HEAT DISSIPATION

       FOR A 36" W X 72" H X 18" D ENCLOSURE
       TOTAL DISSIPATION AREA = 40.5 SQUARE FEET

    2) DIVIDE TOTAL POWER DISSIPATION IN ENCLOSURE BY
       TOTAL NUMBER OF SQUARE FEET OF SURFACE AREA
            656.0
            ───── = 16.2 WATTS PER SQUARE FOOT
             40.5

    3) DETERMINE ENCLOSURE TEMPERATURE RISE FROM GRAPH

          16.2 WATTS/SQ. FT. = 37.5° C TMEPERATURE RISE
```

Fig. 4.13. Enclosure Heat Rise Calculation

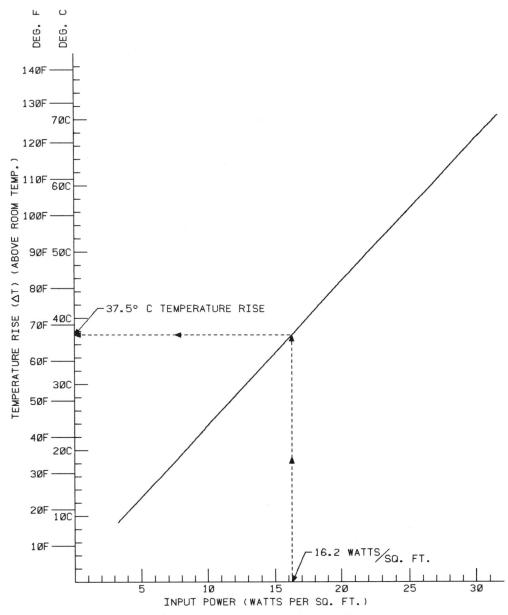

Fig. 4.14. Temperature Rise of Gasketed Enclosures

size should be #12 AWG stranded copper for PC equipment grounds and #8 AWG stranded copper for enclosure backplate grounds. All PC equipment and enclosure backplates should be individually grounded to a central point on the enclosure frame. The daisy chaining of ground conductors is not recommended since it may introduce a "ground loop," which results in faulty equipment operation. Examples of the proper way to ground a PC system and enclosure are detailed in Fig. 4.16.

FOR ENCLOSURE OF FIGURE 4.13 MAXIMUM
AMBIENT TEMPERATURE IS 95° F (35° C)

FOR EQUIPMENT INSTALLED IN ENCLOSURE OF FIGURE 4.13,
MAXIMUM OPERATING TEMPERATURE IS 105° F (40° C)

FORMULA FOR QUANTITY OF AIR REQUIRED TO DISSIPATE
ENCLOSURE HEAT IS

$$Q = \frac{3160 \times KW}{\Delta T}$$

WHERE: KW = ENCLOSURE HEAT LOAD IN KILOWATTS
ΔT = PERMISSIBLE TEMPERATURE RISE OF AIR

CALCULATION OF REQUIRED AIRFLOW

1) CALCULATE PERMISSIBLE TEMPERATURE RISE OF COOLING AIR: ΔT
ΔT = MAX ENCLOSURE TEMPERATURE — MAX AMBIENT TEMPERATURE
ΔT = 105° F — 95° F = 10° F

2) CALCULATE ENCLOSURE HEAT LOAD IN KILOWATTS
656 WATTS = 0.656 KW

3) CALCULATE REQUIRED AIR FLOW IN CUBIC FEET PER MINUTE
$$Q = \frac{3160 \times .656}{10} = 207.3 \text{ CUBIC FEET PER MINUTE}$$

Fig. 4.15.   Enclosure Airflow Calculations

In addition to the many hardware installation requirements and suggestions made by a vendor, numerous wiring recommendations will usually be included. Wiring to the I/O modules should be bundled and each bundle conductor properly labeled with a nonmetallic sleeve-type wire marker. Where possible the wiring should be segregated into ac, dc, and serial communications, with input wiring kept separate of output wiring when possible. Special I/O modules, such as pulse counter modules or analog modules, will require the use of shielded cable for the field wiring. The shielded cable should be run separate of all other wiring in dedicated steel conduits, paying special attention to the shield grounding requirements listed by the manufacturers of the I/O module and the field device.

Other items often overlooked by system designers include such things as the paralleling of I/O outputs to increase the current capacity to a load device. The practice of wiring two output points in parallel to the same load and then operating the output points together by the processor logic is not recommended. Since either a triac, a silicon-controlled rectifier, or a power transistor is commonly used for the output switching device, guaranteed simultaneous switching may not occur every time due to electrical differences in the two switching devices. Also there may be zero crossing differences, or a logic update difference that delays the signal to one output by several milliseconds. This delay is enough for the output switching to first see the full inrush load of the field device, which may cause eventual output point failures at an abnormal rate. Any solid-state device will have a finite ON resistance, which may not match that of other units manufactured at the same time. This ON resistance will cause the currents drawn from each output to be different, and as a result the output point supplying the larger current may fail earlier due to stress overloading.

Transient protection should be considered by the designer whenever he or she must design an output circuit with a switch or other mechanical contact connected between the output point and the field device. Output loads that can be energized from an external switch as well as a PC output point should also be provided with transient protection. All inductive dc output loads should have a transient protector in the form of a diode connected in parallel with the load

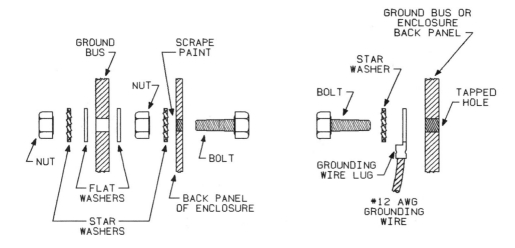

PROPER GROUND BUS MOUNTING   PROPER GROUND WIRE MOUNTING

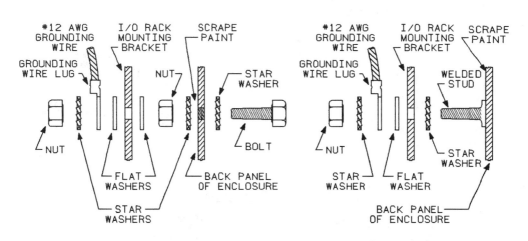

PC HARDWARE MOUNTING AND GROUNDING

Fig. 4.16. Proper Grounding of PC Components

device. Fig. 4.17 depicts several of the more commonly used transient protection schemes used with PC output modules.

Where multiple solid-state input devices are to be connected in series as the signal to an input module, proper operation of the solid-state input device should be tested. Many PC input modules incorporate high-impedance input

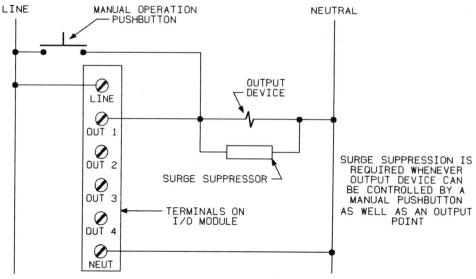

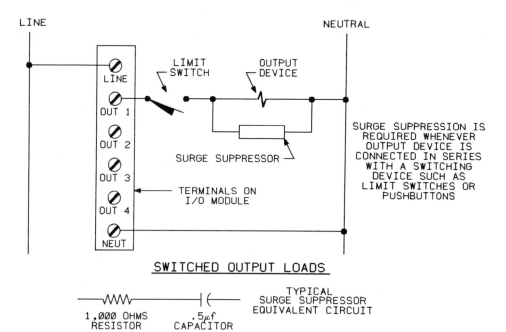

Fig. 4.17. Proper Use or Surge Suppressors

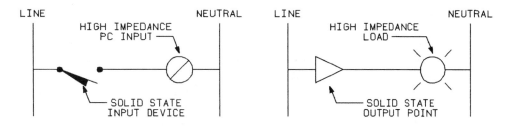

TRIAC, SCR, OR TRANSISTOR SWITCHING DEVICES (PC OUTPUT POINTS, SOLID STATE INPUT DEVICES) PRODUCE A RESIDUAL (LEAKAGE) CURRENT WHEN IN THE OFF STATE. THIS CURRENT CAN RANGE FROM 1 TO 10 MILLIAMPERES. THE RESIDUAL CURRENT CAN BE ENOUGH TO PARTIALLY OPERATE ANY HIGH IMPEDANCE LOAD CONNECTED TO THE DEVICE.

Fig. 4.18. Solid State Switching Devices Used in Input Devices and PC Output Modules

points which work fine for single solid-state input devices, but when multiple devices are connected in series, intermittent operation can result.

Many of the newer sensors and field input devices used with PC-based systems are of solid-state design for reliability. Any electronic-based input sensor which uses a solid-state switch (silicon-controlled rectifier, triac, or transistor) and is wired to a high-impedance PC input point as indicated in Fig. 4.18 can falsely activate the PC input. Equally important, the solid-state switch used in many output modules can create a similar problem with a high-impedance output load device. In order to correct the problem, a loading resistor or impedance must be connected in parallel with the input point sensor or the output point load. The calculation of the proper size and power rating for a loading resistor is illustrated in Fig. 4.19 for input modules and in Fig. 4.20 for output load devices. In lieu of a resistor, a capacitor or resistor/capacitor combination may be substituted. Fig. 4.21 illustrates the calculation of the resistance and capacitance values needed to achieve a loading impedance equal to a given loading resistor value. The resistor/capacitor combination has the benefit of dissipating far less heat than the use of only a resistor for loading purposes. There are several discrete electronic component manufacturers that can supply either off-the-shelf or custom-manufactured resistor/capacitor loading networks.

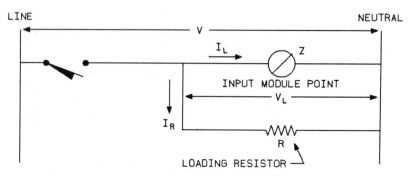

$V = $ SYSTEM VOLTAGE LEVEL

$V_L = $ MAXIMUM VOLTAGE LEVEL AT WHICH INPUT POINT REMAINS OFF

$I_R = $ RESISTOR CURRENT

$I_L = $ INPUT POINT CURRENT

$Z = $ INPUT POINT IMPEDANCE

$R = $ LOADING RESISTOR RESISTANCE

TYPICAL PC INPUT MODULE SPECS
- MAXIMUM OFF STATE VOLTAGE: $V_L = 45$ VOLTS
- INPUT POINT IMPEDANCE: $Z = 25,000$ OHMS
- OPERATING VOLTAGE: $V = 120$ VAC

BY OHM'S LAW: $V = IR$

$$I_L = \frac{V_L}{Z} = \frac{45}{25,000} = 0.0018 \text{ AMPERES} = 1.8\text{mA}$$

RESISTOR IS REQUIRED IF LEAKAGE TO INPUT
POINT IS 1.8mA OR HIGHER

EXAMPLE
IF THE INPUT DEVICE HAS A SPECIFIED LEAKAGE OF 20mA

THEN $R = \frac{V}{I} = \frac{120}{.02} = 6,000$ OHMS

POWER $= V * I = 120 * 0.02 = 2.4$ WATTS

LOADING RESISTOR SHOULD BE 6 KILOHMS, 2.5 WATTS MINIMUM

Fig. 4.19.   Calculation of Input Loading Resistor

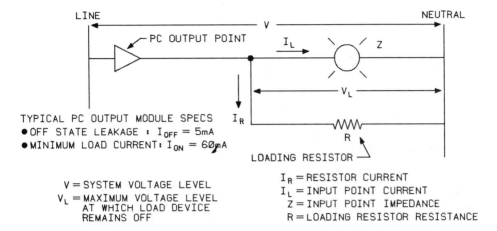

TYPICAL PC OUTPUT MODULE SPECS
- OFF STATE LEAKAGE : $I_{OFF} = 5mA$
- MINIMUM LOAD CURRENT: $I_{ON} = 60mA$

LOADING RESISTOR

V = SYSTEM VOLTAGE LEVEL
V_L = MAXIMUM VOLTAGE LEVEL
    AT WHICH LOAD DEVICE
    REMAINS OFF

I_R = RESISTOR CURRENT
I_L = INPUT POINT CURRENT
Z = INPUT POINT IMPEDANCE
R = LOADING RESISTOR RESISTANCE

OUTPUT POINT WILL LEAK 5 MILLIAMPERES CURRENT WHEN THE OUTPUT
IS OFF. ANY LOAD DEVICE WHICH CAN OPERATE ON 5 MILLIAMPERES OR
LESS WOULD REQUIRE A LOADING RESISTOR TO INSURE THAT THE LOAD
DEVICE IS OFF WHEN THE OUTPUT POINT IS OFF. HOWEVER, SINCE AT
LEAST 60mA MUST BE DRAWN BY THE OUTPUT DEVICE TO INSURE THAT
THE OUTPUT POINT LOAD WILL REMAIN ON, A LOAD RESISTOR MAY
BE REQUIRED FOR ANY OUTPUT DEVICE NOT REQUIRING MORE THAN
60 mA TO OPERATE.

FOR EXAMPLE WHEN AN OUTPUT IS CONNECTED TO AN INPUT

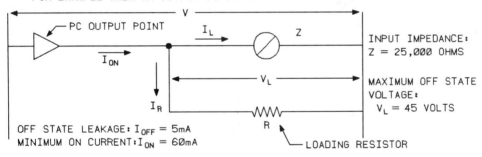

INPUT IMPEDANCE:
Z = 25,000 OHMS

MAXIMUM OFF STATE
VOLTAGE:
V_L = 45 VOLTS

OFF STATE LEAKAGE: $I_{OFF} = 5mA$
MINIMUM ON CURRENT: $I_{ON} = 60mA$

LOADING RESISTOR

INPUT POINT CURRENT: $I_L = \dfrac{V_L}{Z} = \dfrac{45}{25,000} = 0.0018$ AMPERES $= 1.8mA$

1.8 mA IS LESS THAN 60mA MINIMUM ON STATE CURRENT OF OUTPUT MODULE
THE LOAD RESISTOR SHOULD BE

$$R = \frac{V}{I_{ON}} = \frac{120}{.06} = 2,000 \text{ OHMS MINIMUM}$$

POWER RATING $= V * I = 120 * 0.06 = 7.2$ WATTS

Fig. 4.20. Calculation of Output Loading Resistor

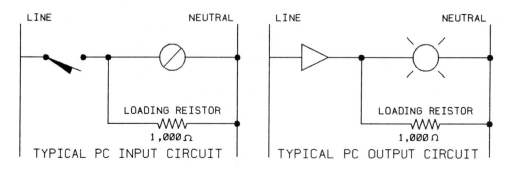

LOADING RESISTOR WILL USUALLY BE 5 WATTS OR LARGER IN POWER RATING.
WHENEVER THE RESISTOR IS ON IT WILL BE GENERATING HEAT. THE RESISTOR
CAN BE REPLACED WITH A RESISTOR/CAPACITOR COMBINATION TO REDUCE HEATING.

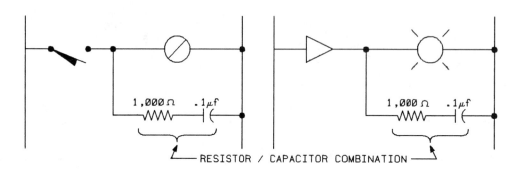

IMPEDANCE OF CAPACITOR $X_c = \frac{1}{2}\pi f c$

f = FREQUENCY IN HERTZ
c = CAPACITANCE IN MICROFARADS
FOR THE COMBINATION
$$X_T = \sqrt{R^2 + X_c^2} = \sqrt{R^2 + \frac{1}{2}\pi f c}$$

FOR RESISTOR VALUE OF $1 \cdot 10^3$ OHMS AND CAPACITOR OF $0.1\mu f$
$$X_T = \left[(1,000)^2 + 1/(2)(\pi)(60)(0.1\mu f)\right]^{\frac{1}{2}}$$
$$X_T = 1013.17 \text{ OHMS}$$

Fig. 4.21. Use of a Resistor/Capacitor Network in Lieu of a Loading Resistor

5

The Central Processing Unit

Every PC, whether it is the smallest unit available in the marketplace or the largest system currently conceived, incorporates a central processing unit (CPU). While these electronic marvels may be referred to simply as CPUs or processors, their functions are all similar, and their design is based on that of a computer. Once the industrial metal covers, the highly filtered and computer-regulated power supplies, and the noise-immune I/O hardware are removed, what remains is a small special-purpose computer. While many manufacturers employ specially designed microprocessors, many PCs are based on the familiar Z80, 8080, 8086, 6800, or 9900 family of microprocessors. In fact at least one PC manufacturer incorporates the same family of microprocessor hardware and the same integrated circuits as those that were selected by the manufacturer of the color computer that ran a word processing program to assist in the writing of this book. The microprocessor family is the 6800 series, and in the PC case, it has been adapted to understand a programming form that consists of relay symbols and industrial control functions. The home computer manufacturer on the other hand has adapted the unit to understand a programming format called *BASIC*. In both cases similar microprocessors are used, but the home computer cannot operate an industrial control system just as a PC is not designed to operate as a word processor. The question to be answered then is: "If the microprocessors used in home computers, minicomputers, and PCs are similar, then what is the difference? After all, at least one manufacturer, Texas Instruments, manufactures the 9900 series of microprocessor, which it uses in its personal computer, its minicomputer systems, and the TI programmable controllers!" The difference is not the microprocessor, but in how the microprocessor is configured and programmed for each application. This chapter will discuss the architecture of the PC processor and how it is both similar to and different from the CPU of a home computer or other computer-based system.

In order to avoid confusion and maintain consistency, the term "processor" will be used in this chapter and the rest of this book to refer to the electronic hardware that makes up a PC's central processing unit. Where it is desired to reference the central processor hardware of a home computer or other standard computer system, the term "CPU" will be used.

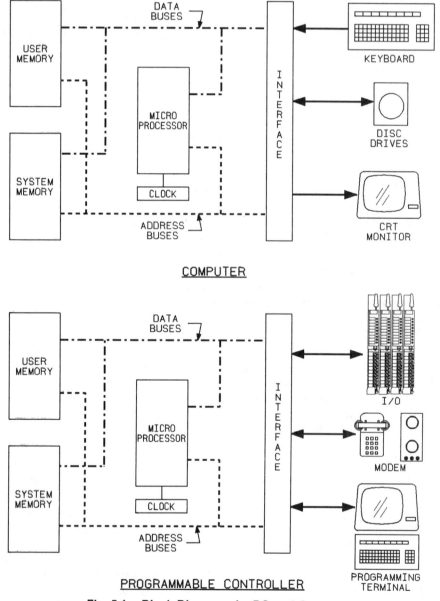

Fig. 5.1. **Block Diagram of a PC and Computer**

PROGRAMMABLE CONTROLLER AND COMPUTER SYSTEM DESIGN

Both the PC processor and the computer CPU are similar in functional components and general design, as indicated in Fig. 5.1. Both systems contain a microprocessor, a user memory, and a system memory. For both systems all components are interconnected by both a data and an address bus system. With the exception of the devices connected to the interface portion of each system, both the PC and the computer systems appear identical. There is, however, a subtle difference in the types of devices and hardware connected to the interface portions of both systems. What makes the arrangements of Fig. 5.1 a computer in one case and a PC in the other, is not only the difference in the devices connected to the interface hardware, but also how the memories, the microprocessor, and the interface hardware all interact with each other.

Before examining how the memory, the microprocessor, and the interface hardware operate to become a PC in one configuration and a computer in another, each part of the processor or the CPU must be examined and understood. The reader should note that the information that will be presented in the remainder of this chapter is not critical for the understanding of a PC. The discussions that will follow are intended to "round out" the reader's knowledge of the overall design and operation of a PC. If the reader understands the general internal construction and operation of a PC, many of the "buzzwords" used by PC manufacturers and users may not seem that mystical to the person who just applies a PC for the first time.

A functional block diagram for a typical microprocessor is shown in Fig. 5.2. The heart of the microprocessor is the arithmetic and logic unit (ALU). Everything a microprocessor does is done by the ALU. As an analogy, the ALU might be thought of as an ultrahigh-powered adding machine. The keys of this adding machine are electric impulses generated in the timing, control, and

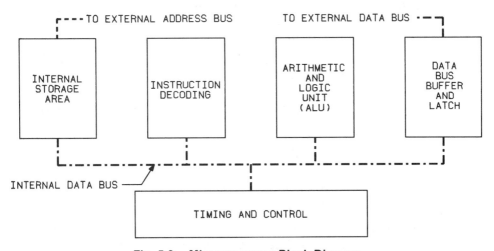

Fig. 5.2. **Microprocessor Block Diagram**

instruction-decoding sections of the system. Instructions and commands programmed in the user memory portion of the CPU or processor are interpreted by the instruction-decoding area of the microprocessor in accordance with various timing and control parameters. Once decoded, the timing and control block of the microprocessor, in conjunction with the instruction-decoding block, issues the proper sequence of electric signals to the ALU and other external components to initiate the proper processing actions for the present instruction being executed. The ALU operates not only as an adder, but also as a decision-making device and a data manipulation device. The ALU is capable of using information stored in the microprocessor's internal storage area as well as in the external memory areas in order to perform the required mathematical or logic operations. Exactly what the microprocessor does, and how it does it, is determined solely by the order and the selection of the instructions that the unit understands.

MICROPROCESSOR LANGUAGES

Every microprocessor has its own instruction set or set of commands which it alone understands. This is not the instruction set that is used when the unit is actually operating as a PC or home computer. The instructions understood by a microprocessor are instructions composed of a series of ones and zeros. The instruction-decoding section of the microprocessor takes a block or instruction composed of ones and zeros, and, depending on the pattern, controls the operation of the ALU and other components of the processor or CPU.

Each microprocessor series, whether it is an 8080, Z80, 6800, or 9900 system, has its own unique instruction set composed of different arrangements of ones and zeros. The microprocessor only understands its own personal set of instructions, whether it is controlling the operation of a PC or a computer. This personal set of microprocessor instructions is called *machine language*. Fig. 5.3 shows a specific pattern of ones and zeros which instruct a microprocessor to add two 16-digit numbers together. The requirement to add together numbers of varying lengths is common to both computer and PC systems. The microprocessor can read, interpret, and carry out the necessary steps to perform the calculation illustrated in 230 microseconds (0.000230 second).

Obviously no one could remember the exact combination of ones and zeros for every instruction available for a particular microprocessor. For simplicity, the microprocessor program can be written in symbolic terms rather than the ones and zeros configuration. The symbols used to represent the ones and zeros are called *mnemonic codes* and are usually two to four letters long. A program written in these mnemonic codes is called an *assembly language* program. A special computer program is then used to convert the mnemonic codes of the assembly language program into the proper configuration of ones and zeros (machine language program) that the microprocessor will understand.

A PC or computer manufacturer develops a special high-level language for his system that is composed of mnemonic symbols and terms he decides upon to represent a complex group of machine language instructions and program-

INSTRUCTION CODE	MNEMONIC	OPERAND	COMMENT
0 0 0 1 0 0 0 1	LXI	D,AUGEND	LOAD AUGEND ADDRESS IN DE REGISTER
0 0 1 0 0 0 0 1	LXI	H,ADDEND	LOAD ADDEND ADDRESS IN HL REGISTER
0 0 0 0 1 1 1 0	MVI	C,8	LOAD C REGISTER WITH "8"
1 0 1 0 1 1 1 1	XRA	A	CLEAR ACCUMULATOR
0 0 0 1 1 0 1 0	LDAX LOOP	D	LOAD AUGEND IN ACCUMULATOR
1 0 0 0 1 1 1 0	ADC	M	ADD ADDEND TO ACCUMULATOR VALUE
0 0 1 0 0 1 1 1	DAA		DECIMAL ADJUST
0 0 0 1 0 0 1 0	STAX	D	STORE SUM TO DE ADDRESS
0 0 1 0 0 0 1 1	INX	H	INCREMENT HL ADDRESS
0 0 0 1 0 0 1 1	INX	D	INCREMENT DE ADDRESS
0 0 0 0 1 1 0 1	DCR	C	CHECK FOR END OF CALCULATION
1 1 0 0 0 0 1 0	JNZ	LOOP	

Fig. 5.3. Microprocessor "Machine Language" Addition Program

ming. For example, the addition routine of Fig. 5.3 might be represented by the symbol "+" for the high-level language developed for both the PC and the computer. However, the mnemonic symbols or terms "equal" and "=" might be used for a machine language routine which sets two values or quantities equal. The user of the PC or computer can then instruct the system to perform certain operations and functions by the proper combination of simple high-level mnemonic symbols which actually represent complicated machine language instruction sequences. Even though both the computer and the PC may use the same microprocessor system, the high-level language mnemonic symbols and the meanings of those symbols are different. Computer programmers can "talk" to a computer in various word-oriented languages, while the PC programmer can "talk" to the PC in an electrical symbol-oriented language.

MEMORY SYSTEMS

The microprocessor is designed to read an instruction and then carry out the meaning of the instruction. The source of the instructions could be a person giving the microprocessor each new instruction, and any data that may be needed for the instruction, as the microprocessor requests it. This operation would be very slow and tedious. The microprocessor is capable of solving instructions at nearly lightning speed, and having to repeat the same sequence of instructions over and over would be highly inefficient. Some sort of memory system is needed that can first store instructions and then allow them to be read as required by the microprocessor.

The memory system is the second major portion of any computer CPU or PC processor. All the instructions that may be needed by the microprocessor are stored in memory along with any data that may also be required. There are different types of memory systems used by both computers and PCs. The selection and use of a particular type of memory is dependent on what information needs to be stored, and how that information is to be used or modified by the microprocessor. Since the architecture of a memory system is usually similar for a PC as well as a computer, no distinction will be made unless necessary.

The best way to visualize a memory system is to think of the "boxes" of a large post office. These mailboxes are usually installed in various walls of the post office in a layout similar to that shown in Fig. 5.4. Each box has its own particular identification according to some layout determined when the boxes were installed. As shown in Fig. 5.4, each mailbox has been assigned a unique identification. For the figure, the scheme of identifying each box has been based on an addressing scheme often used in computer and PC memory systems. Each column of boxes has a unique numerical identification, as does each row of boxes. The identity of each individual mailbox is composed of both the column and the row identifications. For example, box 311 is located in the third column of boxes, the eleventh box from the top. Box 516 is at the bottom of the fifth column of boxes.

Memory is identified in a similar manner. The terms are changed such that each individual box within a column of boxes forms the individual bits of memory and each row of identically numbered column bits form a word. A memory system is composed of words of memory bits, each word having the same number of bits. Computer and PC manufacturers will specify a size for the memory portion of a system based on the number of words of memory designed into the system. The most common memory sizes available for microprocessors, minicomputers, PCs, and large computers is shown in Fig. 5.5. Fig. 5.6 indicates the number of bits per word commonly found for microprocessors, mini-computers, PCs, and large computer systems.

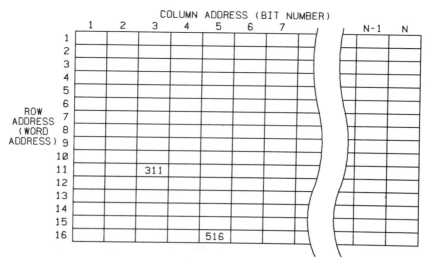

Fig. 5.4. "Post Office" Memory Storage

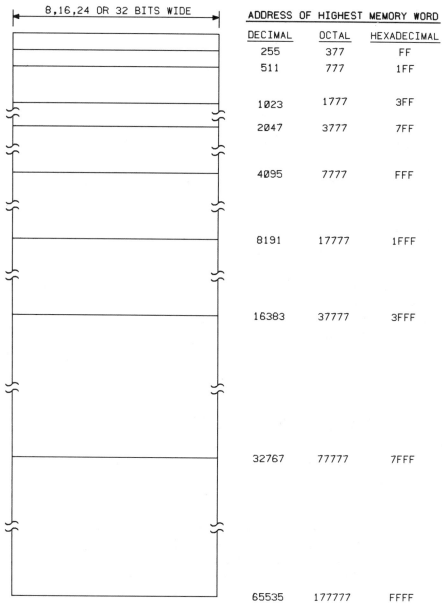

Fig. 5.5. Common Memory Sizes

	ADDRESS OF HIGHEST MEMORY WORD	
DECIMAL	OCTAL	HEXADECIMAL
255	377	FF
511	777	1FF
1023	1777	3FF
2047	3777	7FF
4095	7777	FFF
8191	17777	1FFF
16383	37777	3FFF
32767	77777	7FFF
65535	177777	FFFF

8,16,24 OR 32 BITS WIDE

RANDOM ACCESS
AND READ-ONLY MEMORIES

There are two major types of memory systems available. *Random access memory* (RAM) permits any individual word or bit of memory to be written into or read from an unlimited number of times. Memory systems which, once written into, cannot be further changed or modified but only read from, are

| 1 | 2 | 3 | 4 | 4 BITS/WORD

| 1 | 2 | 3 | 4 | 5 | 6 | 7 | 8 | 8 BITS/WORD   MOST COMMON PC WORD SIZES
                                                      ARE 8 AND 16 BITS PER WORD

| 1 | 2 | 3 | 4 | 5 | 6 | 7 | 8 | 9 |10|11|12| 12 BITS/WORD

| 1 | 2 | 3 | 4 | 5 | 6 | 7 | 8 | 9 |10|11|12|13|14|15|16| 16 BITS/WORD

| 1 | 2 | 3 | 4 | 5 | 6 | 7 | 8 | 9 |10|11|12|13|14|15|16|17|18|19|20|21|22|23|24| 24 BITS/WORD

                                                                          32 BITS/WORD

| 1 | 2 | 3 | 4 | 5 | 6 | 7 | 8 | 9 |10|11|12|13|14|15|16|17|18|19|20|21|22|23|24|25|26|27|28|29|30|31|32|

Fig. 5.6.   Common Memory Word Sizes

termed *read-only memory* (ROM). Both RAM and ROM are used in PCs and computers to meet various memory requirements. The machine language equivalent statements of the high-level operating language commands (BASIC, Fortran, or relay ladder) are stored in ROM since, once entered, they never require changing. The instructions entered by the user of the PC or computer are stored in RAM since these instructions may be changed many times during the life of the system to meet changing conditions and requirements.

CORE-TYPE RANDOM ACCESS MEMORY

There are several styles of RAM memory available for use with a microprocessor. *Core* memory was probably the first style to be used with PCs. It consists of microscopic magnetic donuts wired together, as shown in Fig. 5.7. Each

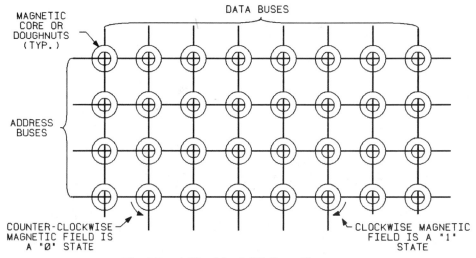

Fig. 5.7.   4-Word by 8-Bit Core Memory

donut is magnetized in either a clockwise or a counterclockwise direction, depending upon the flow of current in the wiring passing through the hole of the donut at the time of programming. Each donut is a bit of memory with rows of donuts equaling a word of memory. Since the individual cores will remain magnetized for long periods of time, even when the memory system is powered down, core memory is said to be *nonvolatile*. Any memory system that maintains its last state when all power is removed is termed *nonvolatile memory*. Core memory does require rewriting whenever it is read. This operation is electrically designed into the memory system, and is transparent to the user.

SEMICONDUCTOR-TYPE RANDOM ACCESS MEMORY

The second style of RAM memory is *solid-state* or *semiconductor* memory. This style of memory is usually *volatile* since it tends to lose all stored information when power is lost, even for a nanosecond.

Semiconductor memories require some sort of standby power source to maintain their contents during power outages. Since this style of memory requires little power to simply retain its data, batteries are usually employed as the backup power source. This is the reason a PC system may incorporate a battery within the processor. The battery does not operate the PC, it supplies standby power for the memory when the main source of power is lost. A discussion on the various types of battery systems employed with PCs is contained in the chapter on power supplies and systems.

STATIC SEMICONDUCTOR RANDOM ACCESS MEMORY

Semiconductor memory can be divided into two classes, *static* and *dynamic*. Static semiconductor memory uses transistors wired in a latch format similar to the latches used in relay control systems. To the electronic engineer these latches may be thought of as D-type flip-flops. A word of memory is composed of a row of flip-flops, each flip-flop being equivalent to a bit of memory. Fig. 5.8 illustrates a static semiconductor memory system composed of flip-flops. The majority of static semiconductor memories manufactured employ field-effect transistors (FETs) since these devices require far less space in an intergrated circuit than the common bipolar transistor, and less FETs are required to make a flip-flop than other transistor types. The static semiconductor memory is one of the fastest memory systems available, and offers other advantages such as low power consumption, small space requirements, and minimal heat generation.

DYNAMIC SEMICONDUCTOR RANDOM ACCESS MEMORY

Dynamic semiconductor memory is the second class of semiconductor memory. It is constructed differently than its static counterpart and requires

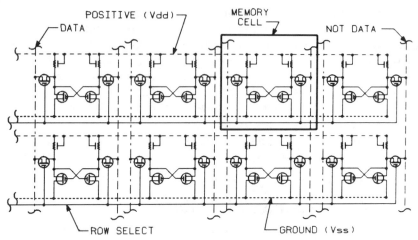

Fig. 5.8. 2-Word by 4-Bit Static Memory

continual "refreshing" in order to maintain its contents. This class of memory uses ultrahigh grade capacitors in conjunction with FETs in order to operate. Fig. 5.9 illustrates the typical internal components and configuration of the dynamic semiconductor memory. This class of memory requires constant refreshing due to the capacitor. In lieu of the latching flip-flop action of the static semiconductor memory, the dynamic semiconductor memory uses a charged capacitor to retain the contents of each bit in memory. Since all capacitors have a finite internal resistance, the charge placed on the capacitor slowly drains away with time. In addition some small amount of charge is continually removed by the FETs since they too have finite internal resistances. Therefore the memory must be continually read and then rewritten back so that the contents are not lost. This class of memory requires more operating power than static semiconductor memory and is slower in terms of accessibility since access to a word or bit must be done between cycles of the refreshing operation.

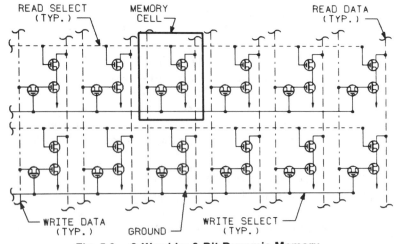

Fig. 5.9. 2-Word by 6-Bit Dynamic Memory

READ-ONLY MEMORIES

The second major type of memory device is the ROM, or read-only memory. This memory type is programmed as part of the manufacturing process and cannot be reprogrammed. ROM is used in applications where the data or instructions for the microprocessor will not be changed often. Standard ROM memory must be discarded and replaced with new units whenever a change is required, while other styles, the programmable types, permit a limited number of changes to be implemented. There are several styles of programmable ROMs generally available. The most common is the *programmable read-only memory* (PROM). While ROM is programmed at the time of manufacturing, PROM memories are user-programmable. *Erasable programmable read-only memory* (EPROM) is available in either of two classes, *ultraviolet erasable* (U.V. EPROM) and *electrically erasable* (E.E. PROM). A new style of ROM that is generally available is *electrically alterable read-only memory* (EAROM).

PROMs are memory systems that, once programmed, cannot be altered. They consist of matrixes of small metal fuses or back-to-back diodes. When the PROM is programmed, the fuse link is blown clear to represent a 0 state, or left intact to represent a 1 state, as illustrated in Fig. 5.10.

When the PROM is constructed of back-to-back diodes, one of the two diodes is overloaded (avalanched) under controlled conditions in order to short-circuit it. The remaining diode is left intact, and depending on which diode remains, the zero or one state of that memory bit is determined. Fig. 5.11 illustrates a partially programmed avalanche diode PROM. The phrase "blowing the PROM" is often heard. It originates from the physical act of blowing out either the fused junction or the diode connections within the ROM during the programming process.

The EPROM incorporates an isolated-gate metal-oxide-semiconductor field-effect transistor (MOSFET) as the memory device. The FET's gate is totally

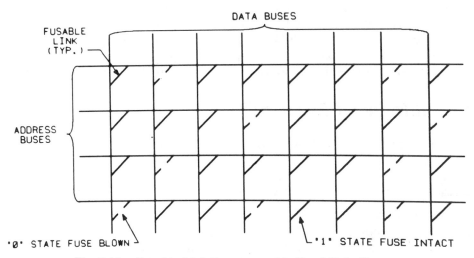

Fig. 5.10. Fusable-Link Programmable Read-Only Memory

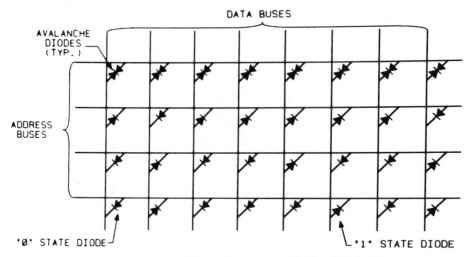

Fig. 5.11.   **Avalanche Diode Programmable Read-Only Memory**

isolated from the electric circuitry of the memory, allowing it to retain any charge placed on it almost indefinitely. If a charge is present, the memory location is at a 1 state; if no charge exists, the memory location is at a 0 state. This style of memory can be erased in either of two ways, depending on its construction. U.V. EPROMs are recognized by their quartz glass covers. Exposure of the U.V. EPROM to ultraviolet light for 30 to 40 minutes causes a photo current to be internally produced, which carries any charges on the FET gate to ground, thereby erasing the ROM. E.E. PROMs are erased by flooding them with an electrical charge in order to clear any charges on the gate of the FET. Both the U.V. EPROM and the E.E. PROM are limited in the number of times they can be erased and reprogrammed. They cannot have selective words erased, but must be completely erased and reprogrammed.

The electrically alterable read-only memory (EAROM) is capable of being erased and reprogrammed without removal from the circuit. These memories are designed similar to the EPROMs discussed earlier, except that they can be erased with lower voltages and more quickly. This style of memory does have the disadvantage of being slow to write or read data or information into or from.

PC MEMORY ARCHITECTURE

All PCs contain both RAM and ROM in varying amounts depending upon the design of the PC. The use of a PC's memory is determined again by the design of the unit. However, all PC memories can be subdivided into at least five major areas. A typical memory utilization map for a PC is depicted in Fig. 5.12.

The operating system or *executive* for the PC is always placed in PROM since, once developed by the manufacturer, it rarely needs changing. The executive of a PC performs several functions. It is the one that actually does the "scanning" in a PC. The PC user programs the PC in a high-level language composed of relay symbols. These symbols are stored in the user portion of the memory in binary format. The executive causes the microprocessor to examine each user instruction, then instructs the microprocessor through the actual conversion of the high-level instruction into its equivalent series of machine language instructions for further action by the microprocessor. Often a user program will contain instructions that require data from other areas of the memory. In this case the executive programming must instruct the microprocessor to gather these data along with the proper machine language instructions for further use by the microprocessor.

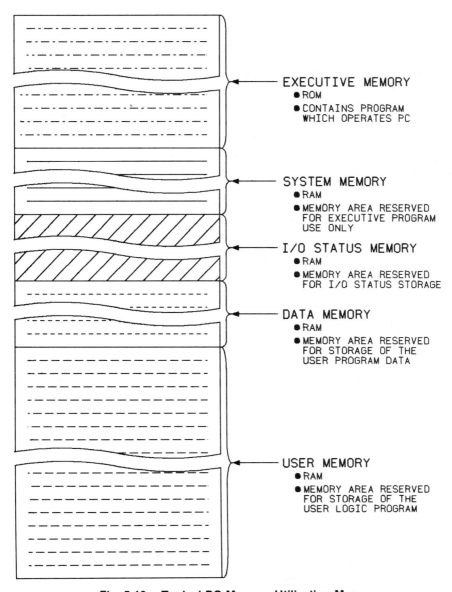

EXECUTIVE MEMORY
● ROM
● CONTAINS PROGRAM WHICH OPERATES PC

SYSTEM MEMORY
● RAM
● MEMORY AREA RESERVED FOR EXECUTIVE PROGRAM USE ONLY

I/O STATUS MEMORY
● RAM
● MEMORY AREA RESERVED FOR I/O STATUS STORAGE

DATA MEMORY
● RAM
● MEMORY AREA RESERVED FOR STORAGE OF THE USER PROGRAM DATA

USER MEMORY
● RAM
● MEMORY AREA RESERVED FOR STORAGE OF THE USER LOGIC PROGRAM

Fig. 5.12. Typical PC Memory Utilization Map

The executive is responsible for providing all system statuses and operating conditions to the microprocessor. The chore of reading all input devices and updating all output devices is also the responsibility of the executive program. In short terms, the executive is a special machine language program that literally runs the PC. It instructs the microprocessor to read each user instruction, helps the microprocessor to interpret user programmed symbols and instructions, keeps track of all the I/O statuses, and is responsible for maintaining a current status of the health of the system and all its components.

In order for the executive to function, a section of the memory is allotted for system administration. As the executive performs its duties, it often requires a place to store intermediate results and information. A section of RAM is installed for this purpose. This area is allotted for use of the executive only and is not available to the user for programming. It might be thought of as a scratch pad for the executive to doodle on as necessary.

Another portion of RAM is allocated for the storage of current I/O statuses. Whenever the executive program instructs the microprocessor to read the current statuses of the inputs to the PC, it stores this information in the *input status* or *image* area. As the executive instructs the microprocessor to scan the user program and interpret the user commands, various output device statuses are generated. These output states and conditions are stored by the micro-processor in the *output status* or *image* area until the end of the logic scan when the output modules get updated to this previously stored information. These status areas of memory are always available for user monitoring and use. They may not, however, be available for user modification except through user programming.

Whenever timers, counters, math, and/or data functions are available, an area of memory must be set aside for *data storage*. The data storage portion of memory is allocated for the storage of such items as timer or counter preset and accumulated values, math instruction data and results, and miscellaneous data and information which will be used by any data manipulation functions programmed in the user memory area.

Some processors subdivide the data memory area into two submemories, one for fixed data and the other for variable data. The fixed data portion can only be programmed via the programming terminal. The processor is not permitted to place data values in this area. The variable portion of the data memory is available to the processor for data storage. The user program assigns a location within this area for the processor to place such information as timer-or counter-accumulated values, the results of a math instruction, or the results of an instruction which generates or manipulates data.

The final area of memory in a PC is allocated to the storage of the user program. It is this memory area that the executive program instructs the microprocessor to examine or "scan" to find the user instructions. The user program area may be subdivided if the processor allocates a portion of this memory area for the storage of ASCII messages, subroutine programs, or other special programming functions or routines.

In the majority of PCs the internal data storage and user program areas are located in RAM. Several systems do offer an option that places both the user logic and the fixed data storage areas in EPROM-type memory. The user can develop his or her program in RAM and run the system to ensure correct operation. Once the user is satisfied that the programming is correct, he or she can duplicate a set of EPROMs from the processor RAM. He or she shuts down the processor and replaces the RAM with the newly programmed EPROM system. Any future changes would require that the EPROMs be reprogrammed. Note that if calculations or data manipulations are part of the user programming, a small amount of RAM may still be required for variable data storage.

In lieu of having to program a separate EPROM, several manufacturers offer an EAROM option with their processors. Processors with this option have both RAM and EAROM internally installed. Once the user has determined that his or her programming performs as desired, on user command the PC internally copies the programming stored in RAM to the EAROM. The PC will then operate in either of two ways. It can execute the user program directly from the EAROM, or the processor can transfer the user program to RAM and execute it from RAM directly. When the PC transfers the user program to RAM from EAROM, the processor will usually verify that the contents of both the RAM and the EAROM exactly match each processor scan. Should a discrepancy occur, the processor can then alert the user to the problem and, if desired, shut down.

In order for the microprocessor to read its instructions from memory, it must communicate with the memory system. This communication is carried out over two groups of electric buses. One bus is designated the *address* bus. It allows the microprocessor to address directly any location in memory. Once a particular memory location is addressed, the contents of the specified location are retrieved over a second bus, termed the *data* bus. The number of individual conductors that make up each bus is a function of the design specifications of the microprocessor and how the memory is arranged for the system. Since a number of the PCs on the market use one of the standard 8-bit microprocessors, the memory address lines may be 16 conductors wide, and the data bus may be 8 conductors wide.

The reader may be asking the question: "If the PC manufacturers use 8-bit microprocessors for the PCs they produce, why or how can they specify the memory system to be 16 bits wide?" The answer to the question lies in the programming of the executive. The executive program is written in machine language statements which are 8-bit instructions to match the 8-bit capacity of the microprocessor. As the executive program instructs the microprocessor to scan the user memory, two consecutive memory locations are retrieved and decoded each time a user instruction is read. This is the equivalent of reading a single 16-bit instruction. Note that many of the newer PCs on the market use larger microprocessors and even multiple microprocessors to operate. These systems may indeed have 16-bit memory systems and multiple data and address bus systems. The combinations of memory, microprocessor, and bus design are as diversified as the number of PCs on the market.

MICROPROCESSOR INTERFACE SECTION

The interface section of the computer CPU or the PC processor is the third major section of a processor system. It is responsible for addressing all of the I/O devices used by the system for the purpose of communicating with the outside world. As indicated in Fig. 5.1, the types of I/O devices are different for a computer system and a PC system. The computer must interface with such devices as keyboards, CRT displays and monitors, disk drives, and magnetic tape storage systems. With the exception of the programmer, a PC strictly communicates with discrete, analog, or computer-based I/O devices and occasionally a BCD-encoded device such as a thumbwheel switch or an LED display. The manner in which a microprocessor that operates a minicomputer or a PC communicates with the external I/O devices is a function of the PC's executive or the computer's operating system, as well as the design of the interface hardware and electronics.

The interface portion of a computer or a PC is responsible for providing information from the microprocessor to the external devices, and for reading the status of any specified external devices for the microprocessor. The interface hardware is addressed by the microprocessor in a similar fashion as the memory. Each I/O device is identified by a unique address, just as each word of memory has its own unique address. For example, if the microprocessor requires the status of a particular input point in a PC system, it addresses that point through the interface hardware. The status of the specified input point is returned to the microprocessor via the data bus and interface hardware. If an output needs to be updated, the microprocessor addresses the selected output via the address buses, along with sending out the desired output state over the data bus. The interface hardware responds to these signals and updates the output point accordingly.

PC I/O ADDRESSING SCHEMES

The PC user will note, when reviewing various PC systems, that each manufacturer uses a different protocol to identify the I/O points and the data memory locations. There are PC systems where the I/O points are identified consecutively regardless of whether they are input or output points, such as by 1, 2, 3, 4, 5, 6, ... Other protocols exist where a distinction is made between input and output points by numbering each kind consecutively in decimal or octal numbering schemes. For example, an input may be labeled I 26 and an output O 26. Other I/O numbering schemes rely on all points being numbered consecutively, but each number preceded by either an I or an O to identify the point as either an input or an output point. This scheme would allow for only a single I/O point with the identity 26, but identified as to type with either the I or the O prefix.

If that is not confusing enough, there are PC systems that block out two groups of consecutive numbers for inputs and outputs. For example, any I/O reference between 0000 and 0999 is an output reference, and any number between 1000 and 1999 is an input reference. There are systems where the I/O point's location in the I/O structure determines its identification. A four-place identification of ABCD, for example, might use position A to indicate an input or output point, B to indicate the I/O channel, C the I/O card's position within the channel, and D the I/O point address on the specified I/O card.

The memory locations used for internal data storage are also labeled in various ways, depending on the manufacturer. Data storage locations may be labeled in both decimal and octal numbering formats. There are data storage areas allocated for specific uses. These areas have specialized identifications, such as "Txxxx" for timer/counter storage areas, "Sxxx" for general storage areas, or even "Mxxxx" for a math storage area where the "xxx" is a numerical identification. The list is varied and goes on.

While it may seem confusing to have all these different numbering schemes, each has its advantages and disadvantages. Since these are subject to the tastes of the user, it is difficult to single out any one method as being the best. The best advice is to examine each identification scheme offered by the various manufacturers and understand its rules and operation. The preference of one system over another can then be incorporated into the evaluation of one PC system versus another. It happens rarely that the selection of one PC system over another is made solely on the basis of the identification scheme used. Rather, the presence for a particular identification scheme will fall under the category of "personal preferences" along with aesthetic designs, functionality, programming ease, and vendor performance evaluations.

PC USER LOGIC SCAN ROUTINES

One item that is often overlooked by the PC user is how the PC solves its logic programming. This oversight may lead to programming problems and improper control system responses. The PC user should review the logic solving operation of the PC being considered with the manufacturer's applications engineer prior to writing a PC program. While most PCs solve their programs sequentially, exactly how the PC solves a single ladder rung of logic may be very important, especially when the PC permits multiple ladder rungs of logic to be grouped together in "pages" or "networks." Some PC systems solve a rung of logic until the first false or OFF condition is encountered. At this point the system immediately updates the rung coil status to OFF, then skips on to the next rung, thereby saving scan time that would be wasted in solving a rung that is going to be false anyway. Many of the newer PC systems solve the user logic in large blocks followed by a partial I/O update. How the processor performs this operation, and in what I/O and user logic block sizes, may be an important consideration in how the user logic is placed in the PC. While the majority of PC applications do not require the user to understand the logic solving rules of the processor, a few minutes spent in review may save later headaches for the programmer. Where timing is critical to the application, this review may expose ways in which the logic solving routines of the processor can be used to advantage to improve scan times and critical timing considerations.

RUNNING ERRORS

Many PCs provide an extreme amount of flexibility in their programming formats and instruction sets. This flexibility is not without a drawback, however. The ability to create a highly flexible program can lead to the creation of a program that the processor cannot solve. As a way to flag the user that the processor has been instructed to do something not permitted, or that a program has been generated that is confusing for the processor to solve, the *running error* has been developed. A running error or *run-time error* is an error indication that the processor has halted its scan because of a user program problem and not a hardware problem.

For example, a user could employ a jump instruction in a program as a means to bypass selected sections of the program under certain conditions. If the PC's jump instruction is formatted to require a label (location) to jump to, multiple identical labels will confuse the processor executive program. When encountering the jump instruction, the processor looks for the location to be jumped to. If two locations have been entered with the same identification, the processor does not know which one is correct. The processor flags a running error and halts its scan. By means of the programming terminal the exact nature of the error is generally displayed for user corrective action. The running error provides a means to flag the user that the program has attempted to send the processor calculating something in never-never land.

MEMORY PROTECTION SCHEMES

For many applications it may be desirable to "lock out" the user and data memory from any future changes without proper access credentials. To accommodate this requirement practically all PCs incorporate *memory protect* keys and/or access codes. Through the memory protect key or processor access identity code those not having the correct access key or code sequence can only monitor a PC program and not change it. When the correct key is inserted in the memory protect lock, or a correct access code combination is entered from the programming terminal, the processor permits the programming terminal to operate in the *program* mode for user logic changes, additions, and deletions.

PROCESSOR MODE SELECTION SWITCH

Many PC processors incorporate two or more keyswitch selectable modes of operation. The first of these modes is the *run* mode, where the processor reads input device statuses, solves the user logic, and updates the output devices. In the run mode the processor scans and solves the user logic programming. A program mode is usually provided to permit the programming of the processor to occur. In this mode the processor may be halted from scanning and solving the user logic program. A *test* mode is often provided for the verification of the user program. In the test mode the processor operates just as though it were in the run mode, except that the output modules remain in the OFF state. The user can activate input devices and watch the operation of the program without having to worry that something might move or start. A final mode often found is an *on-line* program mode. This mode permits the user to monitor and modify his or her programming while the processor is scanning and solving the user program currently stored in memory.

6

Power Systems
and Supplies

The proper selection, design, and installation of a power source for any PC-controlled system is critical for correct system operation. All PC control systems require either singular or multiple sources of power for operation of the control logic and the field control devices. Improper application of a power source can lead to frequent hardware failures or faulty system control. Selection and design of a power scheme for use with a PC begins with the proper distribution of raw power to the control devices and control system. Only then can the proper selection of the power supplies necessary for any PC control system be completed.

Distribution of the raw power for a PC control system begins with the selection of the main source or sources of raw power. These sources can be existing in the facility, or may require installation due to insufficient capacity or availability of existing power sources. The power levels might be anywhere from 120 to 480 volts ac, depending on the particular power distribution system of the facility. For various power generating stations and some railway installations 125 volts dc is often used. Where a PC operates as part of an automatic guided vehicle system (AGVS) or other similar mobile vehicle system, the use of 12 or 24 volts dc is required for system operation. Possible industrial power sources range from lighting distribution panels to large motor control centers (MCCs) and secondary substation switchgear. For those applications relying on dc power, a battery or ac-to-dc converter may be used. Sources of raw power will be required for the PC processor, each functional group of I/O hardware, and each type of I/O device used in the control system. As will be outlined later, it is often advantageous, or possible, to combine multiple PC control system power requirements and thus reduce the number of separate raw power sources required. Some applications even lend themselves to power distribution schemes where process or equipment instrumentation can share the power source with limited PC hardware.

PC POWER SOURCE REQUIREMENTS

Two general classifications of raw power are usually required for any PC system. The first power source must be *clean* for use by the processor and I/O power supplies. A clean power source is one that is free of noise and voltage level fluctuations. The second power source will generally not be as clean as the first and can be termed a *normal* supply. The normal supply will feed I/O devices and their specialized power supplies as required. As an alternate to the clean and normal power sources required for a PC control system, some applications may require an *uninterruptible power source* (UPS) for operation of the control system during general raw power outages. The UPS power may be either ac from a diesel alternator, or a dc source from a battery. The design of a UPS system and its automatic switchover from the raw power source is not encountered in the majority of applications. Design of this type of system will require consultation between the PC applications engineer, the control system designer or engineer, and the UPS hardware supplier.

The best choice of a power source for a PC-based control system is one that is not heavily loaded with power-hungry devices, such as large horsepower motors, welding equipment, or high-power heating devices. Power sources with loads such as those listed may be exceptionally noisy and unsteady due to the cycling of the connected equipment. If at all possible, the power source or sources should be as free from voltage variations and noise as possible. Once a suitable raw power source has been decided upon, the necessary distribution hardware can be selected. This distribution hardware will involve transformers and distribution panels for each power circuit required by the control system.

PC POWER DISTRIBUTION EXAMPLE

To understand better the steps in the design of a power system for a PC-based control system, a hypothetical system will be considered. For this example, the designer is adding additional equipment to a facility for the purpose of expanding production capacity. Review of the present plant power distribution system and the anticipated new power loads indicates that no new power sources to the site will be required since currently there is sufficient spare capacity in the system. A 480-volt ac MCC that is very lightly loaded has been located near the expansion area. This MCC can be expanded to handle the additional new power loads. Since this MCC has very small current loads, and the future loads will also be small, this MCC will serve as the raw power source for the new control system. Fig. 6.1 is a single-line diagram of the existing MCC loads with the future loads also shown. One of the new MCC loads is a transformer for reducing the 480-volt ac power level of the MCC to 240/120 volt ac for the PC system. All PCs manufactured and sold in the United States

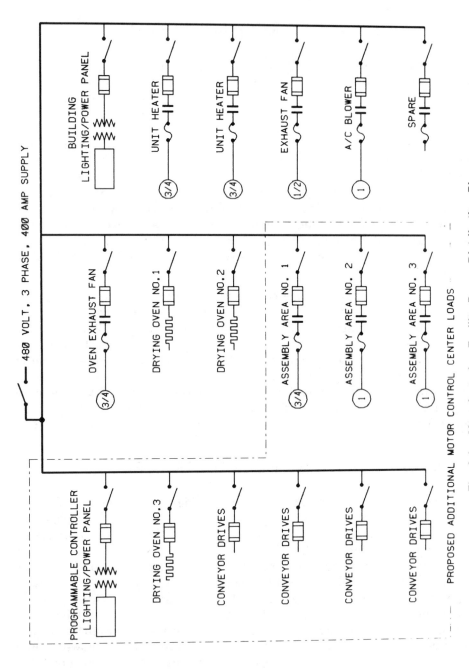

Fig. 6.1. Manufacturing Facility Power Distribution Plan

are designed to operate from either 120- or 240-volt ac power sources. Selection of either voltage is determined by the user to suit his particular requirements. For this example a PC power supply voltage of 120 volts ac will be used.

The PC requirements for this example involve a new PC control system for both the existing drying ovens (retrofit) and a new drying oven. Three new PC-controlled manufacturing machines, located in individual assembly areas, and a new PC-controlled conveyor system will also be added. The three drying ovens will be controlled by a single PC with bumpless backup. Three remote I/O systems will be used with this PC, one remote I/O system for each oven. The three new manufacturing machines will have their own self-contained PC systems. Four remote I/O systems and a single PC processor will operate the new conveyor hardware.

There are numerous ways in which the power sources for this system can be designed. The designs that follow are not necessarily the ones that must be used. The actual system chosen for an application will depend on many factors. Before any scheme is implemented, a review between the system designer or engineer and the manufacturer's applications engineer should take place. This practice will ensure that nothing is overlooked in the final design. Fig. 6.2 is a power distribution diagram for the example area based on the power distribution schemes that will be discussed.

Examination of Fig. 6.2 shows that there are 13 circuits required for the example. Circuit 1 will power the drying oven processors and the auxiliary I/O power supplies assigned for each oven. The I/O field devices for each oven are powered from their own circuits, circuits 2, 3, and 4. Power for the PC control system and the I/O devices of each manufacturing machine is supplied from a dedicated circuit, labeled circuits 5 to 7. The conveyor controls require a dedicated circuit for the processor and one additional circuit for each remote I/O system. The last circuit required provides power for all interior PC cabinet lighting.

Fig. 6.3 is a schematic for circuit 1. The individual loads on circuit 1 are the two PC processors, the remote I/O power supplies for each oven, and the programmer/tape loader receptacles. The two processors are connected through twistlock receptacles so that one operating unit can be removed without disturbing the second one. It should be noted that twistlock receptacles provide a means of quick disconnect as well as line polarity continuity, and good resistance to accidental disconnection from jarring and/or vibration. The three remote I/O power supplies are fed from the same source, through fuses, to permit the removal of one remote I/O power supply without affecting the remaining two. In lieu of fuses, circuit breakers or twistlock receptacles could be utilized for each remote power supply location as an alternate means of disconnecting the supply. The wiring from the isolation transformer to each remote I/O power supply must be sized to minimize any voltage drop in the wiring run. The isolation transformer has been installed to ensure that the voltage fluctuations and noise of the incoming power from the MCC are minimized.

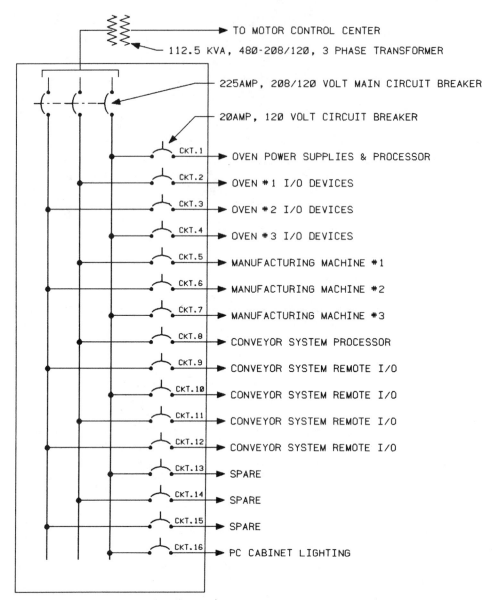

Fig. 6.2. Manufacturing Facility PC Power Distribution

Selection and installation of any voltage stabilization or isolation transformers should be in strict accordance with the manufacturer's recommendations and guidelines. Most PC systems are designed to shut down automatically whenever the supply line voltage drops below a minimum level or exceeds a maximum level. Since many power distribution systems routinely experience variations and even dropouts for a few cycles, the installation of a combination isolation and voltage stabilization transformer is essential if the electronics of the PC are to ride through these nuance interruptions.

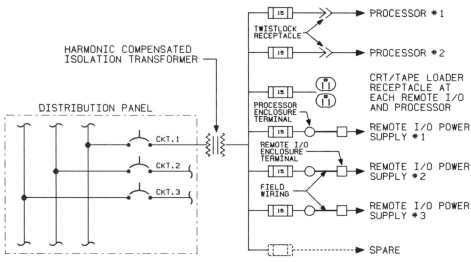

Fig. 6.3. Example Power Distribution Scheme, Circuit 1

There are several styles of stabilization/isolation transformers available. Those units that do not provide filtering of the output harmonics are not recommended. Units are available that incorporate internal output line filters, which attenuate transients having durations of 1 millisecond or less. Another consideration the designer should bear in mind is that the loading of the stabilization/isolation transformer will affect its overall performance. In general the lighter loaded the transformer, the more severe the voltage drop that can be tolerated. For example, a fully loaded transformer can still maintain 100 percent output voltage with only 85 percent of its input voltage. For a system with a 25 percent load the input line voltage can drop to 35 percent of normal with rated output voltage still being maintained.

The designer must bear in mind that the power supplied to the system will remain the same, regardless of the transformer input voltage. When the input voltage drops, the input current will rise to maintain the same power output from the transformer. A transformer operated under reduced input power for extended time periods will heat excessively, and eventually burnout will occur.

The I/O power for each oven is supplied from dedicated circuits of the distribution panel, as indicated in Fig. 6.4. This method of PC power distribution was used because of the types of control loads that exist at each oven. Since the ovens have heating loads which may be solid-state controlled (triacs), better isolation of these noise-generating circuits is recommended. The analog power supplies that may be required for the application will have their own filters to protect them as necessary if they are to be connected from the same circuit.

It is assumed that line voltage variations will shut down the processor before the I/O device senses the disturbance. Most PC power supplies send the processor into an orderly loss-of-power shutdown long before the ON-to-OFF transition threshold of an input module is reached. Alternately, if the possibility exists for I/O power to be lost without the loss of processor power, special

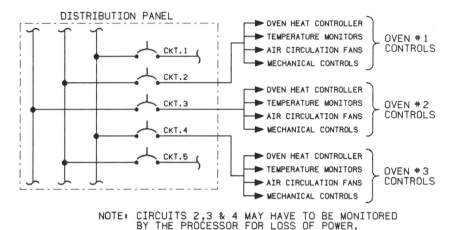

NOTE: CIRCUITS 2,3 & 4 MAY HAVE TO BE MONITORED
BY THE PROCESSOR FOR LOSS OF POWER.

Fig. 6.4. Example Power Distribution Scheme, Circuits 2, 3, and 4

precautions need to be implemented. For the example, the I/O power is fed from a series of circuits different from that of the processor. This will require that the user incorporate sufficient logic within his or her application program to detect the loss of power to one or more remote I/O locations while the processor remains powered. This would ensure that line transients and dropouts do not cause the processor to "believe" that one or more of the input points is OFF.

The power distribution scheme shown in Fig. 6.5 for each of the new manufacturing machines represents the most recommended method of powering the I/O devices and the PC hardware. The output loads are powered directly from the distribution panel in conjunction with an isolation transformer for the processor power and input power. There are no special requirements for powering the output devices, except that the circuit must be sized according to

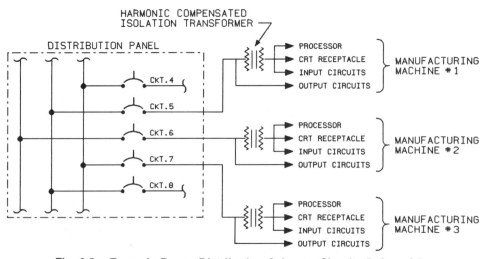

Fig. 6.5. Example Power Distribution Scheme, Circuits 5, 6, and 7

the maximum load requirements of all the output devices and the isolation transformer primary current. Often it will be necessary to supply multiple circuits for the output devices due to current requirements or various voltage differences. In this case the isolation transformer may be placed on its own circuit, and the processor is programmed to monitor the various output device power sources similar to the oven power distribution example just discussed. Again, monitoring of the various output power sources will ensure proper system operation in the event of the loss of one or more output device power sources.

The isolation transformer is the same as was discussed earlier in this chapter. Its prime function is to provide noise isolation and voltage stabilization to the PC processor and input devices. This transformer must be sized to carry the load requirements of the processor as well as the inputs. In most cases the processor power requirements will be greater than the input power requirements. In addition the input devices will be switching the load of the PC input point electronics, usually 50 milliamperes or less, so there will be very little, if any, electric noise generated which could disrupt the processor's operation. Supplying power to the input devices as well as the processor guarantees that in the event of a power interruption, the processor will be shutting down before the input modules have time to react to the loss of power. Once the processor goes into shutdown, it will not react to the later change of state of the input devices, which could cause incorrect logic operations to occur.

The conveyor system represents a third power distribution concept that is often encountered. This application requires that the processor be centrally located with remote I/O systems being implemented to reduce field wiring costs. As was implemented in the oven application, the processor will be powered from a dedicated isolation transformer supplied from a single circuit of the distribution panel. Fig. 6.6 illustrates the power distribution scheme for the processor that is used to control the conveyors. The remote I/O power supplies will not be fed from the isolation transformer, but from the same source that powers the I/O devices, as illustrated in Fig. 6.7.

It is often acceptable to power the I/O devices and a remote I/O power supply from the same source when the I/O devices are low-power devices.

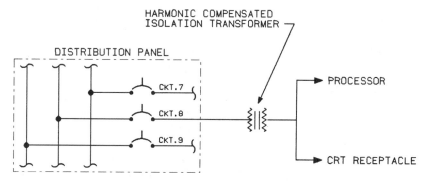

Fig. 6.6. Example Power Distribution Scheme, Circuit 8

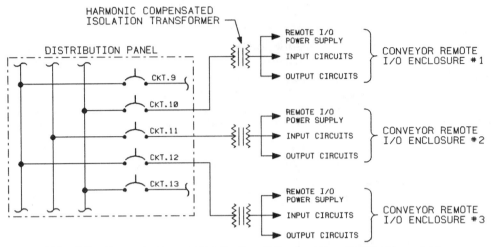

Fig. 6.7. Example Power Distribution Scheme, Circuits 10, 11, and 12

Most conveyor systems require more I/O devices to monitor the conveyor product flow than to actually control the product's flow on the conveyor. The output devices used to control the conveyor's product flow are either solenoid valves or motor starter coils. Usually the motor starter coils are controlled through isolated output modules powered from the starter's control transformer. The noise that may be generated in the motor starter circuits is kept within the circuit and does not affect the remote I/O power source. The remaining control devices are the solenoid valves used to operate stops or actuate transfer devices. These solenoids are generally low-power consumption devices of very few watts. While they do cause small power source variations, the remote I/O power supply is designed with the same noise immunity filtering circuits as the processor power supply, and these variations will not affect the operation of the remote I/O system.

Before applying the conveyor system's power distribution concept, a review of the error-checking design of the PC's remote I/O communications system is recommended. Most manufacturers who offer a remote I/O system provide an error-checking routine between the processor and the remote I/O system to ensure that electric noise does not alter the I/O statuses being passed between the two pieces of hardware. This error-checking routine is both hardware and software designed into the PC by the manufacturer. Several forms of error checking are used, including transmit and verify, multiple transmit and compare, parity encoding, or a combination of these, mixed with special check and verify algorithms.

The purpose of these tests is to ensure that the data being passed between the I/O modules and the processor are correct. When these algorithms indicate a faulty communication of data, the I/O module or channel is deactivated by the processor, and an error condition may be indicated to the operator or user. If an application arises where the remote I/O power supply and the I/O devices are supplied from the same source, a complete review of the remote I/O power supply's noise immunity must be considered. In conjunction with this review the error-checking routines between the processor and the remote I/O system will need to be examined with the manufacturer's applications engineer.

The discussion of power distribution for a PC system is now complete. Three commonly used schemes have been presented. These schemes do not represent the required way to design a PC power distribution system, but rather indicate several of the more common methods used from the numerous possibilities that do exist. The final power distribution scheme employed will depend entirely on the control application, the I/O control devices, the PC system employed by the designer, and the abilities of the designer or engineer.

It should be pointed out at this time that there are two things often overlooked when the design of a PC power distribution system is completed. Many PC enclosures are installed in poorly lit areas, and the installation of a fluorescent fixture in the enclosure will greatly aid the troubleshooting and maintenance of the enclosure hardware. The installation of this light should include a door switch to switch OFF the fixture whenever the door is closed. This light should not be connected to any of the circuits of the control system, but should be powered from its own source or the general building lighting circuit. Fig. 6.8 indicates the lighting scheme that might be employed for the power distribution example.

A second often overlooked item is the installation of a receptacle for powering the PC programming terminal or tape cassette loader. This receptacle should be powered from the same isolation transformer as the processor. Since the programmer and the tape loader hardware communicate directly with the processor, these devices should not be powered from nonisolated sources. Any non-PC processor power source noise or voltage fluctuations would be transmitted directly to the processor, resulting in possible control system operating errors. In addition unintentional ground loops may be created between the foreign and the PC power supplies, which could introduce further electric noise to the system. This receptacle should be labeled "for programmer use only" since the connection of a drill motor or similar device to this

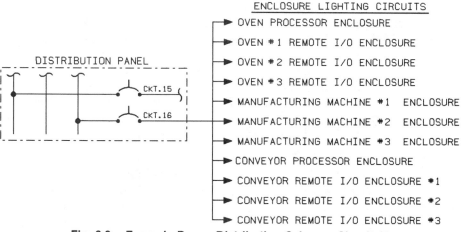

Fig. 6.8. Example Power Distribution Scheme, Circuit 16

receptacle, or another one on the same circuit, could lead to the introduction of noise spikes and voltage fluctuations on the circuit when a programming device or tape loader is being used. Often this receptacle is purchased in a color different from the standard ivory or brown to distinguish it from the others on the site. Changing it from a standard receptacle to a crow's foot or twistlock may be done, but the plugs on all PC support equipment, such as the programmer or the tape loader, will require modification to fit the special receptacle.

TYPES OF PC POWER SUPPLIES

In various chapters throughout this text, reference is made to the power supplies for the processor and the remote I/O system. The power supply or supplies for a PC system are often taken for granted when a system is designed, purchased, and installed. This should not be the case, and therefore a detailed discussion of the operation, selection, and installation of this device is in order.

The power supply often referred to as the *main* supply is responsible for meeting the power requirements of the processor and possibly some small number of I/O modules that are connected in a local configuration to the processor. The I/O modules and I/O hardware that are connected to the processor in a remote configuration usually require the use of a *remote* or *auxiliary* power supply for proper operation. Both main and remote power supplies are designed by the manufacturer to be similar in operation and function. However, additional electronics may be incorporated in the remote supply for the operation of the remote I/O communications lines between the processor and the remote I/O system.

Operation of a PC power supply is the same as that found with any power supply used with electronic equipment. While many of the older PCs on the market use a *linear* design for the power supply, the newer systems are incorporating a *switched mode* (switcher) power supply design when a cost savings is obtainable. As a quick review, Fig. 6.9 is a block diagram representation of a linear power supply. Incoming line power is level-converted with a transformer and passed on to a rectifier for transformation from alternating to direct current. From the rectifier the dc voltage is filtered into pure direct current before being passed to the regulator section. The regulator section contains the necessary electronics to monitor the output voltage continuously with respect to an internal reference and to correct the output voltage as required to maintain the proper level. Before leaving the supply, the output voltage is passed through a protection block that ensures that a short circuit or overload condition in the processor or I/O system does not cause the power supply to destroy itself or the processor.

A block diagram for the switcher power supply is given in Fig. 6.10. Incoming line power is directly rectified and then filtered. This high voltage level direct current is then switched at a very high OFF/ON rate to a filter section. The rate of switching can be anywhere from 20,000 to 50,000 times a second, depending on the load and design requirements of the supply. The filter removes the switching transients from the direct current and also provides a "storage"

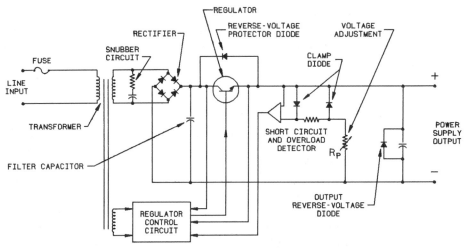

Fig. 6.9. Linear Power Supply

element for the output voltage and current. The effect is to rapidly charge a filter capacitor in very small bursts by the switch block while the load connected to the supply continuously drains off the power in the filter block. The rate at which the switch block "refills" the filter block is a function of the power supply's loading. Before being delivered to the load, the output power is passed through a protection block to ensure that a short circuit or overload does not destroy the supply or load electronics.

The linear type of supply is usually more massive and larger in size than a switcher because of the large transformers, chokes, and capacitors used. Even though they are not complex in electronic circuitry, they are low in efficiency (usually 10 to 50 percent) when compared with that of the switcher (usually 60 to

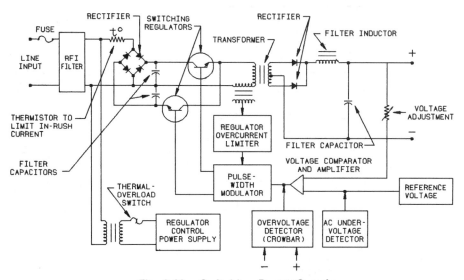

Fig. 6.10. Switching Power Supply

90 percent). With the general reduction in the size of PCs due to increased electronic circuit intergration, the switcher power supply is being implemented more frequently. Table 6.1 provides a generalized comparison of features relating to linear and switching power supplies. The reader may wish to keep this table in mind for future use. Many PC applications may require the user to select a power supply for a specialized I/O application, and Table 6.1 may provide some useful information in the final selection process.

In addition to the standard electronics of the power supply, all PC manufacturers enhance their power supply designs with additional electronics to monitor the incoming line voltage. This is done to ensure that the processor or I/O system has sufficient time to stop operations and go into a "housekeeping" and shutdown mode before total loss of power. This time delay is necessary if the system is to start itself back into proper operation upon power restoration. The housekeeping and shutdown operations of the processor include such functions as completing any communications with the I/O systems, finishing the logic solutions of a particular group of control commands, and storing vital system statuses for reuse upon power up. Directly opposite the power down electronics in function are the power up electronics. Upon restoration of power to the system, the power up electronics inhibit the processor from operating until the power supply has stabilized all of its output power levels. The power up electronics then signal the processor to perform an initialization sequence that returns any stored data to the proper processor areas, and then restarts communications with the I/O systems before the processor actually begins its logic-solving functions.

Table 6.1. Linear and Switching Power Supply Comparison

POWER SUPPLY COMPARISON			
PARAMETER	LINEAR	SWITCHERS	UNITS
SIZE	0.5	2	WATTS/IN
EFFICIENCY	40-55	60-85	%
POWER DENSITY	10	50	WATTS/POUND
RIPPLE NOISE	30	50	MILLIVOLTS PEAK TO PEAK
LINE CYCLE CARRYOVER	2	25-50	
TRANSIENT RECOVERY TIME	25	500	MICROSECONDS
INPUT VOLTAGE TOLERANCE	±10	±20	%
REGULATION	0.1	0.1	%
GENERATED RFI	NONE	HIGH	—

PROPER PC POWER SUPPLY SELECTION

The proper selection of the PC power supply or supplies is probably one of the easiest tasks to perform when designing and purchasing a PC system. A unit must be selected to match the intended supply line voltage level and frequency. In most cases the power supply is multiple rated for more than one line voltage and frequency. The most common PC supplies offered in the United States are for operation on 120 or 240 volts ac at 60 hertz. Often the supply is also rated for 110 or 220 volts ac at 50-hertz operation. Besides these common configurations, some supplies are also available for 125- or 24-volt dc operation. The PC manufacturer should be consulted for any special offerings or other special power supply configurations that might be necessary for a particular application.

Many manufacturers offer several sizes of power supplies from which to choose. The proper size of the supply is determined by the load to be connected to the supply. Many PC models incorporate the power supply as part of the processor. These supplies may be rated for the maximum processor and I/O configuration possible for that particular PC model, so no actual power supply selection need be made. Selection of a particular model of PC automatically incorporates the proper size supply for that model. Other systems are available where the supply incorporated with the processor is not able to handle the full capacity of the PC system. Instead, these PCs have a supply that can handle the processor's maximum load plus a small amount of I/O load. The remaining I/O load must be supplied from auxiliary or remote power supplies. Selection of the additional supplies requires that a load calculation be done prior to the selection of additional supplies.

When additional power supplies are required for a PC system, the manufacturer will usually provide power supply load requirements for all PC hardware, as well as power supply capacities for all power supplies. These capacities are usually in the form of ampere or unit load designations. When ampere designations are supplied, the current capacity of the power supply must exceed the current requirements of all load devices connected to that supply. To simplify the current calculations, some manufacturers specify a unit load value for each device that must be powered from a power supply. The designer must add up the unit load numbers of all PC hardware components connected to the supply, then select a supply that has a unit load capacity greater than the total hardware unit load number. Where a PC processor power supply has a capacity above that required by the processor alone, the difference between the total supply capacity and the required processor load may often be applied to powering additional PC hardware. A word of caution to the user is that it is not permissible to use a PC power supply to power both the PC hardware and some field I/O devices, since any electric noise in the system would have a direct route into the processor electronics. In fact the majority of current PC power supply designs require that a user "plug-in" the power supply via manufacturer-supplied cables. This type of quick power supply installation and connection generally makes the actual dc power terminals not user-accessible.

PC POWER SUPPLY INSTALLATION PRACTICES

Installation of a power supply for a PC system usually requires the mounting and connection of the supply to both the raw power source and the PC hardware. The supply should be firmly mounted with all electrical connections properly covered to prevent accidental electric shock to personnel working around the enclosure. Particular attention should be paid to the manufacturer's heat dissipation instructions. Most power supplies, especially the linear-type designs, dissipate heat while operating. They should be mounted in a free air space in the enclosure, leaving the manufacturer's recommended spacing between the supply and any nearby equipment or enclosure surfaces. Fig. 6.11 illustrates typical spacing requirements for major PC hardware components similar to those suggested by most PC manufacturers as part of their installation

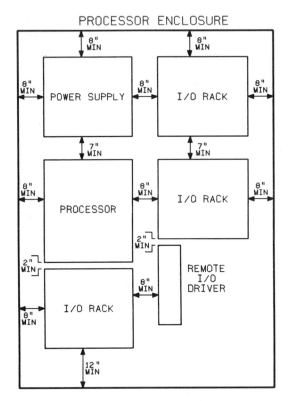

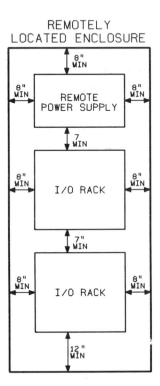

DISTANCES SHOWN BETWEEN COMPONENTS AND ENCLOSURE SIDES ARE RECOMMENDED DISTANCES FOR PROPER AIR FLOW AROUND ENCLOSURE CONTENTS. NO COMPONENT MOUNTED IN ENCLOSURE SHOULD EXTEND CLOSER THAN SIX (6) INCHES FROM THE ENCLOSURE DOOR. MANUFACTURER'S INSTALLATION SPECIFICATIONS AND RECOMMENDATIONS SHOULD BE CONSULTED DURING ENCLOSURE DESIGN FOR SPECIFIC DETAILS REGARDING THE MOUNTING OF COMPONENTS IN AN ENCLOSURE.

Fig. 6.11. Recommended Hardware Spacing

manuals. This figure not only indicates the minimum spacing requirements for the power supplies, but it also indicates the spacing requirements for other PC hardware items, such as I/O racks and remote I/O drivers.

The installation of other heat-producing devices below the power supply should be discouraged. It cannot be overemphasized that the manufacturer's recommendations on operating temperature cannot be ignored. The temperature specifications of the PC system must be considered at the time of system design. Chapter 4 reviews the proper methods used to calculate the temperature rise for enclosures housing PC hardware.

When wiring the supply, proper attention must be paid to the grounding requirements outlined by the manufacturer for the installation of the supply. In addition the wire routing to the PC power supplies should follow the manufacturer's recommendations. Most manufacturers prefer that the power supply wiring be run in dedicated cables from the source to the PC enclosure. Many manufacturers also prefer that the 120-volt ac/dc and lower input and output wiring be kept separate from the 230-volt ac and higher power wiring. If there are any questions concerning the routing of a particular PC cable, the manufacturer's applications engineer should be consulted.

PC BATTERY BACKUP SYSTEMS

A discussion on PC power sources and supplies would not be complete without an examination of the battery systems used by the various manufacturers to maintain the PC memory when the normal source of line power is removed from the unit. Unless the processor memory is of the PROM or core type, the PC manufacturer will be required to use electronic RAM for the processor logic storage function. Use of this type of memory in general requires the installation and maintenance of some form of backup power source to enable the processor memory to maintain its program in the event the normal power source is interrupted. This backup power source is in the form of a battery system. The usual battery systems employed by the various PC manufacturers are carbon-zinc, alkaline, lead-acid, nickel-cadmium, or lithium. Each of these systems exhibits characteristics that should be known to the user, if he or she is to expect uninterrupted service from a PC system using them as the method of memory backup.

The carbon-zinc, alkaline, and lithium systems are the easiest for the manufacturer to install. The battery size is usually chosen as the standard D size. However, when the battery is placed directly on a processor circuit board, the AA size or the button cell is often found. All three of these battery systems are *primary* systems, meaning that they cannot be recharged after having been discharged. Since the carbon-zinc system has very few advantages over any of the other systems, use of it is becoming scarce in the manufacture of PCs. Many PC models incorporate alkaline battery backup because of its low cost and excellent shelf life with respect to the carbon battery. In general the carbon-zinc and alkaline systems are safe for the majority of PC applications.

The lithium battery is becoming the workhorse of the PC industry because of its excellent operating characteristics, long shelf life, and high power density for the package size. Its major drawback is that it can be dangerous if mistreated, and it is expensive. The lithium system of batteries consists of numerous chemical compositions, each composition producing a different character of battery. Table 6.2 lists some of the more common lithium battery systems and the open-circuit terminal voltage for each system. Lithium is the first element of the alkali metal group, metals so chemically active that they are never found as pure elements, but instead as stable compounds. The element will react very violently with as little as 100 parts per million of water. This fierce reaction with water requires that the battery cell structure be hermetically sealed. The battery's strong distaste for moisture, coupled with cathode compositions such as sulfur dioxide (SO_2) under a 2-atmosphere case pressure, resulted in special Department of Transportation regulations (DOT-E 7052) regarding its transportation. While several lithium battery systems meet the requirements of DOT-E 7052, the user should be aware of the regulations and follow their guidelines. Most manufacturers using lithium battery memory backup systems provide detailed instruction sheets with the batteries to inform the user of proper handling, storage, installation, and disposal techniques that should be observed when using the lithium battery. The bottom line is that no matter what the cathode composition, the lithium anode should *not* be taken for granted. If in question, call the manufacturer. Many manufacturers have designed their processor systems to permit the use of multiple-battery chemistry systems. It is often easy to change from a lithium battery system to an alkaline or nickel-cadmium system. The user should be aware that the amount of time that the PC system will retain memory without external ac power may be greatly reduced when changing from a lithium system to another battery system.

The lead-acid and nickel-cadmium batteries are *secondary* battery systems. These two systems allow the battery to be recharged after a discharge. While the primary battery systems listed earlier require replacement at least

Table 6.2. Lithium Battery Chemical Systems

CELL SYSTEMS	ELECTROLYTE	OPEN-CIRCUIT VOLTAGE
LITHIUM/SULFURYL CHLORIDE (Li/SCl_x)	INORGANIC	3.9
LITHIUM/THIONYL CHLORIDE ($Li/SOCl_2$)	INORGANIC	3.7
LITHIUM/VANADIUM PENTOXIDE (Li/V_2O_5)	ORGANIC	3.4
LITHIUM/SILVER CHROMATE ($Li/AG_2Cr/O_4$)	ORGANIC	3.3
LITHIUM/MANGANESE DIOXIDE (Li/MnO_2)	ORGANIC	3.0
LITHIUM/SULFUR DIOXIDE (Li/SO_2)	ORGANIC	2.9
LITHIUM/CARBON MONOFLUORIDE (Li/CF_x)	ORGANIC	2.8
LITHIUM/COPPER SULFIDE (Li/Cu_2S)	ORGANIC	2.1
LITHIUM/IRON SULFIDE (Li/FeS_2)	ORGANIC	1.8
LITHIUM/COPPER OXIDE (Li/CuO)	ORGANIC	1.8
LITHIUM/IODINE (Li/I_2)	SOLID	2.8
LITHIUM/LEAD COPPER-SULFIDE ($LiPb_bCuS$)	SOLID	2.2
CARBON/ZINC = 1.4		

Table 6.3. Battery System Comparison

CHARACTERISTICS	CARBON-ZINC (LECIANCHE)	ALKA-LINE	NICKEL CADMIUM	LEAD ACID	LITH-IUM
ENERGY DENSITY (WH/LB) (WH/IN.^{3})	15-20 1.3-1.6	20-30 2-2.5	9-12 .5-1.0	13-15 1.2-1.5	80-150 6.0-11.5
TYPICAL SINGLE-CELL LOAD VOLTAGE (V)	1.4	1.4	1.2	1.9	2.5
LOW-TEMPERATURE CAPACITY AS PERCENTAGE OF 70°F CAPACITY: +20 °F -20 °F -40 °F	 5 0 0	 20 4 0	 95 75 50	 65 40 20	 95 85 70
SHELF-LIFE STORAGE AT ROOM TEMPERATURE (YEARS)	1-1.5	2-3	5-10	3-8	5

once a year, the secondary systems have much longer operating lives. These systems usually increase the complexity of the processor power supply since a charging circuit is added to ensure that the battery maintains its charge. Both secondary battery chemistries offer excellent characteristics for use with a PC system. The lead-acid battery is usually a gelled electrolyte-type design housed in a plastic case. This battery system is probably the largest physical system employed for a PC. The nickel-cadmium system offers the same excellent characteristics as the lead-acid system, but in the smaller standard AA and D sizes. The nickel-cadmium system is generally used in the majority of secondary battery type memory backup applications. Table 6.3 provides a general comparison between the various battery systems used with PCs.

Any PC system that incorporates a battery system for memory backup must be maintained and monitored. All manufacturers provide instructions on the installation and maintenance of the battery system. The installation instructions usually include step-by-step details on the proper methods and procedures to install the batteries or battery pack in the processor housing. Instructions are also included on the replacement schedule for the battery system. This is probably the single most overlooked maintenance item in a PC system. There will undoubtedly be some readers who will report to work after having read this paragraph and replace all the PC batteries in their facility because they just remembered that the batteries have been in operation for more than a year. It should be noted that many PC batteries should be replaced on a more frequent schedule when there are extended or frequent power interruptions. The designer or programmer should remember that many PC systems offer the capability to monitor a memory location as an alarm when the battery is running low and should be replaced. This memory location can be programmed into the user logic to trigger alarms or indicators so that the battery can be replaced before a total loss of memory occurs due to a low battery condition.

PART III
III
Programming

7

Relay Ladder Programming

When the PC was conceived by the engineers of General Motors' Hydromatic Division, they set forth the very basic requirement that any PC be capable of being programmed in a format not too different from the format used when drafting an electric relay control system schematic. In the industrial control field these schematics are known as *relay ladder diagrams.* In meeting the relay ladder format requirements, PC manufacturers developed a special computerlike language of commands, instructions, and operations that emulate the operation of a standard industrial control relay. This new programming language was called *relay ladder programming.* As will be discussed later in this chapter, it almost directly duplicates the format and functioning of the conventional relay ladder schematic. In fact the PC of today provides many functions and operations not possible to perform with industrial control relays. These nonrelay functions are discussed in greater detail in Chapters 10 to 13.

Prior to an in-depth examination of PC relay-type instructions and a detailed discussion of relay ladder language programming, a quick review of the terminology and operation of the relay is in order. While many readers may find the information presented in the next few pages a review of information they already understand, it is nevertheless presented for those unfamiliar with the industrial control relay, its operation, and its terminology.

GENERAL-PURPOSE RELAYS

There are many types of relays, each one having been developed to meet specific requirements. Two very common relays are shown in Fig. 7.1. They are very simple in construction and provide an excellent example of the basic

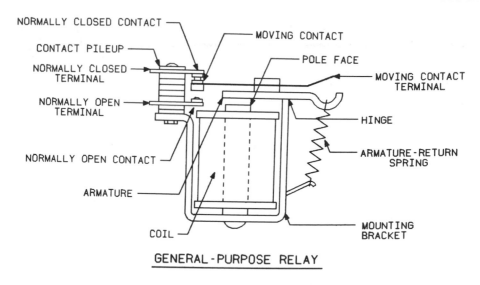

GENERAL-PURPOSE RELAY

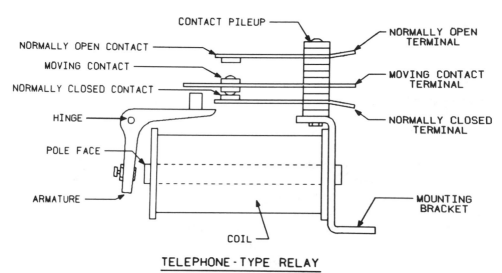

TELEPHONE-TYPE RELAY

Fig. 7.1. Simple Relays

components common to all relays. Every relay contains a metal structure that serves as the means to mount the relay and as a base on which the relay is assembled. This structure is often referred to as the *mounting bracket*. The *coil* of a relay is composed of many turns of very fine wire on a plastic or cardboard core. Over the many turns of wire is wrapped a protective covering of paper, plastic, or other insulating material. The coils of some mass-produced relays are often of an injection-molded design.

Through the center of the core goes a metal rod referred to as the *magnetic pole*. The magnetic pole serves two purposes. It holds the coil to the mounting base, and it acts as a magnet when the coil is energized. The end of the pole piece not fastened to the base is referred to as the *pole face*. The *armature* is also affixed to the base in such a manner that it can be attracted to the pole face whenever the coil is energized. A spring is used to return the armature from the pole face whenever the coil is not energized.

Also attached to the base of a relay is a stacked arrangement of insulating material called the *contact pileup*. The purpose of the pileup is to hold the *contacts* of the relay in position, while electrically isolating them from each other and the remainder of the relay. While the pileup may hold all of the contacts on a relay, some relays, such as the general-purpose relay in Fig. 7.1, may affix one or more contacts directly to the armature with a second contact pileup.

Every relay is composed of at least two contacts. One of these contacts is always called the *moving contact,* or in some cases the *common contact.* The second contact can either be referred to as a *normally open* or a *normally closed* contact, depending on how it operates electrically. Where a relay contains numerous contacts, every contact will be referred to by one of the references just described. However, a number will usually be added in order to distinguish the contact group from others on the same relay. For the sake of simplicity, this discussion will only examine relays with a single normally open/normally closed contact pair.

The name assigned to a relay contact is related to the electrical operation of the contact. A normally open contact is electrically isolated from the moving contact when the relay coil is not energized. Conversely the normally closed contact is mated with the moving contact when the relay is not energized, and power flow is possible between the normally closed contact terminal and the moving contact terminal of the relay. When the relay coil becomes energized, the moving contact is transferred from mating with the normally closed contact to mating with the normally open contact. Power flow is now possible between the moving contact terminal and the normally open contact terminal of the relay. Where the moving contact transfers between a normally open and a normally closed contact, the moving contact may be referred to as a common contact.

The reference terms applied to the contacts and terminals of a relay always describe the electrical operation of the contact when the relay is not energized. A normally closed contact is mated when the relay coil is not energized, and therefore it provides power flow between itself and an associated moving or common contact. The normally open contact description is applied to any contact that does not mate with the moving or common contact when the relay coil is deenergized. Power flow between the normally open contact terminal and the moving or common contact terminal is only possible when the relay coil is energized. Fig. 7.2 illustrates the contact configurations and power flow conditions for both normally open and normally closed relay contacts when the relay coil is both energized and deenergized.

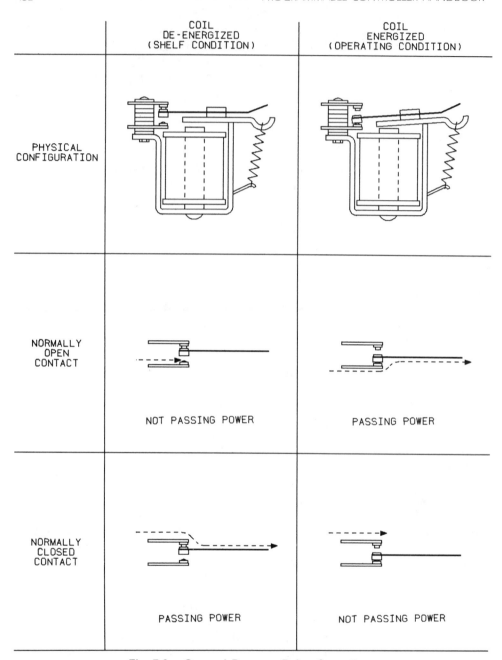

Fig. 7.2. General-Purpose Relay Operation

INDUSTRIAL CONTROL RELAYS

The industrial control relays used for the control of machines and equipment are great-great-great grandchildren of the relay that Morse invented many years ago for his telegraph system. From Morse's simple telegraph relay grew devices designed to meet a host of specialty applications and uses. The industrial control relay is one example of the evolution that took place in the relay field.

The industrial control relay is much more ruggedized than the general-purpose relay. A typical industrial control relay is shown in Fig. 7.3. These relays must be designed for millions of operations in environments of dust, vibration, and temperatures of 100°F and more. Each relay must withstand electromagnetic forces and conditions generated in circuits carrying 600 volts ac or less. Their contact design is somewhat different from that of the general-purpose relay in order to meet the demands of industrial circuits. An industrial control relay is a special-purpose device designed to meet the varied requirements of the industrial control system.

Probably the most drastic difference between the general-purpose or telephone relay contact configuration and the industrial relay contact configuration is the lack of a moving contact terminal, as well as a pair of contact surfaces for each normally open or normally closed contact. As illustrated in Fig. 7.4, the general-purpose relay relies on circuit connection to either the normally open or the normally closed contact and a second circuit connection to the movable contact for electrical operation. The industrial relay contact, shown in the center and bottom portions of Fig. 7.4, does not provide a movable contact terminal, but instead provides a pair of normally open or normally closed contact terminals. This type of construction creates the need for a pair of contact surfaces.

A benefit of providing a pair of contact surfaces with each contact pole is that the arc generated by the opening of the contacts is extinguished twice as fast. The electric circuit is broken in two locations by contacts moving apart at the same rate. The effective distance between the moving contact and either the normally open or the normally closed contact develops twice as fast for the industrial relay contact as it does for the general-purpose relay contact. For every unit distance that the armature moves on the general-purpose or telephone relay, the distance between the contacts changes an equivalent amount. With the industrial relay the distance between the contacts is effectively doubled since one unit distance is generated between each normally open or normally closed contact and the moving contact for each unit movement of the armature.

A second difference that exists between the general-purpose relay and the industrial relay is the quantity of contacts provided. The general-purpose relay typically provides a normally open contact as well as a normally closed contact for a single moving contact. Each moving contact, and its associated normally open and/or normally closed contact, is referred to as a *contact pole*. A general-purpose relay may have numerous poles, each pole consisting of normally open contact configurations, normally closed contact configurations, and/or normally open/normally closed contact configurations. The industrial control relay will have a limited number, usually 12 or less, normally open and/or normally closed contact poles. The normally open/normally closed contact arrangement of the general-purpose relay is not provided with an industrial control relay. Each contact pole is either a normally open or a normally closed contact, but usually not both.

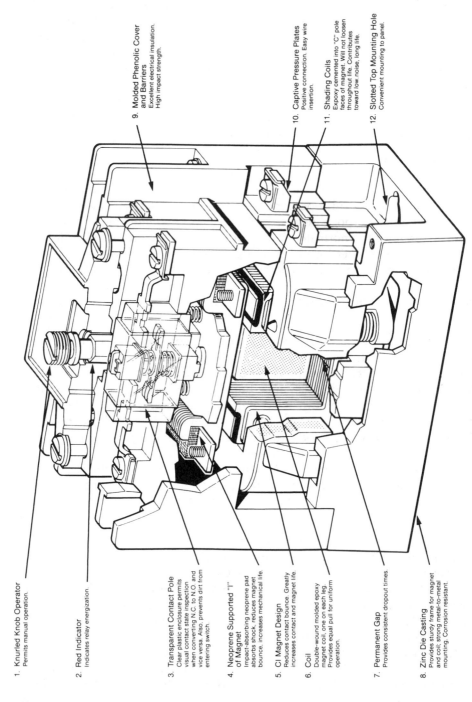

1. **Knurled Knob Operator**
   Permits manual operation.

2. **Red Indicator**
   Indicates relay energization.

3. **Transparent Contact Pole**
   Clear plastic enclosure permits visual contact state inspection when converting N.C. to N.O. and vice versa. Also, prevents dirt from entering switch.

4. **Neoprene Supported "I" of Magnet**
   Impact-absorbing neoprene pad absorbs shock, reduces magnet bounce, increases mechanical life.

5. **CI Magnet Design**
   Reduces contact bounce. Greatly increases contact and magnet life.

6. **Coil**
   Double-wound molded epoxy magnet coil, one on each leg. Provides equal pull for uniform operation.

7. **Permanent Gap**
   Provides consistent dropout times.

8. **Zinc Die Casting**
   Provides sturdy frame for magnet and coil; strong metal-to-metal mounting. Corrosion resistant.

9. **Molded Phenolic Cover and Barriers**
   Excellent electrical insulation. High impact strength.

10. **Captive Pressure Plates**
    Positive connection. Easy wire insertion.

11. **Shading Coils**
    Expoxy cemented into "C" pole faces of magnet. Will not loosen throughout life. Contributes toward low noise, long life.

12. **Slotted Top Mounting Hole**
    Convenient mounting to panel.

Fig. 7.3. Type M Relay Cutaway

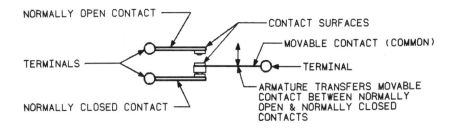

GENERAL-PURPOSE RELAY CONTACT POLE

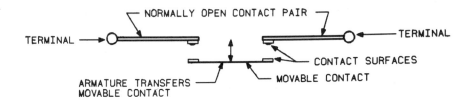

NORMALLY OPEN INDUSTRIAL RELAY CONTACT POLE

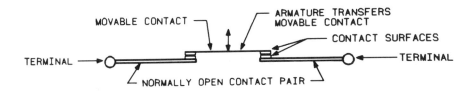

NORMALLY CLOSED INDUSTRIAL RELAY CONTACT POLE

Fig. 7.4. Difference between General Purpose and Industrial Relay Contact Poles

The operation of an industrial control relay contact is identical to that of the general-purpose or telephone relay. Fig. 7.5 details the parts of an industrial relay contact pole and illustrates the operation of both the normally open and the normally closed contact configurations. The relay armature operates the moving-contact actuator to make and break the electric connection between the moving contacts and the normally open or normally closed contacts. The contacts are usually placed in a plastic case to prevent dust and other contaminants from reaching the contact surfaces. One or more springs are provided to ensure the rapid release of the moving contact from the normally open or normally closed contacts.

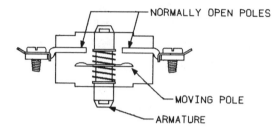

NORMALLY OPEN POLES

MOVING POLE

ARMATURE

WHEN RELAY COIL IS NOT ENERGIZED, NORMALLY OPEN POLE CONTACTS AND MOVING POLE CONTACTS ARE SEPARATED NOT PERMITTING POWER TO BE PASSED THRU THE POLE

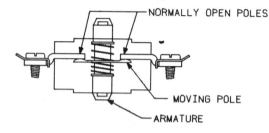

NORMALLY OPEN POLES

MOVING POLE

ARMATURE

WHEN RELAY COIL IS ENERGIZED, NORMALLY OPEN POLE CONTACTS AND MOVING POLE CONTACTS ARE MATED PERMITTING POWER TO BE PASSED THRU THE POLE

NORMALLY OPEN INDUSTRIAL CONTACT

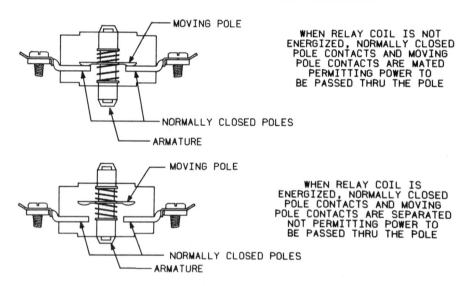

MOVING POLE

NORMALLY CLOSED POLES

ARMATURE

WHEN RELAY COIL IS NOT ENERGIZED, NORMALLY CLOSED POLE CONTACTS AND MOVING POLE CONTACTS ARE MATED PERMITTING POWER TO BE PASSED THRU THE POLE

MOVING POLE

NORMALLY CLOSED POLES

ARMATURE

WHEN RELAY COIL IS ENERGIZED, NORMALLY CLOSED POLE CONTACTS AND MOVING POLE CONTACTS ARE SEPARATED NOT PERMITTING POWER TO BE PASSED THRU THE POLE

NORMALLY CLOSED INDUSTRIAL CONTACT

Fig. 7.5. Industrial Contact Operation

RELAY CONTROL SYSTEM OPERATION

Fig. 7.6 illustrates the basic components of a relay control system. To the left is a limit switch. This device represents a typical input device for an industrial control system. Any device used to supply information to a control system is referred to as an input device. Typical input devices include limit switches, push buttons, pressure switches, photocells, and proximity switches. Part of the limit switch cover has been removed to provide access to the electric contacts of the device. The switch has a pair of normally open and a pair of normally closed contacts. Note that the moving contact is common to both the normally open and the normally closed contact pairs. While this type of moving-contact configuration is not common for industrial relays, many industrial input devices employ this style of contact design.

Centered in Fig. 7.6 is a Cutler-Hammer TYPE M, 600-volt industrial relay. It has been specially modified to provide visual access to two of its eight contact poles. The upper contact pole is configured for normally closed operation, while the bottom pole is configured for normally open operation. This relay operates with a 12-volt ac coil.

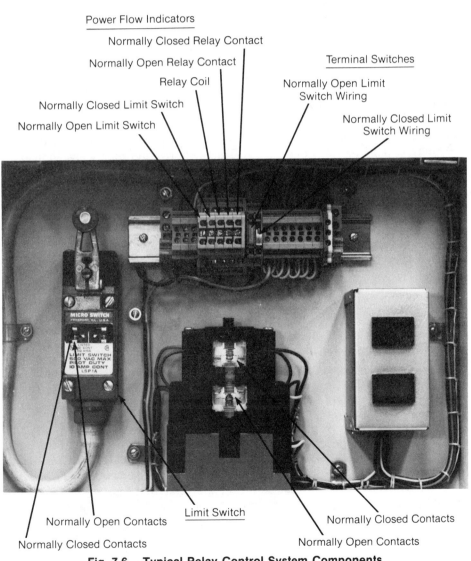

Fig. 7.6. Typical Relay Control System Components

To the right of the relay are a pair of indicator lamps acting as output devices for the control system. Each indicator is wired to an individual contact pole of the relay. The upper indicator is wired to a normally open contact pole of the relay operating exactly like the lower visible normally open contact. A normally closed contact pole, operating like the upper visible normally closed contact pole, is wired to the lower indicator. Either indicator illuminates when the contact to which it is wired is passing electric power.

Both indicators provide status indication of specific conditions relating to the control system. The indicator is one form of control system output device. Other forms of output devices are those that provide a control function or operation external to the control system. For example, a motor might be operated by the control system to move a product or perform another function vital to the overall operation of the hardware or facility being controlled. Motors, indicators, solenoid valves, horns, and alarms are just several of many devices classified as output devices for a control system. Individually or collectively they enable a control system to respond to a changing set of input conditions and states.

Located directly above the relay is a terminal block. Every control system utilizes terminal blocks as a means to connect various signal wires together. The terminal block configuration of the example control system provides several functions. The leftmost three terminals distribute 12-volt ac *common* power to the relay coil and indicator circuits. The next five terminals are indicating terminals. They have been installed to provide a visual indication of the power status of selected control system signal lines. To the right of the indicating terminals are two switched terminals. The left-hand switched terminal, when closed, connects the normally open contact of the limit switch to the relay coil. When closed, the right-hand switched terminal block connects the normally closed limit switch contact to the relay coil. The next seven terminals supply 12-volt ac *line* power to the limit switch and relay contacts. The far right-hand terminal is a ground terminal for the system. To either side of the terminals are retainer clips to fasten the terminals to their metal mounting strip.

Fig. 7.7 provides the electrical relay ladder schematic for the simple control system. The term "relay ladder" diagram arises from the similarity the schematic shows to a ladder. The left-hand power bus or left ladder support is represented by the 12-volt ac *line* bus, while the right-hand 12-volt ac *neutral* bus makes up the remaining ladder support. Each horizontal series of circuit elements forms one rung of the ladder. Each rung is numbered for reference purposes, as is each terminal block and wire.

To the right of the relay coil, outside the ladder diagram, is a cross-reference list of the locations of all normally open (NO) and normally closed (NC) contacts associated with the relay. In ladder rungs 2 and 3 are contacts associated with the relay. These contacts are identified to the relay that operates them by an identification appearing above the contact symbol. Below and to the right of the contact symbol there is a reference to the ladder rung containing the relay that operates the contact.

The operation of the control system is straightforward. In Fig. 7.8 the left terminal block switch has been closed. This switch connects the normally open limit switch contact to the relay coil. Power has been applied to the system. The lower indicator is illuminated because the limit switch is not activated, and the normally open limit switch contact does not provide power flow to the relay coil. Close examination of the limit switch contacts reveals that the moving contact is mated with the two lower normally closed contacts instead of the upper pair of normally open contacts. The normally closed contact of the relay permits power flow to the lower indicator. It is also seen that the relay's upper pole's moving contact is mated with its normally closed contacts. The lower pole's moving contact is not mated to the normally open contacts associated with it. Note that the second and the last terminal block indicators are illuminated, indicating power flow in the normally closed limit switch and relay

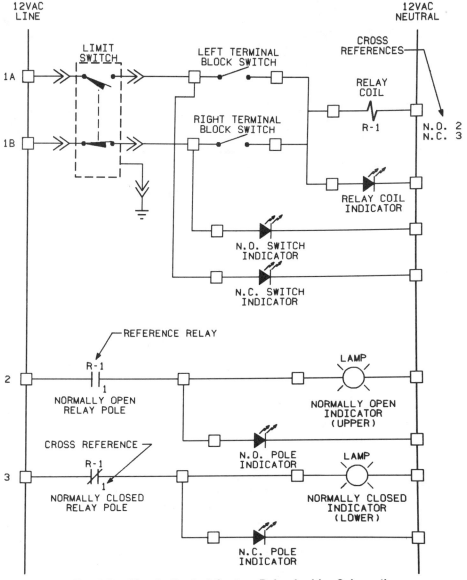

Fig. 7.7. Simple Control System Relay Ladder Schematic

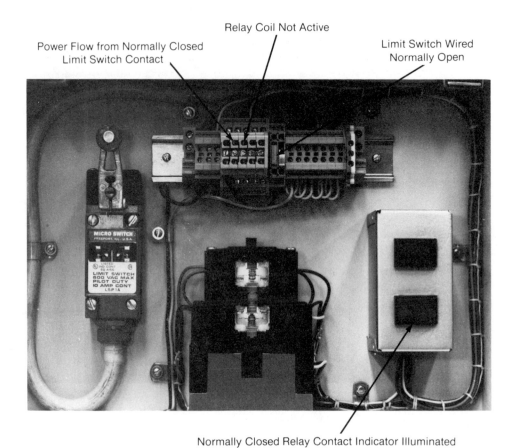

Relay Coil Not Active

Power Flow from Normally Closed
Limit Switch Contact

Limit Switch Wired
Normally Open

Normally Closed Relay Contact Indicator Illuminated

Fig. 7.8. Normally Open Limit Switch Contacts Wired to Control Relay

contacts. Power flow in the electrical schematic of Fig. 7.7 would be through rung 1B to the terminal switch and in rung 3.

When the limit switch becomes mechanically actuated, conditions change, as illustrated in Fig. 7.9. The moving contact in the limit switch transfers from the lower normally closed contacts to the upper normally open contacts. This action is verified by the lighting of the first terminal indicator and the extinguishing of the second terminal indicator. The third terminal indicator is also illuminated, indicating that power is flowing to the relay coil. Close examination of the relay's contacts reveals that the moving contact has separated from the normally closed contacts in the upper contact pole. Conversely, the moving contact now mates with the normally open contacts of the lower pole. The fourth terminal indicator is extinguished (normally closed relay contact), and the fifth indicator is illuminated, indicating power flow through the normally open contact pole of the relay. The final change in Fig. 7.9 is that the lower indicator has extinguished and the upper indicator is illuminated. Returning to Fig. 7.7, power flow will occur through rungs 1A and 2 from the 12-volt ac *line* bus to the 12-volt ac *neutral* bus.

Actuated Limit Switch   Relay Coil Active

Power Flow from Normally Open
Limit Switch Contact

Limit Switch Wired
Normally Open

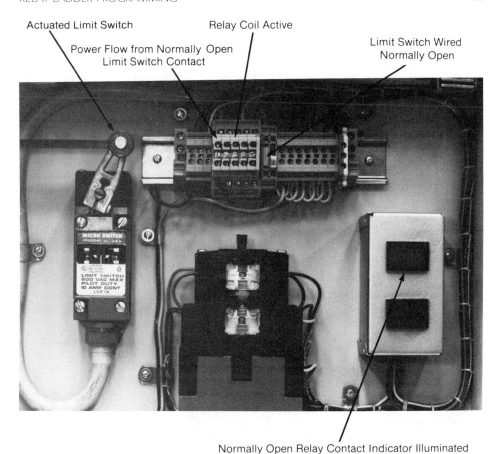

Normally Open Relay Contact Indicator Illuminated

Fig. 7.9. Actuated Limit Switch

When the limit switch is released, the system returns to the conditions of Fig. 7.8. If instead the limit switch is wired to the relay coil in a normally closed fashion, several subtle changes occur. This rewiring is accomplished easily by opening the left terminal switch and closing the right terminal switch. The change from normally open to normally closed limit switch wiring has been made in Fig. 7.10.

At first glance the conditions of this figure appear to be the same as those described in Fig. 7.9. Closer examination reveals only three differences. First and most obvious is the fact that in Fig. 7.10 the limit switch is not mechanically actuated. Second, the left terminal switch is open while the right switch is closed. The final difference can be seen in the terminal indicator lights. The three middle indicators are illuminated. The second indicator shows that the normally closed limit switch contact is closed and passing power. The third terminal indicator shows that the relay coil is energized, while the fourth indicator is illuminated because the normally open contact of the relay is passing power. Referring again to the mini-control system schematic in Fig. 7.7, power flow occurs through rungs 1B and 2.

Conditions change when the limit switch becomes mechanically actuated, as illustrated in Fig. 7.11. The contacts of the limit switch transfer, permitting the normally open contact to pass power. The normally closed contacts break, and power flow is halted to the relay coil. This is verified by the illumination of the first terminal indicator and the extinguishing of the second and third indicators. When the relay coil deenergizes, the normally open relay contact stops passing power to the upper indicator. However, the normally closed indicator becomes illuminated since the normally closed contact of the relay now passes power. The fifth terminal block indicator verifies this condition. Power flow is now occurring in rung 1A to the terminal switch and completely through rung 3.

In summary, anytime an input device is wired to a relay in a normally open manner, the normally open contacts of the relay duplicate the power passing operation of the input device's normally open contacts. The relay's normally closed contacts pass power, exactly the reverse of the input device's normally open contacts. When the input device's normally open contacts are not passing power, the relay's normally closed contacts pass power, and vice versa. Should the input device be wired to the relay in a normally closed manner, the normally open contacts of the relay would duplicate the power passing action

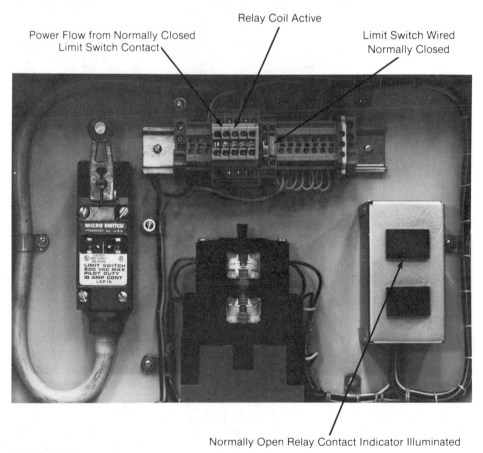

Fig. 7.10. **Normally Closed Limit Switch Contacts Wired to Control Relay**

of the input device. The relay's normally closed contacts will pass power, exactly the reverse of the input device's contacts. Fig. 7.12 provides a tabular summary of the operation just described.

PC CONTACT INSTRUCTIONS

The normally open and normally closed contact instructions are the workhorse instructions of the PC. Since the PC was developed as a means to replace relay control systems, its primary instructions duplicate or emulate the operation of the industrial control relay. Every PC incorporates a normally open and a normally closed contact instruction as well as an output or relay coil instruction. In fact, so solid is the PC's origin as a relay replacer that practically every function that a PC performs is in some form or another initiated by a contact instruction.

In many control systems it is necessary to use the status of a particular input device in numerous relay ladder diagram locations. Many input devices, such as the limit switch selected for the simple control system example, contain a single

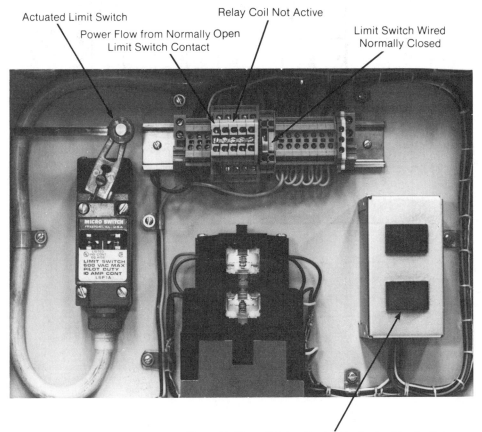

Fig. 7.11. Actuated Limit Switch

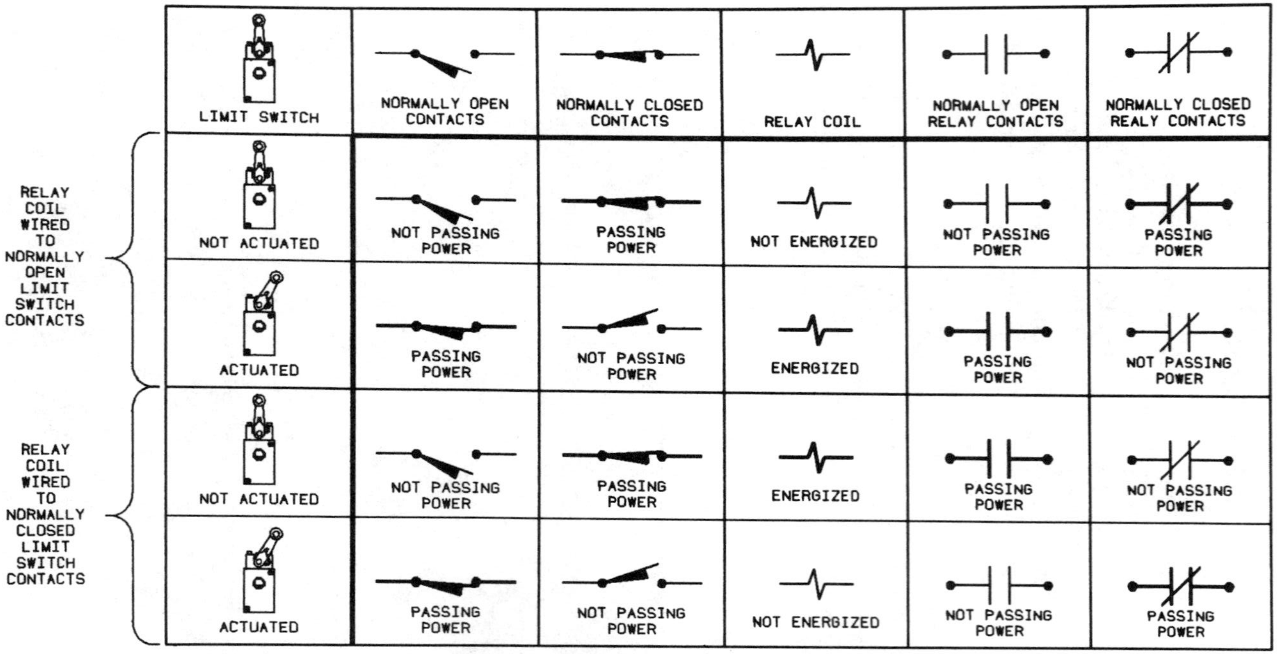

LIMIT SWITCH	NORMALLY OPEN CONTACTS	NORMALLY CLOSED CONTACTS	RELAY COIL	NORMALLY OPEN RELAY CONTACTS	NORMALLY CLOSED RELAY CONTACTS
NOT ACTUATED	NOT PASSING POWER	PASSING POWER	NOT ENERGIZED	NOT PASSING POWER	PASSING POWER
ACTUATED	PASSING POWER	NOT PASSING POWER	ENERGIZED	PASSING POWER	NOT PASSING POWER
NOT ACTUATED	NOT PASSING POWER	PASSING POWER	ENERGIZED	PASSING POWER	NOT PASSING POWER
ACTUATED	PASSING POWER	NOT PASSING POWER	NOT ENERGIZED	NOT PASSING POWER	PASSING POWER

RELAY COIL WIRED TO NORMALLY OPEN LIMIT SWITCH CONTACTS (first two rows)

RELAY COIL WIRED TO NORMALLY CLOSED LIMIT SWITCH CONTACTS (last two rows)

Fig. 7.12. Control System Contact Power Passing Statuses

normally open and normally closed contact. In order to provide the quantity of input device signals required for control system operation, the input device is directly wired to one or more relay coils. Contacts associated with these relays are then used whenever a signal from the input device is required in the control system.

The PC uses the same concept for its input module interface. Any field input device brought to a PC input point can be used in either the normally open or the normally closed configuration, any number of times in the user's program. Probably the most difficult task any potential PC user/programmer has, is learning how to program the correct contact instructions for the desired circuit operation of the field input device.

Anyone associated with relay-based control systems generally thinks of the operation of the relay in terms of the field input device. For example, if the input device is wired normally closed and is actuated, the relay will be deenergized and its normally closed contact will pass power. The same holds true for a device wired to the PC in a normally closed configuration. When the input device is actuated, the input point does not receive power, and any normally closed contact instruction will pass power.

The general pitfall for those first attempting the application of a PC concerns the selection of the proper contact instruction to provide correct input device operation in the user program. The PC does not know, or care, how an input device is wired to its input points. The processor can only respond to the status of the input points. A PC programmer must develop the user program based on the operation of the PC's input points, and *not* on the wiring configuration of the input devices. The general tendency for new PC users is to use the contact instruction that duplicates the wiring configuration of the input device. This is incorrect for some input wiring configurations, as will be indicated shortly.

Perhaps the best way to examine the proper manner to select the correct PC contact instruction is to return to the simple control system example. Fig. 7.13 shows the same hardware as Fig. 7.6. However, the relay and the indicators have had their identities changed. Referring to Fig. 7.13, the relay controls previously discussed can be thought of as PC control system components to illustrate the functioning of relay contact instructions within a PC. For the purposes of discussion, the limit switch can still represent an input device to the control system, in this case a PC-based system. The relay will function like the input point to the PC. The coil of the relay represents the power side of the input module circuitry described in Chapter 3. The contacts of the relay will represent the logic side of the PC input module. The indicators will represent normally open and normally closed contact symbols on a programming terminal.

Close examination of Fig. 7.13 shows that both indicators, now referred to as normally open and normally closed contact instructions, are illuminated. Whenever a contact instruction appears on the programming terminal screen, it can appear in one of two intensifications. The primary reason for providing two levels of intensification relates to the power flow condition of the instruction. Any contact associated with an industrial relay can be checked for power flow in one of several ways. If the type of contact is known (it is usually identified on the

Power Flow Indicators

Normally Closed Contact Symbol

Normally Open Contact Symbol

Input Point

Terminal Switches

Normally Open Limit
Switch Wiring

Normally Closed Limit Switch

Normally Open Limit Switch

Normally Closed Limit
Switch Wiring

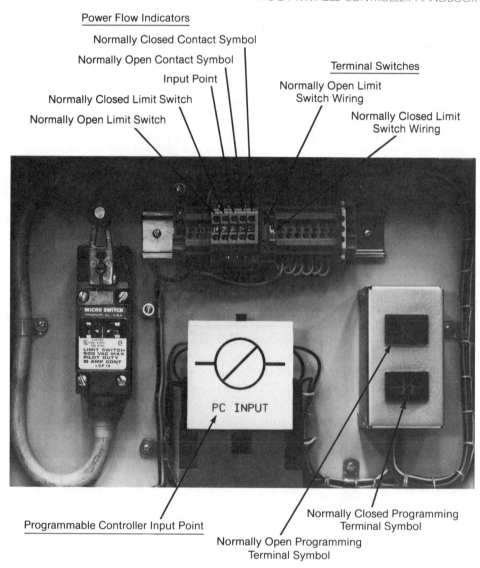

PC INPUT

Programmable Controller Input Point

Normally Closed Programming
Terminal Symbol

Normally Open Programming
Terminal Symbol

Fig. 7.13. Typical PC Control System Components

relay), the state of the relay can be observed. A mechanical indicator is usually provided for this purpose. A second method of determining the power status of a relay contact is to use a volt-ohm meter.

Since the PC is a solid-state computer-based device, it is not possible to use either of these methods to determine the power status of a contact instruction. In fact, one or more bits of memory would have to be interrogated to determine the power status of a particular contact instruction. As a means to indicate the power flow status of a contact instruction, intensification of the instruction is commonly used. In general any contact instruction that is intensified is *passing power.* Those instructions that are not intensified are not passing power. The subject of instruction intensification will be discussed in later paragraphs. For the time being, the intensification convention will be used, with reference to the control system described, for the contact instruction discussion that follows.

The upper portion of Fig. 7.14 is a simplified schematic of the original relay control system described previously. The limit switch is shown on the left feeding the relay coil through the terminal block switches. To the right are the normally open and normally closed relay contacts powering their respective indicators. Directly below the relay schematic is the schematic for the limit switch powering a PC input module. Again, the limit switch contacts are connected through the terminal block switches. To the right are the program-

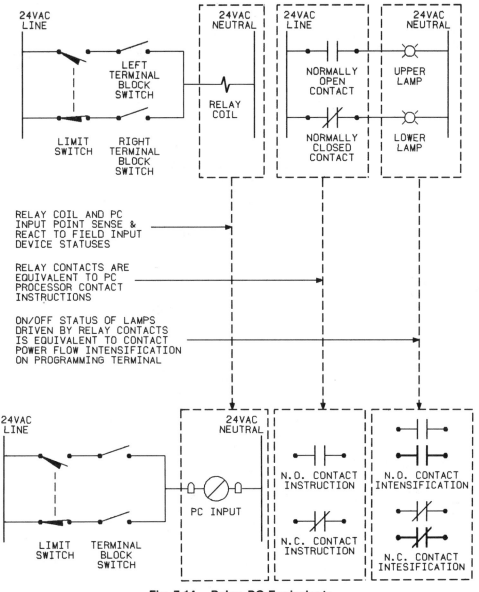

Fig. 7.14. Relay-PC Equivalents

ming terminal instructions for normally open and normally closed contact configurations. The relay of the upper schematic has been "replaced" by the input symbol for a PC in the lower schematic. The relay contacts and indicators are represented in the lower part of the figure by the contact instructions of the programming terminal.

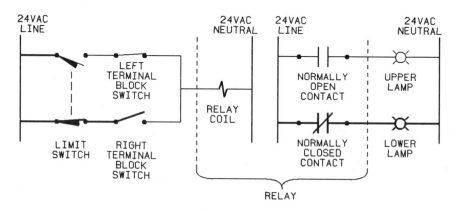

RELAY SYSTEM STATUSES

LIMIT SWITCH	NOT ACTIVATED
LEFT TERMINAL BLOCK SWITCH	CLOSED
RIGHT TERMINAL BLOCK SWITCH	OPEN
RELAY COIL STATUS	NOT ENERGIZED
N.O. RELAY CONTACT	NOT PASSING POWER
N.C. RELAY CONTACT	PASSING POWER
UPPER INDICATOR	NOT ILLUMINATED
LOWER INDICATOR	ILLUMINATED

PC SYSTEM STATUSES

LIMIT SWITCH	NOT ACTIVATED
LEFT TERMINAL BLOCK SWITCH	CLOSED
RIGHT TERMINAL BLOCK SWITCH	OPEN
PC INPUT POINT STATUS	NOT ENERGIZED
N.O. CONTACT INSTRUCTION	NOT PASSING POWER
N.C. CONTACT INSTRUCTION	PASSING POWER

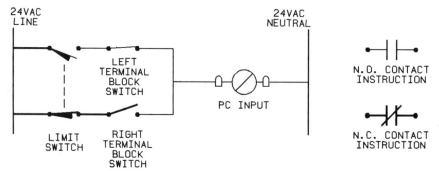

Fig. 7.15. Relay-PC Schematic

From Fig. 7.15 the operation of a PC can begin to be understood. This figure is a duplicate of Fig. 7.14 with the exception that the left terminal block switch connecting the normally open contact of the limit switch to the PC input has been closed, and power has been applied to the circuit. Heavier lines are used to indicate power flow, while an intensified contact instruction indicates that the instruction is passing power.

Examination of Figs. 7.15 and 7.16 indicates that the PC input is not active. The closed terminal block switch effectively wires the limit switch's normally open contact to the input point. Since the normally open limit switch contact is not passing power to the input point, any normally open contact instruction referenced to the input point will not pass power. This is illustrated in both figures by the nonintensified programming terminal instructions. The normally closed contact instruction referenced to the input point is intensified since it would be passing power when the input point is not energized.

The normally closed instruction, passing power exactly like the normally closed contact of a deenergized relay, passes power when its reference coil instruction is not receiving power flow. As the processor scans the user logic, it

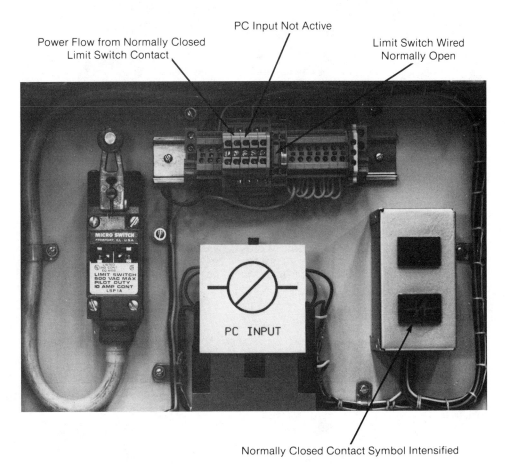

Fig. 7.16. **Normally Open Limit Switch Contacts Wired to PC Input**

first looks to see what type of contact instruction has been programmed. It then examines the input point referenced by the instruction for its OFF/ON status. In our case the input point is OFF; therefore only normally closed instructions referenced to that input are permitted to pass power.

In Figs. 7.17 and 7.18 the limit switch is still wired to the input point by a normally open contact. However, the limit switch has been mechanically

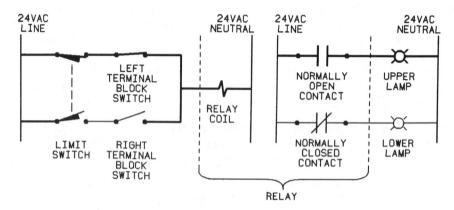

RELAY SYSTEM STATUSES

LIMIT SWITCH	ACTIVATED
LEFT TERMINAL BLOCK SWITCH	CLOSED
RIGHT TERMINAL BLOCK SWITCH	OPEN
RELAY COIL STATUS	ENERGIZED
N.O. RELAY CONTACT	PASSING POWER
N.C. RELAY CONTACT	NOT PASSING POWER
UPPER INDICATOR	ILLUMINATED
LOWER INDICATOR	NOT ILLUMINATED

PC SYSTEM STATUSES

LIMIT SWITCH	ACTIVATED
LEFT TERMINAL BLOCK SWITCH	CLOSED
RIGHT TERMINAL BLOCK SWITCH	OPEN
PC INPUT POINT STATUS	ENERGIZED
N.O. CONTACT INSTRUCTION	PASSING POWER
N.C. CONTACT INSTRUCTION	NOT PASSING POWER

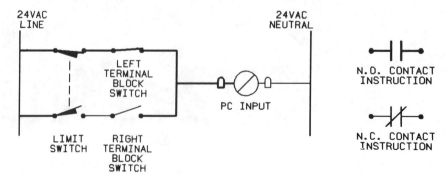

Fig. 7.17. Relay-PC Schematic

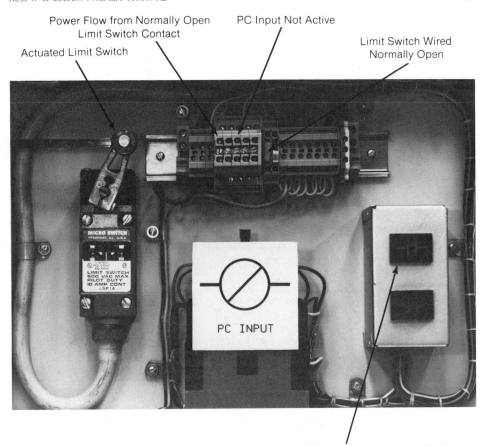

Power Flow from Normally Open
Limit Switch Contact

PC Input Not Active

Actuated Limit Switch

Limit Switch Wired
Normally Open

PC INPUT

Normally Open Contact Symbol Intensified
Fig. 7.18. Actuated Limit Switch

actuated. The normally open limit switch contact now passes power to the input point. When the processor looks at the state of the input point referenced by a programmed normally open contact instruction, and finds the input point ON, it will permit any normally open contact instruction to pass power. This is illustrated in the figures by the intensification of the normally open contact instruction. Note that the normally closed contact instruction referenced to the input is not intensified, since any normally closed contact instruction would not pass power when its referenced input point is active.

Changing the terminal block switches such that the normally closed limit switch contact powers the PC input rather than the normally open limit switch contact, does not change how the processor interprets the passing power status of contact instructions. Figs. 7.19 and 7.20 illustrate the operation of the system when the input point is connected to a normally closed input device. The input point is active with the limit switch not mechanically actuated. Since the input point is on, any normally open contact referenced to that input point will pass power. The processor examines the status of the input module and determines the power passing status of the contact instruction accordingly. Just as was the case with the relay's normally open contact, whenever the PC input is receiving

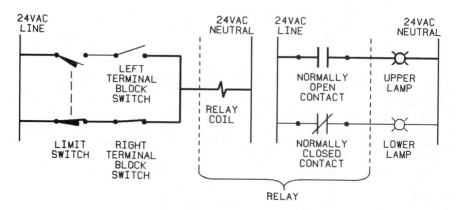

RELAY SYSTEM STATUSES

LIMIT SWITCH	NOT ACTIVATED
LEFT TERMINAL BLOCK SWITCH	OPEN
RIGHT TERMINAL BLOCK SWITCH	CLOSED
RELAY COIL STATUS	ENERGIZED
N.O. RELAY CONTACT	PASSING POWER
N.C. RELAY CONTACT	NOT PASSING POWER
UPPER INDICATOR	ILLUMINATED
LOWER INDICATOR	NOT ILLUMINATED

PC SYSTEM STATUSES

LIMIT SWITCH	NOT ACTIVATED
LEFT TERMINAL BLOCK SWITCH	OPEN
RIGHT TERMINAL BLOCK SWITCH	CLOSED
PC INPUT POINT STATUS	ENERGIZED
N.O. CONTACT INSTRUCTION	PASSING POWER
N.C. CONTACT INSTRUCTION	NOT PASSING POWER

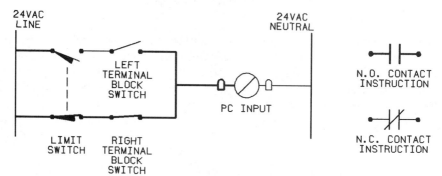

Fig. 7.19. Relay–PC Schematic

power (relay coil), any normally open contact instruction (normally open relay contact) referenced to that input will pass power. Accordingly any normally closed contact instruction would not pass power.

Referring to Figs. 7.21 and 7.22, the statuses of the normally open and normally closed contact instructions change when the limit switch is mechanically actuated. Actuated, the normally closed contact of the limit switch opens, inhibiting power flow to the PC input point. With the input electrically OFF, the normally open contact instruction referenced to the input will not pass power. The normally closed instruction will, however, pass power.

Fig. 7.23 sums up the rules regarding the proper programming of either normally open or normally closed contact instructions with respect to field device wiring. Whenever a device has been wired to an input point in a normally open configuration, and the input device must operate like a normally open connected device in the circuit (user's program), a normally open contact instruction should be programmed. Likewise, if the input device is wired normally closed to the input point, and the input device is to operate normally closed in the circuit, then, again, a normally open contact instruction would be programmed in the user's program. With the normally closed input device not actuated, the input point is ON, and any normally open contact instruction referenced to that input point will pass power similar to the input device.

The normally closed contact instruction is used whenever the input device is wired to the input point in a normally open configuration, and the desired circuit operation of the input device must be normally closed. The input device

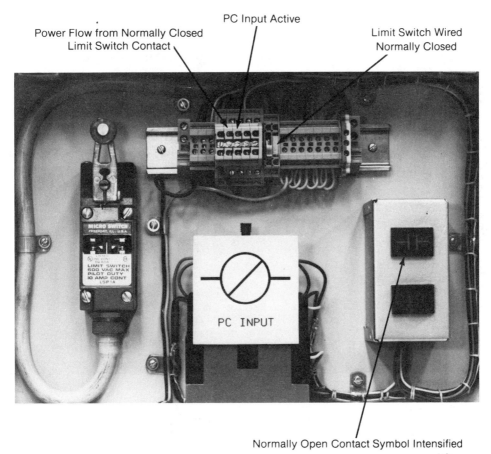

Fig. 7.20. Normally Closed Limit Switch Contacts Wired to PC Input

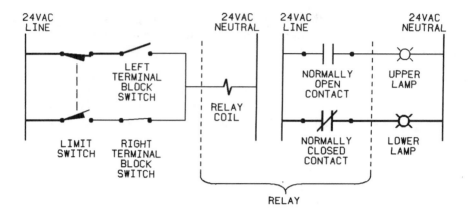

RELAY SYSTEM STATUSES

LIMIT SWITCH	ACTIVATED
LEFT TERMINAL BLOCK SWITCH	OPEN
RIGHT TERMINAL BLOCK SWITCH	CLOSED
RELAY COIL STATUS	NOT ENERGIZED
N.O. RELAY CONTACT	NOT PASSING POWER
N.C. RELAY CONTACT	PASSING POWER
UPPER INDICATOR	NOT ILLUMINATED
LOWER INDICATOR	ILLUMINATED

PC SYSTEM STATUSES

LIMIT SWITCH	ACTIVATED
LEFT TERMINAL BLOCK SWITCH	OPEN
RIGHT TERMINAL BLOCK SWITCH	CLOSED
PC INPUT POINT STATUS	NOT ENERGIZED
N.O. CONTACT INSTRUCTION	NOT PASSING POWER
N.C. CONTACT INSTRUCTION	PASSING POWER

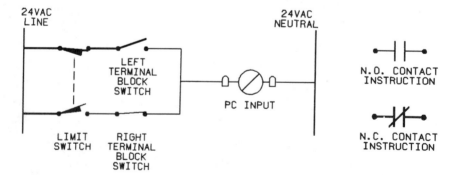

Fig. 7.21. Relay-PC Schematic

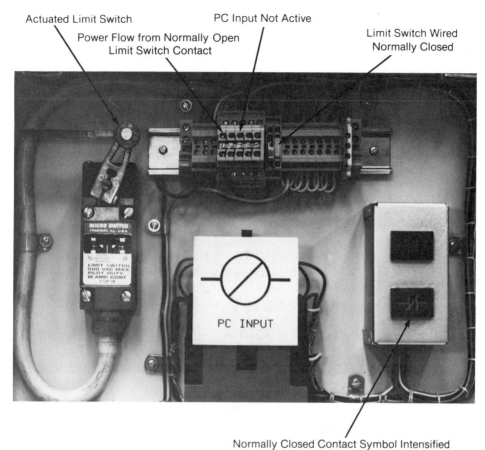

Actuated Limit Switch

Power Flow from Normally Open
Limit Switch Contact

PC Input Not Active

Limit Switch Wired
Normally Closed

PC INPUT

Normally Closed Contact Symbol Intensified

Fig. 7.22. Actuated Limit Switch

wired in a normally open configuration but not mechanically actuated does not pass power to the input point. However, the input point needs to look as if it were wired in the control circuit in a normally closed configuration; therefore the normally closed contact instruction is chosen to meet the need. For the case where the input device is connected to the input point in a normally closed configuration, and the desired operation of the input device in the circuit is normally open, the normally closed contact instruction is used. The input receives power from the normally closed wiring of the input device. Since the desired configuration of the input device in the circuit is a normally open configuration, the normally closed contact instruction is used.

Realizing that the PC responds to the power statuses of its input points, and not to that of the field input devices, is probably one of the most difficult concepts to learn and master, the most common error being the assumption that the proper programming terminal instruction to use is always the one that matches the wiring configuration of the field device. The selection of the proper programming terminal instruction for the circuit operation desired can become more confusing when the field input device's mechanical operation must be considered as well as its electrical operation. Many input devices can be wired in a normally open configuration to the input point, but then are mechanically held closed in normal operation. The fact that the device is released to its normally open state in the operated or tripped condition can add confusion. Fig. 7.24 illustrates the proper programming terminal instruction to select for all possible electrical and mechanical configurations of an input device, depending on the desired operation of the relay ladder diagram device. This figure should be well understood before the reader moves on to the following parts of this chapter.

PROGRAMMING TERMINAL DISPLAY

As the PC scans the user program, it must look up the current status of each contact instruction encountered in either the input status table, the internal coil

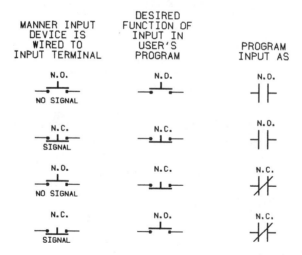

ANY PROGRAMMABLE CONTROLLER ONLY RECOGNIZES THE PRESENCE OR ABSENCE OF A SIGNAL AT AN INPUT TERMINAL. IT IS IMPOSSIBLE FOR ANY CONTROLLER TO KNOW WHETHER AN INPUT DEVICE IS WIRED TO ITS INPUT TERMINAL IN A NORMALLY OPEN OR CLOSED STATE. THE FOLLOWING RULES DICTATE HOW THAT DEVICE MUST BE PROGRAMMED FOR THE VARIOUS STATES.

THE RULES ARE:

1) IF THE WIRED CONFIGURATION OF THE INPUT DEVICE AND THE DESIRED FUNCTION IN THE CIRCUIT ARE THE SAME, PROGRAM THE INPUT AS A NORMALLY OPEN (—||—) INSTRUCTION.

2) IF THE WIRED CONFIGURATION OF THE INPUT DEVICE AND THE DESIRED FUNCTION ARE OPPOSITE, PROGRAM THE NORMALLY CLOSED (—|/|—) INSTRUCTION.

Fig. 7.23. Normally Open/Normally Closed Contact Instruction Programming

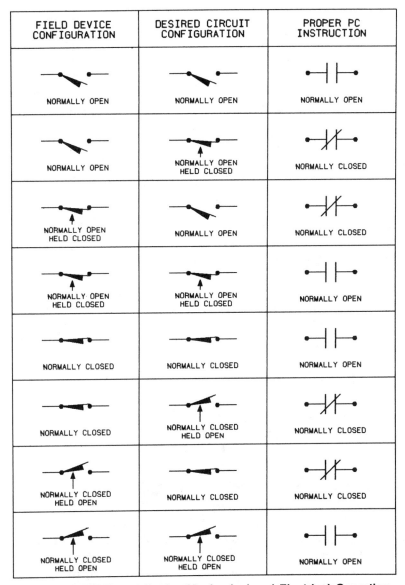

FIELD DEVICE CONFIGURATION	DESIRED CIRCUIT CONFIGURATION	PROPER PC INSTRUCTION
NORMALLY OPEN	NORMALLY OPEN	NORMALLY OPEN
NORMALLY OPEN	NORMALLY OPEN HELD CLOSED	NORMALLY CLOSED
NORMALLY OPEN HELD CLOSED	NORMALLY OPEN	NORMALLY CLOSED
NORMALLY OPEN HELD CLOSED	NORMALLY OPEN HELD CLOSED	NORMALLY OPEN
NORMALLY CLOSED	NORMALLY CLOSED	NORMALLY OPEN
NORMALLY CLOSED	NORMALLY CLOSED HELD OPEN	NORMALLY CLOSED
NORMALLY CLOSED HELD OPEN	NORMALLY CLOSED	NORMALLY CLOSED
NORMALLY CLOSED HELD OPEN	NORMALLY CLOSED HELD OPEN	NORMALLY OPEN

Fig. 7.24. Field Input Device Mechanical and Electrical Operation Programming

status table, or the output point (coil) status table. In short, each relay contact instruction encountered by the processor during its user logic scan must be checked for its passing or inhibiting power status. In order for a ladder rung to be true, at least one continuous path or power flow must be provided from the left side of the programming terminal screen to the right.

Every PC programming terminal provides the same basic arrangement of user program instructions on its display screen. As illustrated in Fig. 7.25, the programming terminal screen will have two vertical lines as part of the display. The left vertical line is usually referred to as the left-hand power rail, and the

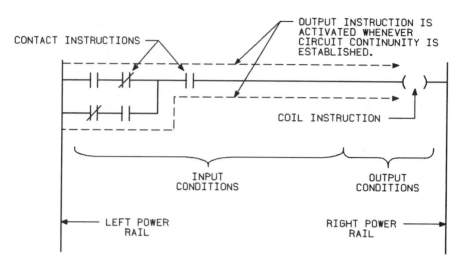

Fig. 7.25. Programming Terminal Display

right vertical line is referred to as the right-hand power rail. The left- and right-hand power rail lines of the programming terminal represent the line and the neutral, or positive and negative, power buses, respectively, of the relay schematic of Fig. 7.7. The rightmost PC instruction within a ladder rung is referred to as an *output instruction*. This instruction will generate some sort of response based on the ON/OFF, TRUE/FALSE status of the ladder rung. This instruction is almost always a relay coil or similar instruction. To the left of the output instruction and the right of the left power rail are the *input conditions* or instructions for the ladder rung. These instructions form one or more possible paths of continuity from the left power rail to the output instruction.

POWER-PASSING CONVENTIONS

A PC user can determine the continuity of any programming terminal ladder rung by observing the intensification level of each contact instruction shown on the screen. Whenever a contact instruction, either normally open or normally closed, is passing power, the instruction is intensified on the programming terminal screen. When a complete path of intensified input instructions exists between the left power rail and the output instruction, continuity is established and the output instruction will be activated. The output instruction can be activated only when a continuous path exists to the instruction from the left power rail.

For example, a normally closed contact is normally shown intensified unless the coil or input point it is referenced to is activated. The normally open contact instruction programming terminal symbol is intensified only when the input point or coil to which it is referenced is activated. Depending on the complexity of the user logic, one or more paths can exist from the left power rail to the output instruction, and there can be multiple paths of continuity at any given time.

It should be pointed out that there are two ways currently used to indicate power flow on a programming terminal screen. Both methods are illustrated in Fig. 7.26. The first of these methods involves simply intensifying only those instructions that are passing power. Any contact instruction not passing power is not intensified on the programming terminal screen. A second method intensifies not only the instructions that are active, but the path or paths that are providing power flow to the output instruction as well. In either case the coil (output) instruction is intensified whenever it is activated. While the ability to intensify those instructions that were passing power was not available with early PCs, user demands have made the use of one of the two methods mentioned a necessity if a PC is to be successful. The ability to follow quickly the power flow of a relay ladder program rung greatly enhances the speed at which a program can be examined for current operation.

Prior to actually developing a PC program based on an existing relay system schematic, a brief look at the terminal used to program a PC is required. While programming terminals will be examined in greater depth in Chapter 14, an overview is required here. For the remainder of this chapter the General Electric Series Six family of PCs will serve as the PC used for many of the programming examples. The Series Six family of PCs provides user flexibility in the application of a PC because of its modular design and construction. Unlike many of the earlier PCs, a unique standardization of hardware and firmware has been woven throughout its development. This provides any Series Six user the opportunity to design a PC-based control system with the knowledge that future changes and/or system expansions to the PC hardware and programming can be implemented with little difficulty. A unique feature of the larger Series Six processors is the ability to interact with a specially designed data processor. Where large amounts of data must be gathered and manipulated, a special data processor can be applied to work in parallel with the PC processor. This coprocessing technique permits faster system response and a flexibility not

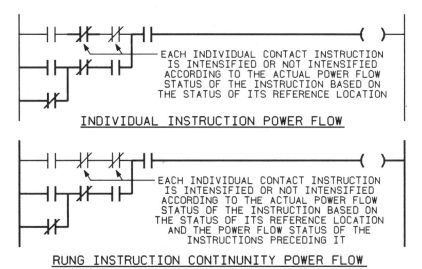

INDIVIDUAL INSTRUCTION POWER FLOW

RUNG INSTRUCTION CONTINUNITY POWER FLOW

Fig. 7.26. Methods of Illustrating Power Flow

generally available with PCs that rely on the PC processor to handle data and logic processing.

Since many of the relay programming examples of this chapter will use the Series Six instruction set and programming protocol, a quick overview of the programming terminal is in order. Fig. 7.27 illustrates the keyboard area of the programming terminal, referred to by GE as the programming development terminal (PDT). Included as part of this figure is a listing of the instruction key groups of the terminal. While the keyboard provides the capability to program two levels of mnemonic instructions, the primary concern of this chapter will be

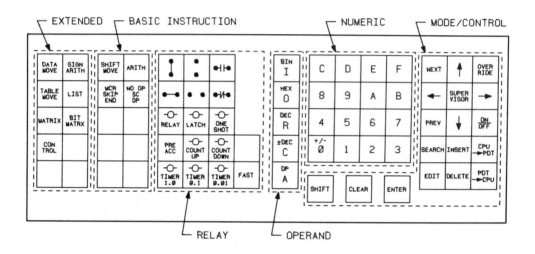

MODE/CONTROL GROUP - USED TO ENTER A PARTICULAR OPERATIONAL MODE OR TO INITIALIZE A DESIRED CONTROL FUNCTION

NUMERIC - USED TO ENTER INSTRUCTION REFERENCES, DATA AND SIMILAR NUMERICAL INFORMATION

OPERAND - THESE KEYS PERFORM DUAL FUNCTIONS: SPECIFYING INPUT (I), OUTPUT (O), REGISTER (R), CONSTANT (C) AND AUXILIARY OPERANDS (A) IN THE UNSHIFTED MODE. IN THE SHIFTED MODE, THESE KEYS DEFINE THE BASE OF ANY NUMERICAL DATA: BINARY (BIN), HEXA- DECIMAL (HEX), DECIMAL (DEC), SIGNED DECIMAL (±DEC) AND DOUBLE PRECISION (DP)

RELAY - USED TO BUILD AND EDIT A RELAY/CONTACT NETWORK

BASIC - SIMPLEST SET OF INSTRUCTIONS WHICH CAN BE USED WITH ANY SERIES SIX PROGRAMMABLE CONTROLLER

EXTENDED - PROVIDES ACCESS TO THE COMPLETE SET OF SERIES SIX INSTRUCTIONS

Fig. 7.27. General Electric Series Six Programming Terminal Functions

the relay, operand, numeric, and mode/control instruction groups. The connection and initialization of the PDT and the PC processor will not be addressed by this chapter. For a detailed discussion on the operation of the PDT, the programming manual for the Series Six should be consulted.

There are numerous ways to begin instructing the new PC user on how a PC can be programmed. The best method to use depends on the background of the individual involved, as well as on the type of hardware available and the experience of the instructor. The PC by its very design is a very flexible piece of hardware. In any application there will be numerous approaches available to solving the needs of the application. The final approach selected will usually provide the desired results to some level of acceptable operation. Likewise there is no best manner to approaching the subject of introductory PC programming, given an audience of varied backgrounds and experience. It is hoped that the methods about to be employed will be successful with the majority of readers.

DEVELOPMENT OF A PC PROGRAM

In lieu of developing a PC program from "scratch," and explaining each step of the process, an existing relay schematic will be converted to PC instructions. This will provide the reader with a concrete example of a system before PC conversion, during each step of conversion, as well as after PC conversion. The system to be converted does not represent an actual operating system. Rather it represents an "out-of-the-blue" relay system design, which can provide many examples for the relay to PC conversion discussions about to follow.

The relay ladder schematic that will be converted is shown in Figs. 7.28 and 7.29. Ladder rungs 1 to 5 contain field input devices wired to relays identified as R1 to R6. As discussed earlier in this chapter, the relays provide a means to obtain additional input device signals for use in the control logic portion of the ladder diagram.

Rungs 6 to 13 contain the actual control logic for the relay-based control system. Four motors and a solenoid valve are controlled by the system. Two relays are required for control system internal use, and have been labeled CR1 and CR2. One timer is used in the relay ladder diagram. This chapter will not address the programming of the timer for the example application. Chapter 9 will address the many forms of PC timers and counters available, their use, and programming. For the purposes of this chapter, discussions on timer programming will be held to a minimum.

The final portion of the relay schematic, Fig. 7.29, contains information concerning the wiring of the motors used in the control system, as well as the power supply design for the relay control system. Since Chapter 6 dealt with the power distribution of a PC-based system, a detailed discussion of this application will not be presented. An overall power distribution scheme will be presented for reference at the end of the programming discussions.

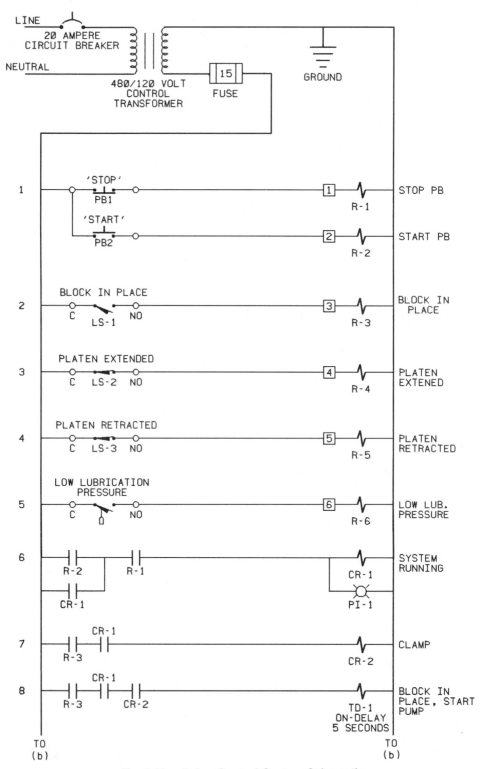

Fig. 7.28. Relay Control System Schematic

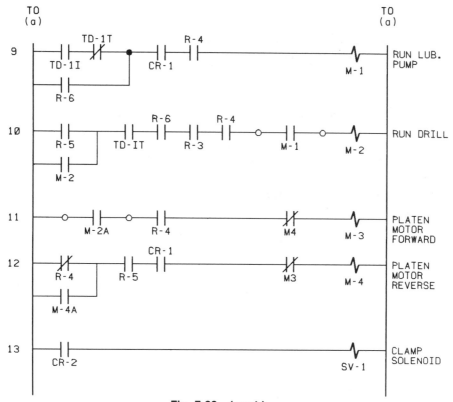

Fig. 7.28 (cont.)

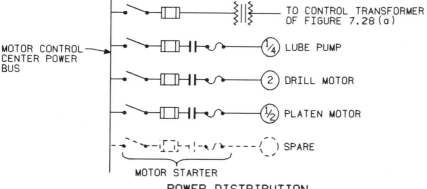

POWER DISTRIBUTION

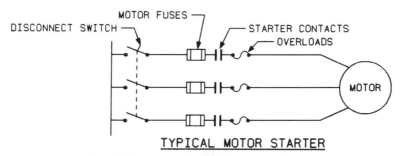

TYPICAL MOTOR STARTER

Fig. 7.29. Relay Example Power Distribution

DETERMINATION OF INPUT/OUTPUT DEVICE OPERATION AND REFERENCES

The first task to be performed is to determine the I/O devices required and allocate I/O reference numbers. This allocation of reference numbers is performed as described in Chapter 4. The example relay diagram contains eight input points, as illustrated in Fig. 7.30. These eight input devices are assigned to eight discrete input module points. Each of the input points is defined by a unique reference identification, indicated directly above the input point symbol. These reference numbers will be used in the PC program to identify particular input devices.

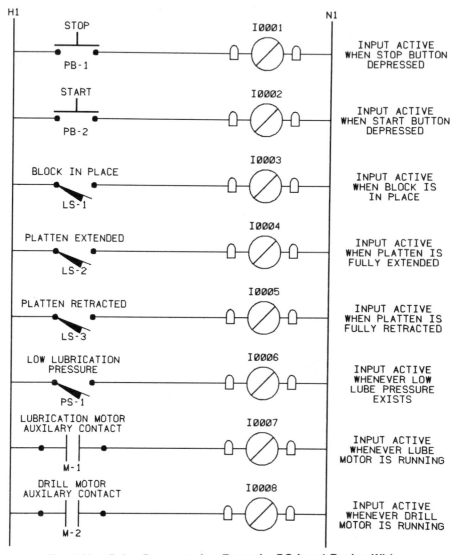

Fig. 7.30. Relay Programming Example, PC Input Device Wiring

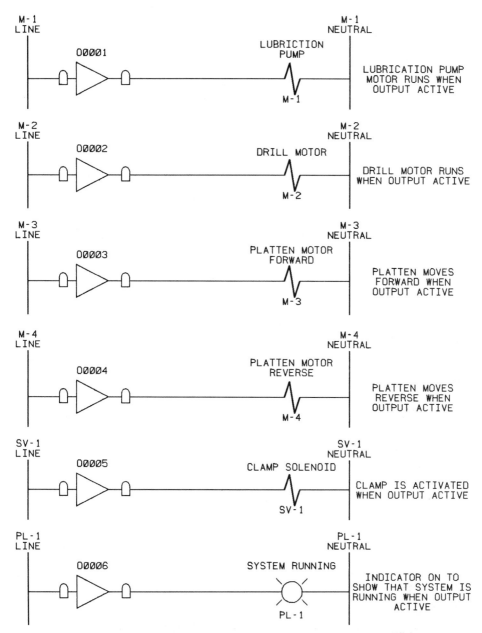

Fig. 7.31. Relay Programming Example, PC Output Device Wiring

The relay control system controls four motors and a solenoid valve. An indicator is also illuminated by the controls to indicate the "running" status of the equipment. These six devices are designated as output devices for the PC. Their output reference designations and wiring are defined in Fig. 7.31. Note that isolated output modules have been selected for this application. In order to keep a foreign voltage out of the MCC starter cubicle, and still use the control transformer in the starter cubicle for starter control power, an isolated output module was selected. In lieu of the isolated output module, an interface relay could have been used, or foreign voltages could have been permitted in the starter cubicle. Fig. 7.32 indicates the power wiring for the PC-based version of the control system.

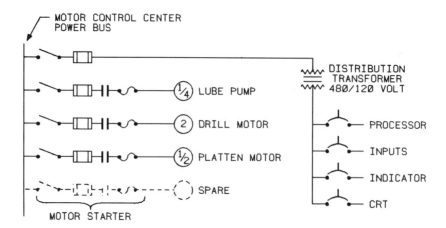

POWER DISTRIBUTION

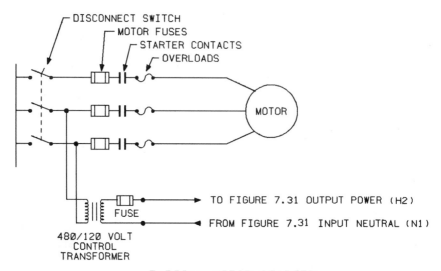

TYPICAL MOTOR STARTER

Fig. 7.32. Programming Example Power Distribution

INTERNAL REFERENCE ALLOCATION

Once the real-world I/O references have been determined, internal PC coil instruction references will have to be allocated. *Internal coil instruction references are analogous to relay control system relays that do not drive real-world devices.* They serve the same function as CR1 and CR2 of the relay ladder schematic. If a PC program requires the programming of coil instructions that will not affect any external devices, the allocation of an internal reference will be advantageous. How the allocation of internal coil instruction references is performed will depend on several criteria. First the architecture of the processor will have the greatest impact on the internal reference allocation. Every PC dedicates a portion of its data memory to the storage of I/O statuses. Another portion of the data memory is also provided for the storage of numerical and/or program data. Depending on its architecture, the processor may or may not allocate a third portion of the data memory for internal relay instruction reference allocations.

Therefore internal coil reference allocation can be performed in either of two ways. First, the internal coil references for the processor could be assigned as required, provided that the processor incorporates a section of the data memory for storage of this type of information. A second way to allocate internal coil references is to assign unused actual output reference points for internal use.

For example, the Series Six PCs provide referencing for 1024 inputs and 1024 outputs, plus the ability to allocate an additional 1024 auxiliary inputs and 1024 auxiliary outputs. Since this application requires very few real-world input and output references to be allocated, the remaining unused output references could be used for internal coil instruction references. For example, references O0001 to O0048 might be allocated for current and future real output points. References O0049 to O0512 might be set aside for future use as either real-world or internal references, and O0513 to O1024 could be designated for use with internal coil instructions. In effect, 512 real-world output points (O0513 to O1024) are being reassigned to internal use only. While an output rack could be installed to address these locations, and output modules installed to respond to the status of these locations, they are essentially unavailable for real-world use. Indirectly the output capacity of the PC has been limited.

Depending on the point of view of the user, this may or may not be detrimental. Some users may object to losing real output points to internal use. They would prefer to have a dedicated internal coil instruction storage area. This is a good argument until all of the internal references have been assigned, and real-world output references must be used as internal reference locations. Other PC users may not prefer the preallotment of internal coil references, arguing that they prefer the total reference allocation flexibility available with processors not providing the special internal coil reference memory area. It is differences in processor design as subtle as this, that often win an order for one manufacturer over another.

Although not used for this example, there is another area of a processor's memory available for the storage of internal coil references. Many processors permit the user to address directly the particular bits of a data memory register or word. While it is customary to address data memory registers or words as a group of bits (such as when programming as timers or counters), some processors permit an extension of the data memory register or word address to be used to specify a designated bit within that register or word. For processors that provide this flexibility, data memory registers or words could be allocated for internal coil instruction reference use. The individual bits of the registers or words would then be assigned to particular coil instructions. For example, a register within the Series Six's data memory, perhaps register R0122, could be allocated for internal coil instruction references. The 16 bits of R0122 could then be assigned as required to 16 unique coil instructions.

DOCUMENTATION OF THE USER PROGRAM

Once the I/O points and internal storage locations have been identified, allocated, and defined, the first steps of documenting the processor program should begin. These steps include the development of tables similar to those of Fig. 7.33. A table should be developed that defines each input, output, and internal coil reference, its use, and the drawings that indicate the wiring and operation of the reference. The tables should list every point available for use, whether it be as an installed spare or a space for future installation. A table should be prepared at this time for internal PC data storage addresses. This table will be filled out during the development of the PC program.

PARALLEL BRANCH PROGRAMMING

Prior to describing the development of a PC program that would duplicate the control provided by the relay control system of Figs. 7.28 and 7.29, several comments need to be directed toward the programming of parallel branches in a PC. Every PC must be able to provide for the programming of parallel branches of instructions within the processor. Fig. 7.34 illustrates a relay ladder rung that is to be programmed in a PC. Depending on the PC manufacturer and how he decided that parallel ladder rungs should be programmed, either of the programming sequences in the figure might be used. Many manufacturers provide separate instructions for branching between subrungs. Where separate instructions are provided, a definite order will occur in the programming of the rung, as indicated in the center portion of Fig. 7.34. The first instruction that must be programmed would be a *branch down* instruction. It indicates to the processor that there is more than one possible path of instructions that might energize the coil instruction. Next, the two contact instructions would be programmed (instructions 2 and 3), followed by another branch down instruction, noted in the figure at 4. The programming of this branch down instruction tells the processor that a subrung is about to be programmed that will operate in parallel to the instructions just programmed.

At 5 in Fig. 7.34 is another branch down instruction. Again this indicates to the processor that there will be another subrung of instructions that will operate in parallel to the instructions about to be programmed. Following the branch down is the contact instruction and another branch down instruction, noted by 6 and 7. A normally open instruction is next programmed as noted at 8. The subrung containing the eighth instruction is closed by a *branch up* instruction as

INPUTS

PC ADDRESS	MNEMONIC	FUNCTION	REFERENCES

OUTPUTS

PC ADDRESS	MNEMONIC	FUNCTION	REFERENCES

INTERNAL COILS

PC ADDRESS	MNEMONIC	FUNCTION	REFERENCES

DATA REFERENCES

PC ADDRESS	MNEMONIC	FUNCTION	REFERENCES

Fig. 7.33.   PC Mnemonic Chart

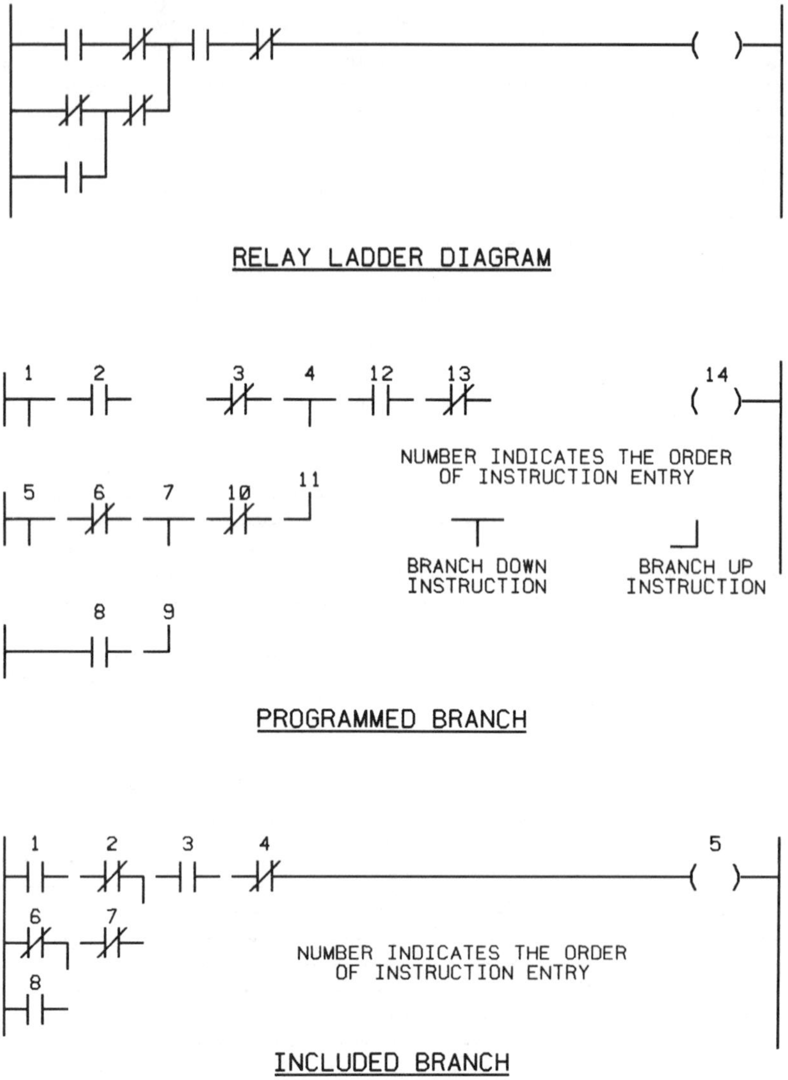

RELAY LADDER DIAGRAM

PROGRAMMED BRANCH

INCLUDED BRANCH

Fig. 7.34. Parallel Branch Programming Methods

noted at 9. The branch up instruction indicates to the processor that all of the instructions for the last subrung have been specified and that programming will continue on the main rung or subrung directly above the current rung.

In order to complete the programming of the middle subrung, a normally closed contact instruction is programmed at 10. Another branch up instruction (11) indicates that the programming of this subrung is now complete. The final two contact instructions of the main rung (12 and 13) can be programmed along with the coil instruction (14). In 14 instructions the relay ladder rung has been converted to an equivalent PC ladder logic rung.

As an alternate to this programming approach, several manufacturers include the branching instructions as part of the contact instructions. As illustrated in the lower portion of Fig. 7.34, the first instruction to be programmed would be the normally open instruction noted as 1. The next instruction would be the normally closed instruction 2. This instruction would have a down branch instruction included with it to signal the processor that the end of a subrung occurs with this instruction. Next the two main rung contact instructions would be programmed along with the coil instruction. The programming cursor would then be repositioned to begin the middle subrung.

Starting the middle subrung would be a normally closed instruction (6) with a down branch to signal again to the processor that the end of a subrung occurs with this instruction. Generally there is no need to place an instruction to indicate the start of the subrung for programming terminals that program in this fashion. Another normally closed instruction (7) is programmed without a branch attached. The cursor would then be moved to start the bottom subrung. After placement of the eighth instruction, a normally open instruction, the rung programming would be complete.

Note that often a set pattern is not required for programming at a terminal using included branch instructions. In lieu of programming the instructions in the order just described, the order could be changed to suit the whims of the programmer. For example, the order of instruction placement might be 1, 6, 8, 2, 7, 3, 4, and 5. The cursor is moved to any permissible location on the CRT screen, and the desired instruction can be placed.

One area of caution should be noted regardless of which programming method is provided by the manufacturer. Each programming terminal and processor has a predefined set of rules as to proper instruction placement and use. These rules are often referred to as the terminal's syntax and logistics rules. They define the proper methods that should be used to develop and enter a PC program. The more common syntax and logistics rules are discussed in Chapter 14 and will not be addressed in this chapter. Prior to program development it is imperative that the programming rules for the equipment being used be reviewed. Where specific programs are developed in this and the following chapters, the programming rules for the PC hardware used in the example program will be followed.

RELAY LADDER DIAGRAM/
RELAY LADDER LANGUAGE CONVERSION

The best approach to developing a PC program from a relay ladder diagram is to understand first the operation of each relay ladder rung. As each relay ladder rung is understood, an equivalent PC rung can be generated. This will require access to the relay ladder schematics and the PC I/O schematics. The amount of trouble that a PC programmer has in the conversion of each relay ladder rung to an equivalent PC ladder rung will probably be directly related to how well he or she understands the programming instructions of the PC being used, and how well the original relay ladder schematic is documented.

The eight control logic rungs of the relay ladder schematic of Fig. 7.28 can be converted to eight rungs of PC relay ladder language commands, as illustrated in Fig. 7.35. Ladder rungs 6 to 13 in Fig. 7.28 have been converted, with minor changes to the PC relay ladder language instructions of Fig. 7.35. The only rung showing any major change is rung 4 in Fig. 7.35 as compared to original rung 9 of Fig. 7.28. Other than this change, only contact references have required altering. Remember that the input point logic contained in Fig. 7.28 has been replaced by the PC input schematic of Fig. 7.30.

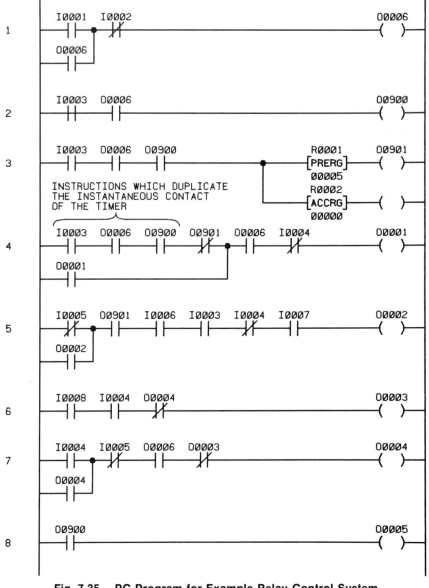

Fig. 7.35. PC Program for Example Relay Control System

In converting rung 6 of Fig. 7.28 to a series of relay ladder instructions, the operation of this rung must be understood. Examination of this rung indicates that it serves as a permissive to operate. A START button, when depressed, assuming that the STOP button is not also depressed, causes internal control relay CR1 to become energized. A normally open contact from CR1 is used as a "seal" or "electric latch" around the START button to provide power flow to the relay coil when the START button is released. A momentary depression of the STOP button will inhibit power flow to the relay coil, causing it to deenergize. By the configuration of the contacts in the circuit, the STOP button overrides the operation of the START button. If both were depressed, the relay would not be activated because of the manner in which the STOP button has been placed in the circuit.

Rung 1 of Fig. 7.35 contains the equivalent logic as performed by rung 6 of Fig. 7.28. The first instruction, a normally open contact instruction with reference I0001, provides the START button status to the logic rung. A normally closed contact instruction, with reference I0002, provides the STOP button status to the circuit. In Fig. 7.28 the STOP button was wired in a normally open manner. However, in Fig. 7.35 a normally closed contact instruction has been used to provide equivalent operation. The reason for the change is PC's response to input module power flow. The STOP button has been wired to the input point in a normally open wiring configuration. This means that the input point will be powered only when the STOP button is depressed. The normally closed contact instruction, reference I0002, passes power unless the input point receives power. Depressing the STOP button powers the input module, thereby causing the normally closed contact instruction to inhibit power flow in the circuit. The PC rung uses a relay coil instruction with reference O0006 to complete the rung. The output designations have assigned output point O0006 to drive the RUNNING indicator. A normally open contact, referenced to relay coil instruction O0006, has been placed in parallel to normally open contact instruction I0001 to provide equivalent operation once the START button has been depressed. Depressing both the START and the STOP buttons together still inhibits the operation of the coil instruction, duplicating the operation of the original relay circuit.

Ladder schematic rungs 7 and 8 in Fig. 7.28 are exactly duplicated as rungs 2 and 3 in Fig. 7.35. Rung 2's pair of normally open contact instructions with references I0003 and O0006 provide equivalent operation to the R3 and CR1 contacts in rung 7 of Fig. 7.28. While the relay schematic uses a relay for internal control system use, an output point (O0900) has been selected to provide the same function in the PC ladder diagram. For the PC program, real-world output references O0900 to O1024 have been assigned for internal references. No output modules will ever be assigned to these output points. These locations will only be assigned for use within the PC ladder program serving functions similar to that of Fig. 7.35, rung 2.

While timers will be covered in depth throughout Chapter 9, a brief description is given here to provide an understanding of the circuit's operation. The contact configurations of rung 8 in Fig. 7.28 are the same as those for rung 3 of Fig. 7.35. The timer is activated by a block being in place (R3), the system

running (CR1), and the clamp activated (CR2). In rung 3 of Fig. 7.35 the block must be in place (I0003), the system must be running (O0006), and the clamp must be activated (O0900).

Timer TD1 of the relay schematic is a 5-second *on-delay* timer. The same type of timer has been programmed in rung 3 of Fig. 7.35. The numbers surrounding the timer instruction indicate the time delay preset that the timer is to have (00005 for 5 seconds) as well as what internal data memory storage locations are to be used to hold the time delay preset period (R0001) and the accumulated time delay period (R0002).

The timer must operate a coil instruction in order to have contact instructions referenced to the timer. Another real-world output location assigned to internal use (O0901) has been assigned to the coil instruction associated with this timer. For purposes of this discussion, the reader should understand that coil instruction O0901 will be activated whenever the timer has completed its 5-second timing period.

Rung 4 of Fig. 7.35 is the only rung not similar in appearance to the rung it emulates in Fig. 7.28. In Fig. 7.28 an instantaneous normally open contact (TD1I), and a timed normally closed contact (TD1T) are wired in series. Together these two contacts act as a momentary signal to activate the lubrication pump. (As soon as the timer is activated, contact TD1I begins to pass power. Contact TD1T normally passes power until the timer has reached its timed-out state.) Provided that the system is still running (contact CR1) and the platen is not extended (contact R4), the timer will time. Once lubrication pressure has been established, the normally open contact R6 will pass power, thereby providing an alternate path of power flow around TD1I and TD1T. Once the timer has timed out, the normally closed timed contact TD1T will change state and inhibit power flow. If contact R6 had not begun passing power by the time TD1T inhibits the passing of power, the lubrication pump would be stopped.

The Series Six PCs do not provide an instantaneous timer output signal to drive a coil instruction. The conversion of rung 9 into an equivalent PC operation requires that some additional programming be done. There are several ways to approach this conversion, and each of them will be discussed in Chapter 9. For this example the simplest method to use is duplicating the instructions that activate the timer wherever an instantaneous timer contact instruction appears. As shown on rung 4 of Fig. 7.35, the three normally open contact instructions that activate the timer of rung 3, have been reprogrammed at the beginning of rung 4 to duplicate the operation of the instantaneous timer contact TD1I of Fig. 7.28.

The remaining instructions of rung 4 in Fig. 7.35 duplicate the operation of their counterparts in rung 9 of Fig. 7.28. Note that the normally open contact R4, wired in rung 9 of Fig. 7.28, has been programmed as a normally closed contact instruction in rung 4 of Fig. 7.35. In the relay schematic the contact functions to stop the lubrication pump when the platen has become fully extended. Since the fully extended platen limit switch has been wired to the input point in a normally open configuration, the normally closed contact instruction must be used to provide equivalent operation of both circuits.

Program rungs 5 to 8 of Fig. 7.35 appear almost identical to their counterparts in Fig. 7.28, rungs 10 to 13. The only noticeable differences are the contact configurations programmed for I0004 and I0005 and the wiring configurations for their counterparts R4 and R5, respectively. Again this is due to the manner in which LS2 and LS3 have been wired to the input modules. The conversion of relay schematic rungs 10 to 13 into programming rungs 5 to 8, respectively, is left as a review exercise for the reader in order to verify his or her understanding of the programming of the relay contact and coil instructions.

PC PROGRAM ENHANCEMENT

The PC program just developed exactly duplicates the relay system wiring it replaces, with the exception of the contact instruction differences from the relay counterparts and the instantaneous timer programming. Often it is possible to enhance a program, deleting redundant operations and instructions. This may require the use of a PC's extended instructions. For example, the relay ladder program instructions of the example might be replaced by a single sequencer instruction by some users. A thorough review of a PC program may reveal duplicate groups of instructions which can be replaced by only a couple of higher order or extended-type instructions.

When examining a PC program for duplicated contact instructions, two forms of duplication should be checked. The first involves the repetition of multiple contact instructions where a single contact instruction referenced to an existing program coil instruction would suffice. This is illustrated in Fig. 7.36. The upper portion of the figure shows three rungs of a program with simple references. Examination of the three rungs reveals that a configuration of three instructions is repeated in all three rungs. These three instructions have been circled. Further review shows that the middle rung contains the three duplicate instructions operating a coil instruction. A reduction of four contact instructions can be made by using a normally open contact instruction, referenced to the coil instruction of the middle rung, in place of the duplicated instructions appearing in rungs 1 and 3. This change has been made in the lower part of the figure.

The second form of redundant contact instruction programming is not as easy to spot. A PC program may contain redundant instructions, but there is not a single occurrence where those instructions simply operate a coil instruction without additional nonduplicate instructions being present. This situation is depicted in the upper portion of Fig. 7.37. Contact instructions 1 through 4 appear in every rung. However, they do not appear without other non-duplicated instructions in any of the rungs, nor do they simply drive a coil instruction without other instructions being present. By programming an additional ladder rung, seven contact instructions can be eliminated in the overall program. As illustrated in the lower half of the figure, the first rung contains the duplicate contact instructions operating an internally referenced coil instruction. The remaining four rungs have had the duplicate instructions replaced by a contact instruction referenced to the coil instruction controlled by the original duplicate instructions.

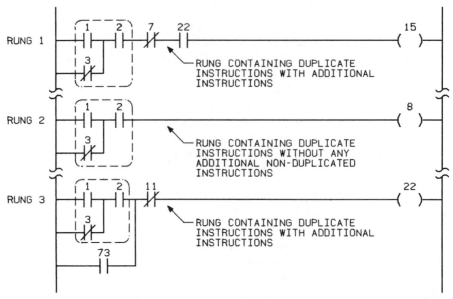

ORIGINAL PC PROGRAM BEFORE REDUCTION

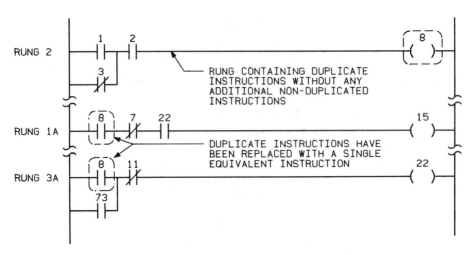

PC PROGRAM AFTER DUPLICATE INSTRUCTION REMOVAL

Fig. 7.36. Reduction of Duplicate Instructions by Instruction Replacement

The reduction of four and seven program instructions in Figs. 7.36 and 7.37, respectively, may not represent much of a reduction. However, a long program can contain many examples of the two types of instruction duplications just discussed. Taken collectively, the amount of extra programming generated by duplication can consume a lot of time. In addition the programming of duplicate instructions can make a PC program difficult to follow and understand.

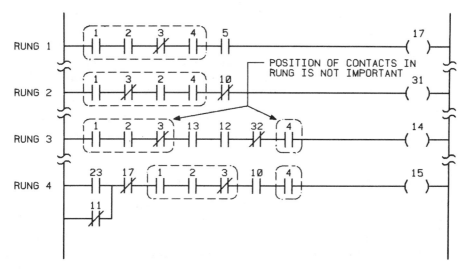

ORIGINAL PC PROGRAM BEFORE REDUCTION

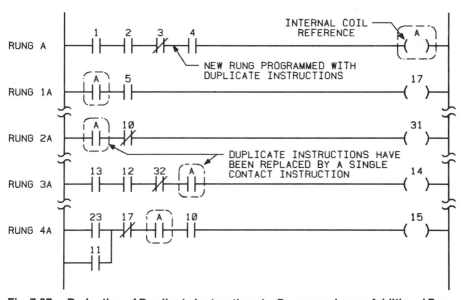

Fig. 7.37. Reduction of Duplicate Instructions by Programming an Additional Rung

Returning to the example conversion program of relay to PC program, examination of Fig. 7.35 reveals three rungs containing duplicate program instructions. The first occurrence is in rung 3 with the pair of normally open contact instructions referenced I0003 and O0006. This same pair of instructions appears in rung 2 as the input conditions for a coil instruction with reference O0900. This means that the two normally open contact instructions with references I0003 and O0006 can be replaced in rung 3 by a single normally open instruction with reference O0900.

Further investigation of the instructions appearing in rung 3 reveal that a normally open contact instruction with reference O0900 already appears. Since the placement of a normally open contact instruction with reference O0900 would be needless duplication, one of the two instructions can be eliminated completely. Rung 3 only requires a single occurrence of normally open instruction O0900.

Rung 4 also contains a pair of normally open contact instructions with references I0003 and O0006. These instructions could be replaced with a single normally open contact instruction with reference O0900. However, since there already is a normally open contact instruction with the reference of O0900, instructions I0003 and O0006 can simply be deleted.

Care should be exercised when making contact substitutions. Even though rung 5 appears to have instruction with references I0003 and O0006, closer examination indicates that what may appear as O0006 is actually I0006. A substitute cannot be made here. Fig. 7.38 shows the PC program with the corrections performed as just discussed. A total of five instructions has been eliminated.

The PC program developed from the relay control schematic shows that another form of duplication has occurred during the transformation. In Fig. 7.28 an internal relay was used as a signal to the rest of the system that the clamp was activated. In rung 13 a normally open contact of this relay, CR2, is used to actually operate the solenoid valve controlling the clamp.

The PC program conversion of Fig. 7.35 contains the same operation in rungs 2 and 8. The duplication appears in the form of an internally referenced coil instruction which is used where an externally referenced coil instruction could be used. The output coil instruction of rung 8 is controlled by a contact from the coil instruction of rung 2. The program can be simplified by deleting rung 8 from the program and changing the reference on the coil instruction of rung 2 from the internal location of O0900 to the solenoid valve's output reference of O0005. Internal reference O0900 becomes available for use elsewhere in the program. All contact instructions with the reference O0900 must also be changed. Fig. 7.39 shows these changes made to the program. Two additional instructions have been eliminated bringing the current total to seven eliminated instructions.

The initial PC program consisted of 39 instructions and eight rungs of logic. After reduction, the program occupies seven rungs and contains 32 instructions. This amounts to an 18 percent reduction in the amount of logic that must be programmed, maintained, and troubleshot. Every PC program will have some amount of duplication in the program. However, once a PC is installed, and the program begins to get expanded with "wouldn't it be nice if" routines, efficiency of instruction use will permit users to get the most from their hardware.

One word of caution is in order concerning the operation of eliminating redundant instructions from a program. Consider the five rungs of logic pulled from a program and illustrated in Fig. 7.40. Rung 1 contains three contact instructions, labeled 1, 2, and 3 for ease of discussion, operating a coil

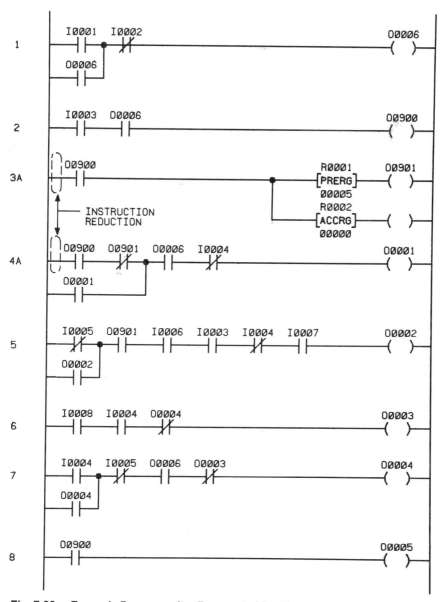

Fig. 7.38. Example Program after Removal of Duplicate Program Instructions

instruction. Rung 2 contains these same three instructions immediately following the branch instruction. These three instructions can be replaced by a single normally open contact instruction, referenced to the coil instruction of rung 1.

The same three instructions appear in rung 3 between the power rail and the branch instruction. Again a single normally open instruction, referenced to the coil instruction of rung 1, can be used to replace these three instructions.

Rung 4 also contains the three instructions. However, they are split on both sides of the branch instruction. In this case a single instruction referenced to the coil instruction of rung 1 cannot be substituted. The reason relates to the operation of the ladder rung. Instruction 2 is always required for the operation of the rung. The power flow status of instructions 1 and 3 is not critical to the circuit's operation due to the parallel instructions around them. A general rule of thumb is that redundant instructions must appear on the same rung or subrung. They must also appear together without being separated by one or more branch instructions if they are to be replaceable by a single equivalent instruction.

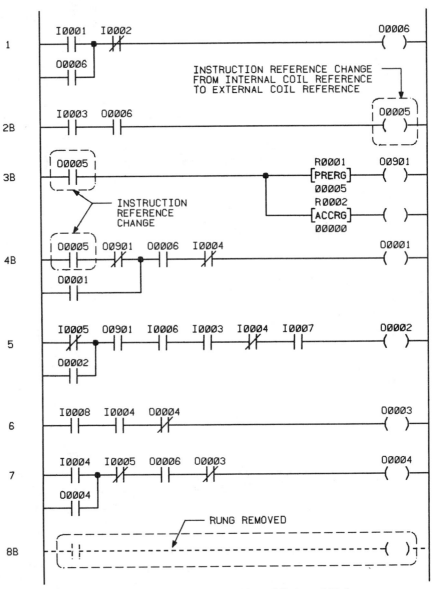

Fig. 7.39. Program Reduction by Use of External Reference

Care should be exercised to ensure that redundant instructions are identical in type and function. Examination of rung 5 in Fig. 7.40 indicates that the redundant instruction references of rung 1 are present, but that the contact instructions themselves are different. Contact instructions associated with references 1 and 3 are programmed as normally open in rung 1 and normally closed in rung 5. The normally closed contact with reference 2 of rung 1 appears as a normally open contact instruction in rung 5 with the same reference. On instinct the programmer may be inclined to use a normally closed contact instruction referenced to coil A, since the contact instructions of rung 5 are programmed as the complementary instructions of rung 1.

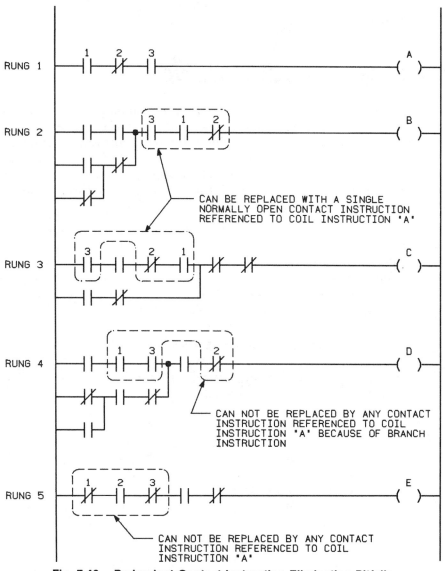

Fig. 7.40. Redundant Contact Instruction Elimination Pitfalls

This is not the case. A truth table can be developed showing eight possible contact-instruction power-flow configurations for rung 1. Included as part of this truth table should be the state of the coil instruction as well as the states of a normally open and a normally closed contact instruction associated with the coil instruction. A similar truth table for the three contact instructions in question in rung 5 can be developed. Examination of the truth table states for the instructions associated with the coil instruction in rung 1 will reveal that neither a normally open nor a normally closed instruction can be substituted for the three instructions of rung 5. When doing redundant instruction replacement, the programmer must be sure that the replacement contact will exactly duplicate the instruction states of the contact group it replaces. This replacement *must be exact*. The programmer *should not* simply consider the operating states of the circuit, but *every possible* state of the contact group to be replaced, as well as the states of the replacement contact.

PC PROGRAMMING EFFICIENCY

Prior to leaving the subject of relay ladder programming to investigate the host of instructions available in the PC marketplace, several topics related to the programming of a PC need to be examined. The first addresses the amount of memory used by an instruction when it is programmed. Each PC on the market has some efficiency level of memory utilization associated with it. This efficiency is not determined by any set standard, but is often referred to in a relative nature by persons involved in the sale and use of PCs.

As an example of the problems associated with determining the efficiency of a particular PC, consider five models of PCs, all given the same program for entry. After the program was verified as to proper operation, data were collected. These data are shown in Table 7.1. The amount of memory required for storage of the program for each PC is noted, as well as the number of instructions required to implement the program.

Table 7.1.   Memory-Efficient Ratings

PC#	MEMORY REQUIRED FOR PROGRAM	NUMBER OF INSTRUCTIONS	MEMORY USE EFFICIENCY	WORDS/INSTR. EFFICIENCY	# OF INSTR./ MEM. REQ. X 100
1	750	640	75%	1.17	85.3%
2	630	625	63%	1.01	99.2%
3	784	710	78.4%	1.11	90.6%
4	693	643	69.3%	1.07	92.7%
5	733	700	73.3%	1.05	95.5%

PC #2 MOST EFFICIENT
IN AMOUNT OF TOTAL
MEMORY USED

PC #2 USED THE LEAST
AMOUNT OF MEMORY WORDS
FOR EACH INSTRUCTION
REQUIRED

PC #2 MOST EFFICIENT
IN OVERALL MEMORY USE

The first question raised is why there is such a difference in the number of instructions required to do the same function. Not all PCs have identical instruction sets. With the variety of PCs available to meet a particular PC application, it is unlikely that a program could be written and programmed identically in two units. The ability of one PC to be more efficient due to a highly developed instruction set can create differences, as can internal design differences in the executive that operates the unit. These variations in instructions affect either directly or indirectly the amount of total memory required to program them. This explains the differences in the memory requirements column. Affecting both columns are the experience and the background of the programmer.

Assuming that all units were given 1000 words of memory for use in implementing the program, PC 2 would appear the most efficient since it requires less memory to implement the required program. This same PC also appears to be the most efficient in terms of the amount of memory required to store an average instruction. PCs 4 and 5 place second and third, respectively, when their memory use efficiencies are considered. These PCs swap positions when the number of words per instruction efficiency is used as the guide. The placement of PCs 1 and 3 as fourth and fifth is just as difficult. While PC 1 is more memory-use efficient, PC 3 is more words per instruction efficient. The last column of the figure expresses the PC efficiencies in terms of the number of instructions per amount of required memory. Examination of this column shows a different lineup of PC efficiencies.

The problem with expressing a PC's efficiency lies in the parameters used to express that efficiency. The data in the table were not generated from any actual tests. A different program would surely produce a different set of results, and probably a different lineup of PC efficiencies. The bottom line is that there are PCs which are more efficient for a given application than others. The actual efficiency of a unit is subject to a host of variables. The only way to determine accurately the relative efficiency of two or more PCs for an application is to program them for the application; to have the programs reviewed by numerous programmers to ensure that each one has been programmed in the best possible manner for that particular PC and the intended application; and then to review the hardware for the job based on the programming efficiencies achieved.

INSTRUCTION MEMORY UTILIZATION

As noted above, one of the areas in which a PC can be rated is the efficiency of its instruction set in using memory. Every instruction programmed in a PC must occupy one or more words of user and/or data memory. Memory is required to store any instruction and its reference in the processor. Depending on the instruction, the amount of user logic and data memory required for a particular instruction will vary. Table 7.2 indicates the amount of memory required for selected Series Six instructions.

Table 7.2. Typical Memory Requirements for Common PC Instructions

INSTRUCTION	MEMORY REQUIRED (WORDS)	INSTRUCTION	MEMORY REQUIRED (WORDS)
GENERAL		**LIST GROUP**	
1. START SERIES BRANCH	1	1. ADD-TO-TOP	4
2. START PARALLEL BRANCH	1	2. REM-FM-BOT	4
3. SERIES CONTACT	1	3. REM-FM-TOP	4
4. PARALLEL CONTACT	1	4. SORT	4
5. END SERIES BRANCH	1		
6. END PARALLEL BRANCH	1	**DATA MOVE GROUP**	
7. OUTPUT RELAY	1	1. MOVE	3
8. OUTPUT LATCH	1	2. MOVE RIGHT 8	3
9. OUTPUT ONE-SHOT	1	3. MOVE LEFT 8	3
10. PHANTOM OUTPUT	1	4. BLOCK MOVE	9
TIMER/COUNTER		**MATRIX GROUP**	
1. UP COUNTER	5	1. AND	5
2. DOWN COUNTER	5	2. IOR - INCLUSIVE	5
3. TIMER	5	3. EOR - EXCLUSIVE	5
4. ADD	4	4. INV - INVERSE	4
5. SUB	4	5. COMPARE	6
6. I/O -- R	3		
7. R -- I/O	3	**BIT MATRIX GROUP**	
8. COMPARE	3	1. BIT MATRIX GROUP	4
9. SHIFT	2	2. BIT CLEAR	4
		3. SHIFT RT	4
SIGNED ARITHMETIC GROUP		4. SHIFT LEFT	4
1. DPADD - DOUBLE PRECISION	4-6		
2. DPSUB - DOUBLE PRECISION	4-6	**CONTROL GROUP**	
		1. DO SUB	3
3. ADD	4	2. RETURN	1
4. SUB	4	3. SUSPEND I/O	1
5. MPY	4	4. DO I/O	3
6. DVD	5-6	5. STATUS	2
7. GREATER THAN	3-5		
		SPECIAL FUNCTIONS	
TABLE MOVE GROUP		1. BIBCD	3
1. TABLE-TO-DEST	4	2. BCD/BI	3
2. SRC-TO-TABLE	4	3. SKIP ALL	1
3. MOVE TABLE	4	4. SKIP N	1
		5. MCR	1
		6. SCREQ	1
		7. DPREQ	1

MEMORY WORDS REQUIRED INCLUDES ALL DATA AND INSTRUCTION
WORDS NECESSARY TO COMPLETE THE INDICATED INSTRUCTION

In addition to the memory required for an instruction, many PCs require additional noninstruction memory to separate one instruction rung from another. The noninstruction memory requirements of a processor, often termed the *programming overhead*, will vary among manufacturers, and in some cases among models of the same manufacturer. Every programming terminal usually provides a method for the user to determine the amount of available memory left in a processor.

DETERMINATION OF PC MEMORY REQUIREMENTS

When evaluating various PCs for an application, the cost of memory will figure into the final system costs for each PC system under consideration. During the evaluation stages the amount of memory required for a particular application needs to be determined for pricing. Unless the program that is going to be used has been developed, and the exact amount of words required for its implementation is known, the amount of memory will need to be approximated. This can be done with the assistance of the applications engineer for each PC being considered.

Every PC manufacturer can provide a set of memory calculation guidelines for use in determining the amount of memory required to meet an application. These guidelines often relate the amount of memory required for an application to the number of I/O points for the application. In most cases the quantity of I/O points can be determined with reasonable accuracy for a given application. Using this information along with the manufacturer's guidelines will produce a memory requirement figure. This figure can then be fine-tuned to suit the feelings of the user and the applications engineer.

A very general guideline that is often used when no manufacturer guidelines are readily available is as follows. The amount of PC user memory required for an application is seven to fourteen times the number of installed I/O points. For example, if an application requires 334 actual I/O devices, and 25 16-point I/O cards must be installed to handle those 334 devices, the PC memory calculation will be based on the 25 cards of 16 points per card, or 400 points. Using the guideline:

400 points $\times$ 7 words per I/O point = 2800 words
400 points $\times$ 14 words per I/O point = 5600 words

The memory requirements for the application will be between 2800 and 5600 memory words. Depending on the number selected, either a 4-K (4096 words) or an 8-K (8192 words) processor would be required. If the program was simple relay-type control with little diagnostics and/or use of extended instructions, then the 4-K memory processor would probably be more than sufficient for the application. If the user was planning on a large portion of the program containing extended instructions and diagnostics, then the 8-K processor should be selected.

The guideline is just that, a guide to a probable amount of memory required to meet the needs of an application. The more time spent in determining the instructions that will be required for the application and the manner in which the PC program will be written, the better will be the information obtained concerning the amount of memory required for that application. It should be noted that the guideline does not consider the amount of memory that might be required for applications involving the storage of large amounts of data pertaining to message-generation routines or other data-oriented applications.

PC RUNG SCAN SEQUENCE

Another area of possible concern to the PC user is the rung scan sequence of the processor. Every processor solves each rung of user logic in a predetermined manner. There are processors that solve the user program one rung at a time, while others solve user logic rungs in multirung user-determined groups, often called *pages* or *networks*. The method used to solve the user logic can affect the overall operation of the program. For example, consider the two rungs of logic shown in Fig. 7.41.

Depending on the PC used, these two rungs could be programmed as a group, forming a page or network, or they might be programmed as individual rungs of logic. Whether they are programmed as individual rungs on a PC that only permits single-rung program entry, or they are programmed in a one rung per page or network manner on a PC that permits multiple ladder rungs to be placed on pages or networks, the processor will solve the first logic rung before going on to the next successive rung or page. This means that the status of coil instruction A is placed in the processor's memory at the conclusion of the solution to the rung. Coil A's latest status is then used in the solving of the next sequential rung, network, or page.

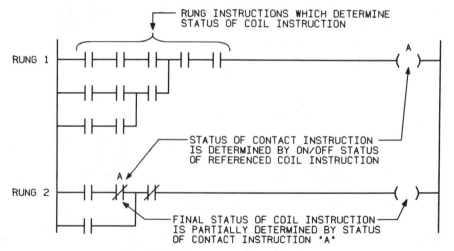

Fig. 7.41. Processor Scan Sequence

PCs permitting multiple logic rungs to appear on a page or network can solve the user logic in several ways. First they could solve the upper rung, followed by the solution of the lower rung. They might also solve both rungs at the same time. The manner that the processor uses during its logic solution can affect the solution of the second rung. Where the rungs are solved sequentially, the coil status of the upper rung will be available for the solution of the second rung. If the two rungs are solved simultaneously, a change in the status of coil instruction A will not be used in the solution of the second rung until the following processor scan. In effect, the second rung of the pair will always be solved with information generated in the previous scan.

Under most circumstances, the manner in which a processor solves the user logic rungs will not be a problem. The overall scan time of the processor is fast enough that the one scan time difference is not noticed. The PC user should, however, be aware of the possible problems associated with programming multiple logic rungs in a page or network format. If a program is highly scan sequence-dependent, possible timing problems could occur. The operating status of multiple rungs programmed on a common page or network should always be checked to verify that the lower rung instructions of a network or page are not dependent on rung instruction results immediately preceding them on the same network or page.

Several PCs have sufficient user program-solving intelligence to solve a rung only as far as necessary to determine the ON/OFF coil instruction status. For example, consider a series of five contact instructions operating a coil instruction. The processor will solve the complete rung and activate the coil instruction only if all five contact instructions are passing power. If during the solution of the rung the processor encounters a contact instruction that is not passing power, it will not solve any remaining contact instructions, but will immediately set the coil instruction to the OFF state. For complex series-parallel combinations of user logic instructions the ability to solve a rung or a subrung only to the first instruction not passing power can greatly reduce processor scan times.

PROCESSOR SCAN TIMES

With some applications the overall processor scan time will be critical for the proper operation of the system it controls. Applications requiring critical scan times include those having I/O devices which must be serviced within some maximum time limit, or internal data generated within certain time constraints. The scan time of a PC can be estimated in either of two ways. It is usually provided by a manufacturer as some unit of time per so many K (thousand words) of memory. This figure is obtained by the manufacturer as an average number during development and testing of the future product. For many applications this number will be sufficient since it reflects an average scan time based on various test programs being run.

Manufacturers often provide detailed data pertaining to the execution time required by a processor to read, interpret, and process each instruction of its instruction set. Table 7.3 provides average scan times for selected Series Six instructions. Note that the scan time for an instruction may vary depending on whether it has to actually be executed. The scan time for a processor can then be determined by adding the quantities of each instruction type, multiplying the quantity of each instruction by its scan period, and summing the totals. The time required for the processor to complete its I/O update scan can then be

Table 7.3. Typical Instruction Execution Times for Common PC Instructions

INSTRUCTION	EXECUTION TIME	INSTRUCTION	EXECUTION TIME
GENERAL		**LIST GROUP**	
1. START SERIES BRANCH	.94	1. ADD-TO-TOP	15.6
2. START PARALLEL BRANCH	.94	2. REM-FM-BOT	11.2
3. SERIES CONTACT	.94	3. REM-FM-TOP	14.7
4. PARALLEL CONTACT	.94	4. SORT	334.0
5. END SERIES BRANCH	.94		
6. END PARALLEL BRANCH	.94	**DATA MOVE GROUP**	
7. OUTPUT RELAY	1.88	1. MOVE	4.7
8. OUTPUT LATCH	1.56	2. MOVE RIGHT 8	6.6
9. OUTPUT ONE-SHOT	2.50	3. MOVE LEFT 8	6.6
10. PHANTOM OUTPUT	.94	4. BLOCK MOVE	17.8
TIMER/COUNTER		**MATRIX GROUP**	
1. UP COUNTER	9.06	1. AND	14.1
2. DOWN COUNTER	9.69	2. IOR - INCLUSIVE	14.1
3. TIMER	9.06	3. EOR - EXCLUSIVE	14.1
4. ADD	2.19	4. INV - INVERSE	12.5
5. SUB	2.19	5. COMPARE	59.7
6. I/O -- R	2.19		
7. R -- I/O	1.25	**BIT MATRIX GROUP**	
8. COMPARE	2.19	1. BIT MATRIX GROUP	12.5
9. SHIFT	2.19	2. BIT CLEAR	12.5
		3. SHIFT RT	18.4
SIGNED ARITHMETIC GROUP		4. SHIFT LEFT	18.1
1. DPADD - DOUBLE PRECISION	10.3		
2. DPSUB - DOUBLE PRECISION	10.3	**CONTROL GROUP**	
		1. DO SUB	29.0
3. ADD	6.9	2. RETURN	1.6
4. SUB	7.2	3. SUSPEND I/O	2.2
5. MPY	13.4	4. DO I/O	75.6
6. DVD	8.44	5. STATUS	5.6 - 7.8
7. GREATER THAN	6.6		
		SPECIAL FUNCTIONS	
TABLE MOVE GROUP		1. BIBCD	1.88
1. TABLE-TO-DEST	12.5	2. BCD/BI	2.19
2. SRC-TO-TABLE	12.1	3. SKIP ALL	1.88
3. MOVE TABLE	12.5	4. SKIP N	1.88
		5. MCR	1.88
		6. SCREQ	2.19
		7. DPREQ	2.19

EXECUTION TIMES ARE IN MICROSECONDS PER INSTRUCTION

calculated in a similar manner and added to the logic scan time to arrive at a final total scan time for the processor.

Perhaps the simplest method to determine the scan time of a processor is to let the processor calculate its own scan time. While some PCs provide a memory location that indicates the amount of time required to complete the last processor scan, the scan time calculation program in Chapter 19 will provide this information for those PCs not incorporating a scan time indication.

THE RAT RACE

Probably one of the most startling incidences that can happen to a PC user is called a *rat race*. A PC-based system may be in operation when a minor program change is suggested to improve system operation. The suggested change only requires a couple of rung changes to implement, so the system is halted and the changes are made. When restarted, everything seems OK, and the modification seems to provide the desired results. While monitoring the system to ensure that everything is correct, all of a sudden things begin to happen that are very unexpected. The system is emergency stopped before any serious damage can occur.

The above events can happen when any program change is not carefully checked out before being put into full operation. The simple change created a *rat race* condition where one event unexpectedly triggers more events. The change in logic has created a loop, or a circle of events.

For example, the control system initiates some action and awaits a signal that the action has been completed. The completed signal is received before the processor has time to react according to the programmed logic. The received signal may trigger other events, again often before the processor has time for its user programming to react to its initial request and response. A rat race condition develops when the processor initiates the next event before completing the first. The scenario is similar to a dog chasing his own tail: He's doing a lot of running, but going nowhere mighty fast.

There is probably no single programming rule that will guarantee that a rat race condition would never occur. Rat races can be caused by any combination of events. They can be caused by the mechanical hardware being controlled, or user logic oriented, or a combination of both. The only sure defense is careful system debugging, simulation, and checkout prior to actual operation.

USE OF DUMMY PROGRAMMING RUNGS

One programming trick often used by PC programmers, especially during the simulation and checkout of a system, is the programming of dummy logic rungs. These rungs consist of normally open contact instructions related to the operation of one or more program conditions. For example, consider a portion of a program being examined which has its operation based on the status of some other portion of the same program. Instead of scrolling the user logic to check the status of a particular contact or logic condition, those conditions are programmed as normally open contact instructions in a dummy rung along with the rung conditions being studied. It now becomes quite simple to know the status of logic conditions that may affect the operation of the monitored rung. Once the monitored rung or rungs have been determined to operate satisfactorily, the dummy rung can be removed. In lieu of dummy status rungs, several PC programming terminals offer alternate screen displays or special status areas for viewing this type of information.

MINIMUM PC INSTRUCTION CYCLE TIME

On each PC scan the processor examines a series of rung input conditions, then sets the status of the associated rung output instruction accordingly. If a set of rung input conditions indicates that an output instruction should be activated, the processor sets the output instruction to the ON state. On the next scan the input conditions for this same output instruction may have changed, now indicating that the instruction should be set to an OFF status. It normally takes a PC scan to generate a change of output instruction state. At best a scan is required to activate an output instruction, and a second scan to deactivate the same output instruction prior to being able to reactivate it. This sequence of events is usually necessary for every output instruction or function available on a particular model of PC.

In certain applications it is desirable, and often necessary, to activate and deactivate an instruction in the same scan. This type of operation is especially desirable for applications involving math or data manipulation instructions, where the application calls for new calculations or the manipulation of data to be performed on each scan. One approach is the programming of duplicate instructions, such that one instruction group performs the desired operation on even processor scans, while the second identical instruction group performs the exact same operation on odd scans. This approach provides the required operation on every scan, but in doing so requires a duplicate set of instructions and good documentation.

Many PCs permit an instruction to be programmed several times in the user program with the exact same references. Processors providing this flexibility often incorporate instructions that control the overall operation of large portions of the user program. Instructions such as *skip* and *master control relay* are typical examples of instructions provided for the control of large sections of the user program. The operation and use of these instructions will be described in the following chapter.

When a PC permits the programming of identical instructions, the ability to deactivate an instruction in the same scan it was activated is often provided. This is done by first programming the output instruction or function to operate when the proper input conditions are met. All user logic that depends on the results of a just completed instruction or function should follow next. Once all instructions using the status of the subject instruction or function are programmed, the subject instruction is again programmed under the control of a single input condition which can never be active. What the processor sees as it scans the user logic is the instruction active, the current status of that instruction used in determining the final status of other user program logic rungs, and finally the instruction in a deactivated state. Since the processor examines the current state of an instruction each time it is encountered, it does not remember that a few ladder rungs ago it set the instruction from OFF to ON. By double programming the instruction, once ON, followed by once OFF in the same scan, the processor is being tricked into causing the instruction to operate every scan, rather than every other scan. As noted before, only some PC models permit the double programming of selected instructions in the same scan. The user is warned to contact the manufacturer for the implications of programming a PC in this manner.

PC PROGRAMMING SAFETY

An area that should be addressed at this time is the safety considerations that must be developed into a PC program. Any PC program will only be as safe as the time and thought spent on the personal and hardware safety considerations of the application. The PC can only operate a piece of equipment or a process with the instructions provided by the user. The greater consideration given to safety by the programmer is reflected in a safer operating system. In addition to the safety considerations that must be applied to a particular application, several others are strictly PC-related.

One such consideration involves the use of internally programmed logic signals in lieu of externally generated PC input signals. At the top of Fig. 7.42 is an electrical schematic for the I/O devices of a typical motor circuit. The auxiliary contact for the starter has been connected to input point I0017. Output point O0010 controls the operation of the starter coil. The START button for the motor is wired to input point I0007, and the STOP button is wired to input point I0009. The desired operation of the motor is that it runs when the START button is momentarily depressed and stops whenever the STOP button is momentarily depressed.

There are two ways that the motor could be programmed. The first method does not use the starter auxiliary contact in the program. For this program a normally open contact instruction referenced to the rung's output instruction O0010 is used to "seal-in" around the START button contact instruction I0007. The alternate method is to use the starter auxiliary contact input point I0017 as the seal for the START button. Both program rungs are shown on the lower portion of the figure.

Of these two programming methods the second is safer, even though it is more costly in terms of field wiring and hardware. The use of the field-generated starter auxiliary contact status in the program is safer because it provides positive feedback to the processor as to the exact status of the motor. This can be seen by considering the following set of conditions. For the majority of PC applications, the processor receives its power from a source separate from the I/O devices. If by some chain of events power was lost to the motor control

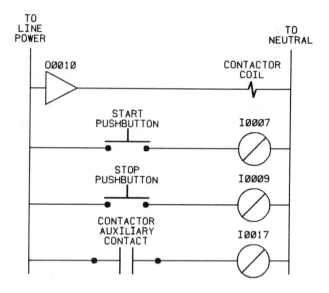

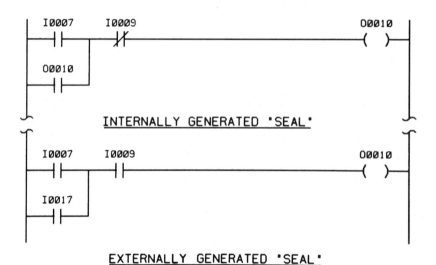

Fig. 7.42. Internally Generated versus Externally Generated Logic Signals

circuitry, the motor would of course stop operation since power would be lost to the starter. In fact, the disconnect switch could be opened to stop the motor.

Under both of these motor power loss events, the processor could still remain operational. If the program was written using a normally open contact instruction referenced to the output coil instruction as the seal for the circuit, the processor would never know when power has been lost to the motor. When power was restored, the motor would restart instantly. Hopefully no one had decided to take advantage of the power outage to investigate or maintain the motor's driven machinery, and the unexpected restarting of the motor would not create an unsafe condition.

Had the program been developed using a normally open contact instruction referenced to the starter auxiliary contact input of the PC, the processor logic would have detected the deenergization of the starter by the loss of power to the auxiliary contact input point. Upon detecting the loss of power on this input, signaling that the motor contactor was not energized, the PC would deenergize the starter output coil instruction and in turn the output point controlling the starter's coil. Providing an input point and the wiring of that input point back to the motor starter auxiliary contact does cost more than simply using an internally generated signal. The PC knows the status of the device it is controlling by means of positive feedback (the starter auxiliary contact and input point), and is not controlling a device by what it thinks real-world conditions are (output point O0010 is ON, therefore the PC assumes that the motor is still operating).

A related safety consideration concerns the STOP button wiring of Figs. 7.30 and 7.42. In both figures the STOP buttons have been wired to the PC input module in a normally open manner. A STOP button is generally considered a safety function as well as an operating function. By the manner in which the button has been wired to the input point, power is required on the input point if the processor is to recognize that a stop action is desired. If by some chain of events the circuit between the button and the input point becomes broken, the STOP button could be depressed forever, but the PC logic could never react to the STOP command since the input point would never become energized. The same holds true if power were lost to the STOP button control circuit.

The safer manner to connect the STOP button to the PC input is in the normally closed wiring configuration. The input point receives power continuously unless the stop function is desired, at which time the input point becomes deenergized. Any faults occurring with the stop circuit wiring, or a loss of circuit power, would effectively be equivalent to an intentional stop operation since power to the input point would be interrupted.

It should be noted that the limit switch wiring in Fig. 7.30 should also be changed to the normally closed switch contacts to ensure operation similar to what was provided in the relay schematic of Fig. 7.28. Fig. 7.43 gives the corrected input wiring for the example problem. Fig. 7.44 provides a finished version of the program developed to duplicate the relay control system.

A final area of safety concern involves the use of multiple PC output points as a means to reduce the possibility of a safety-related accident due to an output point or I/O system failure or fault. As was pointed out in Chapter 2, the possibility of an output point failing either ON or OFF cannot be eliminated due to the solid-state nature of a PC. For critical output devices the use of multiple output points, and the resulting necessary programming, can greatly, if not completely, reduce the impact of a possible output point or a related I/O structure failure. Since the possible I/O and programming configurations have already been discussed in Chapter 2, they will not be repeated here.

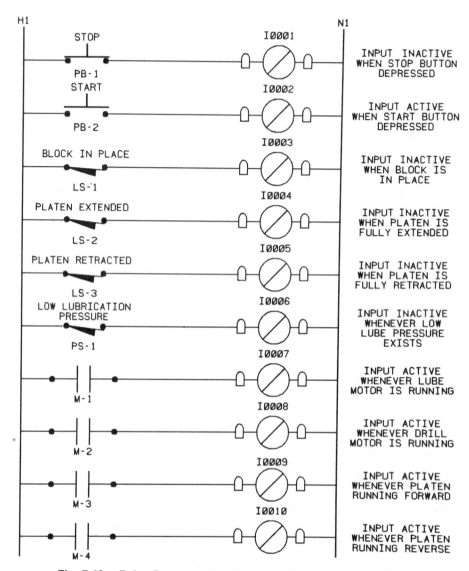

Fig. 7.43. Relay Programming Example, PC Input Device Wiring

END-OF-SCAN INSTRUCTION

One instruction incorporated into every PC is the *end-of-scan* instruction. This instruction is not user-programmable. However, several PC manufacturers provide an equivalent instruction for user use. Since a PC must contain more memory for user logic than is actually needed for a particular PC application, the end of scan instruction acts as a flag to the processor which indicates when all user logic has been scanned and that scanning should return to the beginning of the user program. If a processor were to continue past the end-of-scan instruction, it would find instructions it may possibly not be able to interpret. The end-of-scan instruction is automatically moved whenever the user inserts or deletes an instruction through a programming terminal. Depending on the PC and the manufacturer in question, the instruction may be viewable on the programming terminal for user reference. It should also be noted that the subroutine area of user memory may or may not be located after the formal end-of-scan instruction, again depending on the PC and the manufacturer in question.

Those processors offering a user programmable end-of-scan instruction usually allow this instruction to be placed anywhere in the user program. This instruction is often useful for the initial startup and debugging of a PC program. If the program has been entered into the processor's memory in an order closely related to the actual operation of the controlled machinery or process,

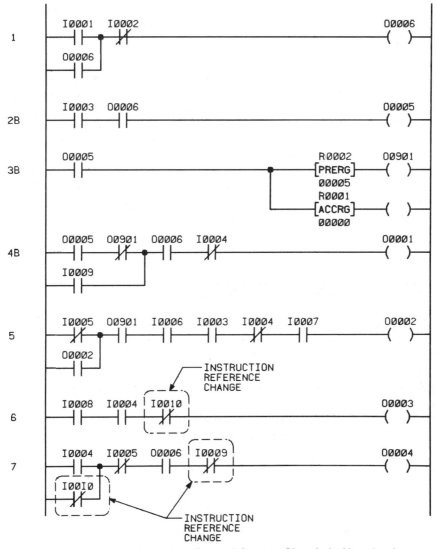

Fig. 7.44. Use of External Control System Signals in User Logic

use of the end-of-scan instruction permits the full program to be broken down into smaller segments for debugging. The user simply programs an end-of-scan instruction at a convenient location in the program. The processor scans from the beginning of the program to the user programmed end of scan instruction. Only the logic in this area is solved. Use of the end-of-scan instruction in this manner allows the user more control in the initial startup and debugging of the logic. Once the user has progressed through most of the program, the user programmed end of scan instruction can be removed and the processor allowed to scan the full amount of user program from the beginning to the manufacturer's end-of-scan instruction. For those processors that do not provide an end-of-scan instruction, often a skip or jump instruction may provide an equivalent form of operation. These instructions are discussed in a later chapter.

EXAMINE-ON AND EXAMINE-OFF INSTRUCTIONS

Two instructions used by at least one major PC manufacturer are the *examine-on* and the *examine-off* instructions. They are exactly equivalent to the *normally open* and *normally closed* contact instructions, respectively. If the reader stops and thinks about the operation of the normally open contact instruction, it will be realized that this instruction passes power whenever the referenced I/O or internal location is active or ON—hence the name for the examine-on instruction. Likewise the examine-off instruction passes power whenever the referenced location for the instruction is OFF, duplicating the operation of the normally closed contact instruction. The same programming terminal symbols that are used for the normally open and the normally closed contact instructions are used for their clone instructions, the examine-on and the examine-off instructions, respectively.

NOT-COIL INSTRUCTION

An instruction offered by several PCs is the *not-coil* instruction. This instruction operates directly opposite the coil instruction discussed earlier in this chapter. The not-coil instruction is energized whenever it is *not* receiving power from its input instructions. As soon as the not-coil instruction's input condition instructions pass power to the instruction, it is deenergized. Any contact instruction referenced to the not coil instruction still passes power according to the operation of the instruction. However, the user is alerted to the fact that overall circuit operations can be reversed. For example, a normally open contact instruction referenced to a not-coil instruction passes power whenever power is not being supplied to the not-coil instruction. This can often confuse someone not familiar with the program and its operation at first glance.

8

Extended Relay Instructions

There are numerous instructions that resemble standard relay symbology in appearance, but perform nonrelay-type operations. The most common of these special relay instructions are the *latch* and *unlatch* instructions. There are contacts called *transitional contacts*, which maintain their power flow or continuity for only a single processor scan. Functions such as *skip* and/or *jump*, *master control relay*, and *zone control relay* allow selected sections of the user logic to be skipped over while the processor either maintains the present coil statuses or turns OFF all coils within the skipped logic area. Often it may be necessary to use an *immediate I/O update* instruction to halt program execution temporarily while specific I/O points are updated. The ability to *jump* forward in a program along with the option to program a *subroutine* gives the PC almost computerlike usability.

In order to illustrate many of the special instructions available, a cross section of PCs will be used for various programming applications. Each instruction will be discussed and an example of its use described. While the examples will refer to specific instructions available from a particular manufacturer, the reader is cautioned that each manufacturer offering an instruction or instructions similar to the one being discussed will usually provide his own version and format of that instruction or operation. The manufacturer cited as the example should not infer that that manufacturer's offering is the only, or best, that is available.

THE LATCH INSTRUCTION

The *latch* instruction can be thought of as a retentive relay which operates identically to its industrial counterpart. Fig. 8.1 shows the electrical operation of

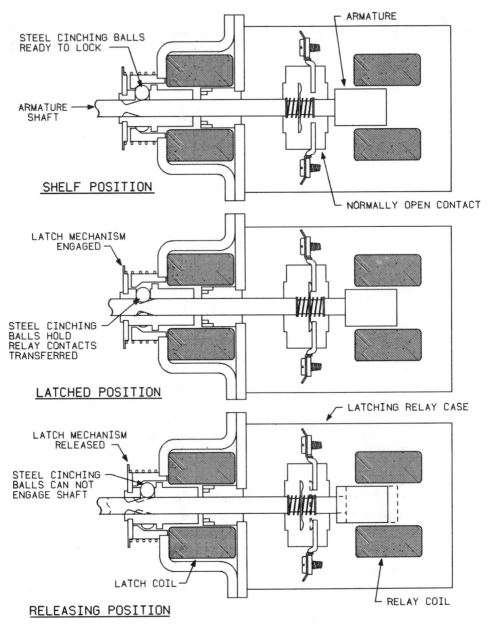

Fig. 8.1. Latching Relay

the standard latching relay. This device consists of two coils designated latch and unlatch coils, respectively. The latching relay can have any mix of normally open and normally closed contact poles, but the maximum number of poles rarely exceeds 12. Whenever the latch coil receives power, the contacts of the relay change state to their nonshelf configuration. All normally closed contacts break to inhibit power flow or continuity, while all normally open contacts make to provide power flow or circuit continuity. Power can be removed and reapplied at will to the latch coil without affecting the pole contact statuses. As soon as the unlatch coil receives power, the contacts associated with the relay return to their shelf state. The normally open contacts reopen to inhibit power flow and circuit continuity, and the normally closed contacts remake to provide power flow and circuit continuity. Again, the unlatch coil may be powered any number of times without affecting the contact status. The latching relay's contacts can only be retransferred by activating the latch coil.

In lieu of the dual coil latch just described, some industrial control systems may incorporate a single-coil mechanical latch. These latches function in an alternate-action mode of operation. The first time the latch's coil is activated, the poles of the latch change state. On the second activation of the coil the poles change state back to their original status. Very few PCs offer a latch instruction which operates similar to the single-coil mechanical latch. The majority of PC latch instructions incorporate the dual-coil concept of operation.

The latch/unlatch instruction is represented in Fig. 8.2 along with a timing diagram for the instruction. The format of the instruction is that of the SY/MAX-20 PC manufactured by Square D. This format is identical to that used by the majority of PC manufacturers offering this instruction pair. Examination of the timing diagram for this instruction reveals that it is identical to that of the latching relay described earlier. The PC user employing the latch/unlatch instruction pair should keep in mind that this is a *pair* of instructions, and one instruction rarely appears without the other. Since the latch/unlatch instruction pair are coil-type or output-type instructions, both the latch instruction and the unlatch instruction have the same reference identifications. Since the PC cannot have two separate mechanical coils, it uses a single reference location within its memory for each latch/unlatch instruction pair. The processor designates this location as a latch/unlatch relay function when it interprets the L or U as part of the instruction. The processor then interprets any contacts referenced to this

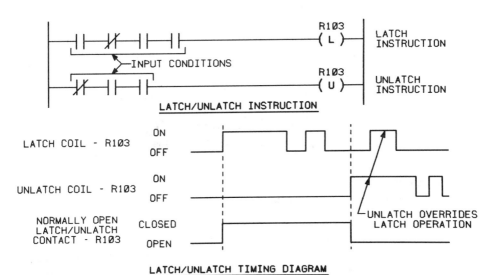

Fig. 8.2. SY/MAX Latch/Unlatch Instruction

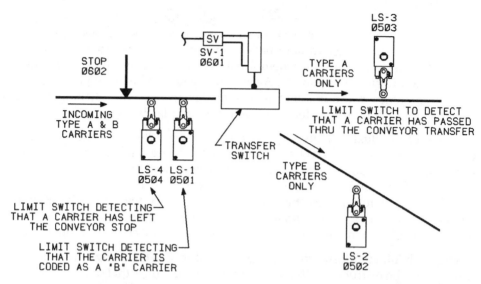

MONORAIL SYSTEM PASSING TYPE "A" CARRIER

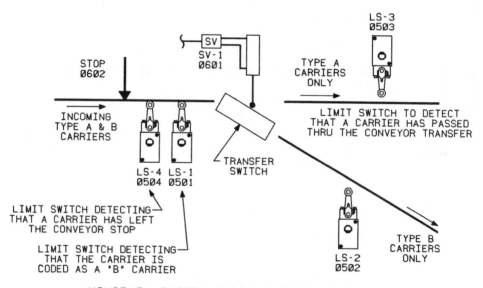

MONORAIL SYSTEM PASSING TYPE "B" CARRIER

Fig. 8.3.   Latch Instruction Example

location as contacts of the latch/unlatch instruction and proceeds to operate them in the customary latching relay format.

The latch/unlatch instruction is a *retentive* instruction. Once the function is either latched or unlatched it maintains its current state until changed. For processors that have volatile types of memory (those requiring battery backup), the latch/unlatch instruction pair will retain their last state through a power outage provided the standby batteries retain the processor's memory. Many processors have a fixed number of available latch/unlatch instructions within their memory areas while others allow an unlimited number of instructions to be programmed.

Often a latch/unlatch instruction can only be referenced as an internal output instruction, one not referenced to a real output point. However, there are also those systems that do allow the latch/unlatch instruction to be directly referenced as an output location. Caution should be observed when a latch/unlatch instruction has been programmed and the process or equipment is started for the first time. The instruction should be preset either ON or OFF for correct process or equipment operation before operating the processor and powering the field devices.

The latch/unlatch instructions can make many programming tasks easier. However, they can also be the basis for many operating headaches. Consider the example of Fig. 8.3. A conveyor system, for this example we will consider it an overhead monorail system, has carriers moving product to various plant locations. Throughout the plant, Y-type switches are installed for the purpose of diverting a carrier from the main route to an alternate route. Two types of carriers may be traveling the main route, and upon reaching the Y-type transfer switch, carrier type A is allowed to pass straight through, while carrier type B is diverted. A limit switch is located prior to the divert switch (LS-1) to sense type B carriers. When a Type B carrier is detected, the switch is activated to divert the carrier. Once the carrier passes through the switch, it trips a second limit switch (LS-2) to reclose the divert switch.

A stop mechanism is located prior to both the divert switch and incoming limit switch that allows only one carrier to be in the divert switch at a time. A limit switch detects any carrier leaving the stop (LS-4), and causes the stop to close. Whenever a carrier clears the divert switch, it is sensed by either one of limit switches LS-2 and LS-3, and the stop is reopened for the release of the next carrier.

Six rungs of logic are required for programming this example, as indicated in Fig. 8.4. The format used is that of the SY/MAX-20 PC. Ladder rungs A to C operate the divert switch, and rungs D to F operate the stop.

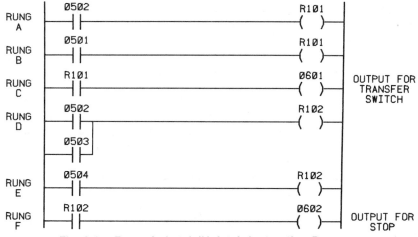

Fig. 8.4. Example Latch/Unlatch Instruction Program

Rung C transfers the state of the latch instruction to the output associated with the divert switch. Rungs A and B control the latch and unlatch instructions associated with the operation of the divert switch. The operation of the latch for the stop is programmed in ladder rungs D and E, while rung F transfers the status of the latch associated with the stop to the output point controlling the stop.

The use of the latch/unlatch instruction pair in this example provides a simple method to control the operation of both the stop and the divert switch. In fact, the use of the latch/unlatch instruction pair greatly reduces the amount of programming required when compared to using other programming techniques such as relay instructions only.

The operating headache arises when the plant operator wishes to take an unsolicited break. He merely has to trip LS-4 manually without a carrier being in the transfer to foul up the operation of the system. What happens is not the result of a problem with the SY/MAX controller. In fact *any* controller programmed as illustrated would react the same way. Tripping LS-4 will activate the latch instruction of rung E. Rung F will then close the entry stop. Carriers will now back up behind the stop, waiting to be released. But since there are no carriers in the divert switch to trip eventually either LS-2 or LS-3, the stop never opens. He could also trip LS-1 at random and send a carrier designated as type A on the alternate route specified for type B carriers. In both cases the use of the latch/unlatch instructions simplified the programming, but in doing so it created potential operating problems.

The solution to both problems is quite simple. A timer instruction could be programmed to sense the tripping of LS-4. It would be set to time out in a period that is longer than the time it takes a carrier released from the stop to travel to either LS-2 or LS-3. If the stop is closed due to any operation of LS-4, the timer will automatically open the stop if a carrier released from the stop does not open it first by tripping either LS-2 or LS-3. The one drawback of this solution is the fact that a carrier hung up in the divert switch will not trip either LS-2 or LS-3. When the timer times out, another carrier will be released. A crash is then inevitable in the switch. In fact if not caught quickly, a sizable number of units could clog the switch. If the carriers are not designed for accumulation, or if the power chain in the track is not designed to release from stuck or accumulated carriers, equipment damage is sure to result. While the use of the timer can solve one potential problem in using a latch/unlatch instruction pair, the operation of the system with the timer could be a bigger problem than the operator anticipates. Since timers will not be covered until the next chapter, the programming necessary to implement the timer will be left as an exercise for the reader after completion of the chapter on timers and counters.

The problem associated with the false tripping of LS-1 can also be corrected with the proper use of a timer. A second timer is programmed to operate from LS-4. This timer sets up a "window" for the operation of LS-1. When LS-4 is tripped, LS-1 must then be tripped within the time period of the timer. Should LS-1 be tripped outside of the time period of the timer, then its signal would be ignored and the divert switch would not be activated. Again this program is left to the reader to do as practice in the application of timers.

In lieu of the latch/unlatch instruction pair, several PCs provide a retentive coil instruction. This instruction operates identically to the latch/unlatch instruction with the exception that it is a single instruction instead of a pair. The retentive coil instruction is used to retain the current status of a coil during a power failure. If the coil was OFF at the time of power loss, upon power restoration the coil will again be OFF. If the retentive coil was ON at the time of power loss, the processor will log the ON condition, and when power is returned, the coil will be returned to the ON state until directed otherwise by the processor logic.

TRANSITIONAL CONTACT, OR ONE-SHOT INSTRUCTIONS

The availability of *transitional contact*, or *one-shot* instructions, in addition to the standard normally open and normally closed contact instructions, can be extremely useful at times. These instructions permit power flow or circuit continuity for a single scan whenever the I/O point or the internal coil location to which they are referenced changes state. The two forms of transitional contact instructions are illustrated in Fig. 8.5 along with a timing diagram for each.

The OFF-to-ON transitional contact instructions passes power for one scan only, whenever the I/O point or the internal coil to which the instruction is referenced goes from an OFF to an ON condition. Should the referenced I/O point or internal coil remain ON longer than a single scan, the instruction will lose power flow until the next OFF-to-ON transition of the referenced coil or I/O point. The ON-to-OFF transitional contact instruction provides the same operation as the OFF-to-ON transitional contact instruction, except that it only passes power for one scan whenever the referenced I/O point or internal coil goes from an ON to an OFF state.

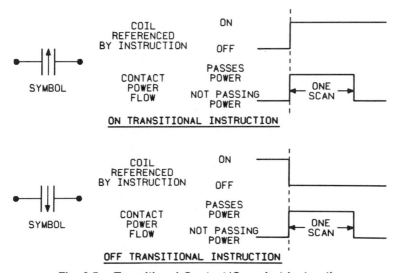

Fig. 8.5.   Transitional Contact/One-shot Instruction

The transitional contact instruction finds excellent use in applications where it is desirable to have an operation occur only once, even though the referenced I/O point or internal coil will be activated longer than for a single scan. In general many of the mathematics, data manipulation, bit manipulation, and special function instructions are designed to operate on every processor scan that the control line for the instruction is true. When these instructions are activated from I/O devices such as push buttons or selector switches, the I/O device can be ON for more than a single processor scan. If the instruction triggered by the I/O device is to operate only once, the use of a transitional contact instruction referenced to the I/O point will guarantee correct instruction operation. The transitional contact instruction may be thought of as a one-shot with a time period equal to that of the processor scan.

MASTER CONTROL RELAY AND SKIP INSTRUCTIONS

Several output-type instructions, which are often referred to as *override* instructions, provide an override control function similar to that provided by the master control relay in a relay-based control system. Instructions comprising the override instruction group are the *master control reset* (MCR), *zone control last state* (ZCL), *skip,* and *jump* instructions. While each of these instructions can provide special control operation for any logic instructions and/or functions programmed within their range, they should not be used as substitutes for hard-wired emergency stop relays. The emergency stop relay or emergency control relay should still be hard wired into the power supply circuits of the processor I/O output modules to ensure a failsafe system shutdown in the event of an emergency situation. The MCR instruction is often referred to by some PC manufacturers as a *master control relay* instruction, and the user should not be confused by the two different names for the same basic instruction.

All master control reset or master control relay, zone control last state, jump, and skip instructions are similar in one respect. They *all* operate over a user-specified range, section, or zone of processor logic. The size of the zone, or the range of the instruction, is specified in some manner as part of the instruction. Fig. 8.6 represents the two commonly used methods to specify the range or zone of this type of instruction. The PC manufacturer will decide on the method he wishes to use, and the user will not be able to change this selection.

The first method involves specifying the number of rungs of user programming to be influenced by the instruction. Whenever the processor interprets the control line of the instruction to be true, the specified number of program rungs indicated by the instruction is controlled according to the type of override instruction selected. A second commonly used method of specifying the amount of user programming controlled by the override instruction requires that the instruction be programmed twice.

The first occurrence of the instruction is in a *conditional* format where the control line for the instruction must be true for the instruction to operate. Once the selected override instruction is activated, all succeeding rungs of the user

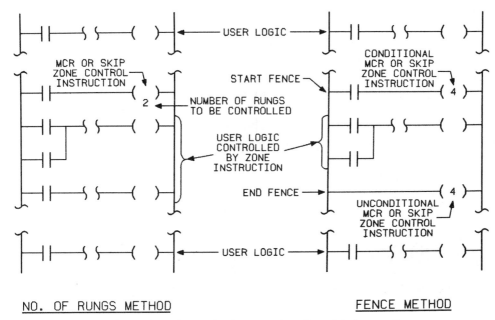

NO. OF RUNGS METHOD   FENCE METHOD

Fig. 8.6.   Indicating the Range of an Override Instruction

program are affected according to the operation of the override instruction in effect. Operation of the override instruction is discontinued when the processor encounters a second identical instruction programmed *unconditionally*. An override instruction that is programmed unconditionally is programmed without any control conditions as part of the control line.

A manufacturer offering the master control reset or master control relay instruction as well as the jump or skip instruction is Struthers-Dunn. Fig. 8.7 represents the programming format and operation of the master control reset instruction for the Director 4001 PC. Whenever the control logic for the instruction is false, the processor will solve the user logic rungs following the control instruction according to the individual control logic conditions and instructions that have been entered by the PC user. If the master control reset instruction is activated, the processor will turn OFF the coils for the specified number of ladder rungs following the instruction.

Note that the master control reset instruction for the Director 4001 specifies the number of pages of logic to be controlled by the instruction, and not the actual number of coils to be controlled. A page of logic within the Director 4001 may contain more than one ladder rung of logic. Those controllers that do not permit paging or networking of the logic, specify directly the number of coils to be controlled, unless the controller is designed to have the instruction programmed twice, once conditionally and later unconditionally.

Fig. 8.8 represents the jump instruction for the Director 4001. Whenever the jump instruction is energized, the specified number of pages will be jumped or skipped over. The jumped coils will not be updated according to their control logic conditions. Coils that were ON during the previous processor scan in which

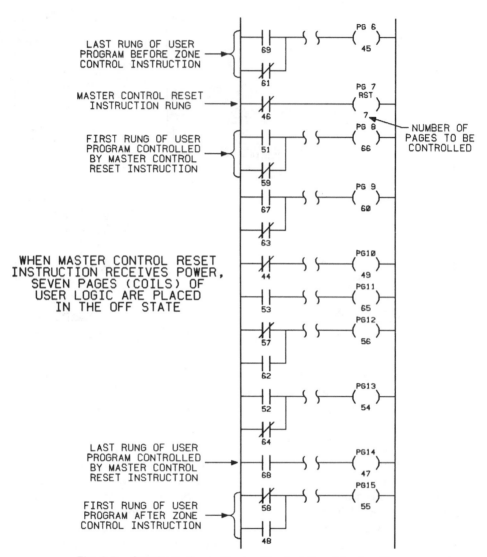

Fig. 8.7. Struthers-Dunn Master Control Reset Instruction

the jump is activated will remain ON, while those that were OFF will remain OFF. Again, the actual number of coils jumped will depend on the number of pages specified in the master control reset instruction since the Director 4001 permits user logic to be paged.

Whenever any of the override instructions are used, the user should pay particular attention to the specific operating characteristics of the instruction offered by the manufacturer. The override instructions may or may not be grouped or nested, depending on the manufacturer's design for the PC being used. Fig. 8.9 depicts the two ways in which override instructions may or may not be overlapped or nested.

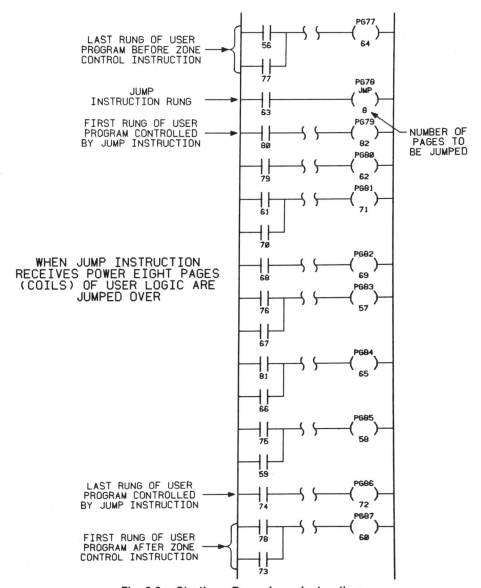

Fig. 8.8.   Struthers-Dunn Jump Instruction

Some processors will actually scan the user logic within the override instruction's range while others may not. The fact of whether or not the processor scans the override instruction's range may be important for the operation of the logic contained within the instruction's range. Many processors will not update timers when the timer instruction is within the range of an active override instruction. Timers may be halted at their current accumulated time, or even reset back to zero, if they are part of a logic zone that is being skipped or jumped.

If the number of output coils to be controlled by the override instruction is greater than the number of coils remaining in the user program, either a processor fault or an error condition could result. The scan time of some controllers may be shortened by the use of a jump or skip instruction. The jump or skip instructions are often used when it is desirable to shorten a processor's scan time by selectively bypassing sections of programming that are not currently in use.

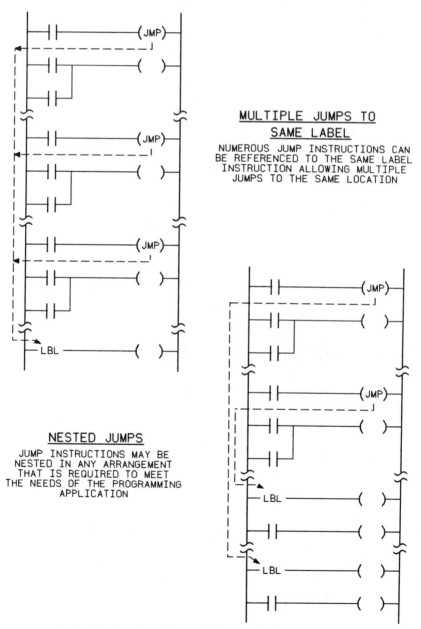

MULTIPLE JUMPS TO
SAME LABEL

NUMEROUS JUMP INSTRUCTIONS CAN
BE REFERENCED TO THE SAME LABEL
INSTRUCTION ALLOWING MULTIPLE
JUMPS TO THE SAME LOCATION

NESTED JUMPS

JUMP INSTRUCTIONS MAY BE
NESTED IN ANY ARRANGEMENT
THAT IS REQUIRED TO MEET
THE NEEDS OF THE PROGRAMMING
APPLICATION

Fig. 8.9. Various Uses of Jump or Skip Instruction

THE HALT INSTRUCTION

One very special override instruction offered by at least one manufacturer is the *halt* instruction. This instruction is in the form of an output coil instruction, and when activated by its control line, it literally stops the processor from scanning. It is in effect a user programmable processor fault. Once halted by this instruction, the processor can be restarted manually through the use of a RESET button located on the processor, provided the user programmed conditions that triggered the shutdown are no longer active.

IMMEDIATE I/O UPDATE INSTRUCTIONS

Applications are often encountered that require that the processor have the ability to examine an input point or a group of input points immediately prior to performing a special control instruction or function. Alternately it may be necessary to provide the results of some processor calculations or decisions immediately to an output point or a group of output points. Other applications may require that a particular control routine be performed using current I/O statuses more than once per processor scan. When any of these conditions exist, a processor having *immediate I/O update* instructions can make programming much easier.

The function of immediate I/O update instructions is to halt the current scanning of the user program for the purpose of performing an update of a selected input image table or of output module points. The scanning of the user program and the I/O module statuses is systematic for all PCs, as shown in Fig. 8.10. The processor scans all input modules to receive input statuses for all field devices. Upon completion of the input update operation the processor begins its scan of the user program. During this operation the processor reads and interprets the various instructions programmed in memory. As these instructions are interpreted, the proper responses are generated. After having solved all user logic the processor begins updating all output modules with the ON/OFF statuses just generated when the user logic was solved. Upon completion of the output module update scan, the cycle repeats with the start of another input update scan.

When the processor reads an instruction that requests the immediate updating of an input point or points, the scan of the user logic is interrupted, as shown in Fig. 8.11. Only the input point or points indicated by the immediate I/O update instruction are updated in the processor's input image table. Once the requested inputs are updated, the processor returns to continue the user logic scan. As shown in Fig. 8.11, most immediate I/O update instructions, which update input data, are normally placed immediately prior to the user programming requiring the most current status of selected input locations.

The immediate I/O update instruction can also be used to provide data and output statuses to the field devices immediately after such information has been produced within the processor. Once data or I/O statuses have been generated within the processor, the user program scan can be interrupted and these

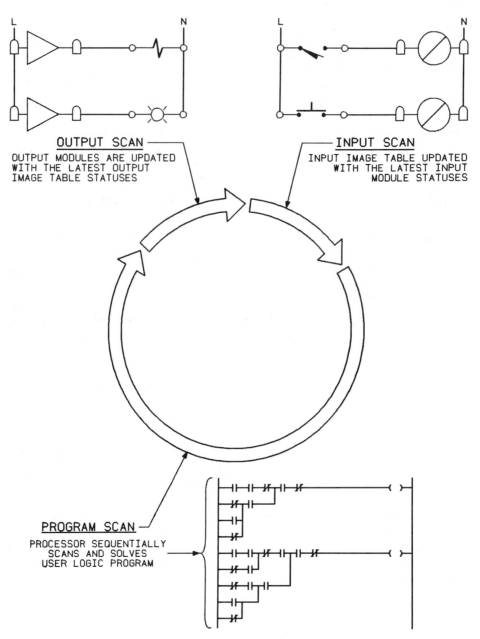

Fig. 8.10. Processor I/O Scan

statuses transmitted from the processor, as indicated in Fig. 8.12. After the processor completes the transmission of data to the output points specified, it returns to the user program where it left off to continue the user program scan. As indicated in Fig. 8.12, most immediate I/O update instructions, which transmit information to the output modules, are usually placed immediately after the user programming that generates the information.

Numerous forms of immediate I/O update instructions are available, depending primarily on the architecture of the processor in question. Many PCs incorporate a single immediate I/O update instruction for both input and output points. The instruction specifies which input and output locations are to be updated as part of the instruction. Other systems provide separate input and output immediate I/O update instructions. Users are given the option of which instruction to use as part of their programming.

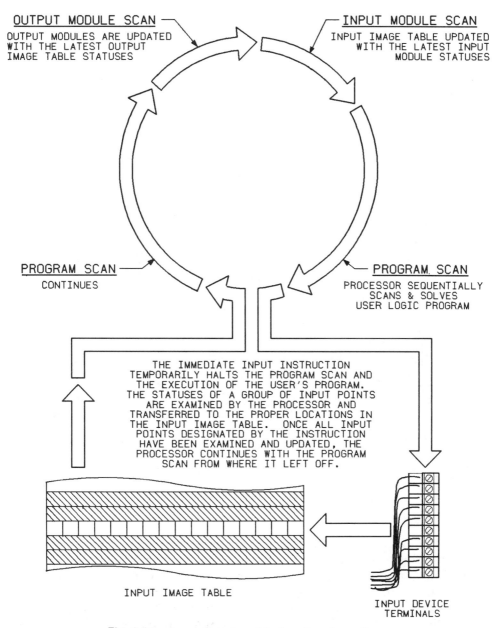

OUTPUT MODULE SCAN
OUTPUT MODULES ARE UPDATED
WITH THE LATEST OUTPUT
IMAGE TABLE STATUSES

INPUT MODULE SCAN
INPUT IMAGE TABLE UPDATED
WITH THE LATEST INPUT
MODULE STATUSES

PROGRAM SCAN
CONTINUES

PROGRAM SCAN
PROCESSOR SEQUENTIALLY
SCANS & SOLVES
USER LOGIC PROGRAM

THE IMMEDIATE INPUT INSTRUCTION
TEMPORARILY HALTS THE PROGRAM SCAN AND
THE EXECUTION OF THE USER'S PROGRAM.
THE STATUSES OF A GROUP OF INPUT POINTS
ARE EXAMINED BY THE PROCESSOR AND
TRANSFERRED TO THE PROPER LOCATIONS IN
THE INPUT IMAGE TABLE.   ONCE ALL INPUT
POINTS DESIGNATED BY THE INSTRUCTION
HAVE BEEN EXAMINED AND UPDATED, THE
PROCESSOR CONTINUES WITH THE PROGRAM
SCAN FROM WHERE IT LEFT OFF.

INPUT IMAGE TABLE

INPUT DEVICE
TERMINALS

Fig. 8.11. Immediate Input Instruction Operation

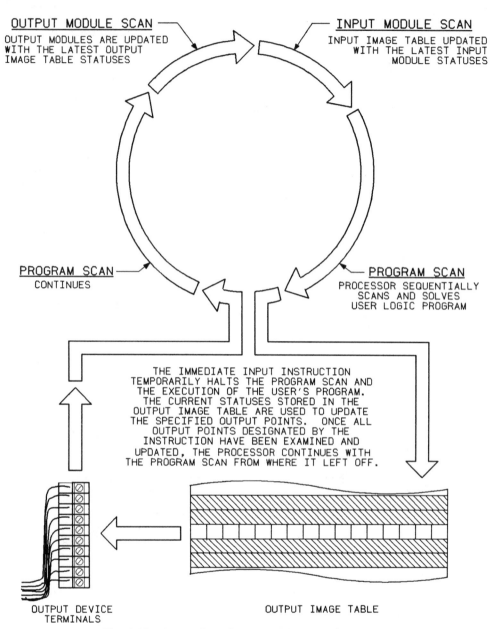

OUTPUT MODULE SCAN
OUTPUT MODULES ARE UPDATED
WITH THE LATEST OUTPUT
IMAGE TABLE STATUSES

INPUT MODULE SCAN
INPUT IMAGE TABLE UPDATED
WITH THE LATEST INPUT
MODULE STATUSES

PROGRAM SCAN
CONTINUES

PROGRAM SCAN
PROCESSOR SEQUENTIALLY
SCANS AND SOLVES
USER LOGIC PROGRAM

THE IMMEDIATE INPUT INSTRUCTION
TEMPORARILY HALTS THE PROGRAM SCAN AND
THE EXECUTION OF THE USER'S PROGRAM.
THE CURRENT STATUSES STORED IN THE
OUTPUT IMAGE TABLE ARE USED TO UPDATE
THE SPECIFIED OUTPUT POINTS.   ONCE ALL
OUTPUT POINTS DESIGNATED BY THE
INSTRUCTION HAVE BEEN EXAMINED AND
UPDATED, THE PROCESSOR CONTINUES WITH
THE PROGRAM SCAN FROM WHERE IT LEFT OFF.

OUTPUT DEVICE
TERMINALS

OUTPUT IMAGE TABLE

Fig. 8.12.   Immediate Output Instruction Operation

Another variable of the immediate I/O update instruction is the number of I/O points that can be updated by the instruction. Many systems require that the instruction specify a single I/O point per instruction. Other systems allow the instruction to update a complete I/O module of points as part of the update sequence. A few processors will update a complete channel of I/O points as part of the instruction. Where it may be necessary to update more than a single I/O point as part of a system's programming, those instructions that allow only single I/O points to be operated upon can get tedious to repeat for each I/O location that requires program scan updating.

Usually the immediate I/O update instructions can be programmed more than once per scan, and often they can address the same I/O locations multiple times within a scan. Note that the immediate I/O update instructions add time to the total scan time of the processor, and that those I/O locations chosen as part of the instruction will be reupdated during the regular I/O scan operation. The user should be aware that numerous update instructions may slow a processor's scan to a point where it might be better to remove all the update instructions and let the processor function according to its normal scan routine.

The NUMA-LOGIC PC manufactured by Westinghouse incorporates the immediate I/O update instruction as part of its instruction set. The format of the instruction is shown in Fig. 8.13. The NUMA-LOGIC instruction updates the I/O modules in pairs beginning with the I/O module that contains the I/O points specified in the instruction. Either discrete I/O points or data I/O points can be updated by the instruction. The instruction does not specify the updated I/O locations as input or output points, but rather updates all input and output addresses specified in the instruction. For example, if the user specified that I/O group 8 was to be updated, the processor updates input addresses 113 through 128 as well as output addresses 113 through 128 since I/O group 8 contains both input and output points with addresses 113 through 128.

JUMP TO LABEL AND JUMP TO SUBROUTINE INSTRUCTIONS

Four instructions which provide programming ease, efficiency, and flexibility are the *jump to subroutine, jump to label, label,* and *return* instructions. The jump to label instruction allows selective jumping between sections of a user's program based on a set of user programmed logic conditions. The jump to label instruction should not be confused with the simple jump or skip instruction discussed earlier in this chapter. The jump or skip instruction simply allows for

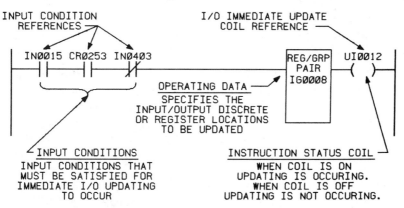

Fig. 8.13. Westinghouse Immediate I/O Update Instruction

selective skipping over portions of the user logic, and not selective jumping to a specified program location from one or more locations within the user program, as is provided by the jump to label instruction. Note that the jump to label instruction can be programmed to work like a simple jump instruction, but the reverse is not true as will be detailed in the following paragraphs.

The jump to subroutine instruction permits the selective jumping to a reserved section of user memory for the purpose of solving a special user developed program entitled a *subroutine*. The jump to subroutine instruction offers the user the ability to conserve memory space by programming repetitive operations once in the subroutine area. These subroutines can then be executed any number of times by the main user program. In some programming applications the jump to subroutine instruction can reduce the processor's scan time.

Both the return and the label instructions are used in conjunction with the jump to label and jump to subroutine instructions. The label instruction can be thought of as the user program target that is jumped to by either the jump to label or the jump to subroutine instruction. The return instruction instructs the processor that the last instruction of the subroutine program has been completed, and that program execution should return to the main user program.

Up until this time the memory area of a PC has been described as being divided into three dedicated zones. Part of the memory is allotted for the storage of I/O statuses, another part stores data values for use in the program, and the third section contains the user programming, as shown in Fig. 8.14. If a PC has the capability of providing subroutines, the user program area of the memory is usually further subdivided into two divisions. The first division consists of the standard user program instructions. A second division is allocated to the storage of program instructions which make up one or more subroutines. During the normal operation of the processor, only the main program area is scanned unless there is a subroutine to be performed. The processor scan would then be directed to continue in the subroutine area by the jump to subroutine instruction. Only at this time would the subroutine area user programming be scanned. Upon reaching a return instruction in the subroutine area, the processor scan would return to the main program area and continue with the main program scan from where it left off. Unless directed by the jump to subroutine instruction, the processor never scans the subroutine area of the memory.

In order to execute a jump to subroutine or a jump to label instruction, a location within the user program must be specified as part of the instruction. In addition the desired location within the main program or subroutine program must be identified if the processor is to know where to jump to. The label instruction provides this identification. It gives a numeric identity for a particular program location. Each label instruction must be unique in its identification since multiple label instructions with identical numerical labels would be confusing to the processor. In addition there may be a limit as to the number of label instruction identification codes that can be used. The label instruction can be used in both the main and the subroutine areas of the user

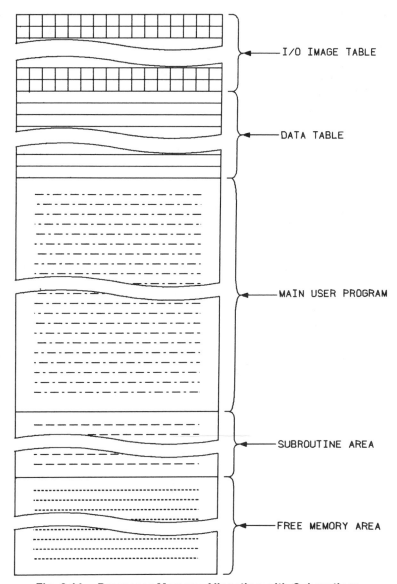

Fig. 8.14.   Processor Memory Allocation with Subroutines

memory since it will be the target of the jump to label and the jump to subroutine instructions.

The Allen-Bradley family of PCs were one of the first to include the jump to label and the jump to subroutine instructions. With the introduction of these instructions, the PC began its evolution into a device that offered traditional relay logic capabilities as well as microprocessor programming power and flexibility.

Whether it is permissible to locate the label instruction in either the main program area or the subroutine area, depends on several criteria. The label instruction cannot be located in the main program areas for a jump to subroutine instruction, nor can it be located in the main program area for a jump to label instruction located in the subroutine area. For a jump to label instruction programmed in the main program area, the label must be located somewhere within that program area, below all jump to label instructions which refer to that particular label instruction, as illustrated in Fig. 8.15. For jump to label instructions, which are located in the subroutine area, the label instruction

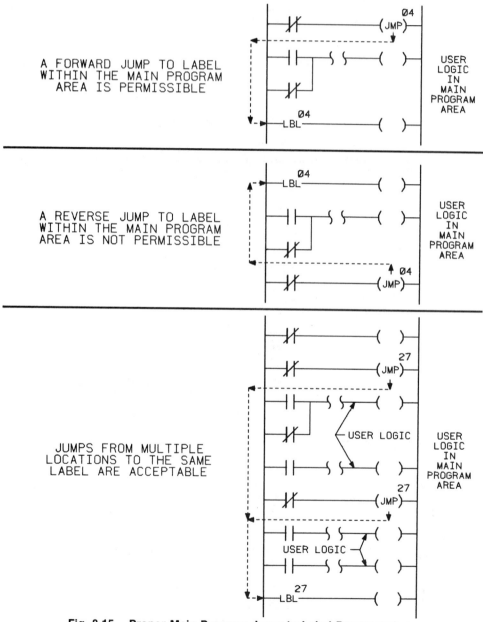

Fig. 8.15. Proper Main Program Jump to Label Programming

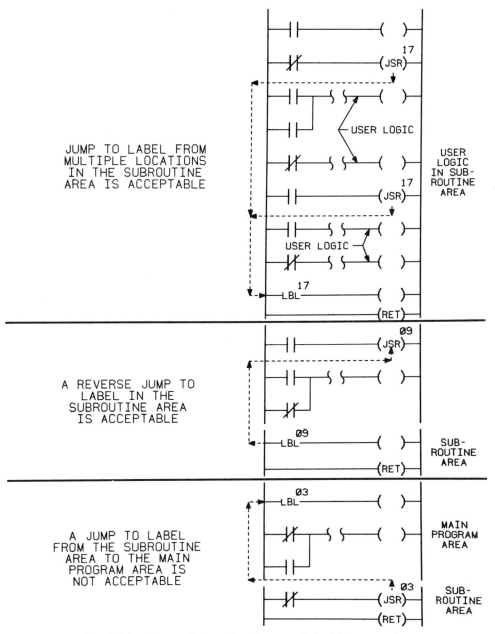

Fig. 8.16. Proper Subroutine Jump to Label Programming

can usually appear anywhere within the subroutine area, as illustrated in Fig. 8.16.

The label instruction associated with a jump to subroutine instruction must be located in the subroutine area of memory. Similar to the jump to label instruction programmed in the subroutine area, the jump to subroutine instruction can refer to a label located anywhere in the subroutine area, but not one located in the main program area. Fig. 8.17 indicates both acceptable and unacceptable locations for the label instruction when associated with the jump to subroutine instruction.

The number of times that a jump to label and/or a jump to subroutine instruction can refer to a particular label instruction is variable, depending on the location of the label instruction and on how it is being used. When the label

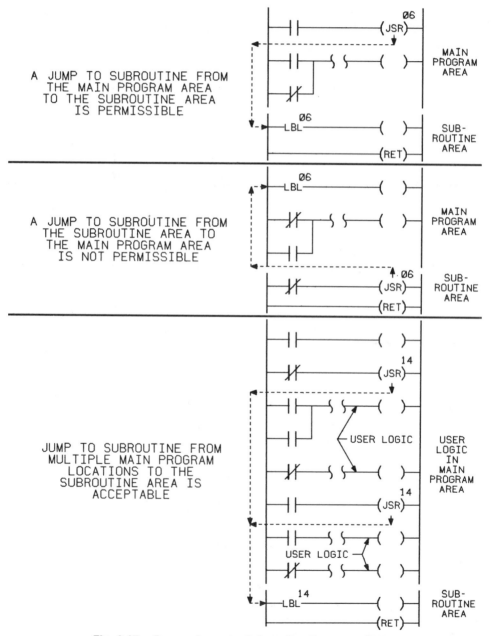

Fig. 8.17. Proper Jump to Subroutine Programming

instruction is located in the main program area, any number of jump to label instructions can reference the same label instruction. As noted earlier, the label instruction will have to be located below all the jump to label instructions which refer to it since jumping backward in the main program area is usually not permitted. Any number of jump to subroutine instructions placed in the main program area can reference a label instruction placed in the subroutine area of the user memory. Any number of either jump to label or jump to subroutine instructions located in the subroutine area can refer to the same label instruction provided that the label instruction is located in the subroutine area of memory.

There is often a limit to the number of times that a group of label instructions can be referenced when they are in the subroutine area. It is often necessary to program multiple *loops* of the same instruction sequence in the subroutine area. This process is called *nesting* and is illustrated in Fig. 8.18. Many processors offering the subroutine instruction permit the subroutine instruction to call itself or other subroutines located in the subroutine area. The number of times the same label instruction, or a number of label instructions, can be "called" is usually dependent on the amount of memory the manufacturer allots the processor for the storage of reference data, such as the user program logic location which originally called the subroutine. Each time the processor interprets a jump to subroutine instruction, it must store internally the location of the jump instruction for later use when it interprets the return instruction indicating that the processor scan sequence should return to the user program from where it originally left. This requirement of logging the originating user logic location uses up memory. The number of times a jump to subroutine instruction can be used without encountering a return instruction is directly dependent on the amount of housekeeping memory allotted for processor use.

The user should be aware that the repetitive calling of subroutines which require long processor times for solution can also lead to problems. Most PC systems require that the I/O structure be updated periodically. In most cases there is a maximum time in which the I/O modules can operate before they must be serviced by the processor. Recalling the standard scanning operations of the processor discussed earlier in this chapter, the I/O is scanned prior to the user logic scan. If the user logic is programmed with multiple calls to a lengthy subroutine, an excessive scan time can develop. If the scan time of the user programming becomes longer than the maximum time permitted by the manufacturer for I/O servicing and refreshing, a processor fault can occur. Reprogramming of the user logic may be necessary to avoid this type of problem.

Use of the jump to label instruction to bypass selected user programming can also result in faulty operation of some instructions that might be jumped over. For example, any timers that might be located in a section of logic that is being jumped selectively could operate incorrectly. When a section of user logic is being jumped, the logic within that area is usually not being solved. Any timer located in that area would not be updated on those scans that bypass the user logic containing one or more timer instructions. In effect the timer would "freeze" at its current time until it is no longer being jumped. When the

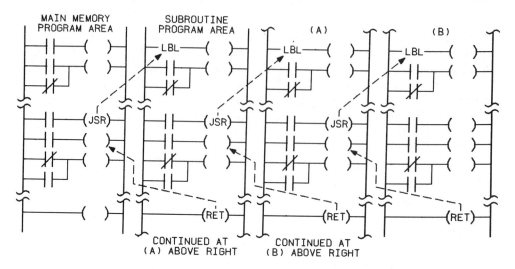

SUBROUTINES CAN CALL OTHER SUBROUTINES
IN THE PROCESSOR SUBROUTINE MEMORY AREA

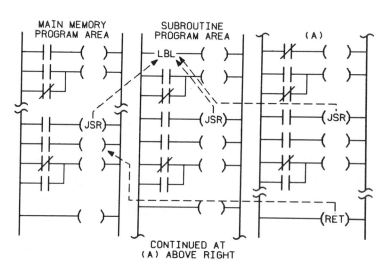

A SUBROUTINE CAN CALL ITSELF A LIMITED
NUMBER OF TIMES IN THE PROCESSOR
SUBROUTINE MEMORY AREA

Fig. 8.18. Subroutine Nesting

processor begins scanning the area again, the timer would begin timing from its last time value.

Since the processor goes directly to the specified label instruction associated with either a jump to label or a jump to subroutine instruction, any programming between the jump instruction and the label instruction will be bypassed. This includes any instructions programmed prior to the label instruction in the same ladder rung as the label instruction. Multiple label instructions may appear in a ladder rung, but the user must be aware of how the rung will be interpreted by the processor.

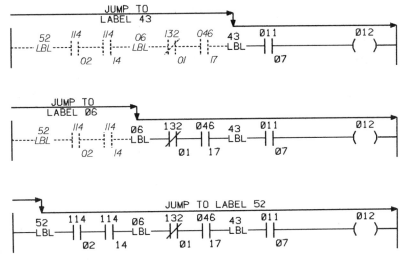

Fig. 8.19. Multiple Label Instructions within a Rung

Fig. 8.19 displays a ladder rung containing multiple label instructions. This rung could appear in either the main or the subroutine areas, and the two labels could be referenced to any number of jump to label or jump to subroutine instructions. Should a jump instruction reference label 43, the instructions to the left would be ignored by the processor, as illustrated in the upper portion of the figure. The only requirement to activate coil 01203 would be that the examine on instruction 01107 be true. Whenever the processor jumps to label 06, illustrated in the middle of the figure, examine instructions 13201, 04617, and 01107 must all be true to activate coil 01203. A label instruction is interpreted by the processor as always active. All the elements of the ladder rung must be true for coil 01213 to be true whenever the jump to label instructions reference label 52, as indicated at the bottom of the figure.

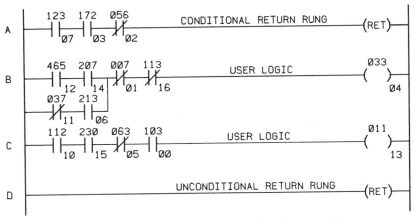

Fig. 8.20. Proper Return Instruction Programming

The return instruction should always be programmed in an unconditional format to ensure proper operation of the subroutine. It is the return instruction that instructs the processor to leave the subroutine area and return to the main user program area, which initially called the subroutine. If the processor interprets a conditionally programmed return instruction as false, it will proceed past the return instruction, looking for an active return instruction, as shown in Fig. 8.20.

When the processor locates the unconditional return instruction, it will return to the area of memory that called the subroutine. If the user only programs conditional return instructions and the processor is scanning the subroutine area, it could get trapped in its scan. Once the processor is in the subroutine area, a return instruction is required for it to exit back to the main program area. If all the return instructions are programmed conditionally, and they all happen to be false, the processor will scan to the end of the subroutine area and stop. A fault would then occur. Extreme caution should be exercised if it becomes necessary to program the return instruction conditionally.

9

Timers and Counters

The *timer* and *counter* instructions are the second oldest pair of PC instructions besides the standard relay instructions discussed in Chapter 7. While most first-generation systems did not include these instructions, they were designed into the majority of second-generation systems. All PCs manufactured today include the timer and counter instructions, but some differences exist between manufacturers in the manner in which a timer or counter is implemented. For the most part timer/counter instructions operate internally within the PC. However, several units require the installation of special hardware timer circuit boards in the processor in order to implement a timer instruction.

There are very few control systems that do not need at least one or two timed functions. The need for a timer can be as simple as providing a startup warning delay of several seconds and as complex as timing the steps of a batch-type process operation. For many years mechanical or clock-driven timers met the majority of industry requirements. With the development of the transistor, solid-state timers became available, offering better accuracies than their mechanical or motor-driven counterparts. While many mechanical timers are still sold today, the electronic version is becoming the timer of choice.

There are numerous styles of industrial timers available to control system designers, and a complete discussion of all available types and their operation is well beyond the scope of this text. However, the more commonly used control system timers have several features in common. Aside from being available in standard voltage ratings and time bases, there are two general classes of timer output contacts, as well as two standard types of timer delays. Many timers offer a retentive operation, allowing them to be used for special applications.

TIMED AND INSTANTANEOUS OUTPUT CONTACTS

Many industrial timers offer both *timed* and *instantaneous* output contacts. The difference in operation between these two classes of output contacts is depicted in Fig. 9.1. Normally when the timer is not activated, the instantaneous and timed output contacts are in their shelf position, regardless of whether they are normally open or normally closed in configuration. Whenever the timer is activated or deactivated, the instantaneous contact output is activated or deactivated in response to the changing input signal. The timed contact will not be activated or deactivated until the timer has waited its preset time delay after timer activation or deactivation. The operation of the output contacts are as described by their class names. The instantaneous output contact changes state immediately upon activation/deactivation of the timer, but the timed contacts transfer state only after the appropriate time interval has passed after timer activation/deactivation.

ON-DELAY AND OFF-DELAY NONRETENTIVE TIMERS

There are two categories of timer delays available for control system use. Fig. 9.2 represents a simplified timing diagram for a timer operating its timed contacts a user specified time interval after having received an input signal. Upon receipt of the input signal, this type of timer, referred to as an *on-delay timer*, begins its timing cycle. The instantaneous contacts on the on-delay timer are transferred upon timer receipt of the input signal. When the timing period is complete, this timer transfers its timed contacts. As its name implies, the on-delay timer transfers its timed contacts from their shelf states some time delay

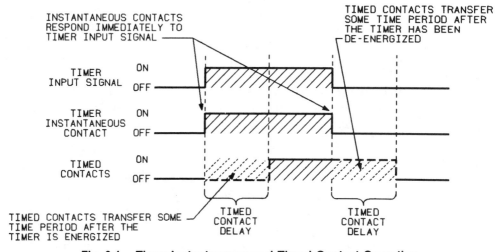

Fig. 9.1. Timer Instantaneous and Timed Contact Operation

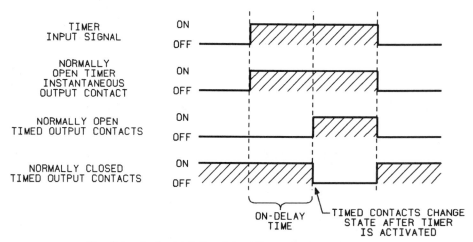

Fig. 9.2. **Industrial On-delay Timer Contact Operation**

period after the timer was actuated or turned ON. Both the instantaneous and the time output contacts of the on-delay timer return to their shelf positions instantly upon removal of the timer input signal.

The *off-delay timer* operates similar to its on-delay counterpart, as indicated in Fig. 9.3. In the case of the off-delay timer, it receives input power continuously. In response to the continuous input signal, the instantaneous and the timed output contacts are in their shelf states. Similar to the on-delay timer, the instantaneous output contact emulates the status of the input signal to the timer. The instantaneous contacts of the off-delay timer transfer to their actuated states immediately upon timer input signal loss and return to their shelf states when the signal to the timer returns. Upon input signal loss, the off-delay timer begins its timing cycle. When the timer's timing cycle has reached completion, the timed output contacts transfer to their actuated states. As with the off-delay timer's instantaneous contacts, the timed contacts return to their shelf states when the input signal returns to the timer.

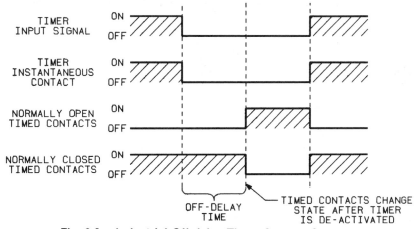

Fig. 9.3. **Industrial Off-delay Timer Contact Operation**

The timers provided with a PC operate similar to those outlined above. The on-delay timer instruction operates as indicated in Fig. 9.4. The instantaneous and the timed contacts referenced to an on-delay timer instruction operate like their hardware timer counterparts. Contact instructions, either normally open or normally closed, which are referenced to the instantaneous output of the timer instruction, transfer state with changes in state of the input conditions to the timer instruction. Likewise, any normally open or normally closed contact instructions referenced to the timed output of the timer instruction operate some user defined time delay period after the timer has been activated, similar to their hardware counterparts.

All PC on-delay timers require a memory location to store the current amount of time elapsed since the timer instruction was activated. The value stored in this location is subject to the timing operation of the timer and changes according to the operation of the timer. As illustrated in Fig. 9.4, this stored value is reset whenever the timer instruction is not activated, and moves toward the timer's preset time delay value whenever the instruction is active. As illustrated, the timed output only operates when the timer has been active for its preset value or longer.

The off-delay timer instruction operates as indicated in Fig. 9.5. The operation of the instantaneous and the timed contacts referenced to an off-delay timer instruction operate like their hardware timer counterparts. Contact instructions, either normally open or normally closed, which are referenced to the instantaneous output of the timer instruction, transfer state with changes in state of the input conditions to the timer instruction. Likewise any normally open or normally closed contact instructions referenced to the timed output of the timer instruction operate some user defined time delay period after the timer has been deactivated similar to their hardware counterparts.

All PC off-delay timers require a memory location to store the current amount of time elapsed since the timer instruction has been activated. The value stored in this location is subject to the timing operation of the timer, and changes according to the operation of the timer. As illustrated in Fig. 9.5, this stored value is reset whenever the timer instruction is not activated, and moves

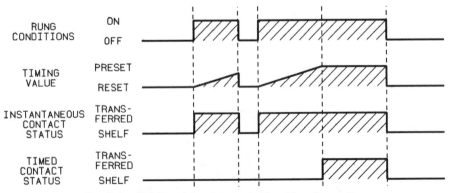

Fig. 9.4. PC On-delay Timer Instruction Operation

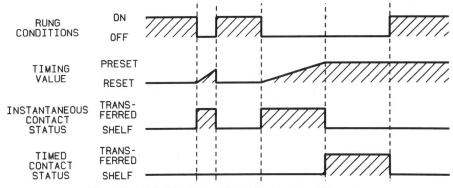

Fig. 9.5. PC Off-delay Timer Instruction Operation

toward the timer's preset time delay value whenever the instruction is active. As illustrated, the timed output operates only when the timer has been active for its preset value or longer.

ON-DELAY AND OFF-DELAY RETENTIVE TIMERS

A retentive operational mode is available with many industrial timers in the event of power loss. Often it is desirable not to have a timer's timed contacts change state immediately upon a change of input state. Both the on-delay and the off-delay industrial timers of Figs. 9.2 and 9.3, respectively, operate in a nonretentive mode. The output contacts for each of these timers are directly affected by the operation of the timer's input signal. The timers of Figs. 9.2 and 9.3 are often referred to as *nonretentive on-delay* and *nonretentive off-delay timers,* respectively.

Fig. 9.6 represents the timing diagram of an on-delay timer similar to the timer of Fig. 9.2. However, the timer of Fig. 9.6 is called a *retentive on-delay timer.* The difference between a nonretentive and a retentive timer is that the

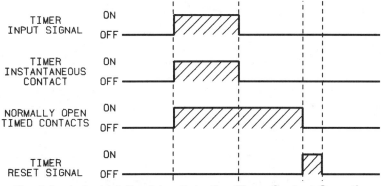

Fig. 9.6. Industrial On-delay Retentive Timer Contact Operation

timed output contacts are permitted to return to their initial states. The nonretentive timer of Fig. 9.2 controls the operation of its timed output contacts in direct response to a changing input signal.

The retentive timer of Fig. 9.6 initially operates its timed contacts in response to the timer input signal. However, a second input signal, commonly referred to as a *reset* signal, is required to return the timer's timed contacts to the pretimer operation state. The retentive timer "remembers" the amount of elapsed time that has passed since it was initially activated, regardless of the timer's input status, unless intentionally reset by a reset signal. A retentive timer retains the amount of timer delay period remaining should the timer input signal change state before it has completed its predetermined timing cycle. When an interruption in timer operation occurs, the timer begins from where it left off in the timing cycle at the next input signal activation. Retentive timers are available both in on-delay types, as shown in Fig. 9.6, and in off-delay types. Note that the instantaneous output contacts of the retentive timer operate in the same manner as the nonretentive timer's instantaneous contacts, transferring in conjunction with the timer's input circuit status.

The retentive on-delay timer PC instruction functions in a way similar to its mechanical counterpart, as illustrated in Fig. 9.7. The retentive timer instruction requires both a control (rung) signal and a reset signal to function. The ON/OFF status of the timer instruction rung controls the timing action of the instruction. A reset signal is required for the reset operation of the instruction.

Whenever the timer instruction is activated by the rung conditions, the timer accumulates time. When it reaches its preset value, any normally open or normally closed contact instructions referenced to the timed output of the timer instruction transfer state. Once transferred, the contact instructions referenced to the timed output of the retentive timer instruction remain in their transferred states until such time as the timer instruction is reset by the reset function. Any loss of control rung power, either part way through a timing cycle

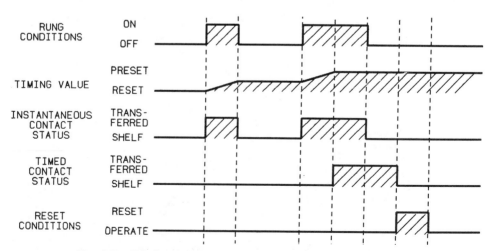

Fig. 9.7. PC On-delay Retentive Timer Instruction Operation

or after completion of a timing cycle, does not cause the timing value of the instruction to be reset. Only a reset instruction can return the accumulated timing value to its initial status. Any normally open or normally closed contact instructions referenced to the instantaneous output of the retentive timer instruction emulate the operation of the timer rung conditions. In addition to the on-delay retentive timer instruction just described, some PCs offer an off-delay retentive timer instruction.

TIMER ACCURACY

The timers available with most PC models give the user the flexibility to implement any type of time delay that may be desired. PC timers are usually crystal oscillator controlled, although several systems use the 50- or 60-hertz ac line frequency as the timer time base oscillator. The most common time base accuracy is 0.1 second. However, most PCs offer a 1.0-second time base as well. Several models permit a 0.01-second time base to be programmed. Use of this time base may place certain restrictions on the use of 0.01-second timers in the PC program due to processor scan time considerations. The cautionary notes included in the programming manuals for PCs offering 0.01-second timers should be understood before using these high-resolution timer types.

Several additional operational characteristics are worth noting for most timer instructions. For most PC timers the reset input overrides the time or control input. If a timer's control input is active as well as its reset function, the timer remains reset and does not time until the reset input becomes inactive.

ELAPSED TIME INDICATION

There are different methods used to indicate the elapsed time of a running timer. Many timers start at zero and accumulate time to their preset value. The time that is displayed is the total accumulated time that the timer has been timing. Activation of the reset function on an *up-count timer,* as they are often called, causes the indicated accumulated time to be reset to zero. The second manner of displaying the elapsed time involves having the timer start at its preset time and time down toward zero. Timers operating in this manner are referred to as *down-count timers.* Activation of the reset function for these timers sets the accumulated value of the timer equal to the preset value for the timer. Whenever the timer is operating, the accumulated time indicated will be the amount of time the timer has to run before it reaches the end of its time delay.

Many PCs permit the user to change the preset value of the timer through mathematics or data manipulation instructions. The flexibility to change the preset time of a timer can be especially handy in many PC applications. The user is warned, however, that care must be exercised when making a timer preset change through either the programming terminal or the user program. If the timer has partially timed through its cycle, and the preset value is set to a value numerically less than the accumulated timer value at the time of the change, erroneous timer operation may occur. The best method to change a timer preset value is to first reset the instruction and then make the required preset time change while the general timer instruction is still held reset.

PC TIMER INSTRUCTION REPRESENTATION

There are two methods used to represent a timer within a PC's ladder diagram program. The first depicts the timer instruction as a relay coil similar to that in Fig. 9.5. The timer is assigned a reference identification as well as being identified as a timer. Several manufacturers add additional information to describe the type of timer, such as on-delay or off-delay. Associated with the relay format of timer representation is the time base of the timer as well as the timer's preset value or time delay period. The final information included as part of the timer symbol is the accumulated value or current time delay period for the timer. The accumulated value for a timer indicates how far the timer has progressed through its time delay period. Whenever the accumulated value equals the preset value for the timer, the timer has reached its timed-out condition, and all timed contacts referenced to the timer instruction have transferred state.

The timer depicted in Fig. 9.8 represents a nonretentive on-delay timer since a reset signal is not included as part of the symbol. The mnemonic "TON" indicates that the instruction illustrated is an on-delay-type timer instruction. Often the mnemonic "TOF" is used to indicate an off-delay timer instruction. Manufacturers offering retentive timers as members of their overall timer instruction set will usually depict them with two relay coils, as indicated in Fig. 9.9. Both the timer and its associated reset coil will have the same reference identification. The reset coil will indicate the preset and accumulated values for the timer it resets to aid program understanding and troubleshooting. The mnemonics "TMR" or "TR" are often used to indicate the reset function for a retentive-type timer.

An area of caution should be noted whenever a coil-type retentive timer is programmed. It involves the locations of the timer instructions and their

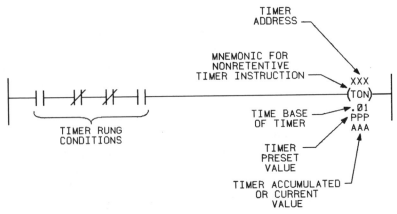

Fig. 9.8. Non-retentive Coil-Formatted Timer Instruction

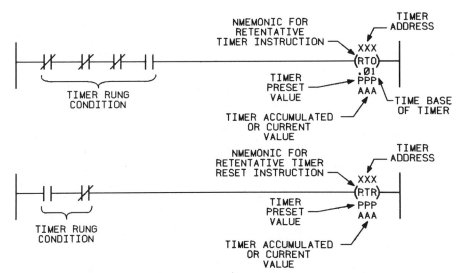

Fig. 9.9.   Retentive Coil-Formatted Timer Instructions

associated reset instructions with respect to the other program ladder rungs which reference the timer's instantaneous and timed outputs. Whenever a timer is programmed to time for a certain preset value and then to reset itself, the timer's reset instruction should be programmed following the last ladder program rung in which a timer output contact (either instantaneous or timed) referenced to the timer in question occurs.

For example, consider a timer that is to time for 60 seconds, increment a counter, then reset itself to start the cycle over again. This type of circuit is often part of the programming necessary to implement an elapsed-time or run-time indicator. Referring to Fig. 9.10, the first instruction programmed (rung A) would be the timer itself. As the processor scans the user program, the accumulated value for the timer will increase by the amount of time that passed since the last time the timer was updated. The counter of rung B is only incremented when the timer reaches its preset value. (Since counters will be

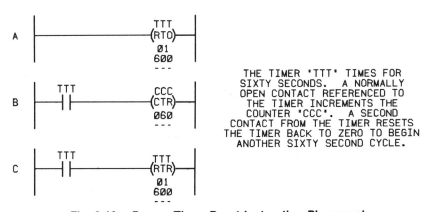

Fig. 9.10.   Proper Timer Reset Instruction Placement

discussed in greater detail later in this chapter, the actual operation of the counter for this example is not significant. The reader should understand that the only function of the counter is to count up one unit every time the timer times out.) The timer reset instruction, rung C, resets the timer each time it times out. Resetting of the timer will occur on the same program scan that the timer reaches its timed-out state. As the example is programmed, the counter is incremented before the timer is reset. If the timer reset instruction had been programmed before the counter (rungs B and C reversed), the counter would never increment, since the timer reset line would reset the timer's output contacts and accumulated time value before the timer's output contacts had the opportunity to increment the counter. Remember that the PC scans the user program instructions sequentially, and since the processor would be instructed to reset the timer immediately after the timer has reached its timed-out state, each time the processor looked for a timed-out timer condition to increment the counter, it would never detect it.

The second timer instruction format is indicated in Fig. 9.11 and is referred to as a *block format*. The block format enables a manufacturer to use one symbol to represent many forms and operations of timers. The timer "block" has two input conditions associated with it. The upper input condition, termed *control line*, controls the actual timing operation of the timer. Whenever this line is true or power is supplied to this input, the timer will time. Removal of power from this input halts the further timing of the timer.

The second input condition for the block-formatted timer is the *reset* function. The reset input resets the timer's accumulated value. It should be noted that there are two conventions used to activate the reset function. Many manufacturers require the reset input to be active, or receiving power flow in conjunction with the timer control input, for the timer to time. Removal of power from the reset input resets the timer to zero (or to its preset value for down-count times). Those remaining manufacturers that do not use this convention do the opposite. They require power flow for the control input

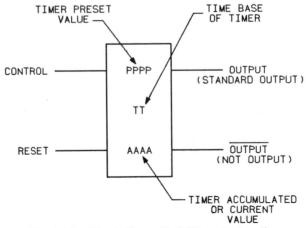

Fig. 9.11. Block-Formatted Timer Instruction

only, and no power flow on the reset input for the timer to operate. For this type of timer operation the timer is reset whenever the reset input receives power. As will be discussed later in this chapter, it is the signal connections to the control and reset inputs of the block-formatted timer that make the timer either nonretentive or retentive.

The timer instruction block for block-formatted timers contains information pertaining to the operation of the timer. The instruction block is identified as a timer, and the time base of the timer is always indicated. Similar to the coil-formatted timer instruction, the block format indicates the preset value as well as the accumulated value for the timer.

All block-formatted timers provide at least one output signal from the timer. However, several manufacturers offer two timer output signals. When a single output is provided, it is used to signal the completion of the timing cycle. For dual-output timer instructions, one timer output signal is activated whenever the timer has completed its timing cycle. The second output signal operates the reverse of the first output signal. Whenever the timer has not reached its timed-out state, the second output is activated while the first output remains OFF. As soon as the timer reaches its timed-out state, the second output is deactivated and the first output is activated.

The location that a block-formatted timer may occupy in a ladder rung is determined by the programming rules of the PC using the block format. Many PC systems require that the timer appear immediately before a coil instruction, as indicated in Fig. 9.12. The identity or reference for timers programmed in this manner is usually the same as the reference assigned to the relay coil instruction(s) operated by the timer. There are PC systems that allow a block-formatted timer instruction to be located anywhere within a program ladder rung, as illustrated in Fig. 9.13. Block-formatted timers, which may be located anywhere within a ladder rung, often permit the user to program other instructions between the timer instruction and the coil instruction, as shown in Fig. 9.13. These instructions do not have to be the relay contacts illustrated, but could be other timers, counters, or applicable PC instructions.

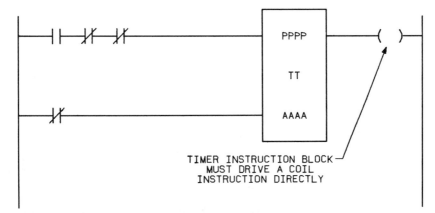

Fig. 9.12. Fixed Timer Program Location

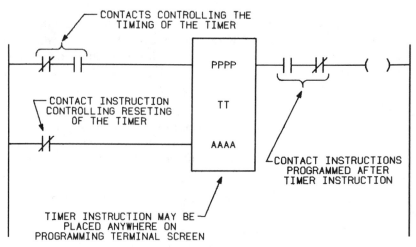

Fig. 9.13. Floating Timer Program Location

PROGRAMMING INSTANTANEOUS CONTACTS

Timers depicted in either the block or the coil format may or may not have an instantaneous output signal associated with them. If an instantaneous output signal is required from a timer, and it is not provided as part of the timer instruction, it may be obtained by programming the conditions associated with the operation of the timer as equivalent instructions operating an internally referenced relay coil instruction first. A contact instruction referenced to the relay (usually normally open) is now used to operate the timer. Instantaneous contacts for the timer can then be referenced to the relay coil instruction, while any time-delayed contacts may be referenced to the timer output coil, as shown in Fig. 9.14.

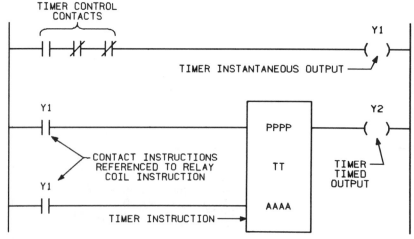

Fig. 9.14. Block-Formatted Timer with Instantaneous Output Programming

The timer instruction format used in Fig. 9.14 is representative of the logic format used in Texas Instruments' TI 530 PC. The TI 530 offers a memory capacity of over 7000 words and a local/remote I/O structure supporting up to 1023 discrete I/O points. Programming of the TI 530 is via Texas Instruments' VPU 200 or VPU 500 CRT programming panels. The TI 530 timer and counter instruction set will be used for all timer examples in this chapter. The TI timers and counters are of the count-down variety. The block format is used by the TI 530 to represent a timer or a counter. It should be noted that even though some of the Texas Instruments' family of PCs use the block format for timer/counter representation, many PCs use the coil format outlined earlier in this chapter. Also many manufacturers use both methods of timer/counter representation. Use of a PC employing the block format for examples of timer and/or counter programming does not infer that either format is preferred. The selection of one model of PC over another is usually determined by cost and hardware comparisons and is generally not based on the manner in which an instruction is depicted on the programming device.

NONRETENTIVE ON-DELAY TIMER PROGRAMMING

Programming a nonretentive on-delay timer is shown in Fig. 9.15. This timer is activated by closing the switch connected to input X1. The preset time for this timer is 10 seconds, after which time output Y1 will be energized. The timing diagram for this example is also shown in Fig. 9.15. When input X1 is activated by the switch closure at A in the timing diagram, the timer begins timing down from its preset value of 10 seconds. After 7 seconds of timing (displayed time equals 3 seconds) the switch is opened at B in the diagram. The only action that results is that the timer is automatically reset back to its preset value (10). Sometime later the timer is reactivated with the switch closure represented at C. The timer again begins its timing cycle, timing down from 10 seconds. At D the timer has been active for 10 seconds (the current time equals zero), and the timer has completed its timing cycle. At this instant the timer activates output Y1 also indicated at D. The timer remains in a timed-out state (the current time remains at zero), with output Y1 activated, until at E the switch is opened. The opening of the switch again resets the timer to its preset value and deactivates output coil Y1. This timer configuration is termed nonretentive since loss of power flow to the timer causes the timer instruction to reset. The timing operation is that of an on-delay timer, since the output coil is activated for a predetermined time after the timer has been activated from the OFF state to the ON state.

RETENTIVE ON-DELAY TIMER PROGRAMMING

The nonretentive on-delay timer can be changed to operate as a retentive on-delay timer by separating the timer's control and reset input signals, such that they operate independently of each other, as illustrated in Fig. 9.16. In this

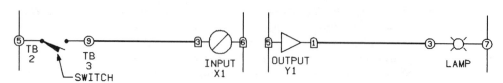

I/O WIRING

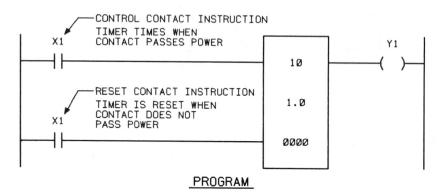

PROGRAM

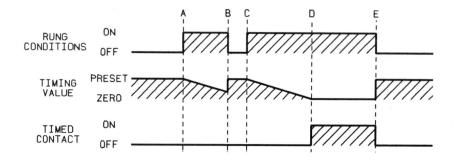

TIMING

Fig. 9.15. Nonretentive On-delay Block-Formatted Timer Example

example two different signals are used to control the timer. A pressure switch is connected to input point X2, and a keyswitch operates input point X3. The operation of the timer is to detect whenever a piping system has sustained a cumulative overpressure or underpressure condition, at which time a lamp is to be illuminated by output point Y2. A keyswitch acknowledges the condition and resets the timer. The timing diagram for this example is also shown in Fig. 9.16.

Operation of the TI 530 timer requires the reset input to be active if the timer is to operate. To meet this requirement the keyswitch has been wired to the

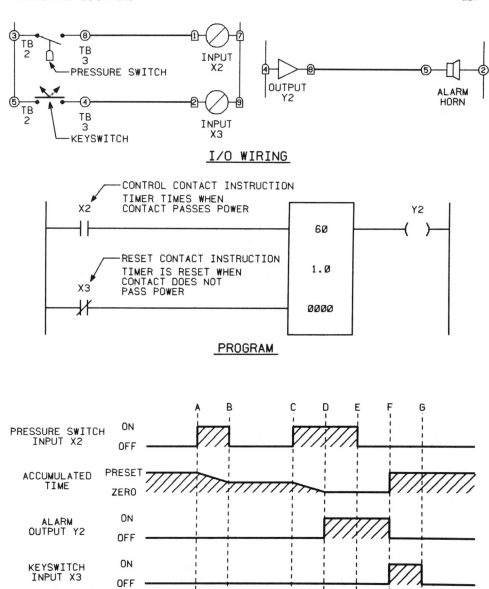

I/O WIRING

PROGRAM

TIMING

Fig. 9.16. Retentive On-delay Block Formatted Timer Example

input point in a normally open configuration. The configuration of the timer instruction with the reset input continuously powered through the normally closed contact instruction and the keyswitch wired normally open to the input point provides for a safer operation. The PC input must receive power in order to reset the timer, and the timer instruction always has the reset input powered due to the normally closed contact instruction. Any time the processor is operating, the normally closed contact instruction will provide a reset signal. Had the keyswitch been wired normally closed, and a normally open contact instruction had been used for the timer reset input any power interruption, even for a few cycles (which may not be enough to shut down the processor, especially if it is on another power source) might inadvertently reset the timer.

Whenever the contacts of the pressure switch close, the control line for the timer is activated via input reference X2. As represented a point A in the timing diagram, closure of the pressure switch initiates the on-delay timing operation. 10 seconds later the overpressure condition is relieved and the control input to the timer is deactivated, as depicted at B. The timer stops timing, but does not reset itself since the reset input to the timer is still active. The timer displays a preset value of 60 seconds and a current value of 50 seconds, indicating that it has operated for 10 seconds. Sometime later at C the condition again activates the timer. Since only 50 seconds remain in the time delay, the timer continues at 50 seconds and times down to zero at D. Since the overpressure condition has not been reduced, the timer times out, and the output lamp connected to output point Y2 is activated. At E, 20 seconds later, the overpressure condition has resided, and input X2 turns off. Since this is a retentive timer, and its reset line is still active, the timer does not reset back to its preset value, and output Y2 is not turned OFF. The timer cannot be reset until the condition is acknowledged at F. When input point X3 is switched ON by the closing of the keyswitch, the timer will be reset to its preset value, and the output lamp connected to output point Y2 will be extinguished. Since the reset input overrides the control input for the TI 530 PC, the timer can be reset back to its preset value at any current time value. The timer does not have to be timed out in order to be reset. The reset line to the timer does have to be active if the timer is to function.

NONRETENTIVE OFF-DELAY TIMER PROGRAMMING

A nonretentive off-delay timer can be implemented as shown in Fig. 9.17. A typical application of the off-delay timer might involve the control of an oven air circulation blower. The blower is to be activated as soon as the heating elements reach a predetermined temperature. Once the heaters are switched OFF, the blower is to continue operation for 60 seconds, allowing the oven to cool below the temperature at which the temperature switch initially operated. Input point X4 is wired to a normally open contact of the temperature switch. Output point Y3 controls the operation of the circulation blower starter. A timing diagram for this example is also shown in Fig. 9.17. The time delay of the timer is set to 60 seconds. Note that output Y3 is programmed as *not coil*. From the chapter dealing with relay programming, the not coil operates in reverse of the standard output coil. Whenever power is not applied to the not coil instruction, contact instructions referenced to the not coil instruction are transferred.

In order to analyze the operation of the off-delay timer it will be assumed that the oven has been in operation for some time, and both the heating elements and the circulation blower are in full operation. In order for the oven to maintain a consistent temperature, the heating elements must cycle ON and OFF. The temperature switch is set near the control temperature for the oven

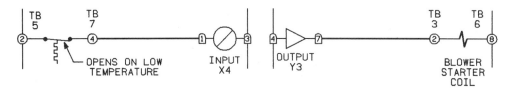

I/O WIRING

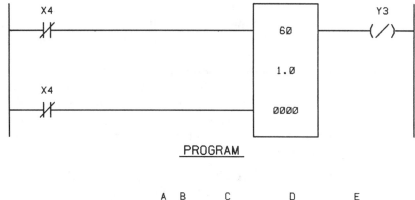

PROGRAM

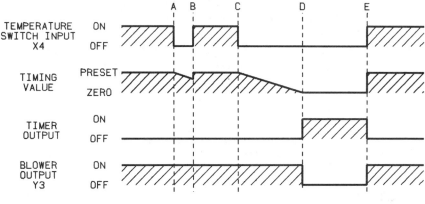

TIMING

Fig. 9.17. Nonretentive Off-delay Block-Formatted Timer Example

such that it senses the normal drop in temperature due to heating element temperature control cycling and deactivates (opens on low temperature) as indicated at *A* in the timing diagram. Whenever input X4 is OFF, the timer will be activated by the normally closed X4 contact references of the control and reset inputs. The timer will start timing down from its preset value toward zero. At *B* in the timing diagram the heaters have come back ON and the temperature switch has detected the increase in oven temperature, causing input X4 to become reenergized by the temperature switch. The timer is reset to its preset value. All during this operation the output from the timer is OFF, which causes the not coil to be active, thereby keeping the blower running.

Sometime later the heaters have been turned OFF, as indicated by a drop in temperature being detected at C. The timer again begins timing down from 60 seconds toward zero. Since the timer is allowed to complete its timing cycle before the heaters are reactivated, output point Y3 is deenergized, as indicated at *D* in the timing diagram. As long as the heating elements remain OFF, the timer remains activated by the normally closed contacts operating its control and reset inputs. As long as the timer is timed out, its output will be ON. The not coil will therefore be OFF. Whenever the oven heaters reheat the oven above the temperature switch's setpoint indicated at *E*, the circulation blower will immediately be started.

It should be noted that the timers of many PCs are retentive through the power-down sequence of the processor. If the timer in the preceding example was timed out at the time of a power-down, the timer would still be timed out when the processor was restarted, provided that the conditions controlling the timer had not changed. For example, if the timer had been timed out when the processor was powered down, the timer would return as timed out on power-up, provided that input X4 was still OFF at power-up. If instead, at power-up, input X4 was ON, the timer would be reset to its preset value on the first scan after the processor was powered up.

Many PCs do not offer the not coil instruction as part of their instruction set. The operation of the circuit in the preceding example may be implemented as indicated in Fig. 9.18, and is preferred by Texas Instruments. A standard coil instruction is used for the timer output. A normally closed contact of this coil is then used to operate output point Y3. The coil instruction is referenced as CR1, indicating that an internal storage or control relay is being used instead of an outputtable coil instruction.

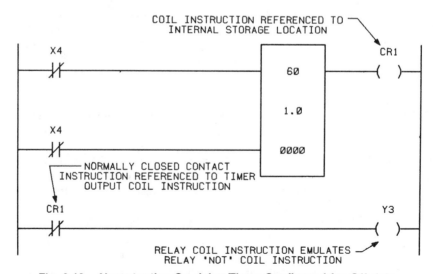

Fig. 9.18. Nonretentive On-delay Timer Configured for Off-delay

RETENTIVE OFF-DELAY TIMER PROGRAMMING

A retentive off-delay timer might be implemented as shown in Fig. 9.19. A field device is connected to input point X5 and causes the off-delay timer instruction to operate whenever power is lost to this input. A second field I/O device is connected to input point X6 for the purpose of resetting the timer. The timer operates not coil CR2, which serves as an internal PC signal that the timer has timed out. The timing diagram for the retentive off-delay timer is also included in Fig. 9.19.

Analysis of the operation of the retentive off-delay timer of Fig. 9.19 begins with both the control input X5 and the reset input X6 energized. At *A* in the timing diagram, input point X5 is deenergized, causing the timer to begin its timing cycle. The timer operates for 10 seconds before input X5 is reenergized at *B*. The timer stops timing and maintains its current time value of 30 seconds. At some later time input X5 again loses power, causing the timer to begin again its timing cycle, as indicated at *C*. At *D* the timer completes its time delay, and as a result internally referenced *not coil* instruction CR2 is energized. Even though input X5 is reenergized at *E*, internally referenced not coil CR2 is still energized since the timer has not been reset back to its preset value. Resetting of the timer to its preset value occurs at *F*, at which time the internally referenced not coil instruction driven by timer CR2 is reenergized.

The retentive off-delay timer, like its cousin the retentive on-delay timer, can be reset at any time during its operation. The retentive off-delay timer does not have to be completely timed out in order to be reset. It should be noted that the reset input to the timer will override the control input of the timer, holding the timer at its preset value even though the control input to the timer is active.

CASCADING OF TIMERS

Often it is desirable to implement a timer that requires a time delay period longer than the maximum preset time allowed for the timer instruction of the PC being used. When this condition arises, it can be solved by simply cascading timers, as illustrated in Fig. 9.20. For this example X7 represents an input signal from some field device. The type of timer programmed for this example is a nonretentive on-delay timer.

Operation of the time delay begins with the activation of the first timer through its control and reset inputs. Since the desired time delay is 120 seconds, the first timer will be programmed as a 90-second timer. When it completes its time delay period, internally referenced coil CR3 will be energized. This action will in turn activate the second timer, which is set for the remaining 30 seconds of the total 120-second time delay. Once the second timer reaches its timed-out condition, internally referenced coil instruction CR4 will be energized to indicate the completion of the full 120-second time delay. The timers will

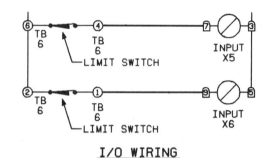

I/O WIRING

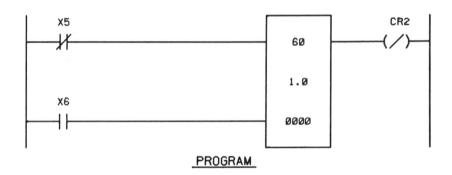

PROGRAM

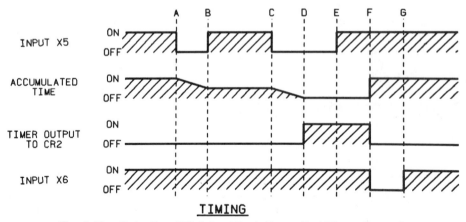

TIMING

Fig. 9.19.   Retentive Off-delay Block-Formatted Timer Example

operate in the nonretentive mode since loss of the X7 input signal to the first timer will reset that timer to its preset value, causing deenergization of CR3. With CR3 deenergized, the second timer is also reset to its preset value, and its output coil will be deenergized.

Similar circuits can be constructed for other types of timers discussed in this chapter when the application requires the timer to have a time delay cycle longer than the maximum allowed for a single timer instruction.

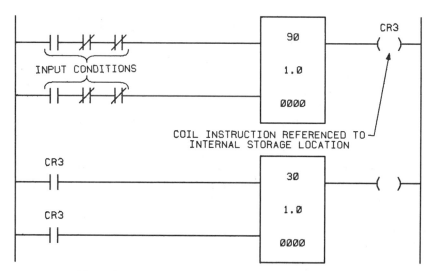

Fig. 9.20.   Cascading of Timers for Longer Delays

Operation of the various types of timers used for any control application will usually involve the proper configuration of the timer examples presented in this chapter. Occasionally the timers presented in this chapter will require additional counter instructions or other basic programming instructions in order to meet the application requirement. An example of a timer and counter combination will be presented later in this chapter. Additional examples involving the use of timers are included in the chapter dealing with applications examples.

WATCHDOG TIMERS

One type of timer that is not user programmable is the *watchdog* timer. This timer is a hardware and software timer built in as part of the processor. It is used to monitor the scanning of the processor and the associated I/O electronics. This timer system is in effect the "heartbeat" of the PC. During the processor scan, at times predetermined by the manufacturer, a specific memory location is switched either OFF or ON. Some finite time later in the scan, the same memory location is set to the opposite state. The ON/OFF state of each of these memory locations (some PCs incorporate more than one for added security) is monitored by dedicated circuits consisting of electronic one-shots. The changing state action of these memory locations serves to keep the one-shots from timing out. Should any one-shot actually time out, a fault is indicated, and the processor is shut down immediately. While these timers are not available for user use, most PCs incorporate them as part of their design.

Many PC users wish to provide additional monitoring of the processor and its associated I/O hardware. One method often employed is to program an oscillator circuit within the user logic area of the PC. This circuit is a simple normally closed contact instruction programmed to operate a coil instruction. Both the coil and the contact instructions are referenced to the same output point. As the processor scans, the output point will toggle ON or OFF. A dc output point is used at the referenced output location to drive the contacts of a motion detector. A dc output point is selected over an ac output point due to the quick response of the dc point. Motion detectors are often used to ensure that a shaft rotates between a pair of user determined speeds, where the shaft's rotation is monitored by a pulse pickup. In lieu of the pulse pickup, the output of the PC is used. Since the PC scan will vary between an upper and a lower limit, the motion detector can be set with an upper and a lower limit outside the PC's actual maximum and minimum scan times. As long as the PC scans, the output will oscillate. As long as the output oscillates, the input to the motion detector is satisfied. Should the output point stop oscillating due to processor shutdown or other problems relating to the circuit (loss of power, output point failure, broken wiring, etc.) the motion detector will trip. The trip signal can then be used for whatever requirements the user has.

COUNTERS

The counter instructions available for use in PC programming exactly emulate the electronic and mechanical counters found in relay-based control systems. While the majority of counters used in industry are probably *up-counters,* numerous applications require the implementation of *down-counters* or of combination *up/down counters.* Practically every model of PC offered has some form of counter instruction included as part of its instruction set. While many PC systems strictly offer the up-counter instruction, increasing numbers of systems are including the down-counter instruction in conjunction with the up-counter instruction. Many manufacturers have combined these two separate instructions into a single instruction, called the up/down-counter.

In conjunction with the counter instruction is a *counter reset* instruction, which permits the counter to be reset. Up-type counters are always reset to zero. However, the down-counter may be set to its preset value, or reset to zero depending on the design of the PC instruction in question. Many manufacturers include the reset function as an integral part of the general counter instruction, while others dedicate a separate instruction to the task of resetting the counter.

Caution should be exercised when implementing a counter, as not all counter instructions "count" in the same manner. Many up-counters count only to their preset values. Additional counts above the preset value are ignored by the counter instruction. Other up-counter instructions permit the counter to keep track of the number of counts received above the counter's preset value. Conversely some down-counters strictly count down to zero and no further. Other down-counters may count below zero, or "wrap around" and begin counting down from the largest preset value that can be set for the PC's counter instruction. For example, a PC offering up/down-counters that have a maximum counter preset limit of 999, may count up as follows: 995, 996, 997, 998, 999, 000, 001, 002,. . . . Conversely the same counter would count down in the following manner: 004, 003, 002, 001, 000, 999, 998, 997, 996, ...

COIL-FORMATTED COUNTER INSTRUCTIONS

As was discussed earlier with regard to the timer instruction, there are two methods used to represent a counter on the programming terminal. Similar to the timer, a coil programming format is employed by many manufacturers, as illustrated in Fig. 9.21. The counter function is represented as a coil containing a mnemonic descriptor, such as "CTU," indicating whether the counter functions as an up- or a down-counter, respectively. Associated with the symbol is an identification reference number used to identify contact instructions programmed elsewhere in the program as operating from the counter instruction. Also displayed as part of the counter symbology are the counter's preset value as well as the current accumulated count for the counter. Most counter instructions permit the user to "preload" the accumulated count to a predetermined value at the time of counter instruction programming. In addition many PCs permit the user program to alter the counter preset and accumulated values as required in order to meet the needs of almost any application.

The coil-formatted counter reset instruction is similar to the coil-formatted timer reset instruction, with the exception that the coil contains a mnemonic identification that labels the instruction as a counter reset function, for example, "CTR" or "CR." The identification number associated with the reset instruction indicates which counter will be reset when the reset instruction is activated.

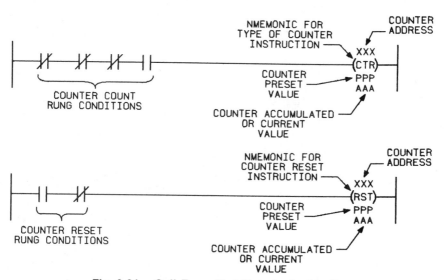

Fig. 9.21.   Coil-Formatted Counter Instruction

BLOCK-FORMATTED COUNTER INSTRUCTIONS

In lieu of the coil format, many PC manufacturers have adopted the block format to represent the counter on the programming terminal. Fig. 9.22

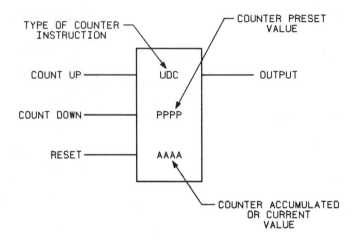

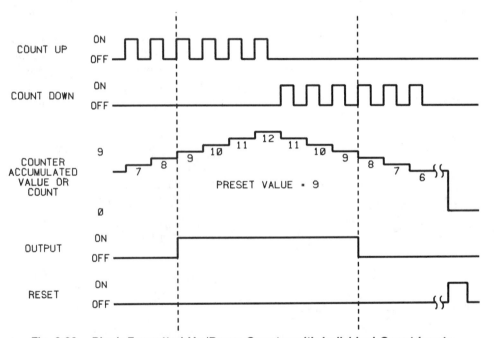

Fig. 9.22. Block-Formatted Up/Down-Counter with Individual Count Inputs

illustrates a typical block format used for an up/down-counter, as well as a function diagram illustrating the counting action of the instruction. Another form of the up/down-counter block format is illustrated in Fig. 9.23. The major difference between Figs. 9.22 and 9.23 lies in how the instruction operates. While both counter representations count identically, one format, Fig. 9.22, has a separate count-up and count-down inputs. As an alternate to this representation, several manufacturers employ a single count input to operate the counter (Fig. 9.23). A second counter input registers whether the counter counts up or down as determined by the absence or presence of that input when the count

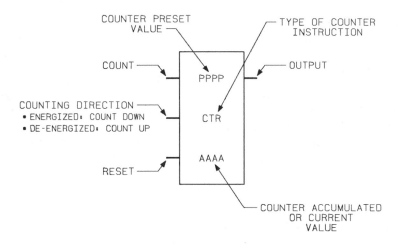

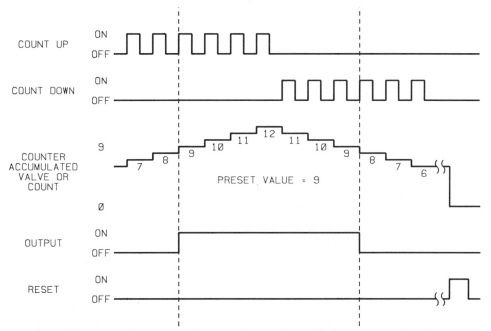

Fig. 9.23. Block-Formatted Counter Instruction with Count Direction Input

input is toggled. Both up/down-counter representations employ a reset input as the third input to the instruction block.

Those manufacturers offering separate instructions for the up-counter and/or the down-counter will usually represent the counter instruction as depicted in Fig. 9.24. The instruction block will indicate whether the counter functions as an up- or a down-counter. The counting diagram for an up-counter represented in this manner is included as part of Fig. 9.24.

All PC counters operate or count on the leading edge of the input signal. The counter will either increment or decrement whenever the input condition to the counter instruction transfers from a zero or OFF state to a one or ON state. The counter will not operate on the trailing edge of the control signal. A one, or ON, to zero, or OFF, transition of the input conditions will not cause the counter to operate in either an increasing (up) or a decreasing (down) count mode.

Care should be exercised with respect to the operation of the RESET signal of a counter. Some counter instructions require the reset input conditions for a block-formatted counter instruction, or the reset instruction of the coil-formatted counter instruction, to be OFF except when it is desired to actually

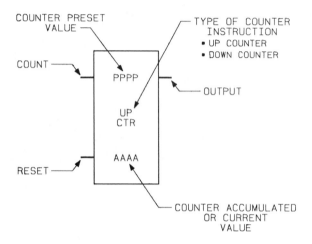

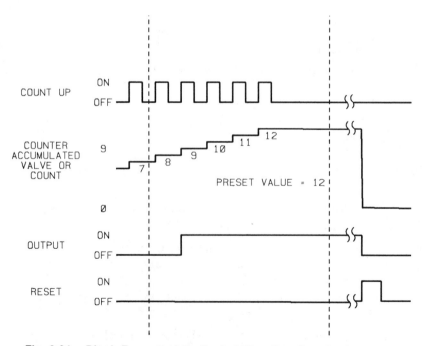

Fig. 9.24. Block-Formatted Dedicated Direction Counter Instruction

reset the counter. Other manufacturers require that the user ensure that the reset input conditions for a block-formatted counter instruction, or the reset instruction for a coil-formatted counter instruction, be active, or ON, for the counter to operate, except when it is desired to actually reset the counter instruction.

Most PC counters are retentive through the power-down and power-up sequence of the processor. Whatever count was contained in the counter at the time of a processor shutdown will be restored to the counter when the processor resumes operation. The counter may be reset, however, if the reset instruction is activated at the time of system power restoration.

In order to illustrate the operation of an up/down-counter, the programming format for the TI 530 PC will again be used. As was indicated earlier in this chapter, the selection of the TI 530 PC does not indicate a preference for the TI method of representing or operating a counter.

UP/DOWN-COUNTER EXAMPLE

As an example of how an up/down-counter might be used, consider an application where it is necessary to store or hold a manufactured product temporarily until quality control tests can be completed. Once the product has been produced, it is placed on a carrier which transports it through other manufacturing steps. The carriers are moved by either an in-floor towline or an overhead power and free monorail system. Before final inspection and packaging of the product, it must be stored for several hours pending the outcome of some quality control tests. Since the product is manufactured in lots, there is no need to sort the product, each lot just needs to be kept separate from the rest.

In order to store the product during the time it takes to perform the quality control tests, a transport conveyer arrangement similar to that depicted in Fig. 9.25 might be used. The PC controlling the transport conveyers must keep track of the number of product carriers loaded into a storage zone, and will divert incoming carriers to the next available storage zone whenever the zone currently being filled is full. Once cleared for final packing, the PC will release product carriers from the specified storage zones, one carrier at a time, until all specified zones are emptied.

Due to the nature of the programming required, and the fact that only the simplest PC instructions have been discussed at this point of the text, the programming necessary to determine which storage zones shall be filled or emptied, and the order in which these zones are to be acted upon, will not be discussed. The selection of which storage zones are available for filling, as well as the order for releasing product from the zones, is simple to accomplish with a PC. However, an understanding of several advanced PC instructions is necessary in order to keep the programming simple. Once the reader has completed reading this text and many of the advanced instructions described herein, he or she may wish to consider how this task might be accomplished with the aid of several of these advanced instructions.

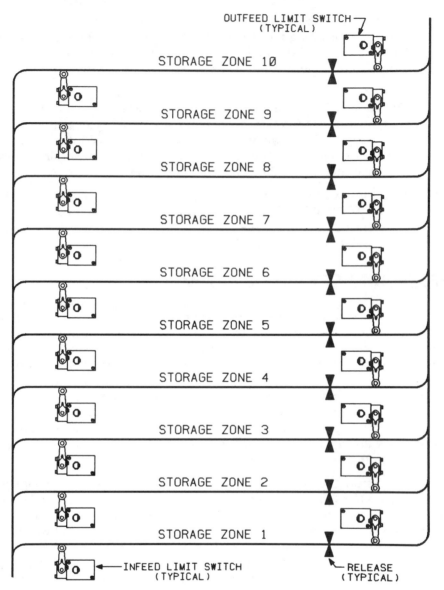

Fig. 9.25. Counter Application Example, Storage Area

Each storage zone is assigned a dedicated up/down-counter to keep track of the number of carriers stored in that particular zone. Each time a carrier is diverted into a zone from the infeed conveyer system, the counter associated with that zone counts up one unit. Release of a carrier from a zone decrements the counter assigned to that zone. Carrier detectors are used to sense the passing of a carrier and increment or decrement the proper counter. An example of the up/down-counter programming that would be used for each storage zone is shown in Fig. 9.26.

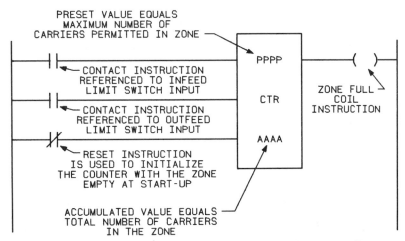

Fig. 9.26. Counter Application Example, Typical Program Rung

COUNTER/TIMER INSTRUCTION EXAMPLE

Many control systems incorporate a time-of-day clock for the logging of data pertaining to the operation of the process. The implementation of a clock as part of a PC's program is straightforward and simple to accomplish. A single timer instruction along with a pair of counter instructions are all that is necessary to implement a 24-hour time-of-day clock.

The three ladder rungs of programming necessary to implement the time-of-day clock are illustrated in Fig. 9.27. A timer instruction is programmed first with a preset value of 60 seconds. This timer times for a 60-second period, after which internal coil CR10 is activated. The energization of CR10 causes the counter of rung B to increment one count. On the next processor scan the timer is reset and begins timing again. The counter of rung B is preset to 60 also. Each time the timer completes its time delay, the counter of rung B is incremented. When the counter reaches its preset value of 60, internal coil CR11 is energized. Energization of CR11 increments the second counter programmed in rung C. Rung C's counter is preset for 24. Note that whenever CR11 is activated, it also resets the counter in rung B to begin the 60-count sequence again. The counter of rung C is designed to operate internal coil CR12. The function of CR12 is to reset the counter of rung C after it reaches the count of 24. The time of day is generated by examining the current count or time for each counter or timer in the three ladder rungs of Fig. 9.27. The counter of rung C indicates the hour of day in 24-hour military format. The current minute is represented by the count value of the counter in rung B. Rung A's timer displays the seconds of a minute as its current time value.

If the 24-hour format is not desirable, the program of Fig. 9.27 can be modified as shown in Fig. 9.28 to show the time in a 12-hour format, along with an A.M. or P.M. indication. The ladder rungs of Fig. 9.27 are duplicated for the 12-hour clock with the exception that the preset value for the counter in rung C is set to 12 instead of 24. A fourth rung is also required for the A.M./P.M.

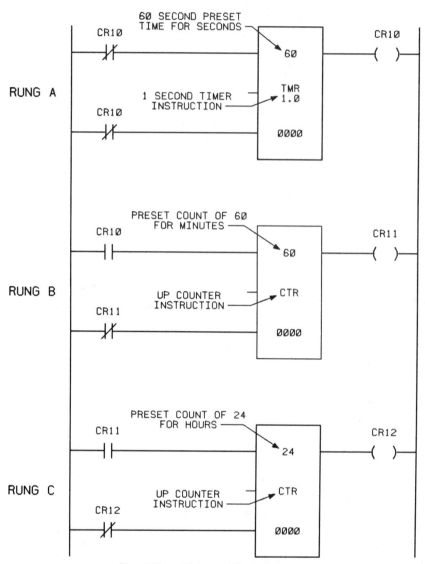

Fig. 9.27. 24-hour Clock Program

indication, as shown. Internal coil CR12 operates the counter of rung D. This counter is programmed to count to two and energize internal coil CR13. The energization of CR13 resets this counter back to zero. Whenever the counter of rung D contains a zero, the 12-hour time indicated by the timer/counters of rungs A to C will be understood to represent the A.M. A one in the counter of rung D will indicate a P.M. time. Note that the counter of rung D never reaches a value of two more than a single scan. This fact allows this counter to indicate the A.M. as a count of zero, and the P.M. as a count of one. Additional examples of timer and counter program applications are contained in the chapter on applications programs.

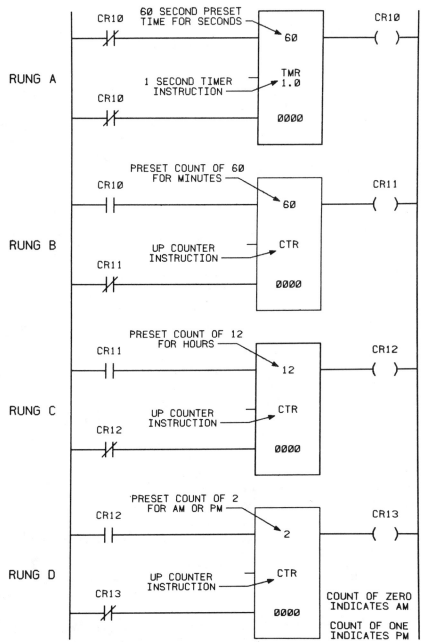

Fig. 9.28. 12-hour Clock Program with A.M. and P.M. Indication

SCAN COUNTERS

One form of counter offered with several PC models is a *scan counter*. The scan counter is activated by standard relay-type instructions, but does not function like the counters described above. The scan counter does not require the OFF-to-ON toggling of the control input to count. Once the control input is activated, the scan counter counts the number of processor scans that occur while the control input is active. The scan counter has a preset value, which is similar to the preset value of a standard timer or counter except that the scan counter activates an output whenever the current number of completed scans equals the scan counter's preset value. The use of the scan counter instruction simplifies the programming necessary to synchronize the operations of a complex program with the processor's scanning. For example, many error-checking routines may be referenced to the scan of the processor. The user program may be designed to detect the result of a complex calculation to be varying within some tolerance level for a number of processor scans before the result of the calculation is used. Scan counters with a preset value of one (1) will operate like a one-shot.

TIMER/COUNTER PROGRAMMING CAUTIONS

One final note before leaving the subject of timer/counter programming concerns the operation of timers and counters that have just been loaded or programmed into a processor. Most PCs record the preset as well as the current values of all timers and counters implemented in a user program when a tape is made of the processor's memory. These recorded values are reloaded into the processor from the tape during the tape load operation. Whenever a tape is loaded into a processor, the program timer/counter current values should be verified to ensure proper system operation. For example, the clock programs described earlier will be recorded with the timer/counter accumulated values equal to the time of day when the tape was made. Unless the tape is reloaded on some later occasion at the exact same time of the day as it was made, the time-of-day calculation will be in error.

Many processors also reset all timer current values to zero when entering the program mode. When the processor is returned to the run mode, the timers begin their time delay at their preset values instead of where they left off. While this action usually only occurs with nonretentive timers, the user should verify the operation of the PC's timers prior to programming in order to avoid later problems and/or confusion.

Finally, the user should know the format used to store a preset or accumulated value within the processor's data memory. Many processors store these values in a BCD number format, while others use a binary format. The actual format used may be of importance for some applications, especially where manipulation of either a preset or accumulated value must be performed by additional user programming. The user is cautioned to check how timer/counter preset and accumulated values are stored within the processor prior to programming. Information relating to the various number systems and formats commonly used with PCs is contained in the following chapter.

10

Equality
and Mathematics

When PCs were first introduced, their instruction set was limited to relay, latch, timer, and counter functions. As the PC gained increased acceptance in the industrial marketplace, additional demands and applications began to tax the limited capabilities of the first- and second-generation units. In order to meet the increasing demands, math functions began to appear as part of the instruction sets for the third- and fourth-generation systems. The ability to perform addition and subtraction became commonplace along with the ability to perform the comparison functions of less than, equal to, and greater than. Multiplication and division were not far behind in rounding out the mathematical set of instructions. Today these functions are usually offered as standard features with double precision and floating-point capabilities being standard in larger top-of-the-line systems.

EQUALITY AND INEQUALITY

The functions for mathematical comparison come in many forms, depending on the manufacturer's system design. In some form or another every manufacturer who includes math instructions as part of the standard instruction set includes the capability to compare two numbers for equality or inequality. The ability to compare is provided through the use of the three basic functions of *less than, equal to,* and *greater than.* While many manufacturers provide these three instructions directly, others combine them in various arrangements to provide the same capability of comparing two numbers for equality or inequality.

255

A PC offering the math equality functions is Square D's SY/MAX-300 PC. Four basic instructions are offered, as shown in Fig. 10.1. The *equal* instruction examines two data memory locations for equality, while the *not equal* instruction provides an indication of inequality. The *greater than or equal to* instruction examines a data memory location's contents for being numerically greater than or equal to the contents of a second data memory location. A final instruction, the *less than* instruction, provides an indication that the contents of a selected data memory location are numerically less than the contents of a second data memory location. With these four instructions any desired comparison can be performed simply by the proper combination of instructions and numerical values.

In conjunction with the equality instructions, the SY/MAX PC utilizes either the conjuction *if* or the transitive verb *let* to indicate whether an *equality compare* or a *data move* is to be performed. The two equality instruction blocks for the SY/MAX controller are illustrated as part of Fig. 10.1. The compare instruction compares two data memory locations for a specified equality or inequality, and indicates whenever the comparison is true. The data move (transfer) instruction can be used in two ways. First it can be used to set a data location equal to a fixed value. A second use for the data move instruction is to set the contents of a specified data memory location equal to the value contained in a second data memory location.

Fig. 10.2 indicates the necessary SY/MAX-300 instructions for comparing a variable number against a constant number for a condition of either equality or inequality. Whenever the values contained in data memory locations S11 and S12 are equal, coil instruction 2-02 will be ON. Likewise, whenever the values contained in memory locations S15 and S33 are not equal, coil instruction 2-11 will be ON. Both the equal and the not equal instructions will be performed only when they receive power through contact instructions 1-01 or 1-02, respectively.

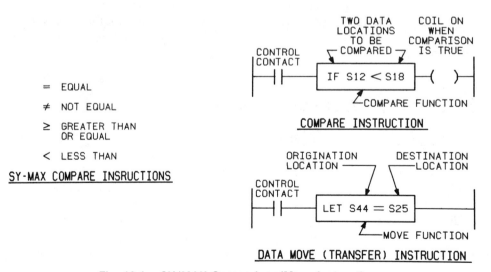

Fig. 10.1. SY/MAX Comparison/Move Instructions

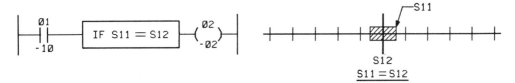

EQUAL TO INSTRUCTION

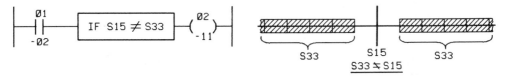

NOT EQUAL TO INSTRUCTION

Fig. 10.2. SY/MAX Equal To and Not Equal To Instructions

Often the need arises to examine a number's magnitude relationship with a numerical value stored in another memory location. When the prime concern is that the content of one memory location is either strictly less than or greater than another numerical value located in another data memory address, the functions in Fig. 10.3 might be employed. This figure illustrates how two data memory locations, S66 and S21, can be compared for less than and greater than status. The value placed in data memory location S66 is a fixed value to which the content of data memory location S21 is to be compared. The upper portion of the figure illustrates the instruction configuration required to test for the content of S21 being numerically larger than the value in S66. Coil 2-09 will be ON whenever this condition exists. The lower portion of the figure depicts the same inequality function, less than, being used to indicate whenever the content of S21 is less than the value in S66. The SY/MAX PC, like many other available PCs, does not provide both a dedicated greater than instruction and a

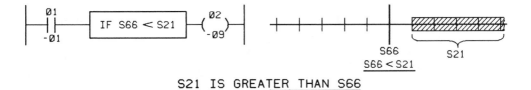

S21 IS GREATER THAN S66

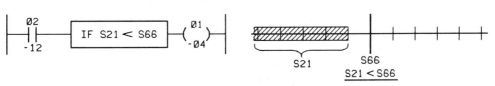

S21 IS LESS THAN S66

Fig. 10.3. SY/MAX Greater Than and Less Than Instructions

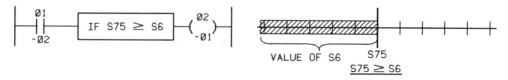

S6 GREATER THAN OR EQUAL TO S75

S6 LESS THAN OR EQUAL TO S75

Fig. 10.4.   SY/MAX Greater Than/Less Than or Equal To Instructions

dedicated less than instruction. Instead, a single instruction is provided, and the user can interchange the two data memory locations as required to generate the necessary greater than or less than function. Whenever the less than condition is met in the example figure, coil 1-04 will be activated. The contact instructions 1-01 and 2-12 permit the greater than and less than comparisons to occur.

When the numerical value in question can be equal to *and* either less than or greater than a second numerical value, the simple programming of Fig. 10.4 might be used. This figure illustrates the instruction programming necessary to provide an indication of either greater than or equal to, or less than or equal to. Again, the SY/MAX PC provides only the greater than or equal to function. The proper placement of the two data memory locations to be acted upon determines whether the operation will be greater than or equal to, or less than or equal to.

Another numerical comparison that is often required involves evaluating a numerical value for its inclusion or exclusion within a specified numerical range. Fig. 10.5 shows the programming necessary to determine whether a value falls inside or outside a specified range of numerical values. Programming to provide this function may require the use of multiple comparison functions. Note that it may also be necessary to program multiple ladder rungs in order to achieve the proper comparison information as indicated in Fig. 10.5.

Fig. 10.6 displays several examples of various comparison instructions combined together to perform a complex comparison. Rung A consists of a pair of greater than or equal to instructions which energize coil 1-06 whenever the numerical value in data memory location S69 is greater than or equal to the numerical contents of S23, but less than or equal to the numerical contents of S31. Ladder rung B provides an indication via coil instruction 1-07 that the content of S69 is greater than the numerical value stored in S45. Finally the coil instruction of rung C, coil 1-08, provides an indication that the value is S69 is greater than the numerical value stored in S45, or that the content of S69 is in the range between the values contained in S23 and S31.

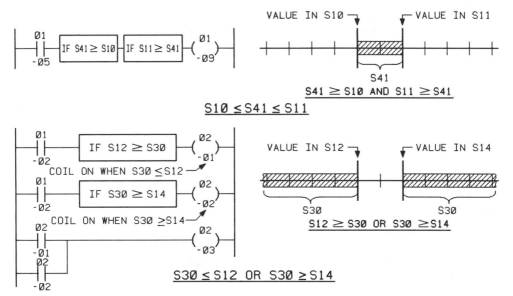

Fig. 10.5. **Numerical Zone Comparisons**

Use of the comparison instructions is generally straightforward in application. However, one error very common in occurrence involves the use of comparison instructions for applications where a raw material or product is being loaded into a container or vessel. The receiving container or vessel is continuously weighed, or a level indicator is being continuously monitored by the PC as the vessel or container fills. When the weight or level reaches a preset value, the supply is to be cut off. While the vessel or container fills, the PC performs a comparison between the vessel or container's current weight or

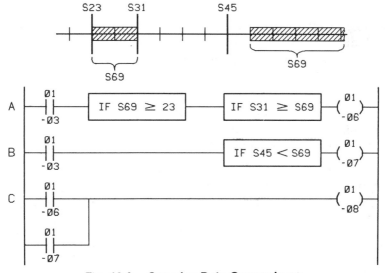

Fig. 10.6. **Complex Data Comparisons**

level and a predetermined constant programmed in the processor. If the programmer uses only the equality instruction, problems may result. As the vessel or container fills, the comparison for equality will be false. At the instant the vessel or container's level or weight reaches the desired preset constant of the equality instruction, the instruction becomes true. The usual control action is then to stop the flow of product to the container or vessel.

The problem arises when the supply system "leaks" additional product to the vessel or container. The equality instruction could go false again if there is sufficient leakage to raise the vessel or container's weight above the constant value set in the processor's memory. Once the instruction goes false, this time from the vessel or container being overweight or overfilled, it could instruct the supply system to restart the filling operation. Unless this is discovered in time, the vessel or container could overfill. The simplest solution is to program the comparison instruction as greater than or equal to. This way any excess product entering the vessel above the preset limit will not affect the filling operation. It may be necessary, however, to include additional programming to indicate a serious overfill condition.

It should be pointed out that several PC manufacturers do not provide dedicated equality/inequality functions as illustrated for the SY/MAX PC. Instead, they combine the equality/inequality functions as part of another instruction or instructions. For example, one model of PC uses various status indicators associated with the subtraction instruction to provide the equality/ inequality function. Another PC line provides the independent equality/ inequality functions; however, they must be used in conjunction with any of several other programming instructions in order to operate. While it will be the user's opinion as to his or her preference for implementing an equality/in-equality function, most PCs provide these functions in some form or another.

MATHEMATICS

The standard math operations available for the majority of PCs on the market today include the ability to add, subtract, multiply, and divide. Special functions, such as square and square root, are offered with several processors, while fewer yet offer such transcendental functions as sine, cosine, or log.

The size of the numerical data that can be manipulated varies between manufacturers, with the most common ranges being 0–999 and 0–9999. Systems that allow for multiplication and division usually allow these functions to handle numbers as large as 999,999 or 99,999,999.

While most systems allow only integers to be used for the math operations, they do permit decimal and/or fractional remainders to be provided as part of the solution for the division function. The ability to indicate a negative result in subtraction is normally available; however, additional programming may be necessary to retrieve this information.

The SY/MAX-20 PC offers four math operations in addition to the math comparison instructions previously discussed. These instructions include *addi-*

tion, subtraction, multiplication, and *division.* The format for each of these instructions is shown in Fig. 10.7. Control of the math functions for the SY/MAX-20 PC is the same as is standard in the PC industry. Execution of the function requires that continuity be present on the control line to the instruction block if the instruction is to be performed. For every processor scan for which the control line of the instruction is true, the instruction will be performed. Where it is desirable to perform the instruction only once, the use of a one-shot or transitional contact instruction must be used. Resetting and refiring of the one-shot or transitional contact instruction will cause the mathematics instruction to repeat its operation.

SYMBOL	FUNCTION	OPERATING RANGE		
+	ADDITION	5 PLACE AUGEND	+ 5 PLACE ADDEND	= 5 PLACE SUM
—	SUBTRACTION	5 PLACE MINUEND	— 5 PLACE SUBTRAHEND	= 5 PLACE DIFFERENCE
×	MULTIPLICATION	3 PLACE MULTIPLICAND	× 5 PLACE MULTIPLIER	= 5 PLACE PRODUCT
/	DIVISION	5 PLACE DIVIDEND	/ 5 PLACE DIVISOR	= 5 PLACE QUOTIENT & 5 PLACE REMAINDER

MATH INSTRUCTION RANGE: - 32,767 TO + 32,767

SY/MAX MATH INSTRUCTION RANGES

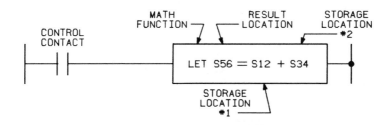

SY/MAX MATH INSTRUCTION FORMAT

Fig. 10.7. SY/MAX Mathematics Instructions

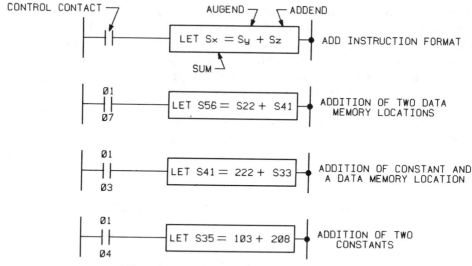

Fig. 10.8. SY/MAX Addition Instruction

The addition instruction for the SY/MAX-20 is shown in Fig. 10.8. The top part of the illustration depicts the general format for the SY/MAX addition instruction. The remaining portion of Fig. 10.8 illustrates how the addition instruction can be used to add together the numerical data contained in two data memory locations or two numerical constants, or how the addition instruction can be used to add the numerical value contained in a data memory location to a numerical constant.

Where a PC can add only numbers that have a result less than 999 or 9999, a double-precision addition program can provide the necessary results. A double-precision addition routine is illustrated in Chapter 19.

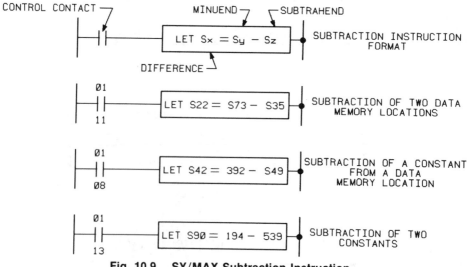

Fig. 10.9. SY/MAX Subtraction Instruction

The subtraction instruction is shown in Fig. 10.9. It is similar to the addition instruction in appearance, with the exception of the sign change. Control of this instruction is exactly the same as that of the addition instruction. A routine for performing double-precision subtraction is also provided as part of Chapter 19.

The multiplication instruction is shown in Fig. 10.10. The SY/MAX PC can multiply any two numbers together provided the product does not exceed 32,767. Should the product exceed 32,767, a flag bit is set by the controller to indicate this condition.

Fig. 10.11 illustrates the format and several examples of the division instruction. The user should be aware that the quotient of a division consists of an integer number and a remainder. Depending on the PC manufacturer, the remainder will be expressed as either a decimal number or a whole number. While the SY/MAX PC records the remainder of a division operation as a whole number, the user should be aware of the format in which a remainder is represented. The result of 30 divided by 12 can be displayed as either a 2 and a 6 (6 being the whole number remainder of the division) or a 2 and a 5 (5 being the decimal equivalent of ½).

GET/PUT AND PULL/PUSH

Before leaving the subject of mathematical instructions, two specialty or semimath instructions need to be discussed. It is often necessary to move the numerical contents of a particular memory location to another memory location. Depending on the PC manufacturer, the movement of single memory locations of numerical data can be accomplished in either of several ways.

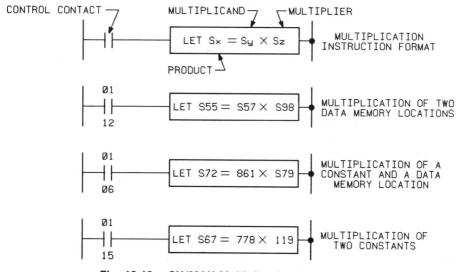

Fig. 10.10. SY/MAX Multiplication Instruction

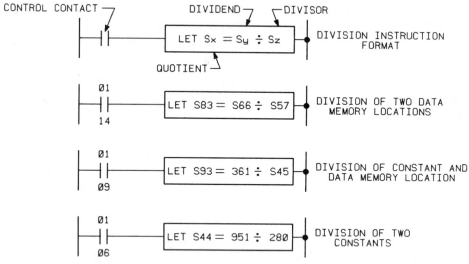

Fig. 10.11. SY/MAX Division Instruction

First, a simple addition instruction can be used. The numerical content of the memory location from where the data are to come from is added to zero (0) through the use of a math addition instruction. The results of the addition are then stored in the data memory location to which the numerical data are to be moved. In effect, the numerical contents of a specified data memory location are placed in a second specified location through the use of an addition instruction.

Use of the addition instruction has drawbacks in that the instruction may have to be reset before it can be reactivated to move the numerical contents of a specified data memory location to another location. This may mean that data cannot be transferred every processor scan, or at least on those scans that may require the transfer to occur. Furthermore the addition instruction may require several words of memory to implement. As an alternate approach, several manufacturers have developed the *get/put* and *pull/push* instruction pairs.

These instruction pairs permit the user to move the numerical contents of a specified data memory location to another location without the use of the addition instruction. The get and pull instructions of the instruction pairs retrieve the numerical data from the data memory location specified by the instruction. The put and push instructions of the respective instruction pairs then place the numerical data in the data memory location specified as part of the instruction. The movement of data from one location of the data memory to another is accomplished by getting (*pull*ing) the numerical data from the specified location, and then *putt*ing (*push*ing) them into the specified destination location.

The get or pull instructions are usually represented as normally open contact instructions, while the put or push instructions are usually represented by coil-type instructions. Both the get/put and the pull/push instruction pairs will have special reference configurations as part of their respective normally open or coil instruction formats to indicate the specified data memory location used by the instruction, and the numerical data currently contained in that location.

NUMBERING SYSTEMS

No discussion on mathematics would be complete without a discussion on numbering systems. There are five numbering systems found in common use with PCs. These numbering systems are decimal, binary, octal, hexadecimal, and binary-coded decimal (BCD). Depending on the particular PC, any one or more of these systems may be used.

The numbering system used for a particular PC function can vary depending on the PC in question. For example, a PC can label its I/O points in the decimal system and record numerical data within the data memory in the binary system. The programming terminal may have the ability to display a stored value in several number systems, either one system at a time or in multiple systems at a glance. It will be advantageous to the PC user to understand the number system used by a particular PC, as well as how to convert between the various numbering systems.

The decimal numbering system uses a number set consisting of 10 elements or digits. Each of the 10 digits is unique, and is represented by the symbols 0, 1, 2, 3, 4, 5, 6, 7, 8, and 9. Any number expressed in the decimal numbering system can be divided into individual subdivisions or places, each subdivision being a power of 10. The value of the decimally represented number is determined by multiplying each digit by its corresponding power of 10 and adding these results together. Fig. 10.12 shows the number 49,365 expressed in a decimal manner.

The binary numbering system uses a number set that consists of two digits. These digits are zero (0) and one (1). Each digit of the binary number system represents a particular power of 2. The decimal equivalent of any binary number can be computed by multiplying the binary digit by its corresponding power of 2 and adding up the results. Fig. 10.13 illustrates the binary equivalent of the decimal number 49,365.

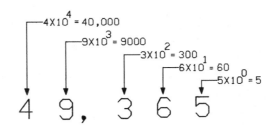

$$-4 \times 10^4 = 40,000$$
$$-9 \times 10^3 = 9000$$
$$-3 \times 10^2 = 300$$
$$-6 \times 10^1 = 60$$
$$-5 \times 10^0 = 5$$

4 9, 3 6 5

```
THE DECIMAL NUMBERING SYSTEM IS
  BASED ON VARIOUS POWERS OF TEN.
THE VALUE OF A DECIMALLY REPRESENTED
NUMBER IS DETERMINED BY MULTIPLYING
EACH DIGIT BY ITS CORRESPONDING POWER
   OF TEN, AND SUMMING THE RESULTS.
```

Fig. 10.12.   Decimal Numbering System

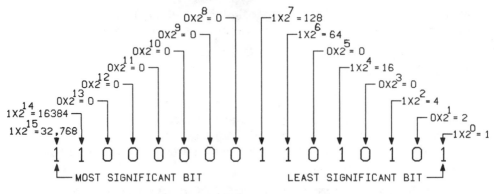

1100000011010101 BINARY EQUALS 49,365 DECIMAL

THE BINARY NUMBERING SYSTEM IS BASED
ON VARIOUS POWERS OF TWO.   THE DECIMAL
EQUIVALENT OF A BINARY NUMBER CAN BE
COMPUTED BY MULTIPLYING THE BINARY
DIGIT BY ITS CORRESPONDING POWER OF
TWO, AND SUMMING THE RESULTS.

Fig. 10.13. Binary Numbering System

It is the binary number system that is employed by the microprocessor of all PCs and computer/microprocessor-based equipment. The two digits of the binary system, one and zero, easily correspond to the two possible states of a digital signal. Generally the OFF or FALSE state or condition is represented by the binary digit zero (0). The ON or TRUE state or condition is represented by the binary digit one (1). The ON/OFF state of any signal can be represented by the binary numbering system digits. In addition any numeric data can be represented in a binary form as outlined above. Once in the binary form, the data or digital states can be logically compared and operated on, as well as manipulated by digital means. For example, if the number 3 is to be doubled (decimal 3 equals 011 in binary), the binary digits representing the number 3 can be shifted to the left one position to produce the result of 6 (decimal 6 equals 110 in binary). This is in fact a simplified example of how any digital device can use a specially designed algorithm to manipulate information that has been stored in a binary format.

The octal numbering system consists of eight digits, 0 through 7. It assigns each digit position a power of 8, as illustrated in Fig. 10.14. The decimal equivalent of any number represented in an octal format is obtained by multiplying each octal digit by the power of its place and summing the results.

Occasionally the hexadecimal numbering system is encountered when working with digital devices. The hexadecimal system incorporates 16 digits represented by the numbers 0 through 9 and the letters A through F. The decimal equivalent of any hexadecimal number can be calculated by multiplying each digit of the hexadecimal number by the power of its position, a power of 16, and summing the results, as indicated in Fig. 10.15.

The final numbering system commonly used by computers, microprocessors, and PCs is the binary-coded decimal numbering system. This system relies on

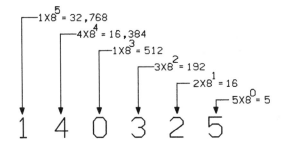

$$1X8^5 = 32,768$$
$$4X8^4 = 16,384$$
$$1X8^3 = 512$$
$$3X8^2 = 192$$
$$2X8^1 = 16$$
$$5X8^0 = 5$$

1 4 0 3 2 5

140325 OCTAL EQUALS 49,365 DECIMAL

THE OCTAL NUMBERING SYSTEM IS BASED
ON VARIOUS POWERS OF EIGHT.   THE
DECIMAL EQUIVALENT OF AN OCTAL NUMBER
CAN BE COMPUTED BY MULTIPLYING THE
OCTAL DIGIT BY ITS CORRESPONDING
POWER OF EIGHT, AND SUMMING THE RESULTS.

Fig. 10.14.   Octal Numbering System

groups of four binary digits to represent each digit of a decimal number. Four binary digits are used to represent the decimal numbers 0 through 9, as indicated in Fig. 10.16. Any decimal number is broken down into individual four-digit binary equivalent numbers for each of its decimal digits.

NUMBER SYSTEM
CONVERSION INSTRUCTIONS

The majority of PCs available display numerical data on the programming terminal in a decimal format. Howver, they may actually store the displayed data and manipulate data within the processor's memory in either a binary or a BCD format. In addition, numerical data may be input to or output from a PC in either a binary or a BCD format. To ensure that the PC user does not have to write software programs to convert binary data to BCD-formatted data and vice versa, most PC manufacturers provide an instruction or instructions that simplify the conversion of numerical data between the binary and BCD number systems.

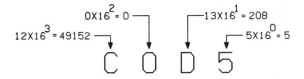

$$0X16^2 = 0$$
$$13X16^1 = 208$$
$$12X16^3 = 49152$$
$$5X16^0 = 5$$

C O D 5

COD5 HEXADECIMAL EQUALS 49,365 DECIMAL

THE HEXADECIMAL NUMBERING SYSTEM IS
BASED ON VARIOUS POWERS OF SIXTEEN.
THE DECIMAL EQUIVALENT OF A HEXADECIMAL
NUMBER CAN BE COMPUTED BY MULTIPLYING
THE HEXADECIMAL DIGIT BY ITS
CORRESPONDING POWER OF SIXTEEN,
SUMMING THE RESULTS.

Fig. 10.15.   Hexadecimal Numbering System

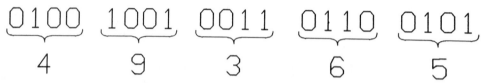

BCD	
0000	0
0001	1
0010	2
0011	3
0100	4
0101	5
0110	6
0111	7
1000	8
1001	9

0100 1001 0011 0110 0101

4   9   3   6   5

Fig. 10.16. Binary-coded-decimal Numbering System

There are two number system conversion instructions required to change numerical information from one system to another. The first of these instructions is the *binary to BCD conversion* instruction. As illustrated in Fig. 10.17, the binary to BCD conversion instruction converts numerical data stored in a binary number system format into an equivalent number represented in the BCD number system format. Many PCs store and manipulate numerical data in a BCD format within the processor. Data being received by the PC from an external device may be in a binary format. The use of the binary to BCD conversion instruction will permit the user to convert the incoming numerical data from a binary format to a BCD format prior to use and/or manipulation by the processor.

Once data have been manipulated or generated by a PC that uses a BCD format, it may be necessary to convert the data into a binary format for use by a device driven by the PC. This operation can be performed by using a *BCD to*

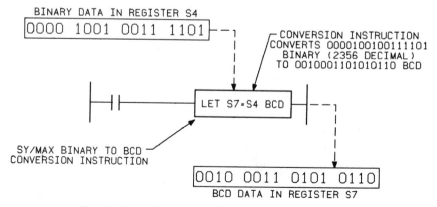

BINARY DATA IN REGISTER S4

0000 1001 0011 1101

CONVERSION INSTRUCTION
CONVERTS 0000100100111101
BINARY (2356 DECIMAL)
TO 0010001101010110 BCD

LET S7=S4 BCD

SY/MAX BINARY TO BCD
CONVERSION INSTRUCTION

0010 0011 0101 0110

BCD DATA IN REGISTER S7

Fig. 10.17. Binary-to-BCD Conversion Instruction

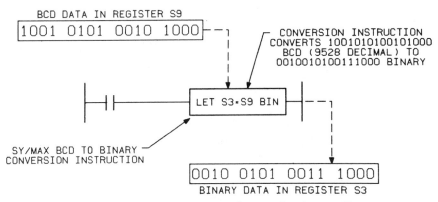

Fig. 10.18. BCD-to-Binary Conversion Instruction

binary conversion instruction. This instruction operates in reverse of the binary to BCD conversion instruction previously discussed. Numerical information represented in a BCD format can be transformed into its binary equivalent representation prior to output by the output modules. Fig. 10.18 illustrates the BCD to binary conversion for the SY/MAX PC.

The user should be aware of one major pitfall when changing numerical data from one number system format to another within a PC. The programming terminal may be designed to assume that the numerical information stored in the data memory of the processor is in either BCD or binary format. This means that should the user convert any numerical data, or place any numerical data in a data memory location in a format other than the format that the programming terminal is designed to interpret, the programming terminal display of these data could be in error. For example, the programming terminal of the SY/MAX PC assumes that any numerical data stored in a data memory location is in a binary format. When the programming terminal must display the contents of a particular location, it assumes that the data stored in this location are in a binary format and thus converts them internally from the binary number system into the decimal number system. If the number had been placed in the storage location in question in a BCD format, the programming terminal would convert the BCD series of zeros and ones as if they were a binary series of zeros and ones. The BCD storage of the number 1234 (0001 0010 0011 0100) is not the same as its binary counterpart (0000 0100 1101 0010). The same warning applies to the use of the hexadecimal numbering system. A processor storing numerical information in a BCD format may not understand what 0110 1101 1110 0101 (hexadecimal 6DE5) really means, especially if it is the preset value for a timer.

11

Data Manipulation

The data manipulation instructions allow the movement, manipulation, or storage of data in either single- or multiple-element groups, from one data memory area of the PC to another. The data being manipulated may be in the form of numbers, bit configurations, and binary codes, representing such information as bar code information or ASCII character codings. Data movement can be performed in a single processor scan, or distributed over numerous processor scans. There are four classifications of data manipulation instructions including *move, shift, search,* and *FIFO/LIFO.* Use of these instructions in PC applications that require the generation and manipulation of large quantities of data will greatly reduce the complexity and quantity of programming that would otherwise be required.

The data manipulation instructions are generally designed for operation on words of memory contained in the internal data storage area of the processor memory. However, these operations may be performed directly on input and/or output image table storage words. When the data being used by the data manipulation instruction are addressed to the input or output image area, the generally preferred procedure is to buffer the information in the internal data memory storage area prior to using them in the user program. This procedure gives the user a chance to verify the correctness and completeness of the data prior to program use.

WORDS OR REGISTERS
AND FILES OR TABLES

Each data manipulation instruction requires at least two or more words of data memory for operation. The words of data memory in singular form may be referred to as either *registers* or *words,* depending on the terminology chosen by the manufacturer. The terms *table* or *file* are generally used when a consecutive group of data memory words is implied, again the term chosen being determined by the manufacturer. Fig. 11.1 shows a portion of a processor's data memory area with representations of a single word or register of data memory and a file or table of data memory words. The number of individual data words or registers available within a processor is totally dependent on its memory configuration. The number of files or tables that can be designated within the data memory area of a PC will be determined by the amount of data memory available for use and the number of data words required for each table or file. Data tables and files allocated in a PC may be stand-alone, overlapping, or subsets of other files or tables. Fig. 11.2 shows the various ways in which multiple files or tables can be implemented within the data memory area of a PC.

To simplify the explanation of many of the various data manipulation instructions available, the instruction protocol for the Allen-Bradley PLC-2 family of PCs will be used. Since Allen-Bradley uses the terms "word" and "file" to designate single and groups of data memory words, respectively, these terms will be used in this chapter's examples to avoid confusion when referring to program examples involving the PLC-2 family instruction protocol. Allen-Bradley's standard symbols and data memory location reference numbers will also be used to maintain continuity.

DATA REPRESENTATION IN THE PC

The data contained in the files, tables, registers, or words of data memory will ultimately be in the form of a series of ones and zeros. The meaning conveyed by the pattern of ones and zeros will be determined by the format used to generate the binary pattern. For example, information stored as data in the data

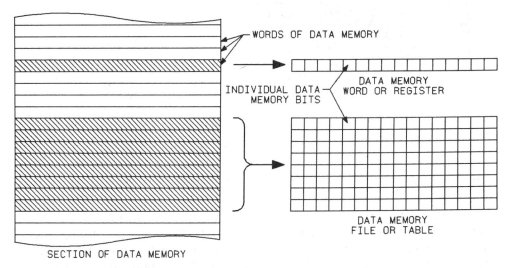

Fig. 11.1. Difference between a Word or Register and a File or Table

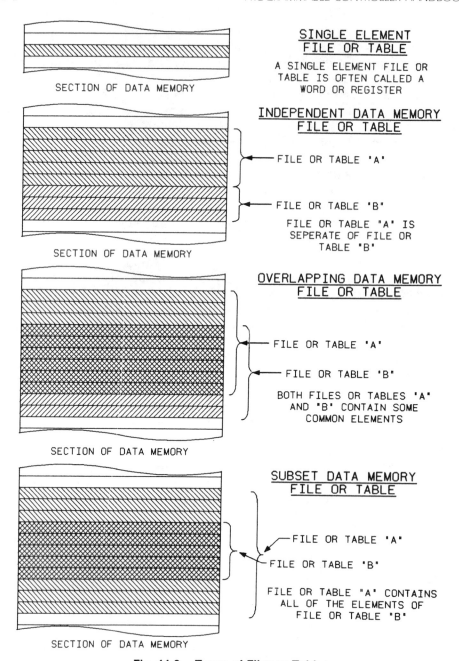

Fig. 11.2. Types of Files or Tables

memory may represent numerical information to be used in the control of a process or machine.

Several encoding methods are used to store numerical data in a PC. The numerical data may be stored in pure binary, octal, or possibly BCD format. When messages are stored in a PC memory, the ASCII character format is usually employed. Fig. 11.3 shows several numerical encoding representations

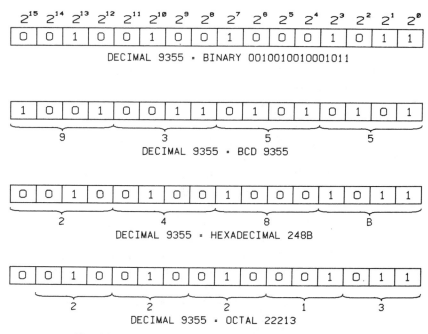

Fig. 11.3. Numerical Data Storage in a PC Memory

of data stored in a processor's data memory, while Fig. 11.4 indicates the storage of ASCII data in the form of ones and zeros in a processor memory.

Regardless of the form in which information is stored in memory, the data manipulation instructions act upon each data word as a unit, and not as individual ones and zeros. This permits these instructions to be ideally suited for use in PC programs which require the movement and manipulation of data values instead of a series of ON/OFF signal statuses.

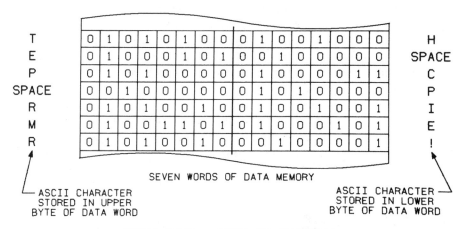

Fig. 11.4. ASCII Data Storage in a PC

MODES OF DATA INSTRUCTION OPERATION

Many of the data manipulation instructions may be implemented on the basis of a word per scan, a specified number of data memory words per scan, or a complete group of data memory words per scan. The advantage of exercising control over the quantity of data memory words acted upon per scan is indirectly related to the control of the scan time of the processor.

As an illustration consider a data manipulation instruction that is occasionally activated to move 500 data words from one location of the processor's memory to another. If the instruction can be distributed over several scans, it will have very little overall effect on the total scan time of the processor in comparison with the same instruction being completed in a single scan. The ability to move 50 data memory words over 10 processor scans will lengthen the processor scan time slightly for the 10 scans required to complete the instruction, instead of having a single scan almost 10 times longer than normal if the instruction were completed in a single scan. The actual time to perform the instruction will be about the same, just its impact on the processor's scan rate will be different. This difference could be important for applications requiring quick responses by the PC for varying control conditions.

If a data manipulation instruction is performed on a single data memory word or register per scan basis, the instruction is operating in an *incremental* mode. When the instruction totally completes all required data manipulations within a single processor scan, the mode of operation is termed *continuous* or *complete*. A *distributed* mode of operation involves the breaking up of the instruction over several processor scans. The total number of instruction operations that will be performed in a single scan is specified by the software programmer, and can be any number less than the total length of the table or file of data that is being acted upon. Fig. 11.5 shows the operation of the incremental, distributed, and continuous or complete instruction modes in graphic form.

Once an instruction is started, it may not be possible to interrupt its operation. For example, once a data manipulation instruction is started in the complete mode, loss of power flow to the instruction may not halt the instruction and prevent it from completing its cycle. Special care must be exercised when programming a data manipulation instruction to ensure that toggling of the control input to the instruction will not affect the overall operation of the instruction. In most cases, once the data manipulation instruction is completed, the control input line for the instruction will require an OFF-to-ON transition for the instruction to be restarted. Care must also be exercised when using data generated by a data manipulation operation. Where distributed or incremental modes of operation are selected, care must be exercised to ensure that the data instruction is complete before the data generated by the instruction are used elsewhere in the program.

DATA INSTRUCTION COUNTER/POINTERS

Data manipulation instructions that have an incremental, distributed, or complete mode of operation incorporate a *counter* or *pointer* to indicate the

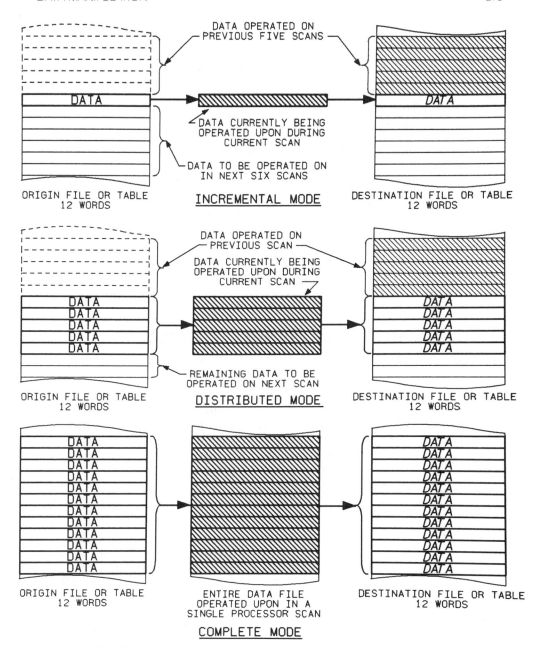

Fig. 11.5. Three Modes of Data Instruction Operation

current area of the data file or table that is being acted on. This pointer or counter will either indicate the next file or table position *to be acted upon,* or the last file or table position that *was acted on.* Particular attention should be paid to exactly which instruction location is indicated by the pointer, since various PC manufacturers infer different meanings to the counter or pointer value.

The user may wish to monitor the counter or pointer through other program instructions to ensure that the data he or she wishes to use later have been operated upon by the data manipulation instruction. It may also be possible to change the value of the pointer or counter by other instructions to permit selected jumping around to various file or table locations. The ability to "force" the indicator to various instruction locations adds greater flexibility to the instruction than would otherwise be possible with a fixed indicator. Care must again be exercised to ensure that the value of the counter or pointer is not forced beyond the total length of the file or table.

Some data manipulation instructions require that the indexing of the counter or pointer be done by additional instructions in the program, while others automatically increment the pointer or counter from file or table location to location each time the data manipulation instruction is activated. Exact operation of the pointer or counter can differ among manufacturers, and proper counter/pointer indexing should be considered at the time of programming. For instructions that offer the distributed and/or complete mode of operation, a counter or pointer is usually required for operation of the instruction. This counter or pointer is used internally by the instruction for operation, and is usually not user-accessible for external program control.

DATA INSTRUCTION ENABLED/DONE INDICATORS

If the PC manufacturer provides the ability to regulate the operation of the data manipulation instructions by the use of either the incremental or the distributed mode of control, two status indications are provided, as illustrated in Fig. 11.6. Once the instruction has been activated, an *enabled* indicator is energized by the instruction to alert the user that the instruction has been activated but not yet reached completion. When the instruction has been completed, a second status indicator, usually called the *done* or *complete* indicator, is activated to inform the user that the instruction has completed its

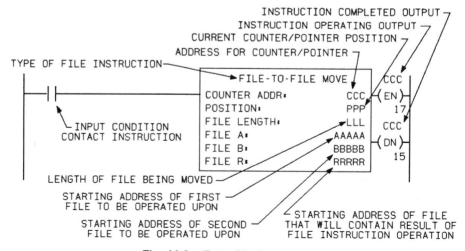

Fig. 11.6.   Data File Instruction

function. Through the proper use of the enable and the done or complete indicators, and the monitoring of the counter or pointer, the user can know the current status of any data manipulation instruction that is operating in either an incremental or a distributed mode.

DATA MOVE INSTRUCTION

As was noted earlier in this chapter, there are four classifications of data manipulation instructions. Three instructions comprise the *move* class of instructions. Two of these instructions move the contents of a single data memory word to or from a specified location in a larger group of data memory words. The third instruction copies or moves the contents of a particular location in one group of data memory words to the same location in a second group of data memory words.

The instructions that move the contents of a single data memory word to or from a larger group of data memory words have several instruction names that may identify them. Where the manufacturer uses the terms "table" and "register" to identify groups and single data memory words, respectively, these instructions will usually be called *table to register* and *register to table* moves. The titles *file to word* and *word to file* will be used when the manufacturer designates single and multiple data words as words and files, respectively. Figs. 11.7 and 11.8 indicate the difference between the two types of move instructions.

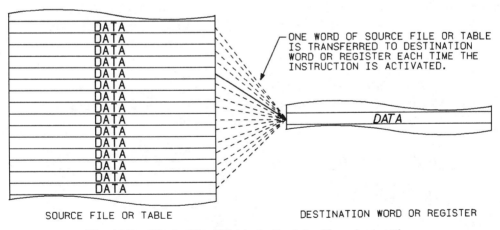

SOURCE FILE OR TABLE   DESTINATION WORD OR REGISTER

Fig. 11.7.   File to Word/Table to Register Move Instruction

The file-to-word or table-to-register move instruction copies or moves the contents of a specified location in a table or file of data memory words to a single register or word location in the data memory area of the processor. The contents of a single file or table location are moved each scan. In order to move the total contents of the table or file, as many processor scans as there are table or file locations will be required. In every processor scan during which the instruction is active, the previous contents of the single data memory word or register are replaced with the current contents of the specified table or file location. Operation of this instruction usually does not alter the contents of the source table or file, it only copies the contents to the destination data memory word or register. Once the instruction has indexed through the source file or table, it must be reset back to either the start of or another location in the source file or table, prior to being reactivated.

A counter or pointer is used with this instruction to specify the location in the source table or file that is to be moved to the destination word or register. Depending on the PC in question, this counter or pointer may be user set to increment automatically from location to location each time the instruction is executed. It may also be set to remain stationary, or it may be externally manipulated by the user program to any desired table or file location.

Often a program will generate data and information that must be stored for later use. The word to file or register to table move instruction provides the ability to move the contents of a single data memory word or register location into a specified location in a larger table or file of data memory words. Again, whether the instruction is termed word to file or register to table is a function of the terminology of the manufacturer.

Whenever data are moved from a data memory word or register location to a specified table or file location, the previous contents of the table or file location are replaced with the contents of the source word or register. Movement of data may not clear the source data memory word or register to zero.

Record keeping of the current location to be loaded in the data memory table or file is done through the use of a counter or pointer, which is associated with the instruction. Depending on the PC, this counter or pointer may be user set to index automatically through the table or file with each data move, or it may be externally controlled or modified by the user program to select any table or file location. When the complete data memory table or file has been fully loaded, any additional data that are to be loaded are usually lost or loaded into the last file or table location, until the counter or pointer is reset to the start or to another location in the destination file or table.

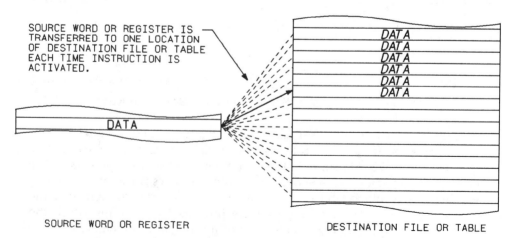

SOURCE WORD OR REGISTER IS TRANSFERRED TO ONE LOCATION OF DESTINATION FILE OR TABLE EACH TIME INSTRUCTION IS ACTIVATED.

DATA

SOURCE WORD OR REGISTER

DESTINATION FILE OR TABLE

Fig. 11.8. Word to File/Register to Table Move Instruction

It may at times be necessary to transfer data from a PC to another device, or between PCs, in order to meet the requirements of a particular application. The use of the two move instructions discussed can make this operation quite simple to perform. As an illustration of the use of these two instructions, consider an application where numerical data must be transmitted from one PC to another. Fig. 11.9 displays in block form the manner in which the data may be transmitted.

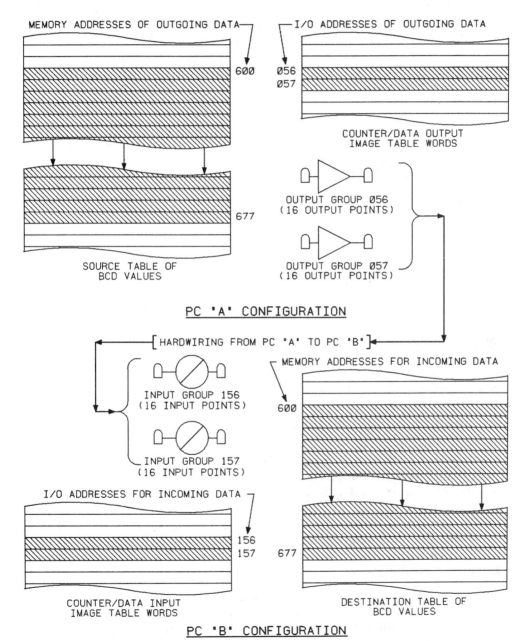

Fig. 11.9. **Data Transmission between Two PCs**

The information to be transmitted is stored in PC A data memory locations 600 through 677 octal. The Allen-Bradley PLC-2 PC family uses the octal numbering system to identify memory locations and addresses. The data are to be placed in the same data memory locations in PC B, addresses 600 through 677 octal. The I/O systems of the two processors will be wired together in order that the data may be transferred. Eight I/O modules will be required for the transmission of data between the processors. Four of the I/O modules will be output modules placed in the fifth I/O rack of PC A, while the remaining four I/O modules will be input modules located in I/O rack 5 of PC B. Two output modules with address 057 will output the data from PC A to two input modules of PC B addressed 157. The information concerning the current location being transmitted will be output from PC A through the pair of output modules with address 056, to the input modules of PC B with address 156. The type of I/O modules selected for the application is up to the user. However, dc-type modules should be considered since they offer fast signal response.

This example will require the use of both the file to word and the word to file move instructions. Fig. 11.10 represents the programming necessary to perform the transmission of data between the two processors. Three rungs of programming are required in PC A to move the data from the data memory locations 600 through 677 octal to the data output modules at address 057. Rung A is a scan counter with a preset value of 64, the number of locations to be transferred (octal 77 equals decimal 64). This scan counter is set to reset itself via the examine off instruction every time its accumulated value reaches its preset value. The function of the scan counter is to specify the current location in the data file to be transmitted.

In order for PC B to know which location is being transmitted, the current value of the scan counter must also be transmitted to PC B. Rung B functions to get the current value of the scan counter and place this value in the output address 056. The use of the get instruction in PC A retrieves the current scan counter accumulated value located in address 400, and places it in output address 056 via the put instruction. Output address 056 contains two output modules, which are wired to input modules on PC B.

The final logic rung in PC A's program, rung C, is responsible for actually moving the data from addresses 600 through 677 octal to the output address 057. A file to word move instruction is used to perform this operation. Examination of the block form of the instruction indicates that the address used for the counter is the same address that was assigned to the scan counter. The function of the counter in this instruction is to indicate which file location is being acted on. The scan counter accumulated value is the current location to be moved. By the file to word move instruction and the scan counter having the same address, the selection of which file location is to be moved is directly tied to the current value of the scan counter.

The length for the file in the file to word move instruction is 64 words. The source file is located from word 600 through word 677 octal inclusive, and the memory word to which the contents of the selected file location will be moved is address 057.

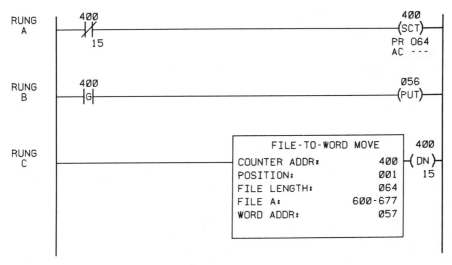

PC "A" PROGRAM

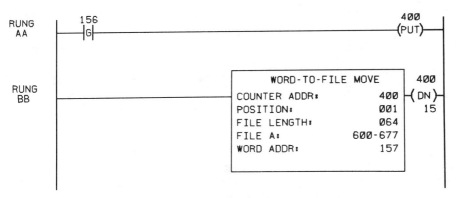

PC "B" PROGRAM

Fig. 11.10.   **Program for Passing Data between Two PCs**

Note that the length of the file and the preset value of the scan counter are both 64. Since the file-to-word move instruction's counter address and the scan counter's address must be the same if the instructions are to maintain correct operation, the length of the file must be equal to the preset value of the scan counter. Should the file length be greater than the scan counter's preset value, any file positions above 677 octal would never be transmitted. Likewise if the length were chosen less than the scan counter's preset value, a run-time error could result when the file to word move instruction's position value exceeded its length value.

Only two rungs of logic are required in PC B to receive the transmitted data. Rung AA receives the current scan counter accumulated value and places it in memory location 400. The get instruction of rung AA takes the BCD-formatted data input to address 156 from PC A and places it in address 400 via the put instruction.

A word to file move instruction comprises rung B. The address of the counter for the word to file move instruction is address 400, the same as was used for the put instruction of rung AA. By using the same addresses, the position in the file being filled will be determined by the current count of the scan counter in PC A.

Reviewing how this happens starts with remembering that the current accumulated value of the scan counter in PC A is moved to output address 056, which is wired to input address 156 in PC B. The numerical value placed in address 156 is moved to location 400, the position counter of the word to file move instruction. Again, the length of the file is 64 words beginning with address 600 and ending with address 677 octal. The memory location selected as the source of data to be placed in the file is input address 157, the address of the input modules wired to the output modules of PC A which transmit data from PC A.

Once both processors are running, the transference of data between them will begin. The example has assumed that the data being transmitted are in a BCD numerical format. Since this is often not the case, additional programming may be necessary to ensure correct data transmission. Additional programming may also be desirable to ensure that PC A is operating before the data are actually used in PC B. Whenever PC A is not functioning, its outputs will be OFF, a zero state to PC B. If PC B is still operational, its programming will attempt to load its data file with zeros. This is undesirable, and the user may wish to inhibit PC B from operation whenever PC A is not properly transmitting acceptable data.

The final instruction that comprises the group of three move instructions is the *file to file* or *table to table* move instruction. Either title may be used to describe an instruction that copies or moves the contents of a specified data memory location in a consecutive group of data memory words to the same location in a second consecutive group of data memory locations. Fig. 11.11 displays in graphic form the operation of the file to file or table to table move instruction. On each scan that the instruction is active, the location specified by the pointer or counter associated with the instruction is copied from the source table or file to the destination table or file. Once all data have been duplicated, the instruction is completed, and the counter or pointer will require resetting along with an OFF-to-ON transition of the instruction's control logic before the contents of the source table or file will again be duplicated to the destination table or file.

The file to file or table to table move instruction is operable in the incremental mode, where the control logic for the instruction must be cycled to cause a specified memory location to be copied. This instruction may also be operated in the distributed mode. The distributed mode of operation for the table to table or file to file move instruction moves tables or files of data, on a specified number of locations per scan basis, controlled by the OFF-to-ON transition of the control logic.

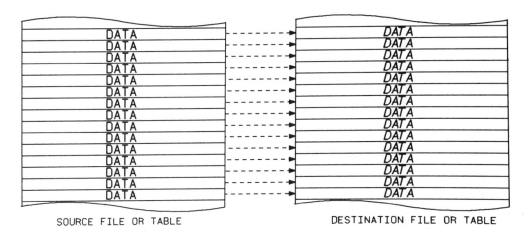

SOURCE FILE OR TABLE   DESTINATION FILE OR TABLE

ENTIRE SOURCE FILE OR TABLE IS
MOVED TO DESTINATION FILE OR
TABLE IN ONE OR MORE PROCESSOR SCANS.

Fig. 11.11.   File to File/Table to Table Move Instruction

When the instruction operates to copy the entire contents of the source table or file to the destination table or file in a single scan, the instruction is operating in the complete mode. Several manufacturers have renamed this instruction, when it operates in the complete mode, the *block move* instruction. The reason for the different designations is that in the complete mode of operation, the instruction is effectively copying a complete block of data memory words into another block of data memory words.

An excellent use of the file to file or table to table move is in a program that controls some sort of batch process or operation. Many programs that control batching operations operate the same sequence of events no matter which recipe is being run. The difference between recipes lies in the dwell times for each step, the quantity of raw material used, or the temperature or pressures used in a particular step. Table 11.1 represents the major parameters for a batch recipe program. The logic required to perform the various steps of the batch process or operation is usually the same for any recipe being run. The quantity of recipes and the number of parameters that must be specified for each recipe change from recipe to recipe. The actual logic required to control the batch process may require less processor memory than the storage of the parameters for each recipe.

An alternate method to control the process would be to duplicate the control logic for each recipe and substitute the correct recipe parameters for each recipe that must be produced. This would be a time-consuming task and would be a very inefficient use of processor memory. The concept represented by Table 11.1 is definitely a more efficient and simpler approach to the problem. The use of the file to file or table to table move instruction provides for the movement of a particular recipe's parameters to the proper locations within the control logic program of the processor.

Table 11.1. Batch Recipe Parameters

PARAMETER	RECIPE 1	RECIPE 2	RECIPE 3	RECIPE 4	RECIPE 5
POUNDS INGREDIENT 'A'	50	25	15	8	0
POUNDS INGREDIENT 'B'	0	25	15	2.7	0
POUNDS INGREDIENT 'C'	5	0	15	9	0
POUNDS INGREDIENT 'D'	3	8	17	28	0
GALLONS 'A'	0	4	3.5	36	5
GALLONS 'B'	0.	4	9	6	8.3
GALLONS WATER	8	0	2.2	8	75
MIX SPEED	2	3.5	9	9	10
MIX TEMPERATURE	65	70	96.8	93	100
MIX TIME(MIN)	30	22	10	42	60
SETTLING TIME	10	15	0	17	0
MOLD QUANTITY	1	2	2	3	5
BAKING TEMPERATURE	350	275	410	390	500
OVEN CONVEYER SPEED	1	4	2	4	5
COOLING TIME	30	20	45	35	0
COATING NUMBER	4	1	2	3	0
COATING TEMPERATURE	80	92	73	103	50
COATING PRESSURE	5	3	7	8	0
DRYING TIME(HRS)	48	36	48	20	0

A sample of the logic necessary to copy each recipe's parameters into the control logic for the process is shown in Fig. 11.12. If the selected recipe is to be designated by the use of thumbwheel switches and loaded with the use of a push button, these devices might be wired to the processor's I/O system as shown in Fig. 11.12. The program logic necessary to transfer a recipe's parameters can then be performed in a single rung of ladder logic for each recipe to be transferred. The processor logic to perform the movement of a particular recipe's parameters could then be written similar to the ladder logic rung also shown in Fig. 11.12.

In order to transfer the parameters of a particular recipe, three conditions must be met. These conditions are programmed as permissives for the file to file move instruction. An examine on instruction for the "load" push button begins the ladder rung. Following the examine on instruction with I/O address 11217 is a get instruction and an equal instruction that examines the current thumbwheel switch value for a match. The equal instruction contains the address of a memory location which stores the identification value for the recipe that will be moved by the file to file move instruction. The get instruction preceding the equal instruction causes the current value of memory location 122, the input image table address for the thumbwheels, to be compared with the memory location value of the equal instruction.

If the value of the thumbwheels matches the value stored in the memory location indicated in the equal instruction, the file to file move instruction is activated. The file to file move instruction can be operated in either the distributed or the complete mode, depending upon the impact the instruction might have on the processor scan time. The location specified for file A would be the memory addresses that contain the various recipe parameters for a particular recipe. The memory addresses for file R would be the same for every

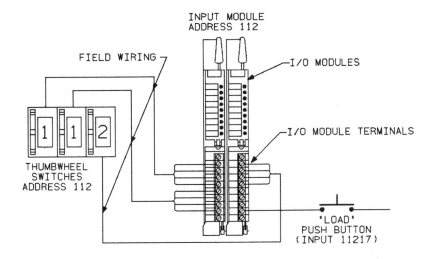

BATCH RECIPE I/O WIRING

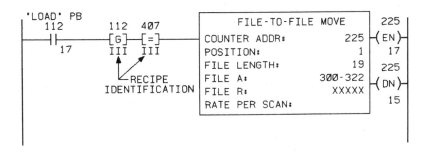

TYPICAL RECIPE PARAMETER MOVE INSTRUCTION

Fig. 11.12. Batch Recipe Program

file to file move instruction required, and would indicate the memory location that holds the operating parameters for the particular recipe being run by the program. The *length* parameter for the instruction would be the same for all file to file move instructions. The length would be equal to the number of parameters that must be specified for operation of the recipe program. The *position* specification would indicate which memory location is being moved when the instruction is active. A *counter* address for use by the instruction in keeping track of the current location being transferred would need to be specified, along with a *rate per scan* value. The ladder rung depicted in Fig. 11.12 would be programmed as many times as necessary, one time for each recipe required by the application. The only parameters that would be changed are the memory location specified by the equal instruction and the memory locations specified as file A.

DATA SHIFT/ROTATE INSTRUCTIONS

The second classification of data manipulation instructions are the *shift/ rotate* instructions. There are two instructions that comprise this class of instructions. The first shifts the contents of each word or register of a data memory file or table to the next higher data memory address location. This instruction is called the *data shift-up* instruction. The *data shift-down* instruction is the second instruction in the shift/rotate class of instructions. The data shift-down instruction moves the contents of each word or register of a specified data memory file or table to the next lower data memory address or location. Both of these instructions usually provide a means to input new data into the shifted or rotated file or table, as well as providing a location for any data being shifted or rotated from the specified file or table. When the shift/rotate instructions are programmed in such a manner that new data can be entered into the data memory file or table while existing data is moved out of the data memory file or table, the instruction is termed a *shift* instruction (information is being "shifted" *through* the data memory file or table). When the shift/rotate instructions are programmed in such a manner that no new data can be entered into the data memory file or table and any existing data cannot be moved out of the data memory file or table, the instruction is termed a *rotate* instruction (information is being "rotated" *around within* the file or table. When operated as a rotate instruction, the location allocated for both the inputting of new data and the outputting of existing data is designated as the same address. Fig. 11.13 displays the operation of the data shift-up and data shift-down instructions graphically.

Programming of the data shift instructions requires the specification of a data memory word for the inputting of data to the instruction, a data memory word for the outputting of data from the instruction, and the location and length of the file or table to be shifted either up or down. The data shift-up instruction places the contents of the highest numerical memory location of the specified file or table in the memory location specified as the output address. It then proceeds to move the contents of each memory location of the file or table to the next numerically higher memory location. The instruction concludes by moving the data value located in the memory address specified as the input address to the numerically lowest memory location designated as the start of the file or table.

The data shift-down instruction operates in reverse of the data shift-up instruction. The instruction starts by moving the data contained in the numerically lowest memory location of the file or table to the location designated as the output address. The data contained in each memory location of the file or table are then shifted down to the next numerically lower memory location. The final operation of the instruction is to move the data contained in the instruction's input address location to the highest numerical memory location of the file or table.

The data shift instructions require a pointer or counter for internal use to keep track of which memory location is being shifted either up or down. A rate

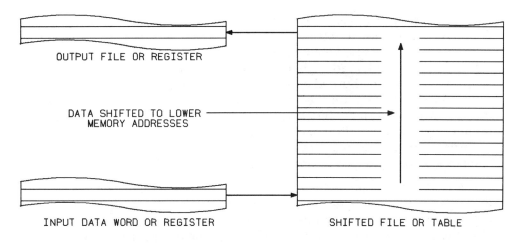

OUTPUT FILE OR REGISTER

DATA SHIFTED TO LOWER
MEMORY ADDRESSES

INPUT DATA WORD OR REGISTER

SHIFTED FILE OR TABLE

DATA TABLE OR FILE SHIFT DOWN OPERATION

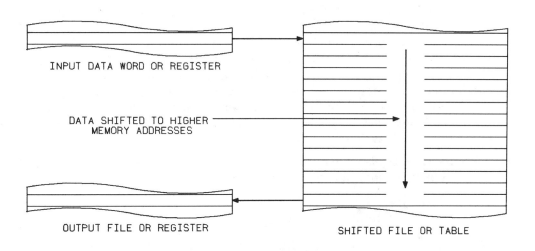

INPUT DATA WORD OR REGISTER

DATA SHIFTED TO HIGHER
MEMORY ADDRESSES

OUTPUT FILE OR REGISTER

SHIFTED FILE OR TABLE

DATA TABLE OR FILE SHIFT UP OPERATION

Fig. 11.13. Data Shift Instructions

per scan value is also required. However, the instruction does not have an incremental mode of operation. A distributed mode may be available should the user be concerned about processor scan time. Should the user wish to specify the same memory location for both the input and the output addresses, the data shift instruction operates as a data rotate instruction.

An excellent example to illustrate the use of the data shift instructions involves an application where an indexing table moves parts past several test stands. Each test stand performs a functional test on the part, and the results of each test must be recorded with the serial number associated for the part tested. A printout of the test results is provided for shipment with each part, and a permanent record is also kept by the manufacturer. An example of where this type of test practice is used might be the power supply industry. Once a power supply has been built, burned in, and quality control tested, the manufacturer often runs several tests on the unit to inform the customer of the excellent performance of the unit. Test results that may be provided include percent output ripple under no-load and full-load conditions, safety circuit trip points for overvoltage and overcurrent conditions, and data on the supply's ability to compensate for rapid load or line supply variations.

Fig. 11.14 is a diagram of the test facility. A 12-position indexing table carries the parts to be tested to each of four test stands. For ease of discussion each position on the stand is numbered, and the four test stands are labeled. Operation of the facility involves indexing the table in a clockwise direction. A tested part is removed from the table and held for label printing at the end of an index cycle. An untested part is also loaded on the table from a load conveyor after completion of an index cycle. The four test stands are placed around the table as shown in Fig. 11.14.

In order to retain the test stand results with each part tested, five files will be required in the data memory area of the PC. These files are listed in Fig. 11.15. The first file is 12 words long and will be used to keep record of the serial number associated with each part as it is indexed on the table. The remaining files are of varying lengths and will be used to record the results of each test as it is performed. These four files do not have to be 12 words long since they are required to store the test results for the remaining index positions of the table only. Each of these files has been numerically identified with the same number as the test stand to avoid confusion.

The first word of each file has been designated as the counter or pointer for the shift-up instruction. Note that each test stand file is equal to the number of

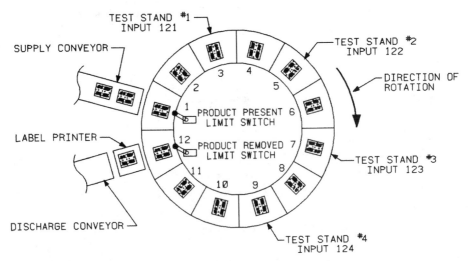

Fig. 11.14. Power Supply Test Facility

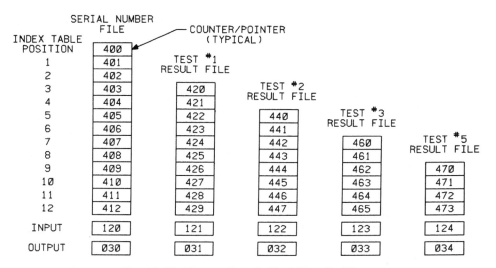

Fig. 11.15. Power Supply Test Results Files

index table positions remaining from the position of the test stand to the discharge conveyor. This is because the table is designed to index prior to the loading of the file by the file shift operation. For example, test stand 2's file is seven words long since there are seven index positions after test stand 2. The part will be tested in position five, and the input word for file 2's shift-up instruction will be loaded with the test results. The table will index at the completion of all the tests, and the shift-up instruction will load the file with the value contained in the shift-up instruction's input address. When the value contained in the input address is loaded into the file, the part will have already moved to index location 6 since the shift-up instruction will be triggered in conjunction with the indexing of the table. The file addresses, shift-up input and output addresses, and shift-up instruction counter addresses are listed in Fig. 11.15.

The actual logic required to perform the record keeping for the test facility is quite straightforward and simple to implement. Fig. 11.16 displays the various ladder logic rungs necessary for this application. Rung A is comprised of a latch triggered by the successful completion of all tests and the load and discharge operation. The function of the latch is to start the loading of the files with the test results, and the loading of the first file with the serial number associated with the part placed on the index table. Each test stand supplies a signal to the PC as to the status of the test. Unless a test is in progress, the test stand will signal the PC that it is available to test another part. The "complete" signal will be low or false whenever a test is in progress. Proper loading of the test results into the PC is accomplished by directly wiring the test results in BCD format to the proper I/O modules. It is assumed that the test stands have the capability to output the test results in a BCD format.

Ladder rung B is comprised of a shift-up instruction triggered by the latch instruction. The shift-up instruction retrieves the serial number for the part (location 120) and loads it into the serial file. The serial number of the part being passed from the discharge conveyor to the printer is placed in output location 030 for use by the printer and possibly a numerical display device for operator monitoring.

Ladder rungs C to F are additional shift-up instructions, which load the results of a test in the file associated with the test being performed at that particular test stand. Each shift-up instruction is triggered by the latch signal generated in rung A. The counter address, file addresses, and input and output addresses for each instruction are indicated as part of the instruction. The input addresses specified

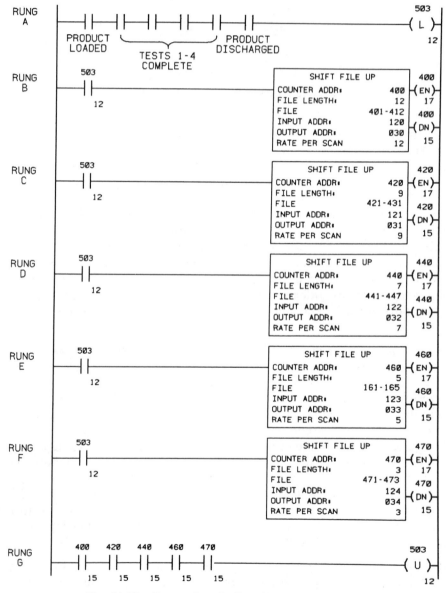

Fig. 11.16.   **Power Supply Test Facility Program**

are input image table addresses representing the I/O module which supplies the BCD test values to the processor from the test stand. The output addresses specified are output image table addresses which could have an operator display device connected for monitoring purposes. These output image addresses would also be used by the printer when the test results are printed.

The remaining rung of the program, rung G, resets the latch when the files have been loaded, shifted to the next higher memory location and the discharged part's data output. The unlatch instruction of rung G is triggered by the done bits of the five shift-up instructions.

Since it will take several processor scans for the table to physically index to its next location, the distributed mode of operation has been chosen for each shift-up instruction. The complete mode could have been used for each instruction. However, since it will take longer for the table to rotate then for the PC to complete its functions in the distributed mode, the scan time of the processor can be conserved without negative effects on the overall operation of the system. Since rung G monitors the done address of all the shift-up instructions, the latch will only be unlatched when all instructions have been completed. Each shift-up instruction requires a false-to-true transition of the control line to begin operation. Any instruction that is enabled in the distributed mode remains internally enabled until it completes its function. Once complete, the instruction must be reenabled in order to start over.

The logic for the indexing test table is complete. The shift-down instruction could have been substituted for the shift-up instruction used in the example. The shift-down instruction could also be used in this application if the table had the ability to index backward under special conditions. Additional programming would be required to keep track of the number of reverse steps, as well as a method to extend the files used for data storage in order to keep track of the test values for each step backward.

DATA SEARCH INSTRUCTION

A single instruction comprises the third class of data manipulation instructions. The *search* instruction provides a PC user the ability to search a specified table or file of values for a particular data value. Fig. 11.17 illustrates the operation of the search instruction graphically.

The search instruction requires the specification of a table or file that is to be searched. An address that contains the data value to be searched for in the file or table is also required for operation of the instruction. The final information necessary for proper operation of this instruction is an address for the counter or pointer.

Operation of the search instruction can only be implemented in the complete mode. When a false-to-true transition of the control line occurs, the search instruction begins searching each consecutive word or register of the specified file or table for a match between the data contained in each file or table location and the data contained in the instruction's source word or

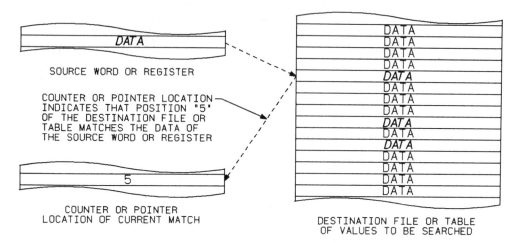

Fig. 11.17. Search Instruction

register. If a match is found, the counter or pointer is updated to indicate the location of the match, and any further searching of the table or file is halted. The user can monitor the pointer or counter for the location of the match as desired. On the next false-to-true transition of the control line the instruction can either begin its search from the start of the table or file, or continue the search from the next sequential memory location in the file or table following the location where the last comparison was discovered.

The search function will search until the end of the table or file is reached. The instruction then provides the user with an indication that the end of the table or file has been reached. If a search instruction is initiated on a file or table, and no matches are found, the instruction will simply indicate that the end of the table or file has been reached.

An excellent use of the search instruction involves applications where a manufactured part or item must be inspected or released for additional work. For example, many products must "cure" during their manufacture before going on for further processing. A table or file of acceptable product codes or serial numbers can be kept in the data memory of a PC. As a product is released for additional processing, its product code, serial number, or date code can be checked and verified against the acceptable numbers stored in the processor's data memory. Those units matching the file or table listings can be processed, those that do not match could be checked against other lists in the processor for other types of matches, or simply set aside for later retries or operator verification.

FIFO/LIFO INSTRUCTIONS

Another class of data manipulation instructions are the *FIFO/LIFO* instructions. FIFO is an acronym for the words "first in, first out," which describes the operation of a particular type of data storage queue. The acronym LIFO stands for "last in, first out," which is a second form of a common data storage queue.

Fig. 11.18 represents the operation of the FIFO queue and the two instructions that comprise the FIFO class of instructions.

Operation of the FIFO queue can be compared to that of a cup dispenser. Cups are loaded into the dispenser from the top and fall to the bottom in the order in which they are loaded. The dispenser can be loaded with cups up to the maximum dispenser capacity. Cups can be unloaded from the dispenser bottom in the same order in which they were loaded; they can be removed until the dispenser is empty.

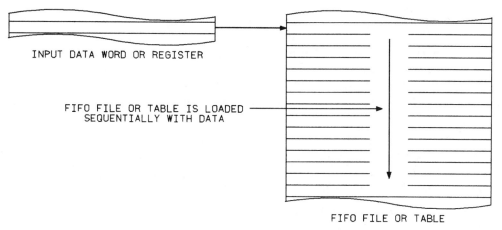

INPUT DATA WORD OR REGISTER

FIFO FILE OR TABLE IS LOADED
SEQUENTIALLY WITH DATA

FIFO FILE OR TABLE

FIFO LOAD OPERATION

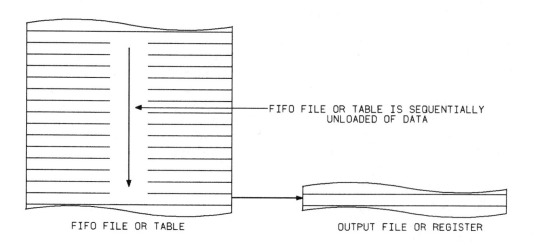

FIFO FILE OR TABLE IS SEQUENTIALLY
UNLOADED OF DATA

FIFO FILE OR TABLE

OUTPUT FILE OR REGISTER

FIFO UNLOAD OPERATION

Fig. 11.18. FIFO Instructions

The FIFO queue created in the data memory of the processor operates in a manner similar to the cup dispenser just described. The FIFO queue is filled by the *FIFO load* instruction to the maximum number of data memory words specified for the file or table that comprises the FIFO queue. The data that are loaded by the FIFO load instruction are stored in the queue file or table in the exact order in which they were loaded. Removal of the data from the FIFO queue is performed with the *FIFO unload* instruction. A FIFO queue can be loaded and unloaded at will, with the only restriction being that the maximum capacity of the queue cannot be exceeded. A FIFO queue does not have to be completely filled before being unloaded, nor does it have to be completely empty before it can be reloaded.

The first of the FIFO instruction pair is the FIFO load instruction, which loads the FIFO queue. Data that have been placed in a data memory location designated as the input address for the FIFO load instruction are moved into the FIFO queue whenever the FIFO load instruction is activated. Only one piece of data can be loaded into the FIFO queue per scan, unless multiple FIFO load instructions are programmed in the PC to load the same FIFO queue.

Operation of the FIFO unload instruction places the data currently located at the "bottom" of the FIFO queue in the data memory location specified as the output address for the unload instruction. Again, only one data word can be unloaded from the FIFO queue unless the user chooses to program multiple FIFO unload instructions addressing the same FIFO queue.

Every FIFO load or FIFO unload instruction requires the specification of a counter or pointer for operation. The purpose of the counter or pointer is to keep track of the number of items stored in the FIFO queue. With every file or table specified for use as a FIFO queue, a corresponding counter or pointer must be specified. Each FIFO instruction referencing a particular FIFO queue must also use the counter or pointer assigned to that FIFO queue. Failure to reference the counter or pointer associated with a specific FIFO queue will result in faulty FIFO load and/or FIFO unload instruction operation.

All FIFO instructions usually provide status indicators, which can be monitored to tell the current status of the FIFO queue. An empty indicator provides information that the queue has been completely emptied by the FIFO unload instruction. If the FIFO queue is full, a second status indicator provides this information as a warning that no additional data can be loaded into the queue until data already placed in the FIFO are removed. Note that many FIFO unload instructions may not actually clear the contents of the FIFO file or table during the FIFO unload operation, but will simply copy the data to the output address and internally log the queue file or table location containing the original data as available for new data.

An excellent application for use of the FIFO data manipulation instruction is depicted in Fig. 11.19. Many manufactured items are labeled with specific information concerning the contents of the container or package, the manufacturing codes and dates, or even information concerning the use of the product or item. The manufactured items are often generated in a random pattern and accumulated on an accumulating conveyor or other type of

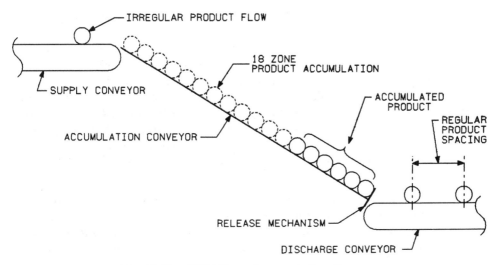

Fig. 11.19. FIFO Type Accumulating Conveyor

accumulation device prior to delivery to the labeler. Fig. 11.19 shows the product being supplied on a conveyor to an accumulating device. The product is supplied at irregular intervals, but the average supply rate is equal to the rate in which the product is supplied to the labeler.

For this example it will be assumed that the hardware supplying the labeler will provide the correct data for the labeling process. The accumulation device is designed to accumulate a maximum of 18 items. The product is released from the accumulation area to the labeler at a specified rate.

Fig. 11.20 shows the data memory locations that will be used for this application. Data memory input image table address 173 will receive the data from the production hardware. The FIFO queue will be 18 data memory words long beginning with word 700. The data that are to be unloaded from the FIFO

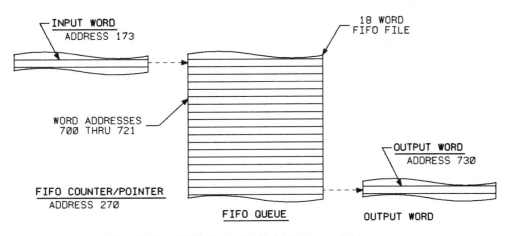

Fig. 11.20. FIFO Example Data Memory Assignments

queue will be placed in address 730. Address 730 could also be the input address for a data shift instruction, which would track the product on the supply conveyor to the labeler. The counter address associated with the queue file is address 270. Fig. 11.21 illustrates the actual instructions for the application. Ladder rung A contains a FIFO load instruction which loads the product information into the FIFO file. Whenever the production hardware delivers an item to the accumulation device, it toggles I/O input address 17400. A false-to-true transition of this input loads the current data values placed at word input address 173 by the production hardware into the FIFO queue. The queue's counter will be incremented each time another item is released to the accumulation conveyor.

Rungs B and C physically release the product from the accumulation conveyor and unload the product information from the FIFO queue. Every time the discharge conveyor is ready to accept another item, I/O input address 17401 is toggled. Rung B causes the accumulation conveyor's release device to actually

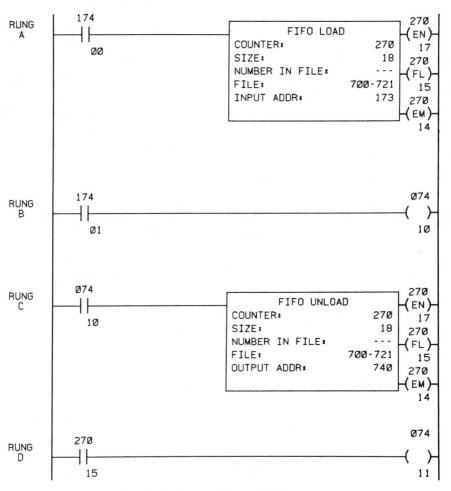

Fig. 11.21. FIFO Example Program

release the item by activating I/O output address 07410. Activation of this address enables the FIFO unload instruction of rung C. The FIFO unload instruction takes the data stored at the "bottom" of the FIFO and places them in address 730 for later movement into a data shift operation designed to track the product as it moves down the conveyor toward the labeler.

The final rung of the example program provides a halt signal to the production hardware in the event the accumlation conveyor becomes full. The FIFO queue FULL status indicator is monitored, address 27015, and used to activate I/O output address 07411. I/O output address 07411 can be used to halt the production machinery and/or as an alarm indication to operating personnel.

A variation of the FIFO instruction is the LIFO instruction. It differs from the FIFO instruction in that the last item loaded into the file or table is the first item removed from the file or table. As an illustration consider the washing and storing of dishes in a kitchen cabinet. As each dish is washed and dried, it is placed on top of the stack in the cabinet for storage. When a clean dish is needed, it is removed from the top of the cabinet stack instead of from the bottom. Hence the term LIFO, or last in, first out. The LIFO is synonymous with the "stack" of a microprocessor where data are "pushed onto" the stack for storage and later "popped off" during retrieval. Fig. 11.22 illustrates the operation of the LIFO instruction.

The first of the LIFO instruction pair is the *LIFO load* instruction, which loads the LIFO queue. Data that have been placed in a data memory location designated as the input address for the LIFO load instruction are moved into the LIFO queue whenever the LIFO load instruction is activated. Only one piece of data can be loaded into the LIFO queue per scan, unless multiple LIFO load instructions are programmed in the PC to load the same LIFO queue.

Operation of the LIFO *unload* instruction places the data currently located at the "top" of the LIFO queue in the data memory location specified as the output address for the unload instruction. Again, only one data word can be unloaded from the LIFO queue unless the user chooses to program multiple LIFO unload instructions addressing the same LIFO queue.

Every LIFO load or LIFO unload instruction may require the specification of a counter or pointer for operation. The purpose of the counter or pointer is to keep track of the number of items stored in the LIFO queue. With every file or table specified for use as a LIFO queue, a corresponding counter or pointer may be specified. Each LIFO instruction that references a particular LIFO queue must also use the counter or pointer assigned to that LIFO queue. Failure to reference the counter or pointer associated with a specific LIFO queue will result in faulty LIFO load and/or LIFO unload instruction operation.

Many LIFO instructions provide status indicators which can be monitored to tell the current status of the LIFO queue. An empty indicator provides information that the queue has been completely emptied by the LIFO unload instruction. If the LIFO queue is full, a second status indicator provides this information as a warning that no additional data can be loaded into the queue,

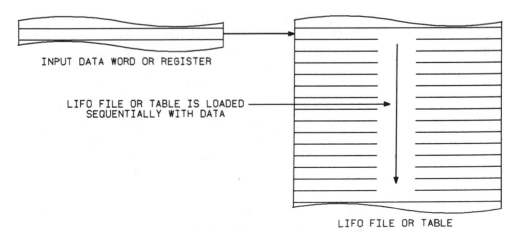

INPUT DATA WORD OR REGISTER

LIFO FILE OR TABLE IS LOADED
SEQUENTIALLY WITH DATA

LIFO FILE OR TABLE

LIFO LOAD OPERATION

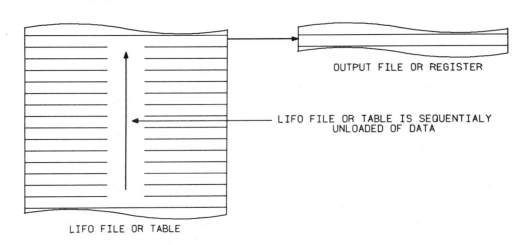

OUTPUT FILE OR REGISTER

LIFO FILE OR TABLE IS SEQUENTIALY
UNLOADED OF DATA

LIFO FILE OR TABLE

LIFO UNLOAD OPERATION

Fig. 11.22.   LIFO Instructions

until data already placed in the LIFO are removed. Note that many LIFO unload instructions may not actually clear the contents of the LIFO file or table during the LIFO unload operation, but will simply copy the data to the output address and log the queue file or table location containing the original data as available for new data.

Before leaving the subject of FIFO and LIFO instructions, a word of caution is in order regarding how data are stored in a FIFO/LIFO queue. All FIFO/LIFO instructions have a file or table associated with them for the storage of data. The user cannot assume that the instruction will store and remove incoming and outgoing data in a specific order within the instruction's file or table. The only guarantee the user has is that the data will be stored and retrieved according to the operation of the instruction being used.

For example, a FIFO queue of 100 words or registers may not place the fifty-third data value entered into the queue, into the fifty-third memory location of the FIFO's file or table. The user cannot assume that since there are 52 other queue entries currently indicated on the instruction's counter or pointer, the next one will be placed in the FIFO instruction's fifty-third file or

8888	9843
7111	9354
9000	9000
4309	8888
6500	7695
2222	7111
1378	6500
9843	5791
1000	4309
7695	3464
2012	2222
9354	2012
5791	1689
1689	1378
3464	1000

FILE OR TABLE BEFORE SORTING   FILE OR TABLE AFTER SORTING

DATA TABLE OR FILE DESCENDING SORT OPERATION

8888	1000
7111	1378
9000	1689
4309	2012
6500	2222
2222	3464
1378	4309
9843	5791
1000	6500
7695	7111
2012	7695
9354	8888
5791	9000
1689	9354
3464	9843

FILE OR TABLE BEFORE SORTING   FILE OR TABLE AFTER SORTING

DATA TABLE OR FILE ASCENDING SORT OPERATION

Fig. 11.23. Sort Instructions

table location. Likewise the removal of 10 queue entries may not mean that the instruction has automatically moved that fifty-third entry to the forty-third file or table location. The FIFO/LIFO instructions are generally not designed to permit the user to "steal" data from an intermediate queue location. An experienced PC user may be able to experiment with a particular PC model and learn how the FIFO/LIFO instruction operates internally, but if the reason for obtaining this type of instruction operation information is to "steal" data from the middle of a FIFO/LIFO queue, then the user should consult a qualified PC applications engineer to obtain suggestions on alternate programming methods to meet his or her required PC programming needs.

SORT INSTRUCTIONS

Many PCs provide instructions that enable the user to sort lists of numerical data in ascending and/or descending order. These instructions are referred to as *sort* instructions. The *sort ascending* instruction takes a file or table of specified length and reshuffles the contents so that the numerical data contained in each successive file or table location are numerically greater than the value of the data contained in the previous location. The *sort descending* instruction takes a file or table of specified length and reshuffles the contents so that the data contained in each successive file or table location are numerically less than the value of the data contained in the previous location.

The sort instructions may or may not employ a duplicate data memory file or table for operation. Where a duplicate file or table is provided, it is used to list the original position of the data prior to the sorting operation. Fig. 11.23 illustrates both the ascending and the descending sort instruction operation.

It should be noted that a PC may or may not sort by numerical sign. Some PCs will sort negative numbers as less than positive ones, while others do not take into account the sign of the number. When the PC does not sort according to sign, it will sort the file or table contents by absolute value.

12

Bit Manipulation

The instructions that compare and manipulate individual bits of a processor's memory are probably among of the most useful sets of instructions. There are several classes of these instructions available, including the *logic* group of operations, the *bit shift/rotate* group of operations, and the *bit examine*, *set*, and *reset* group of operations. The logic operations provide for the *ANDing*, *INCLUSIVE ORing*, *EXCLUSIVE ORing*, and *COMPLEMENTing* of memory bits. Bit shift instructions include the shifting and/or rotating of memory bits to the right or left, while the *bit examine*, set, and *reset* instructions provide for the examination of a particular bit for its OFF or ON status, and the ability to set or clear any specified bit in memory.

The bit instructions are generally designed for operation on words of memory contained in the internal data area of the processor memory. However, these operations may be performed directly on input or output image tables under some circumstances. When it is desired to use input or output statuses as data for a bit instruction, the generally preferred procedure is to move the data from the input image table to the internal data area, or from the internal data area to the output image table as required.

DATA MEMORY FILES, REGISTERS, TABLES, AND WORDS

Each bit instruction requires at least one or more words of data memory for operation. In singular form these words of memory may be referred to as *registers* or *data words*, depending on the terminology chosen by the manufacturer. When these memory words are referred to as a consecutive group of data memory words, the group may be referred to as either a *table* or a *file*, again the term used being determined by the manufacturer. Fig. 12.1 shows a section of a processor's data memory area with representations of a single word or register of data memory, and a file or table of data memory words.

The number of data words or registers available within a processor is totally dependent on the design of the PC. The number of files or tables that can be designated within the data memory area of a PC will be determined by the amount of data memory available, and by the number of data words required for each table or file. Data tables or files of data memory bits may be stand-alone, overlapping, or subsets of other files or tables of data memory bits. Fig. 12.2 shows the various ways in which a file or table of data memory bits can be implemented within the data memory area of a PC.

DATA MEMORY WORD BIT LABELS

In order to understand the operation of the bit instructions better, a detailed discussion on the conventions of labeling the bits of one or more data memory words is required. Many manufacturers designate the group of bits belonging to a word or register, or a file or table, as a *matrix*. For clarity the term matrix will be used in this text to reference a group of data memory bits formed from a word or register, or a file or table. Fig. 12.3 shows a section of the data memory with the individual memory bits labeled for both a word or register and a file or table.

The memory architecture for any PC determines the number of bits available in a particular memory word. The total number of bits in a matrix will be the number of bits in a word or register of the data memory times the number of words that make up the matrix. However, several PCs do not follow this convention since they allow users to specify the number of bits they desire. This

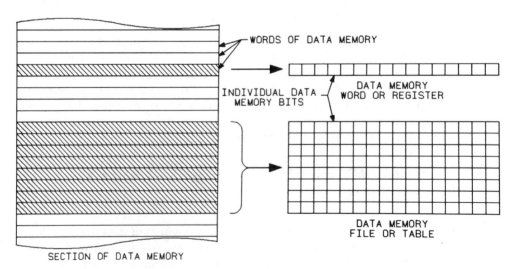

Fig. 12.1. Difference between a Word or Register and a File or Table

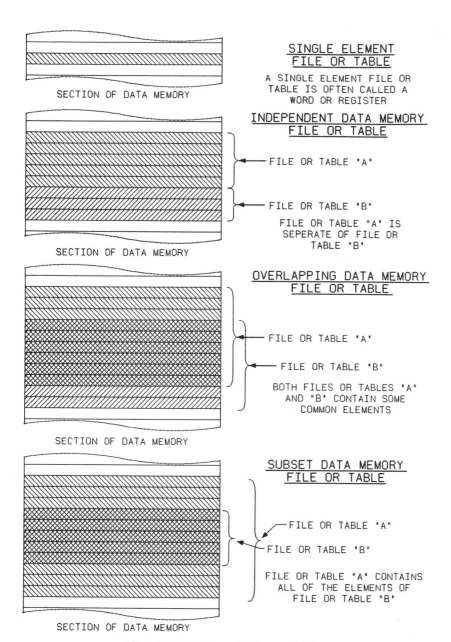

SINGLE ELEMENT
FILE OR TABLE

A SINGLE ELEMENT FILE OR
TABLE IS OFTEN CALLED A
WORD OR REGISTER

SECTION OF DATA MEMORY

INDEPENDENT DATA MEMORY
FILE OR TABLE

FILE OR TABLE 'A'

FILE OR TABLE "B"

FILE OR TABLE "A" IS
SEPERATE OF FILE OR
TABLE "B"

SECTION OF DATA MEMORY

OVERLAPPING DATA MEMORY
FILE OR TABLE

FILE OR TABLE 'A'

FILE OR TABLE 'B"

BOTH FILES OR TABLES 'A"
AND "B' CONTAIN SOME
COMMON ELEMENTS

SECTION OF DATA MEMORY

SUBSET DATA MEMORY
FILE OR TABLE

FILE OR TABLE 'A"

FILE OR TABLE 'B"

FILE OR TABLE "A" CONTAINS
ALL OF THE ELEMENTS OF
FILE OR TABLE "B'

SECTION OF DATA MEMORY

Fig. 12.2. Types of Files or Tables

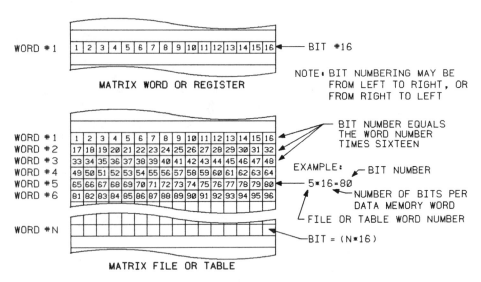

Fig. 12.3. Data Memory Matrix Addressing

number can be anything up to a manufacturer specified maximum. Once the user specifies the number of bits he or she wishes to have in a particular matrix, the processor automatically allocates the number of data memory words required to meet the need. The processor will always allocate complete words, and any bits remaining as part of a partially allocated word are "lost."

It should be noted that some PCs start with the number 0 to identify the first matrix bit, while those PCs not following this convention identify the first matrix bit as number 1. Also, the order of numbering the bits in each matrix word may be from right to left or from left to right. The method of numbering has no effect on the operation of the PC or the bit instruction, but to avoid confusion the user should be aware of the manufacturer's matrix bit identification conventions prior to applying any bit instruction.

MATRIX LOCATION CONTENTS

Each matrix bit location will contain either a one (1) or a zero (0) value, which may represent an ON or true condition as a one and an OFF or false condition as a zero. The bit instructions affect directly the status of each bit within a matrix. The individual matrix bits can be sensed, shifted, compared, set, cleared, or inverted at the discretion of the user, in units as small as a single bit or as large as the largest matrix file or table allocated in the processor's data memory.

BIT INSTRUCTION OPERATING MODES

In addition some processors permit the bit operations to be implemented on the basis of a word per scan, a specified number of words per scan, or a complete group of words per scan. The advantage of having control over the number of words, and individually the number of bits acted upon in a scan, is indirectly related to controlling the total scan of the processor. Consider a bit instruction that occasionally operates on a matrix of 800 bits or 16 words or registers. If the operation performed by the bit instruction can be distributed over several scans, it will affect the instantaneous scan time to a lesser degree than when it does the complete operation in one scan. The ability to perform a matrix instruction on the 800-bit matrix in groups of 160 bits will lengthen the processor scan time slightly for the five scans required to complete execution of the instruction, instead of a larger scan time increase for a single scan. The actual time to perform the instruction will be the same, just its impact on the processor scan rate will be less for the distributed operation.

When bit instructions are performed on a single data memory word or register per scan basis, the instruction is operating in an *incremental* mode. If the instruction is totally completed in a single processor scan, the mode of operation for the bit instruction is termed either *continuous* or *complete*. A *distributed* mode involves the breaking up of the instruction over several processor scans. The distributed mode can break the instruction down into any number of smaller operations, with the number of operations per scan less than the total number of operations to be completed. Fig. 12.4 graphically illustrates the operation of the incremental, distributed, and continuous or complete instruction modes.

Once a bit instruction is started, it may not be possible to interrupt its operation. For example, loss of power flow to the instruction's control input may not halt the instruction from completing its cycle once it has started. Special care should be taken when programming the bit instructions to see whether toggling the control input to the instruction will affect the operation of the instruction once it has been initiated. In most cases, once a bit instruction is complete, the control input line for the instruction will require an OFF-to-ON transition for the instruction to be restarted. Care should also be taken when using the data generated by a bit instruction. Where distributed or incremental operating modes are selected, the programmer should ensure that the bit instruction is complete before using the data elsewhere in the program.

To simplify the explanation of some of the various bit instructions available, the instruction protocol for the Modicon 584 PC will be used for the examples of this chapter. Since Modicon uses the term "register" to designate a single word in the data memory, and the term "table" to designate a consecutive group of data memory words, these terms will be used in this chapter to avoid confusion when discussing programming examples involving the 584 instruction set. Modicon's standard symbols and data memory location reference numbers will also be used to maintain continuity.

Since the Modicon 584 utilizes a 16-bit memory word structure, each data memory word contains 16 bits. The bits of a Modicon 584 matrix table are numbered consecutively from left to right, starting with 1 for the leftmost bit of the first data memory word of the table. Bit 16 is the rightmost bit of the first data memory word, with the leftmost bit of the matrix table's second memory data word designated as bit number 17. This numbering convention is continued throughout the remainder of the matrix table. Since the total

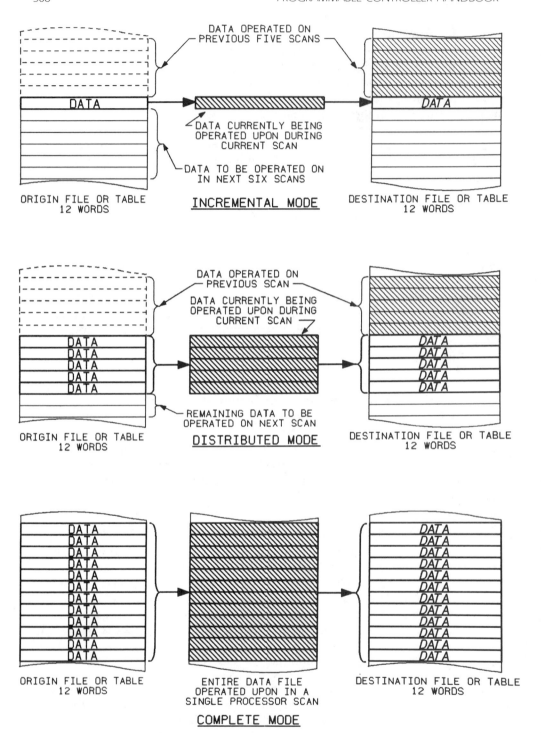

Fig. 12.4. Three Modes of Bit Instruction Operation

number of bits in any Modicon 584 matrix must be equal to the number of bits in a memory word times the number of words allocated for the table, a 584 matrix table must contain 16 times the total number of table words.

LOGIC BIT INSTRUCTIONS

As noted earlier, there are four types of logic bit instructions. These instructions provide for the ANDing, INCLUSIVE ORing, EXCLUSIVE ORing, and COMPLEMENTing of data memory matrixes. All logic bit instructions, except the complement instruction, require two *source matrixes* upon which to operate. The results of all logic operations are placed in a *destination matrix*. Depending on the PC manufacturer, the destination matrix may be a separate matrix or one of the source matrixes. The complement instruction requires a single source matrix, and will return the result to a destination matrix. Most PCs attempt to leave at least one of the source matrixes intact to erase user programming.

Those PCs that offer logic bit instructions require that each matrix used for a logic operation be arranged in either of two possible configurations. The source matrixes may both be file or table matrixes, or one of the source matrixes can be a word or register matrix while the remaining source matrix is a file or table matrix. As illustrated in Fig. 12.5, unless one of the source matrixes is a word or register matrix, each of the source matrixes and the destination matrix must contain identical quantities of bits. When one of the source matrixes is permitted to be a word or register matrix, this matrix must occupy a full data memory word. Whenever a logic bit instruction is performed between a source file or table matrix pair, like addressed bits of each source matrix are operated upon, with the result being placed in the equivalent destination matrix location. When one of the two source matrixes is a word or register matrix, like bit positions

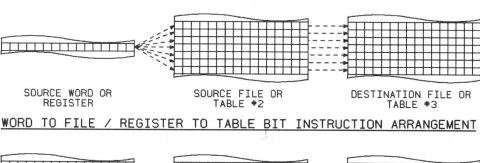

SOURCE WORD OR   SOURCE FILE OR   DESTINATION FILE OR
REGISTER   TABLE #2   TABLE #3

WORD TO FILE / REGISTER TO TABLE BIT INSTRUCTION ARRANGEMENT

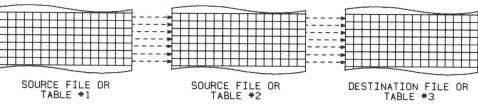

SOURCE FILE OR   SOURCE FILE OR   DESTINATION FILE OR
TABLE #1   TABLE #2   TABLE #3

FILE TO FILE / REGISTER TO TABLE BIT INSTRUCTION ARRANGEMENT

Fig. 12.5. Logic Bit Instruction Matrix Operation

within each source matrix word are operated upon, and the result is placed in the like bit position of the destination matrix word.

When a PC manufacturer provides the ability to regulate the operation of the logic instruction by the incremental mode of control, a counter or pointer is provided to indicate the current matrix position that is being operated upon. The user may monitor this indicator through other program instructions to ensure that the data he or she wishes to use later have been operated upon by the logic instruction. In addition two status points are usually provided as indicators. Once the instruction is activated, an *enabled* indicator is energized to alert the user that the instruction has been started, but has not reached completion. Upon completion a second status indicator, usually called the *done* or *complete* indicator, is activated to let the user know that the instruction has been successfully completed. Through the proper use of the enable and done indicators, and the monitoring of the counter or pointer, the user can determine the current status of any logic instruction that has an incremental mode of operation.

For those processors incorporating the distributed or complete modes of operation, a counter or pointer is also required. However, this counter or pointer is used internally by the processor to keep track of how far the instruction has progressed through the matrix file or table. This pointer should not be changed by the user programming, even though it might be possible to do so. The user may wish to monitor the value of the counter or pointer as an indication of how far the logic instruction has progressed on any given program scan operating in the distributed mode.

It is often desirable to manipulate the counter or pointer current position value externally for an incremental logic bit instruction. While the majority of PCs offering bit manipulation instructions permit program manipulation of the counter or pointer indicators, extreme care should be exercised when doing so. It will be the user's responsibility to ensure that the logic instruction's counter or pointer value is being adjusted to the correct new matrix word or bit location. Should the new counter or pointer value exceed the number of matrix words or registers that make up the matrix file or table for the logic instruction, improper processor operation could result. Many PCs are designed to detect this condition and shut-down automatically, or to reset the counter or pointer automatically back to the beginning of the file or table.

It is worth noting that the value indicated by the counter or pointer of a logic instruction can indicate either of two different locations in a matrix. Some PCs are designed such that the counter or pointer indicates the *last* matrix bit position acted *upon* by the logic instruction. Other PCs are designed such that the counter or pointer indicates the *next* matrix bit location to be acted on. The user should be aware of exactly what location in the matrix file or table the logic instruction's pointer indicates before performing any user program manipulation or monitoring of the counter or pointer value.

The AND instruction provides for the logic ANDing of two matrixes. Fig. 12.6 shows the truth table, English statement, and Boolean equation for the AND instruction as well as the logic diagram symbols. Referring to the truth table of

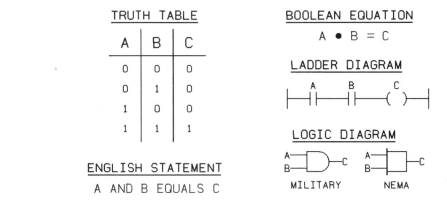

TRUTH TABLE

A	B	C
0	0	0
0	1	0
1	0	0
1	1	1

ENGLISH STATEMENT

A AND B EQUALS C

BOOLEAN EQUATION

$A \bullet B = C$

LADDER DIAGRAM

LOGIC DIAGRAM

MILITARY   NEMA

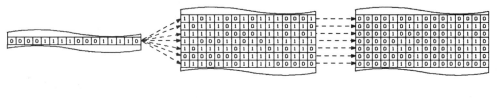

SOURCE MATRIX #1   SOURCE MATRIX #2   DESTINATION MATRIX

WORD TO FILE / REGISTER TO TABLE 'AND' INSTRUCTION

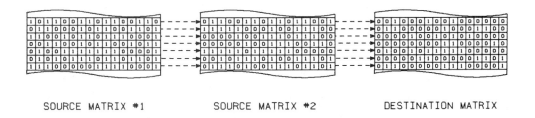

SOURCE MATRIX #1   SOURCE MATRIX #2   DESTINATION MATRIX

FILE TO FILE / REGISTER TO TABLE 'AND' INSTRUCTION

Fig. 12.6.   Logic And Instruction

Fig. 12.6, the logic AND instruction requires both source matrix bit statuses to be a one in order for the result to also be a one. Fig. 12.6 illustrates the operation of the AND function on two source matrixes with the result being placed in the destination matrix.

An example of a typical use of the logic AND instruction is shown in Fig. 12.7. When handling information in bit form it is often advantageous to create a *mask* to screen out certain bits within a larger group of data matrix bits. The first source matrix is programmed as the mask matrix for the AND operation. Those bits in the second source matrix, which we will call the data mask, are to retain their current status, have their corresponding bits in the mask matrix programmed to a one. Those bits of the data matrix that are not desired have their corresponding mask matrix bit set to a zero. The proper setting of the mask matrix bits must be done when the instruction is initially programmed in the processor by the user who addresses the mask matrix data memory words with the programming terminal and then sets each mask bit to its individual one or zero state.

For our example the mask matrix has been placed in 584 data memory locations 40010 through 40019. The data that are to be acted upon by the AND function are placed in data memory locations 40100 through 40109. The resultant data generated by the AND operation will be returned to locations 40100 through 40109. The Modicon 584 PC series designates the second source table and the final destination table as always having the same data memory addresses for many of the bit and data manipulation instructions. A normally open contact with identity 10022 will activate the instruction, and coil 00021 will signal that the instruction is operating for that scan. The Modicon 584 AND instruction operates to completion every scan that power flow is received to the control input of the instruction. The function block format for the 584 instruction (mask matrix). The AND instruction operates to place the result of the operation back in the location of the second source matrix. Both the second source matrix (data matrix) and the destination matrix have the same data memory addresses, which are placed in the middle position of the instruction. The bottom part of the instruction contains the length of the matrixes. Power flow through contact 10022 causes the instruction to AND the two source matrixes together logically and place the result back in the location of the second source matrix now acting as the destination matrix. Fig. 12.7 shows the bit configurations for the mask matrix and the data matrix, with the result of the operation shown as a separate matrix.

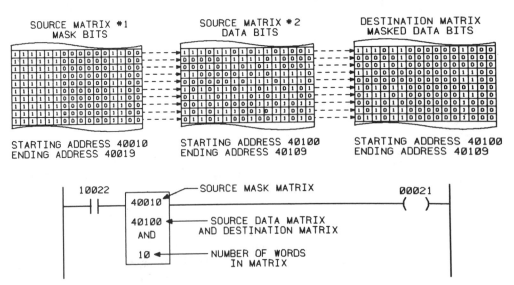

Fig. 12.7. **Masking with the Logic And Instruction**

The logic OR instruction, often referred to as an INCLUSIVE OR, provides for the ORing of individual matrix bits. Fig. 12.8 shows the truth table and Boolean equation for the logic INCLUSIVE OR function as well as representations for the OR operation between matrixes. The OR function examines corresponding bits of two source matrixes for a one condition. If either matrix or both of the matrixes contain a one, then the corresponding bit of the destination matrix is also set to a one status. Only when both source matrixes contain a zero in

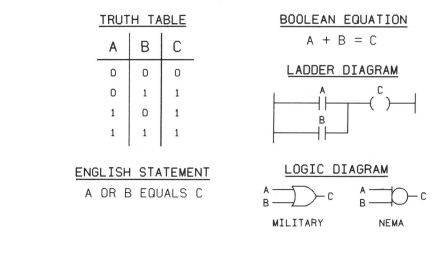

TRUTH TABLE

A	B	C
0	0	0
0	1	1
1	0	1
1	1	1

BOOLEAN EQUATION

A + B = C

LADDER DIAGRAM

ENGLISH STATEMENT

A OR B EQUALS C

LOGIC DIAGRAM

MILITARY   NEMA

SOURCE MATRIX #1   SOURCE MATRIX #2   DESTINATION MATRIX

WORD TO FILE / REGISTER TO TABLE "INCLUSIVE OR" INSTRUCTION

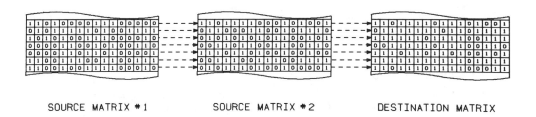

SOURCE MATRIX #1   SOURCE MATRIX #2   DESTINATION MATRIX

FILE TO FILE / REGISTER TO TABLE "INCLUSIVE OR" INSTRUCTION

Fig. 12.8. Logic Inclusive Or Instruction

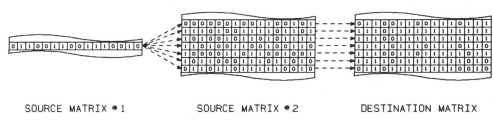

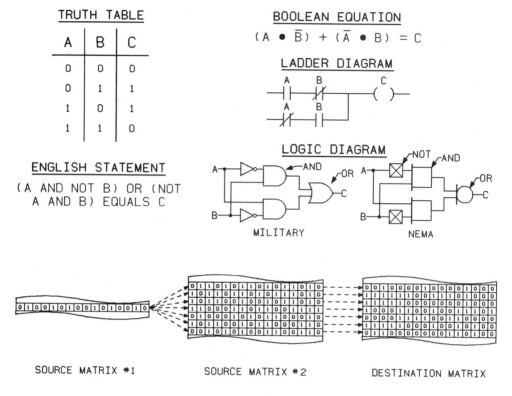

TRUTH TABLE

A	B	C
0	0	0
0	1	1
1	0	1
1	1	0

BOOLEAN EQUATION

$$(A \bullet \overline{B}) + (\overline{A} \bullet B) = C$$

LADDER DIAGRAM

ENGLISH STATEMENT

(A AND NOT B) OR (NOT A AND B) EQUALS C

LOGIC DIAGRAM

MILITARY   NEMA

SOURCE MATRIX #1   SOURCE MATRIX #2   DESTINATION MATRIX

WORD TO FILE / REGISTER TO TABLE "EXCLUSIVE OR" INSTRUCTION

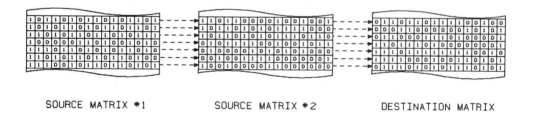

SOURCE MATRIX #1   SOURCE MATRIX #2   DESTINATION MATRIX

FILE TO FILE / REGISTER TO TABLE "EXCLUSIVE OR" INSTRUCTION

Fig. 12.9 Logic EXCLUSIVE OR Instruction

corresponding locations will the destination matrix also contain a zero. The logic OR operation often finds use as a means to check for one statuses in multiple data memory words. An example of the OR instruction will be combined with the EXCLUSIVE OR instruction.

Fig. 12.9 shows the Boolean equation and logic truth table for the EXCLUSIVE OR instruction. As can be seen from the diagrams, the EXCLUSIVE OR operation examines two matrixes for opposite bit statuses in corresponding matrix bit locations. If the corresponding bits in the source matrixes are the same, a zero is set in the corresponding destination matrix bit location. A one is set in the destination matrix only when the two bits in the same locations of the source matrixes differ.

To illustrate the operation of both the INCLUSIVE OR and the EXCLUSIVE OR instructions better, consider an application where 32 reversible motors are to be monitored for operation. Information on the direction of rotation of each motor is not as important as the fact that a motor is operating when instructed by the control system. One possible method to provide this type of monitoring capability is with the use of both OR-type instructions. The fact that a motor is running either in the forward *or* in the reverse direction is first obtained. This information is then compared with the current motor starter auxiliary contact status by an EXCLUSIVE OR instruction to determine when an instruction to run is not accompanied by the fact that the motor is operating. One advantage of this routine is that only one input point is required for each motor. The forward and the reverse auxiliary contacts of a motor starter are wired in parallel to a single input point.

Fig. 12.10 shows the wiring of 32 motors to the input and output modules of a Modicon 584 PC. The I/O references have been provided for later use with the programming instructions. Fig. 12.11 shows the instructions and data table statuses for each of the function blocks in the program. Note that an intermediate storage or work matrix is required for this application. This work matrix provides a location to store temporarily the results of the OR operations and the position of any motor that operates incorrectly. The work matrix has been designated for data memory locations 40625 through 40656.

The input and output addresses for all of the motor I/O devices can be used directly with the logic instructions. There is no need to move the I/O statuses into the data memory area prior to instruction operation. The Modicon instruction set permits the use of source data from either the input and/or the output image table areas as well as from the data memory area. Destination results from an instruction may be placed directly in data memory or in output image table memory.

Node 1's instruction block in Fig. 12.11 clears the work matrix to all zeros. This EXCLUSIVE OR instruction is necessary if the data gathered in the previous scan are not to affect the data being calculated in the current scan. The use of an EXCLUSIVE OR instruction provides the clearing of the work matrix through both source matrixes being the same. The EXCLUSIVE OR function will always detect the same bit statuses in corresponding source matrix locations, and will therefore place a zero in the same destination location.

A pair of OR instructions comprise nodes 2 and 3. Node 2's OR instruction ORs the current forward output statuses with the recently cleared work matrix. Any motor operating in the forward direction will have a one bit set in the work matrix. Those motors that are also operating in the reverse direction have their output status ORed with the forward statuses already stored in the work matrix by the OR instruction of node 3. For memory efficiency, nodes 1 and 2 could have been replaced by a word to word move instruction. The forward motor

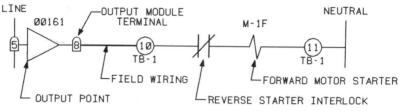

TYPICAL FORWARD MOTOR STARTER WIRING

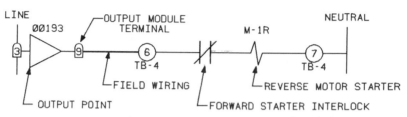

TYPICAL REVERSE MOTOR STARTER WIRING

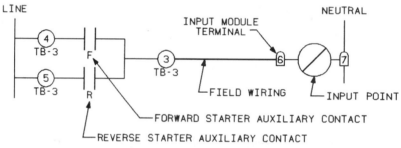

TYPICAL MOTOR STARTER AUXILIARY CONTACT WIRING

MOTOR NUMBER	FORWARD OUTPUT ADDRESS	REVERSE OUTPUT ADDRESS	AUXILARY CONTACT ADDRESS
1	00161	00193	10129
↓	↓	↓	↓
16	00176	00208	10144
17	00177	00209	10145
↓	↓	↓	↓
32	00192	00224	10160

I/O MODULE ADDRESSES

Fig. 12.10. Motor Monitor Circuit I/O Wiring and Addresses

statuses would be moved, and the work matrix cleared in one opertion, instead of the two shown in Fig. 12.11.

Comparison of the statuses of the auxiliary contacts with the current running status of each motor is provided by another EXCLUSIVE OR instruction in node 4. This EXCLUSIVE OR instruction places a one in the work matrix location where the corresponding bit of the input matrix (the ORed result of the forward

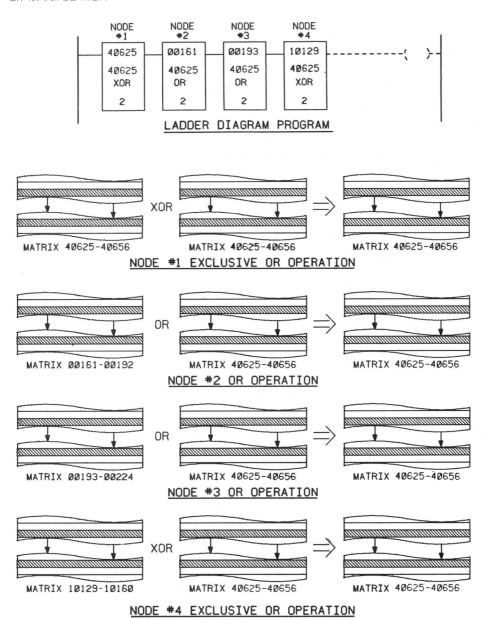

Fig. 12.11. Motor Monitor Program and Table Assignments

and reverse output signals) does not match. The use of additional instructions such as bit sense will allow the user to identify the offending motors for corrective action.

The remaining logic bit operation to be examined is the COMPLEMENT instruction. It differs from the AND, INCLUSIVE OR, and EXCLUSIVE OR functions in several ways. Only one source matrix is required for the COMPLEMENT instruction, and this matrix can consist of either a single memory word or a group of data memory words, as shown in Fig. 12.12. The truth table for the COMPLEMENT instruction shows that this function takes the current zero or one status of a matrix location and places its complement or opposite state in the equivalent destination matrix. For every one in the source matrix, a zero is placed in the corresponding destination matrix location. Where a zero occupies a source matrix location, a one is placed in the same location in the destination matrix.

BIT SHIFT/ROTATE INSTRUCTIONS

There are two bit shift instructions commonly found as part of the instruction set for PCs offering bit manipulation instructions. The *right shift* instruction moves all bits in a matrix word or register to the next bit location or position right of its current location or position. Bits that occupy the rightmost location of a matrix data memory word are shifted to the leftmost location of the next numerically greater matrix data memory word.

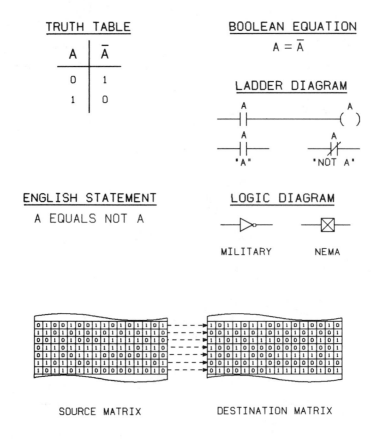

FILE TO FILE / REGISTER TO TABLE "COMPLEMENT" INSTRUCTION

Fig. 12.12. Logical "Complement" or "Not" Instruction

A matrix can also be operated on by a *left shift* instruction. The left shift instruction moves all bits in a matrix word or register left one bit location or position. Bits in the leftmost position of a matrix data word or register, are shifted to the rightmost location or position of the next numerically higher matrix data word. Fig. 12.13 shows the bit movements for both the right shift and the left shift bit instructions. Fig. 12.14 displays the before and after matrixes for a right shift and a left shift instruction.

Every matrix shift instruction requires an input bit and generates an output bit. The input bit and output bit locations are labeled in Figs. 12.13 and 12.14 for each shift instruction. The output bit is bumped from the matrix as a result of the shift, while the input bit is set to a one or cleared to a zero by the instruction after the instruction completes the shift operation. When the PC manufacturer permits a matrix to be specified as a certain number of bits long instead of a number of words long, the location of the output bit for a right shift, or the input bit for a left shift, may change from the leftmost position of the last matrix data word to any of the other bits of the last matrix data word. Those manufacturers that require the matrix to be specified as a number of data memory words do not provide for movement of bits in and out of the middle of the last matrix data word with the shift instruction. For these PCs the use of additional instructions will be required, such as *bit sense* and *bit clear* or *bit set*.

The bit shift instructions can be transformed into *bit rotate* instructions by designating the same address for both the infeed and the outfeed bits. The number of bit positions rotated will be equal to the number of bit locations in the matrix plus one, the extra position coming from the infeed/outfeed location. The matrix can be rotated either to the left or to the right by selecting the proper direction desired. In lieu of using the bit shift instruction as a bit rotate instruction, several PCs offer a dedicated bit rotate instruction. Those units offering the bit rotate instruction automatically move a bit from the outfeed location of the matrix back to the infeed location.

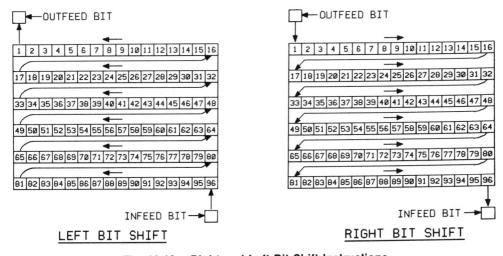

Fig. 12.13. Right and Left Bit Shift Instructions

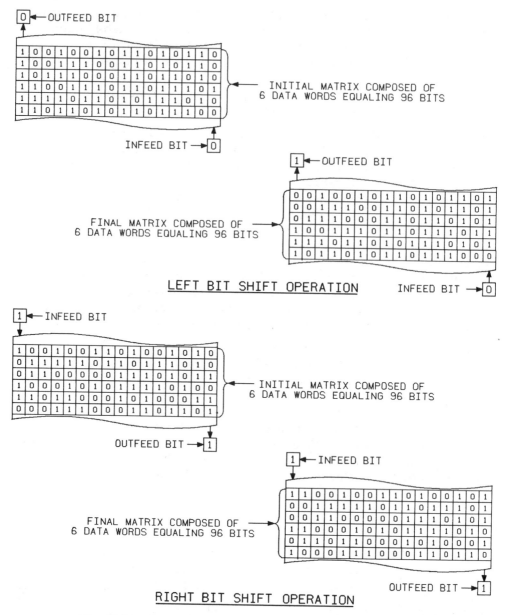

Fig. 12.14. Examples of Right and Left Bit Shift Instructions

The bit shift/rotate instructions are very useful for the construction of ring counters and shift registers. An example use of the shift instruction will be combined with the bit set and bit clear instructions.

MATRIX COMPARISON

Many processors permit any matrix of bits in a PC's data memory to be compared with another matrix of bits located in either the PC's input or output status memory, or in another area of the data memory. This comparison is performed by the *matrix compare* instruction. This instruction examines two matrixes bit by bit for a mismatch between corresponding locations, and indicates the position of that mismatch. Fig. 12.15 shows 48 input devices grouped together as a matrix. The compare instruction continuously monitors these inputs against a predetermined set of standard values. As the machine moves from step to step, a new set of standards is used for the comparison. If any step is detected to have a bit location miscomparing, the machine will be stopped immediately and the discrepancy indicated. Fig. 12.16 shows part of the logic necessary to perform this monitoring routine in Modicon format.

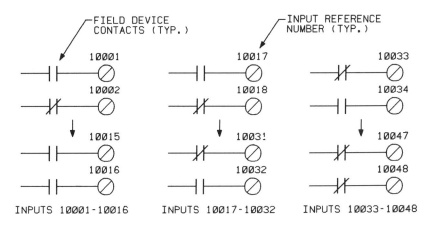

INPUT DEVICE CONNECTIONS

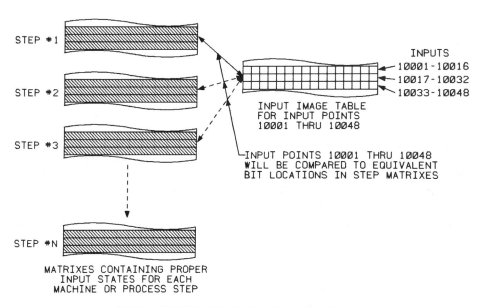

DATA MEMORY COMPARISON FILES

Fig. 12.15. Matrix Compare Example—Operation

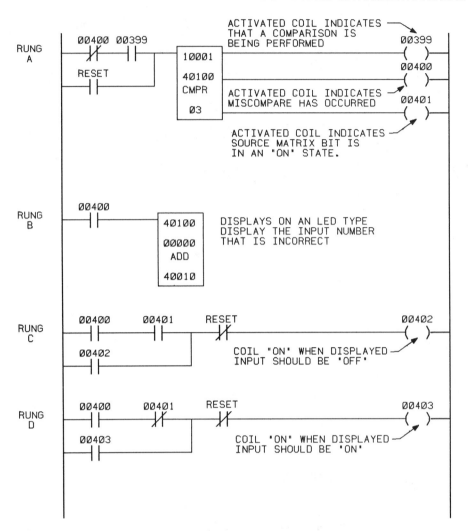

Fig. 12.16. Matrix Compare Example—Programming

The first ladder rung of Fig. 12.16, rung A, contains the actual compare instruction. The upper portion of the instruction block points to the first input point to be monitored. 48 inputs will be monitored since the lower portion of the instruction block contains a 3, representing three input status words of 16 points per word. The middle section of the instruction block contains the address of the pointer that indicates the numerical matrix location of the input bit which does not match the predetermined status. The three data memory words which contain the standard input point statuses are designated as the three next higher data memory words above the address assigned for the miscompare pointer. Movement of the correct standard values for the miscompare is not shown as part of this logic. A data memory move instruction will be required to move the current step's standard input point values into the comparison memory area at the proper time.

Three status indicators are provided as part of the Modicon compare instruction. The uppermost output indicator indicates that the comparison is active. The middle output status indicator is programmed to energize coil instruction 00400. The function of this status indicator is to signal that a miscompare has occurred. Contact instructions referenced to the coil instruction driven from the middle status indicator can be used elsewhere in the control logic to inhibit operation of the machine. A normally closed contact instruction from coil instruction 00400 is also used to inhibit the compare from continuing. The bottom output status indicator reflects the current status of the source matrix bit that is miscompared. This indicator is connected to coil instruction 00401.

Ladder rung B of the example program is an addition instruction that moves the current pointer value, which indicates the location of the miscompare, to an output location. The addition instruction is used here simply to move information from one data memory location to another. This rung is activated by a contact instruction from coil instruction 00400. The output location specified would have a digital display connected to it for the purpose of displaying the input address of the faulty bit. The number that is displayed is the actual input point reference number that has miscompared times 10,000. Note that if a different group of input addresses were involved, the addition instruction could provide the proper offset correction during the moving of data between locations.

The remaining two ladder rungs indicate the manner in which the input miscompares. Rung C is activated when there is a miscompare (coil instruction 00400 ON) and the input point should be OFF (coil instruction 00401 ON indicating that the input bit is also ON). Ladder rung D operates in a manner opposite to that of rung C to indicate that the input point should be ON. Note that rungs C and D require sealing in since the status indicators of the compare instruction are active only when the instruction is active. The use of the reset contact clears the indicator, and also restarts the compare instruction along with the operation of the machine.

DIAGNOSTIC INSTRUCTION

Another instruction that is often offered in place of, or in addition to, the compare operation is the *diagnostic* instruction. This instruction can be used in a similar manner as the compare instruction to verify the status of large amounts of data, or to perform machine diagnostic routines. Operation of the diagnostic instruction may require the programming of a logic EXCLUSIVE OR function to first generate a data memory table or file of miscompares between a file or table of actual system statuses and a file or table of standard statuses. Once this file or table of miscompares is generated, a diagnostic instruction can be performed to determine the location and type of error.

When the diagnostic instruction is executed, a search of the miscompare file or table is performed to find the first location of a miscompare. A fourth file or table is then required for the purpose of listing the location and nature of the error. When the diagnostic instruction is being performed on I/O points, the exact I/O point that is faulty is listed in the error file along with an error number to identify the quantity of the miscompares found by the instruction to this point in the instruction's operation. When internal data are being verified, the data memory word and the bit in question are displayed. Once an error is located in the miscompare file or table, the diagnostic instruction is halted so that the error information can be placed in the error file or table. Unless the diagnostic instruction is then stopped from further operation on the next scan, the instruction will pick up with the next bit from where it stopped. The search for another miscompare will continue, and if another miscompare is located, a new address and numerical identifier code will be placed in the error file or table in place of the previous error information. When the diagnostic instruction reaches completion, an indication that the instruction has been completed is signaled. If the instruction is still active on the next processor scan, the search may be reinitiated at the beginning of the miscompare file or table, or the instruction may wait for the next false-to-true transition of the control input.

BIT MODIFY INSTRUCTIONS

There are two instructions which comprise the bit modify group of instructions. The first of these instructions is the *bit set* instruction, which permits the setting of any bit in a matrix to a one state. The ability to set any bit in a matrix to a zero state is provided through the use of a *bit clear* or *bit reset* instruction. The bit modify instruction group acts on only one bit at a time, requiring the use of multiple instructions to change more than one matrix bit per scan.

The bit modify group of instructions has several modes of operation, depending on the manufacturer of the PC. The simplest mode is the single-shot mode. This form of operation specifies a particular bit in a matrix, and this bit is either set or cleared every scan that the command is activated. If the matrix specified for the bit set or bit clear operation is later shifted by one of the bit rotate instructions, the new data that are shifted into the location specified by the bit modify instruction will be altered on the next scan that the modify instruction is activated.

Some manufacturers design the bit modify instruction to progress continuously through a matrix and change the status of each bit of the matrix. This automatic mode of operation requires more synchronization between the current position to be modified, and whether the bit in that location is to be set or cleared.

Operation of a bit modify instruction in the automatic mode requires the use of a counter or pointer to indicate the current bit location that is to be or has been modified. Operation of the bit modify instruction in the automatic mode requires that the counter or pointer be set to the desired starting bit in the matrix, and that the instruction be activated. Each bit location is either set or cleared, depending on the instruction, at a rate of one bit location per scan. The instruction will continue through the matrix at a one bit per scan rate until the end of the matrix is reached. At the end of the matrix, the pointer or counter may be reset to the start of the matrix, so that on the next scan the first bit of the matrix will be modified. Note that in lieu of resetting the counter or pointer automatically to the beginning of the matrix, some bit modify instructions require that the counter or pointer be reset to the beginning of the matrix by the user logic before the instruction can be reactivated.

When the automatic mode of operation is offered, a submode is also usually available. This mode does not allow the pointer or counter to increment automatically, but instead holds it at a set value or bit location. The instruction then operates in a manner similar to the single-shot mode, except that it is possible to force the pointer or counter to different matrix locations on each scan prior to a bit modification. Use of this submode allows the user to program a single instruction and use it each scan to change selected bit statuses. However, if more than one bit location must be changed in a single scan, multiple instructions will still be required.

BIT SENSE INSTRUCTIONS

Two additional bit instructions that are available are the *bit sense* or the *bit examine* instructions. These instructions allow the examination of any bit in a matrix for either a one or a zero condition. They only check the status of a particular bit in a matrix and do not alter the current status of the bit being examined.

As was the case with the bit modify instructions, two operating modes are available. The single-shot mode of operation specifies a particular matrix bit as part of the instruction, and every scan the instruction receives power the specified matrix location is examined. In the automatic mode a counter or pointer is employed to increment automatically the instruction from bit to bit location in the matrix. Again an automatic submode is usually offered that permits the incrementing of the counter or pointer to be halted. The counter or pointer can then be externally changed by the user program, enabling the user to jump around within a matrix, sensing any matrix location's status.

To show a possible application which uses several of the commands discussed in this section, consider the conveyor system depicted in Fig. 12.17. This application uses a conveyor to move parts along a test bed for automated testing and inspection. Parts are placed on the conveyor at the left and indexed past four different test stations. At the end of the conveyor the parts are removed and sorted as to good or bad. As each part moves down the conveyor, the status is to be kept indicating whether the part is good or bad. If a part is found to be faulty, it is tracked for the length of the conveyor so that no additional tests are performed past the last test which the part failed. Sorting of the parts at the end of the conveyor is done according to the status of the part. The conveyor indexes 16 times in order for a part to move from one end to the other. Parts placed on the conveyor are verified at index position 1. If a part fails to be

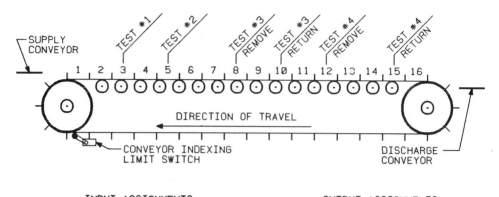

INPUT ASSIGNMENTS

10036 - CONVEYOR INDEX COMPLETE
10096 - TEST #1 COMPLETE
10097 - TEST #2 COMPLETE
10098 - TEST #3 COMPLETE
10099 - TEST #4 COMPLETE
10100 - TEST #1 FAILED
10101 - TEST #2 FAILED
10102 - TEST #3 FAILED
10103 - TEST #4 FAILED
10104 - PART FEED
10105 - PART AT POSITION #1

OUTPUT ASSIGNMENTS

00584 - INDEX TO NEXT POSITION
00585 - EXIT PART STATUS
00203 - PERFORM TEST #1
00205 - PERFORM TEST #2
00207 - PERFORM TEST #3
00209 - PERFORM TEST #4

MATRIX ASSIGNMENTS

40121 - 16 BIT MATRIX

Fig. 12.17. Test Conveyor Example

placed on the conveyor, a faulty part is indicated, so no tests are performed on a vacant location.

Three types of instructions will be used to implement the program for this application. The three instructions are bit rotate, bit sense, and bit modify. The Modicon 584 format for each of these instructions is shown in Fig. 12.18. The Modicon bit rotate instruction is capable of either shifting or rotating a matrix of bits in either a left or a right shift motion. The bit modify instruction is capable of modifying any designated matrix location to either a zero or one state, while the bit sense instruction can examine the status of any designated matrix bit.

The logic required for this application is shown in Fig. 12.19. Rung A initiates the indexing of the conveyor from position to position. The signal to index is generated whenever all four test stands have completed their respective tests, and a new part has been fed onto the conveyor system. Once generated, this signal is sealed until the index cycle is complete, as indicated by limit switch 10036.

The function of rung B is to provide the shifting of the matrix, one bit position to the right, every time the conveyor mechanically indexes. The matrix indexing operation is initiated by the transitional (one-shot) action of the contact referenced to coil instruction 00584 at the beginning of rung B. The bit rotate instruction is programmed as a right bit shift instruction since neither the second nor the third instruction inputs are powered. Use of the middle output function of the bit rotate instruction to drive coil instruction 00585 provides an indication as to whether the part exiting the conveyor is good or bad.

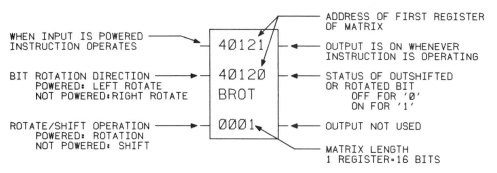

BIT ROTATE/SHIFT INSTRUCTION

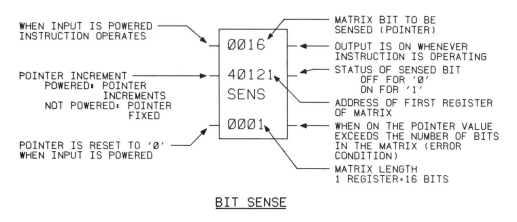

BIT SENSE

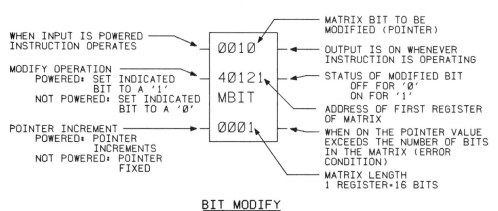

BIT MODIFY

Fig. 12.18. Conveyor Example Bit Instructions

Program rungs C, E, G, and I permit a test to be performed. Rung C initiates test 1; rung E initiates test 2; rung G initiates test 3; and rung I initiates test 4. Since a test can only be performed when the conveyer is stopped, a normally closed contact instruction is used to initiate the bit sense instruction. The function of the bit sense instruction is to examine the corresponding matrix

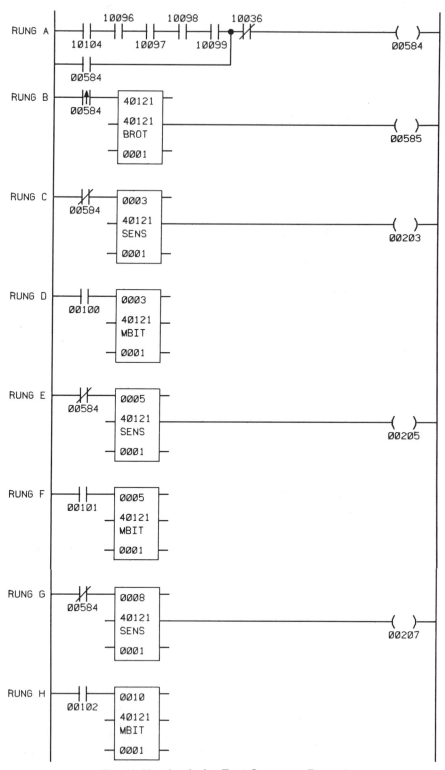

Fig. 12.19. Logic for Test Conveyor Example

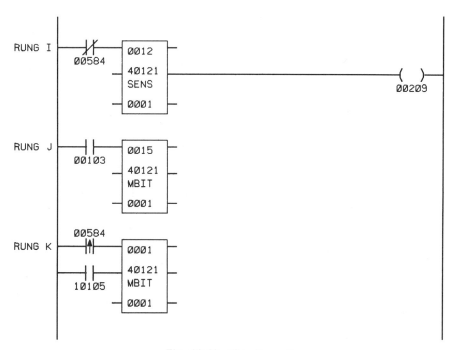

Fig. 12.19. *(Continued)*

location for a good/bad indication of the part currently at the test stand. If the part is good, and therefore available for the test, the output coil instruction for that test stand will be activated signaling the start of another test. Should the output coil instruction not be energized, the test will not be performed.

The remaining ladder rungs, rungs D, F, H, and J, are used to set a matrix bit to zero should a part fail a particular test. The bit modify instruction is used to perform this operation. It is initiated by any part failing a particular test. Since the middle input to this instruction is not used, whenever the instruction is activated, the matrix location corresponding to that position on the transport conveyor is set to zero. If the part passes the test, no matrix modification is required, and the test complete signal indicates that the conveyor can be indexed whenever the remaining tests are completed.

The last rung of logic for this example, rung K, involves sensing position 1 for the presence of a part for testing. A bit modify instruction is used for this operation also. The upper instruction input is operated from a transitional contact referenced to the index coil instruction 00584. Whenever the index operation is complete, the bit modify instruction is activated for a single processor scan. Matrix location 1 is modified according to the status of contact instruction 10105. Contact instruction 10105 is driven from a field input device, indicating whether a part has been placed on the conveyor in location 1 for testing.

13

Special Languages
and Instructions

Practically every PC model has one or more features that make it different from the rest of the units it competes against. While many of these features are associated with the overall design of the hardware, others are software-oriented and/or hardware/software-oriented. This chapter will examine many of the "special" instructions, functions, and alternate programming languages that seem to have become more "standard" among the special offerings. While this chapter will not examine each and every special instruction, function, and/or language that is available, it should give the reader an excellent overview of what is offered.

Any special instruction, function, and/or language can be either a hardware, a hardware and software, or strictly a software enhancement that gives a particular model or models of PC the ability to perform a function or operation more easily. Early PCs incorporated special instructions that enabled the unit to perform proportional/integral/derivative (PID) control of continuous processes. Enhancements in PID control schemes have combined special-purpose I/O modules with more user-friendly programming instructions. The ability to program in computer languages such as BASIC provides special system flexibility in addition to the more standard relay ladder language. Many PCs program in Boolean formats and symbols, permitting yet another dimension to their programming flexibility.

Many special-purpose I/O modules have been developed to expand the application horizon of the PC. Where accurate positioning is required, a host of *servo, axis positioning,* and *stepper* modules have been introduced. *Machine diagnostic modules* allow a PC to "learn" the operation of a machine, then relentlessly monitor its every action for a hint of trouble in order to alert the operator in time to avoid serious problems. Through the use of *ASCII I/O* modules, RS 232C communications are available at every I/O rack to serve numerous application requirements. In fact a new class of I/O modules has appeared in conjunction with the introduction of specialty instructions, functions, and languages. This new class of I/O module has been termed the "intelligent" I/O module, since many incorporate microprocessors which prehandle the I/O information prior to PC processor use. In some cases the intelligent I/O module has the ability to operate in a reduced capacity or manual mode without PC processor support.

PROPORTIONAL/INTEGRAL/DERIVATIVE

Many control applications involve some form of closed-loop control to maintain a process characteristic such as a temperature, pressure, flow, or level at a desired value. As time progresses, the process characteristic will change as a result of other process and system variations. Every continuous process characteristic has a desired setpoint for ideal conditions. Under operating conditions the process characteristic in question is measured, and an error value is calculated for the amount that the measured value deviates from the desired value. The error that is detected is then used to modify the control devices in the loop to compensate for the amount of error.

Fig. 13.1 provides an equation that represents the standard PID control action. The equation consists of a proportional term, an integral term, a derivative term, as well as a desired setpoint term. The PID equation includes an error term generated from a second equation also shown in the figure. Most PCs solve the integral and derivative portions of the PID equation as an approximation to the actual terms of the standard PID control equation. The PID equation in its PC approximated form is shown in the lower portion of the figure.

$$V = (V_S) + (K_P * E) + (K_I \int_0^t E \, dt) + (K_D \frac{dE}{dt})$$

V=CONTROL VARIABLE $\quad$ K_I=INTEGRAL GAIN
V_S=OUTPUT SET POINT $\quad$ K_D=DERIVATIVE GAIN
K_P=PROPORTIONAL GAIN $\quad$ t=TIME
E=ERROR

PID CONTROL EQUATION

$$E = SP - PV$$

E=ERROR $\quad$ SP=DESIRED SET POINT
PV=ACTUAL PROCESS CHARACTERISTIC

PID ERROR EQUATION

$$V = (V_S) + (K_P * E) + (K_I \sum_0^t E \, \Delta t) + (K \frac{\Delta E}{\Delta t})$$

WHERE $\sum_0^t E \Delta t = (E_1 * \Delta t) + (E_2 * \Delta t) + \ldots + (E_N * \Delta t)$

PC PID APPROXIMATION EQUATION

Fig. 13.1. PID Process Control Equations

Probably the best way to examine the operation of the PID functions of a PC is to explore a typical process characteristic and how the PID function acts to control that characteristic. Controlling the level of a liquid in a tank or vessel represents a classic example of a process control application requiring typical PID control. As illustrated in Fig. 13.2, the closed-loop control components consist of an analog input signal from the level sensor and an analog output signal from the PC to the fill control valve. The closed-loop operation of the system results in the tank level being measured, this measurement being compared against the desired setpoint by the PID controls, and a corresponding output signal being sent to the fill control valve to adjust for any error between the actual and the desired liquid levels. Depending on the amount of liquid flowing from the tank, the amount of liquid entering the tank will have to be adjusted to prevent the tank from becoming empty or running over.

The simplest approach to controlling the liquid flow to the tank is a simple ON-OFF type of control action. Whenever the tank level drops below the desired level, the fill control valve is fully opened until the desired level returns. This form of control will usually result in the control valve being quickly cycled. The low level condition will be detected, and the valve opened. If the infeed liquid rate is much greater than the outflow rate, the tank will fill rapidly and the infeed flow stopped as soon as the desired level is reached. Depending on the various rates of infeed and outfeed liquid flow, the actual tank level will vary some amount from the desired value. The infeed control valve will either cycle at a slow rate, causing large variations in the level in the tank, or cycle at a fast rate, keeping the level fairly constant but imposing undue wear on the control valve. Fig. 13.3 illustrates in graphic form the action of the control valve with respect to the liquid level in the tank over time.

Closer examination of the system reveals that there will always be some flow rate of liquid from the tank which can be replaced by a similar infeed flow rate to the tank. A control system can be devised that maintains a constant infeed

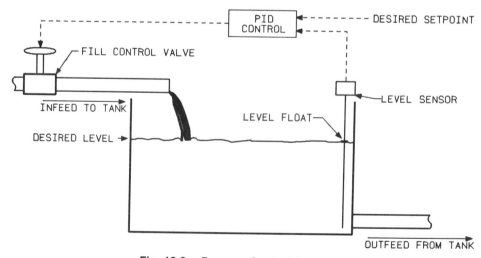

Fig. 13.2. Process Control Example

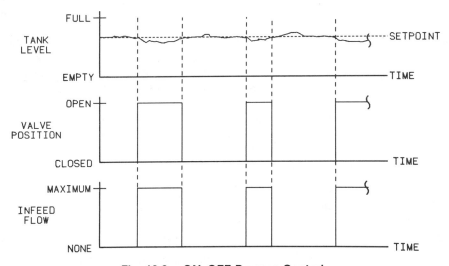

Fig. 13.3. ON–OFF Process Control

rate and adjusts this rate according to the difference between the desired level and the actual level in the tank. This type of control is often referred to as *proportional control*. As the liquid level changes, the controller changes the infeed valve position in proportion to the difference, or error, in the actual versus the desired level of fluid in the tank. Fig. 13.4 illustrates proportional control for the tank.

Examination of the graphs of Fig. 13.4 illustrates that the tank's liquid is maintained much closer to the desired setpoint than is achieved with simple ON-OFF-type control. The proportional controller does not permit the level to return to the setpoint unless the outfeed flow rate returns to the same value as was present at the time the setpoint was determined. Proportional controllers

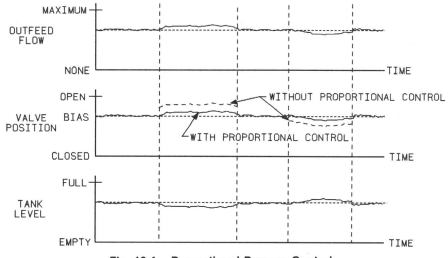

Fig. 13.4. Proportional Process Control

only generate a signal in relationship to the amount of error in the actual versus the desired process characteristic. A proportional controller will not compensate for long-term excesses or deficiencies in the process characteristic with respect to time. The position of the infeed valve that allows the tank level to be maintained at the setpoint is termed the *bias* point for the valve. The infeed valve will be adjusted above or below the bias point in relationship to the amount of actual versus desired error in the closed loop, and will not be changed to compensate for any long-term excesses or deficiencies which may occur.

With the addition of integral action the tank level can be returned to the setpoint level over time as flow conditions vary. Integral action or control, in conjunction with proportional control, evaluates the amount of long-term error that exists between the desired and the actual tank levels, and changes the valve bias position to return the system to the desired setpoint level. The tank liquid level is maintained much closer to the setpoint level than with proportional control, as illustrated in Fig. 13.5. With proportional control the magnitude of the process characteristic is compensated for, while integral control compensates for the time that the error persists.

Often a process system will be slow to respond to changes made by the control system. These delays in system response are usually due to numerous factors relating to the design of the system. For example, in our example system if the tank is reasonably large, it may be some time before the integral action of the controls can return the tank level to the setpoint level. The tank infeed flow would surely be designed to handle the worst-case average outflow from the vessel, but the control system would not quickly compensate for any sudden outflow surges that might occur. The introduction of derivative control permits the control system to respond more quickly to sudden process characteristic changes.

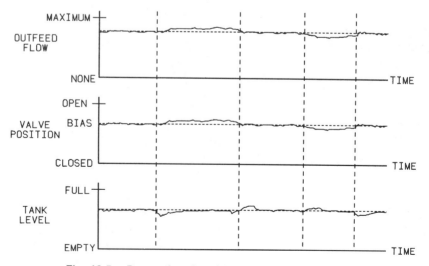

Fig. 13.5. Proportional and Integral Process Control

The derivative action of a PID controller responds to the rate at which the actual versus desired error is generated. Where this error is generated slowly, the derivative action of the control provides little compensation to the setting of the control valve. When a large error develops over a relatively short time, the derivative action of the control is such that the system responds more quickly.

In our example, should a large flow of liquid occur from the tank, a large error would be generated in a relatively short period of time. The proportional action of the controls would begin to compensate for the increase in outflow with an increase in infeed flow. Since the level would have dropped in the tank, the integral action of the controls would provide some additional amount of infeed valve control to return the tank level to the setpoint level. Since the outfeed change occurred over a short period of time, the derivative action of the control would provide additional control to the integral signal so that the tank level could return to the setpoint value sooner than if only integral action were operational. Fig. 13.6 illustrates the quick response of full PID control.

The PID control of any process characteristic provides for the characteristic to be corrected under normal operating conditions in three ways. The magnitude of the difference in desired versus actual characteristic level or rate is affected by the proportional component of control. Integral control examines and compensates for the time duration of a desired versus actual process characteristic difference. Finally the rate at which the actual versus desired process characteristic difference develops is compensated for by the derivative component of the control system.

The tank filling example used to explain PID control of a process characteristic was quite simple in operation. Often process control loops will be dependent on other process loops. For example, if the tank of the previous example were a water-holding tank for a small community, the control might be much more complex. The infeed water might come from deep wells and require filtering and chemical treatment. The filtering equipment might be

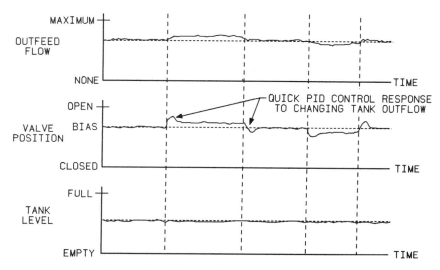

Fig. 13.6. Proportional, Integral, and Derivative Process Control

monitored for efficiency which would affect the infeed flow rate. If the incoming water needed treatment for hardness, minerals, or similar undesirable conditions, the chemical treatment equipment might monitor the infeed water flow as well as the overall tank water conditions. Chemical treatment would be added to the incoming water flow as required to ensure that the outfeed water incoming from the tank met community standards.

A PC that performs PID calculations is Texas Instruments' PM550 PC. The PM550 includes special loop access modules (LAM) for the monitoring of process loops programmed in the PC. The LAM provides an operator with the ability to selectively change process loop variables, such as setpoint, gain, or other tuning parameters. Any loop parameter can be displayed, loop control can be changed from manual to automatic, and all alarms and warnings can be displayed. The PID programming for the PM550 can be configured for ratio control, cascade control, and feedforward control, to name several.

LOOP SPECIFICATION SHEET

Loop Description:

PM550 Loop Number_____

MEMORY ALLOCATION
Tuning Constants in____(C or V)

	Beginning Address	Tune C	Tune V	Ending Address
Constant Table	_____	_____	_____	_____
Variable Table	_____	_____	_____	_____

Are loop flags for alarms and mode switching allocated in the image register?____ If yes, give beginning address:_____

PROCESS VARIABLE   Address:_____
20% Offset?_____   Square root?_____
Special calculation?_____   If yes, give address:_____
Low Range =_____   High Range =_____
Engr. Units_____   -Transmitter_____

CONTROL CALCULATIONS   Sample Time =_____
Remote Set Point?_____   If yes, give address:_____
Special Calculation?_____   If yes, give address:_____
LOCK - Set Point?_____   Auto/Manual?_____Cascade?_____
Error Squared?_____   Error Deadband?_____
Gain =_____   Reset Time =_____
Rate =_____   Reverse Acting?_____
Output Address: _____   20% Offset?_____

ALARMS
Process Variable
       Low =_____   High = _____
Deviation
       Yellow =_____   Orange =_____

Fig. 13.7. Loop Specification Sheet

In order to provide the reader with an example of the programming necessary for implementation of a PID control loop, a quick description of the PM550 program will be reviewed. The programming of the PID program for the PM550 is well explained in the Texas Instruments manuals, and should be reviewed when using the PM550 for process control. The actual PC program has been divided into four categories.

Seven programming responses are required to allocate a portion of the processor memory for the storage of loop configuration and loop data parameters. Once the processor has been configured for the control loop, six additional programming responses provide information relating to such items as high and low ranges, whether or not certain options are to be in effect, and if any special process characteristic calculations must be performed. Programming responses 14 through 21 specify information concerning the PID control action in the processor. Such information as setpoint, gain, reset time, and derivative time are specified during these responses, as are the control locks (setpoint, auto/manual, cascade, etc.) and special calculation data locations. Finally four program responses are devoted to the programming of high and low alarm limits and two deviation limits.

The programming sequence of the PM550 PID program is completely prompted. The programmer simply answers the questions on the programming terminal as they are asked. The questions asked by the programming terminal during PID programming are backed up with a detailed description and discussion in the PM550 manuals. The programmer need only answer each program terminal question with either numerical data, a memory address, or a *yes* or *no* answer. A special loop specification sheet is provided with the PM550 manuals to assist and document the programming of the control loop. An example of a loop specification sheet is shown in Fig. 13.7. Once the loop has been programmed, 60 error conditions can be recognized by the PID programming and displayed to the user.

AXIS POSITIONING AND STEPPER MOTOR CONTROL

Another group of intelligent I/O modules are those that can directly control a drive or stepper motor. These modules are designed to receive *move* commands from the user program, and convert those commands into pulse trains for use in operating or controlling a drive or stepper motor. Each drive or stepper motor that can be controlled is often referred to as an *axis*, since the drive or stepper will usually cause motion about some machine axis. Once the move commands are complete, the module signals the user program that the requested move or moves are complete and the module is ready to receive additional commands. While many I/O modules can only control a single drive or stepper motor, others permit multiple drives or steppers to be used for applications requiring multiple axes of positioning.

Generally two modes of operation are available. The first mode is often referred to as the *single-step mode*. In this mode of operation one or more move commands may be sent to the I/O module. Each move command is executed by the module in the order received, as illustrated in Fig. 13.8. When a move is complete, the module signals the processor with a status signal. The user program must then signal the module when to begin the next move. This mode of operation is ideal for applications that require dwell times or other system functions to occur between program moves.

When a dwell time between moves is not required, a *continuous mode* of operation may be available. This mode permits the module to cycle through a complex sequence of programmed moves without user program intervention. The module carries out the directed moves independent of the processor scan. Any moves programmed in the continuous mode blend into a smooth control action of programmed accelerations and decelerations.

When two or more axes are provided with a positioning module, they may be controlled independently or synchronized. When an axis is independently programmed, it operates completely independent of any other axis. Should it be necessary to coordinate the motion of two or more axes, a synchronized mode can be used. In this mode each axis must complete its given move prior to any axis being allowed to continue, as illustrated in Fig. 13.9.

Many axis positioning/stepper motor I/O modules can be configured for either open-loop or closed-loop control. In open-loop configuration the module simply sends control signals to the drive or stepper and does not monitor the axis's true position. When closed-loop control is used, additional position-sensing hardware, such as encoders, are connected to the module to provide positive axis positional information. In addition a module may provide other I/O points for end of travel, jog, and home position signals.

Any processor of the General Electric Series Six PC family is capable of communication with the Series Six axis positioning module (APM). This module permits a PC user to apply PC-based control to any application involving axis

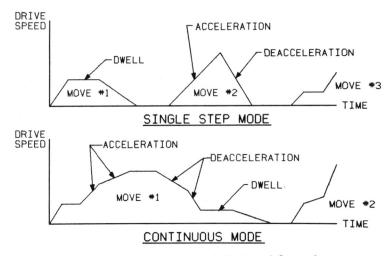

Fig. 13.8. Axis Positioning Modes of Operation

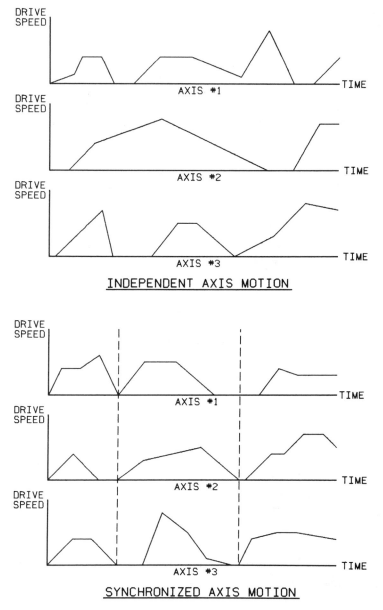

Fig. 13.9. Independent and Synchronized Axis Positioning Control

positioning. The Series Six APM permits the control of three axes per module, with over 60 modules being controllable from a single Series Six processor. The APM can store 10 profiles (axis position movement routines), and any or all profiles can be changed by the user program at will through a simple download of the new profile from the PC's data memory. Any module is capable of providing axis control over a distance of 1400 feet with a resolution of 0.001 inch. The resolution can be increased to 0.0001 inch in an overall maximum distance of 140 feet. The velocity of motion can reach 600 inches per minute at 0.0001-inch resolution or 6000 inches per minute at 0.001-inch resolution.

The APM commands fall into four major groups. The move commands specify the characteristics of a move such as velocity, acceleration, deceleration, dwell time, and positive positioning. Ten axis set-up commands define the various characteristics for each axis. These commands include instructions for setting the maximum velocity and acceleration, jog velocity, end-of-travel positions, and various servo constants. A third group of APM commands instruct the module on specialized tasks, such as percent of rate override, move at constant velocity, set axis position, and run a canned cycle. The final group of commands permit the APM to be directly controlled from ladder logic coil instructions. These commands permit the APM to be enabled and disabled, to abort all pending moves, to return to home, to execute any one of the 10 possible user programmed canned cycles, and to jog in either the forward or the reverse direction.

During each processor scan the APM is capable of returning data to the processor. These data can include error codes, numerical data, and discrete information relating to the current status of the module. Indicating lights on the module as well as approximately 60 error conditions can be examined and/or monitored by the user to pinpoint hardware and/or software faults. Numerical data relating to the current position and velocity of an axis are available for user program monitoring and use. Discrete signals relating to the module status include such items as servo ready, error condition detected, at position, waiting next move, and APM disabled.

While the actual programming of the APM is quite simple and will not be reviewed here, the reader should realize that the features of the Series Six APM permit any PC user to integrate motion and/or servo control easily into any PC application. The use of intelligent axis positioning/servo control I/O modules such as GE's APM permit the development of more flexible, higher productivity machines. These modules also extend to the user significant cost savings over conventional computer or numerical control systems for many applications. The ability to maintain and troubleshoot easily a servo system programmed in simple relay ladder instruction formats can produce extensive cost savings during the life of a system.

COMPUTER PROGRAM LANGUAGE MODULES

The PC has been generally marketed as a noncomputer means to control an industrial complex or piece of hardware. It has been designed to be programmed in a high-level language of symbols and brief statements, instead of a computer-based language of words and statements. Early PC users wanted nothing to do with computer languages such as BASIC, Fortran, Cobol, and Pascal. Somewhere along the line the idea of being able to program the PC in computer languages as well as relay ladder language became an option. Today more and more PCs are permitting the user to do programming in languages other than relay ladder language. Multiple language capability is being offered as a function of the programming terminal, or as a stand-alone intelligent I/O module.

One example of an intelligent I/O module dedicated to computer language programming is Industrial Solid State Control's model 386 COP. The COP operates independently of the ISSC IPC-300 PC, but interfaces to the PC through the PC's I/O structure. The COP module can provide either 16 K or 32 K of user RAM or ROM for user programs. Two serial interface ports are available for connection of the COP to external devices, such as color terminals and printer/displays. The COP programming language is a special form of BASIC, called *COPBASIC*. In effect the COP can be thought of as a small personal computer that is PC compatible.

Use of an intelligent I/O module such as the ISSC COP can greatly enhance the operation of a PC. String variables and arrays (alphanumeric data instead of numerical data) can be handled with relative ease. Messages such as "today is February 11, 1983" are stored easily in the COP and printed out by PC command. Note that many PCs permit the equivalent message to be stored. However, storing and manipulating it in the PC can often become quite a programming task. When an application requires complex calculations, such as log, sine, cosine, tangent, and exponent, the transcendental functions offered in COPBASIC are available. Random numbers can be generated as needed, the greatest integer less than or equal to a specified value can be provided, as well as the absolute value, and the sign (+, −, or 0) of a value can be determined by simple commands. Where necessary special user functions can be written in machine language and called upon either from the COPBASIC program or indirectly through the PC ladder language program.

Generation of a COPBASIC program is accomplished in the same manner as a BASIC program is generated for any computer. COPBASIC is an interpretive language, meaning that the COP microprocessor must interpret each statement prior to execution. The COPBASIC applications program is written and then entered through a terminal connected to one of the module's two access ports. Once entered, the program can be run and monitored through the terminal.

Communications between the COP and the PC are exchanged in a portion of the COP's memory, which is accessible to the PC processor. A 256-word data table in the COP can be written to and read from by either the COP or the PC processor. When the PC wishes to send data to the COP, the data are placed in a designated area of the COP data memory addressed in a manner similar to an I/O module. The COP reads this data memory to receive information and commands from the processor. The reverse operation occurs for information that must be sent from the COP to the PC. The COP places the outgoing data in its data memory where the PC will retrieve them on the next processor scan.

Fig. 13.10 illustrates both a PC program and a COPBASIC program for the generation of alarm messages. The PC program sets a number value in a specified data memory location (7) depending on the error message that needs to be generated. The COPBASIC program examines this data memory location (PEEK (X) command) and prints the proper message depending on the numerical value found in the location.

The use of computer languages similar to BASIC and Fortran greatly enhances the operation of a PC. The ability to generate a computer language program

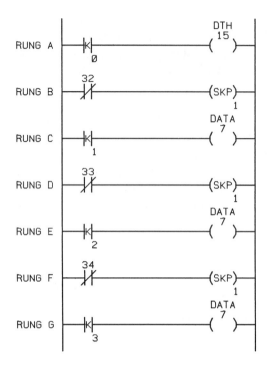

RUNG A INITIALIZES COP DATA
TABLE LOCATION 0 FOR THE TRANS-
MISSION OF INFORMATION BETWEEN
THE PC PROCESSOR AND THE COP
PROCESSOR. RUNGS C,E, AND G
INDICATE WHICH MESSAGE OR
MESSAGES ARE TO BE GENERATED
ACCORDING TO THE SKIPPING
ACTION OF RUNGS B,D, AND F.
RUNGS C,E, AND G ARE SKIPPED
UNLESS NORMALLY CLOSED CONTACT
INSTRUCTIONS 32,33, AND 34 ARE
NOT PASSING POWER TO THE
RESPECTIVE SKIP INSTRUCTION IN
RUNGS B,D, AND F.

ISSC PC PROGRAM CALLING MESSAGES FROM COP

```
10 LET X=0
20 IF PEEK (X)=0 THEN GO TO 20
30 IF PEEK (X)=1 THEN PRINT 'CAUTION
   - MOTOR OVERLOAD EXISTS'
40 IF PEEK (X)=2 THEN PRINT 'LOW
   AIR PRESSURE'
50 IF PEEK (X)=3 THEN PRINT 'EMERGENCY
   STOP CORD PULLED'
60 POKE (X,0)
70 GO TO 20
```

STATEMENT 20 CONTINUALLY
EXAMINES COP DATA LOCATION 0
FOR A NON-ZERO NUMBER. WHEN
A NON-ZERO NUMBER IS PLACED
IN LOCATION 0, STATEMENTS 30
THRU 50 PRINT THE INDICATED
MESSAGES, STATEMENTS 60 THRU
70 RESET THE MESSAGE GENERATION
ROUTINE

COP BASIC PROGRAM FOR MESSAGE GENERATION

Fig. 13.10. Example ISSC PC and COP Programs

which operates independently of the PC permits large volumes of data to be
handled without needless PC program complication or the need for a
computer. Intelligent computer I/O modules can provide another dimension to
the overall flexibility of a PC.

BOOLEAN EQUATION PROGRAMMING

Many PCs program in Boolean equation formats using Boolean operators
such as *AND, OR, NOT, NAND, NOR*. The use of Boolean operators and programming
formats is probably the second most widely used PC language other than the

relay ladder programming language. The popularity of Boolean is due to its similarity to the relay ladder language. An excellent example of a PC which programs in Boolean as well as in relay ladder language is the Eagle Signal EPTAK series of PCs. The EPTAK PCs provide the user with an extensive group of instructions and commands. A partial listing of the EPTAK Boolean commands, called EPTAK CONTROL LANGUAGE by Eagle Signal, can be found in Table 13.1.

Table 13.1. Eagle EPTAK CONTROL LANGUAGE Commands

LOGIC INSTRUCTIONS

LOAD	LOAD DATA FROM SOURCE INTO ACCUMULATOR
CLOAD	LOAD INVERTED DATA FROM SOURCE INTO ACCUMULATOR
STO	STORE ACCUMULATOR CONTENTS AT SPECIFIED LOCATION
AND	LOGIC AND ACCUMULATOR DATA WITH SPECIFIED LOCATION DATA
CAND	LOGIC AND ACCUMULATOR DATA WITH SPECIFIED LOCATION INVERTED DATA
IOR	LOGIC OR ACCUMULATOR DATA WITH SPECIFIED LOCATION DATA
COR	LOGIC OR ACCUMULATOR DATA WITH SPECIFIED LOCATION INVERTED DATA
XOR	EXCLUSIVE OR ACCUMULATOR DATA WITH SPECIFIED LOCATION DATA
XCOR	EXCLUSIVE OR ACCUMULATOR DATA WITH SPECIFIED LOCATION INVERTED DATA
COMP	COMPARE ACCUMULATOR DATA WITH SPECIFIED LOCATION DATA
BINV	INVERT ALL EIGHT BITS OF THE ACCUMULATOR
INV	INVERT ACCUMULATOR
NOP	PERFORM NO OPERATION
JMP	UNCONDITIONAL JUMP TO SPECIFIED STATEMENT LABEL
JEQ	JUMP TO SPECIFIED STATEMENT LABEL IF EQUALITY CONDITION IS SET
JLT	JUMP TO SPECIFIED STATEMENT LABEL IF LESS THAN CONDITION IS SET

ARITHMETIC INSTRUCTIONS

STET	LOAD 4 BCD DIGITS INTO ACCUMULATOR FROM SPECIFIED LOCATION
STOD	STORE ACCUMULATOR DATA IN SPECIFIED LOCATION
PLUS	ADD ACCUMULATOR DATA TO SPECIFIED LOCATION DATA
MINUS	SUBTRACT SPECIFIED LOCATION DATA FROM ACCUMULATOR DATA
MULT	MULTIPLY ACCUMULATOR DATA BY THE SPECIFIED LOCATION DATA
DIVID	DIVIDE ACCUMULATOR DATA BY THE SPECIFIED LOCATION DATA

SPECIAL PURPOSE INSTRUCTIONS

CALL	TRANSFER TO SUBROUTINE PROGRAM
RET	RETURN TO MAIN PROGRAM FROM SUBROUTINE PROGRAM
PRINT	PRINT SPECIFIED MESSAGE
KEYIN	ENTER DATA INTO PROCESSOR MEMORY FROM KEYPAD

The workhorse instructions of a Boolean programming language such as the EPTAK CONTROL LANGUAGE consist of the *load, store, AND, OR,* and *complement* instructions. The use of these elementary instructions is illustrated in Fig. 13.11. The first and second examples of Fig. 13.11 illustrate the use of simple instructions such as AND, OR, load, and store.

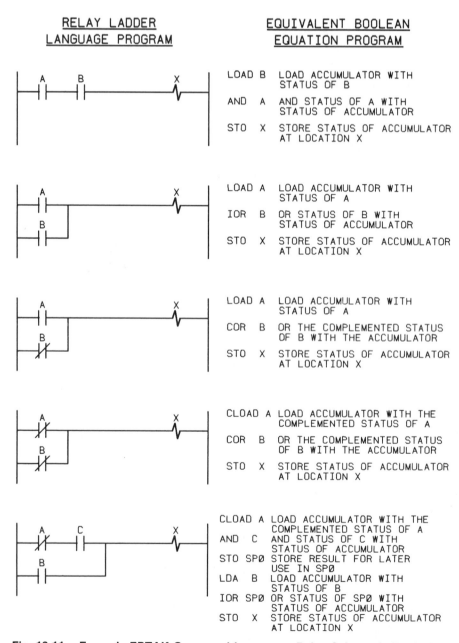

RELAY LADDER
LANGUAGE PROGRAM

EQUIVALENT BOOLEAN
EQUATION PROGRAM

LOAD B LOAD ACCUMULATOR WITH
        STATUS OF B
AND A AND STATUS OF A WITH
        STATUS OF ACCUMULATOR
STO X STORE STATUS OF ACCUMULATOR
        AT LOCATION X

LOAD A LOAD ACCUMULATOR WITH
        STATUS OF A
IOR B OR STATUS OF B WITH
        STATUS OF ACCUMULATOR
STO X STORE STATUS OF ACCUMULATOR
        AT LOCATION X

LOAD A LOAD ACCUMULATOR WITH
        STATUS OF A
COR B OR THE COMPLEMENTED STATUS
        OF B WITH THE ACCUMULATOR
STO X STORE STATUS OF ACCUMULATOR
        AT LOCATION X

CLOAD A LOAD ACCUMULATOR WITH THE
        COMPLEMENTED STATUS OF A
COR B OR THE COMPLEMENTED STATUS
        OF B WITH THE ACCUMULATOR
STO X STORE STATUS OF ACCUMULATOR
        AT LOCATION X

CLOAD A LOAD ACCUMULATOR WITH THE
        COMPLEMENTED STATUS OF A
AND C AND STATUS OF C WITH
        STATUS OF ACCUMULATOR
STO SP0 STORE RESULT FOR LATER
        USE IN SP0
LDA B LOAD ACCUMULATOR WITH
        STATUS OF B
IOR SP0 OR STATUS OF SP0 WITH
        STATUS OF ACCUMULATOR
STO X STORE STATUS OF ACCUMULATOR
        AT LOCATION X

Fig. 13.11. Example EPTAK Command Language, Relay Schematic Equivalents

Note that these examples, as well as the remaining ones in the figure, begin with a load instruction. Boolean programmed PCs utilize special memory locations to assist in the solution of the user program. The most fundamental of these locations is one usually called the *accumulator,* a term common to microprocessor machine language programming. The accumulator should be thought of as a current status storage location that is used for the solving of the user program instructions. The accumulator contains the OFF/ON status for a particular group of user instructions currently being solved. In the first example program of Fig. 13.11 the LOAD B instruction loads the accumulator with the current OFF/ON status of contact B. This status, the current OFF/ON status of contact B, is used by the next instruction, AND A, to arrive at a new OFF/ON status for the accumulator. In effect, the current status of contact B is logically ANDed with the current status of contact A, and the result is stored in the accumulator. Upon execution of the STO X instruction, location X is set to the OFF/ON status of the accumulator.

A similar operation occurs in the second example program of Fig. 13.11, except that the two contact instruction states are logically ORed together in lieu of being ANDed together. The remaining examples of Fig. 13.11 illustrate the use of the complement instruction in conjunction with the AND, OR, and load instructions. Note that the complement instruction is appended to the AND, OR, and load instructions simply as a C.

The final example program illustrates the programming for a more complex combination of instructions. This example also illustrates the need for using an intermediate storage location in a Boolean program. The ANDed A and C contact instructions must be ORed with the B instruction prior to the storing of the result in output X's location. This is accomplished by first ANDing the statuses of contacts A and C in the accumulator. This resultant status is then stored in a temporary location, SP0, so that the current status of contact B can be loaded into the accumulator. Once B is loaded into the accumulator with the LDA B instruction, the accumulator status is ORed with the status previously stored in the temporary storage location, SP0. The accumulator now contains the status of (A and C), or B. The STO instruction can then update location X's status to the current accumulator status.

With the use of Boolean program commands, any relay ladder schematic can be converted to a Boolean program. Many people, especially those with instrumentation and/or computer backgrounds, prefer to design PC systems and user programs with Boolean logic equations rather than with relay ladder schematics. As can be seen from the figure, the conversion from Boolean equations to Boolean programming is simple to perform.

DIAGNOSTIC MONITORING HARDWARE

A problem faced by nearly every PC user is how to efficiently troubleshoot a PC-based control system. New diagnostic monitoring I/O modules are providing an answer to this question. These modules, such as Allen-Bradley's

1774-DM1, provide a PC user with the ability to identify quickly the possible sources creating a problem with an operating system, in order to reduce downtime and increase system utilization.

Modules such as the 1774-DM1 actually "learn" the operation of the process or machine during a *learning* mode of operation, where the user operates the process or machinery in a known-correct sequence while the module "looks on." Once the user is satisfied that the module has learned the correct sequence of machine events and operations, it is placed in the *monitor* mode, enabling it to watch over the PC and the system hardware being controlled.

The module is given the task to monitor up to 1500 I/O status changes of the operating machine, and to compare them to the standard I/O statuses it learned earlier. The Allen-Bradley module is capable of detecting three types of faults. The first is an I/O condition that changes but was not learned. An example of this form of fault is a motor that stops due to loss of power, overload trip, or similar conditions. The module monitors for I/O points that fail to activate and provides this information to the user as the second type of fault. Finally the last type of fault involves a multiple I/O change where a point changes more often during a cycle than it should.

The module is preprogrammed with messages to inform the user of the nature of the fault, the location and/or address of the fault, as well as the OFF/ON state of the fault. This information is conveyed to the user through an RS-232C port on the module, which can be easily configured (baud rate, stop bits, parity, etc.) to communicate with a host of printers and intelligent displays.

Very little user programming is required to implement the Allen-Bradley diagnostic monitoring module. In fact, with three to eight rungs of user program the module can be set to learn or monitor as well as ignore any I/O point desired. In addition the module can be used in a LIFO mode for operations that do not repeat in an exact cyclic fashion. In this mode the module monitors the last I/O to change and provides this information to the user for diagnostic purposes. To enhance the module's capabilities further, it can communicate over Allen-Bradley's Data Highway communications network to other devices for improved fault diagnostic information.

SEQUENCER INSTRUCTIONS

A special instruction that has become a workhorse instruction is the sequencer instruction. Sequencer instructions are named after the mechanical sequencer switches they clone. These switches are often referred to as drum switches, rotary switches, stepper switches, or cam switches, in addition to the sequencer switch identification.

Illustrated in Fig. 13.12 is a typical mechanical sequencer switch. The switch consists of a series of poles (labeled A through V), which are activated by pins located on a moving drum. As the figure illustrates, the pins can be placed at random locations around the circumference of the drum to activate one or

Fig. 13.12. Sequencer or Drum Switch

more selected switch poles. Since the switch poles are designed with a combination of normally open and normally closed contacts for each pole, a pin can be set to either interrupt or complete the flow of current in a switch pole.

A motor is used to drive the drum, as pictured to the lower right of the assembly. The sequencer switch is driven with a block motor that ensures the drum will complete one revolution every minute. The numbers located on the left circumference of the drum indicate seconds. The sequencer illustrated can have a switch pole closed for a minimum of 0.5 second and as long as 59.5 seconds. Note that the switch is designed so that the drum is in continuous motion. Many sequencer switch drums are operated by a solenoid ratcheting the drum through discrete steps each time the solenoid is activated.

Sequencer switches are used whenever a repeatable operating pattern is required. An excellent example is the sequencer switch used in washing machines to pilot the washer machinery through a washload. Once loaded with clothes, the sequencer begins running. Water is brought into the proper level. The sequencer is incremented and the wash cycle begins. After a predetermined time the wash cycle stops and the tank is pumped out. A spin rinse is provided prior to tank refill for a formal rinse. Once the rinse is complete, a spin dry ensures that the water is completely removed. The cycle is always the same, and each step occurs for a specific time. The domestic washing machine is an excellent example of the use of a sequencer, as are similar devices such as dishwashers, dryers, and time clock-controlled devices.

A typical PC sequencer instruction is illustrated in the upper portion of Fig. 13.13. The instruction permits the user to specify the number of switch poles to be operated upon by the instruction. Some sequencer instructions have an upper limit on the number of switch poles that can be operated upon by a single instruction. However, multiple instructions may be used to gain the required number of poles. The instruction also includes a specification of the number of steps on the sequencer drum. Again an upper limit may be imposed by the instruction, but multiple instructions can be used as necessary.

For an actual sequencer instruction a state table must be specified, as illustrated in the lower portion of Fig. 13.13. This state table consists of

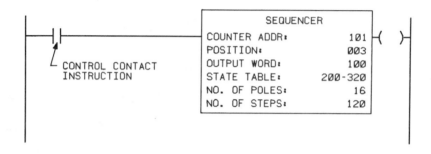

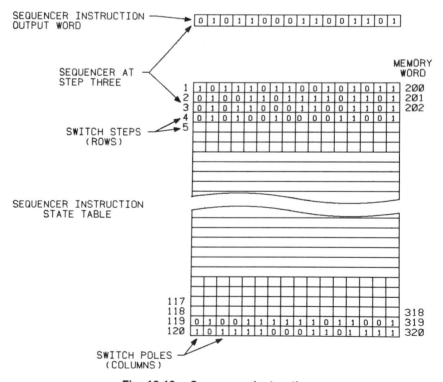

Fig. 13.13. Sequencer Instruction

consecutive data memory words programmed with either zero or one states. Each zero or one state represents the state of a predetermined switch pole for a particular sequencer step. For example, the mechanical sequencer of Fig. 13.13 has 120 discrete steps (60-second drum rotation time, times two steps per second). Therefore the state table for the PC instruction will require 120 words of data memory. Since there are 19 switch poles on the mechanical sequencer, each word (step) of the state table will contain 19 bit locations or positions. Each of the 19 bit locations of a particular state table word represents the ON (one) or OFF (zero) condition of a predetermined switch pole for the step represented by the state table word.

Note that a PC using 16-bit data memory words requires two words for each program step. This makes the sequencer state table 240 words long. The first 120 words contain the information for switch poles 1 through 16. State table words 121 through 240 contain the step statuses for switch poles 17 through 19. Since only three bit positions would be required in words 121 through 240, the remaining 13 bit locations would be unused. If at a later time additional contacts were required, a maximum of 13 switch poles could be controlled by the sequencer instruction without the necessity of allocating additional data memory words. Most sequencer instructions allocate data memory according to the user's requirements. The user specifies the number of steps required, the number of switch poles to be controlled per step, and the starting address for the state table. The programming terminal and processor then allocate the proper amount of memory area to meet the application. The user should verify that the area designated is virgin data memory area and will not be used for other program functions or calculations.

The user must designate an output status word for the sequencer instruction. This is a memory location to which the contents of a specified step are copied while that step is active. It is the individual bits of the output status word that are used as the reference addresses for the contact instructions used later in the user program. The output status word may be addressed as an internal storage word where the switch pole statuses are to be used for internal program use, or an output image table word if the user wishes to control output devices directly. The output status word can usually be displayed on the programming terminal to indicate the current sequencer switch pole statuses.

Finally, a counter or pointer location must be specified to indicate the current sequencer step being performed. The counter or pointer is incremented by the processor each time the sequencer instruction is activated. Many sequencer instructions reset the counter or pointer automatically to the beginning of the sequencer state table upon completion of the last sequence step. Other instructions provide an individual reset control line to return the counter or pointer to the beginning of the state table under user control. In addition some sequencer instructions permit the user program to manipulate the counter or pointer value. This permits the user to skip selected steps as well as to jump forward or backward in the state table to other steps. The user should be aware that changing the counter or pointer to a location outside of the state table can cause incorrect control operations and/or run time errors.

PART IV

Peripherals and Accessories

14

Programming Terminals

One accessory that no PC system can be without is the programming terminal. This unit accepts the commands of the user, converts them into a form that can be understood by the processor, and then places each command in the proper memory location within the processor's logic storage area. The programming terminal allows the software engineer to examine various parts of the PC memory, as well as to enter new instructions or modify existing ones. As optional functions, these units often have the ability to produce a hard-copy printout or a magnetic tape or disk recording of the information and commands stored in the processor memory. The programming device always has some form of keyboard for instruction and data entry, but various styles of information displays are available. Many manufacturers incorporate a display that shows only one instruction at a time with the keyboard, while others have a cathode ray tube (CRT) for displaying requested data. Those programming devices that do not contain CRTs are generally called *hand-held* programmers, while those with CRTs are generally referred to as *CRT programmers.*

HAND-HELD PROGRAMMERS

The hand-held programming unit usually contains no more than several dozen program instruction and data entry keys. A display is incorporated with the keypad to display such things as numerical data and processor memory location addresses. A series of indicators may be incorporated alongside selected instruction keys to indicate the status of that particular key to the user. Additional indicators are employed for the display of processor operating modes and status. The hand-held programmer is generally small in size and

351

weight, often the size of an electronic office calculator. Most hand-held programmers plug directly into the processor through a multiconductor cable and plug, and in most cases require no external power source for operation. For complicated control programs hand-held programmers can be unwieldy to use for the initial processor instruction entry and debugging because of their limited capacity to examine and enter only one user command at a time. They can be useful for the quick examination of data or instruction statuses during system operation and maintenance, provided that the control program is well documented. The hand-held programmer should not be confused with a remote processor loader/monitor/access panel, which will be discussed in Chapter 16. Most hand-held programmers are only usable on models of PCs that do not have a large extended instruction set consisting of data moves, word and bit manipulations, or similar specialty commands. Fig. 14.1 illustrates a typical hand-held programmer.

Many hand-held programmers have a single numeric display divided into two areas. One area displays the numeric designation for the particular memory location being programmed or examined. The second area of the display indicates the numerical identifier or data value being programmed in the memory location being examined. Other hand-held programmers incorporate a liquid crystal display designed to show individual relay contact instructions or other programming data. In lieu of the single numeric display, several manufacturers incorporate multiline liquid crystal or LED readouts for user use. Hand-held programmers having multiline displays provide more information to the user, and in several cases the display is capable of displaying a portion of the relay ladder diagram.

The keypad areas of most hand-held programmers are the same. Some use pressure-sensitive keypads, while others incorporate the type of keys that are commonly found on calculators or typewriters. Due to the limited quantity of keys available on a hand-held programmer, they all tend to offer the same capabilities. For data entry there are keys labeled 0 through 9. Both ENTER and INSERT keys are provided for the initial entry of data or commands into the current memory location or between two memory locations. The INSERT key moves all existing data and commands from the addressed memory location to the next higher location in memory before the automatic entry of new information. To allow the quick examination of sequential memory locations, a STEP key automatically indexes the hand-held programmers from one memory location to the next. To display the current contents of a particular memory location, a READ or DISPLAY key is used. To make corrections or remove the current command from the hand-held programmer's display, the CLEAR key is used. Some hand-held programmers provide a DELETE key for the removal of commands or data from the processor memory while others use a combination of keys such as CLEAR/ENTER.

The remaining keys of a hand-held programmer depend on the characteristics of the PC or PCs with which the unit will work. There will usually be keys to enter a timer instruction, the timer's time base, and its preset value. If counter instructions are available, the necessary keys for the selection of the type of counter and the counter's preset value are included. Since many PCs program

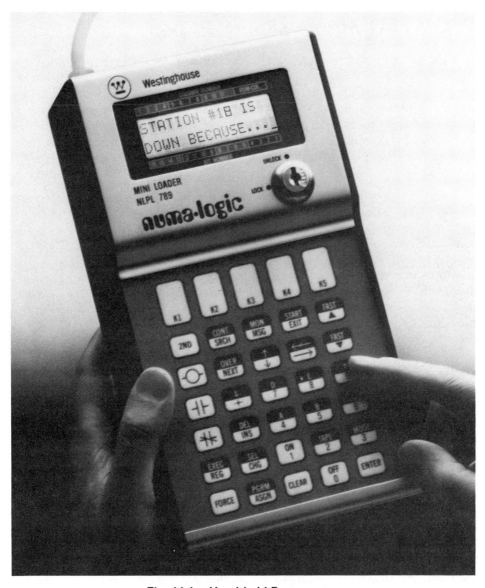

Fig. 14.1. Hand-held Programmer

in a language of Boolean equations as well as with relay instructions, many hand-held programmers include dual-purpose keys designed for programming in both formats. The keys used for relay symbology include open and closed contact configurations, branch instructions, and assorted coil symbols and functions. Those programmers with Boolean capability have the standard AND, OR, and NOT functions along with the available output device instructions. Each of the relay, timer, counter, Boolean, and output key functions may incorporate an indicator adjacent to the key to inform the user that that particular function or instruction is part of the command or data that is stored in the memory location currently being examined.

PROM PROGRAMMERS

A special version of the hand-held programming device is the programmable read-only memory (PROM) programmer. These devices do not offer the same capabilities as the hand-held unit previously described, but are designed strictly for the programming of PROM-type memory chips and therefore are not connected to a processor in a programming manner. The PROM programming unit allows the user to place the control program and data in a permanent form of memory device. The Westinghouse PROM programmer is illustrated in Fig. 14.2.

Several versions of the PROM programmer are available, depending on the model of PC being used and whether that model offers a PROM memory option. One version of this type of programmer requires that the user manually enter each memory location's instruction or data directly into the PROM. A second version is more automatic in that it works with a processor and copies the data stored in the processor's memory to the PROM memory device automatically. The use of this style programmer usually means that the data and instructions to operate the PC system can be stored and tested in the PC processor or the programming unit prior to the actual transfer to the PROM memory units.

The PROM programmer can be more cumbersome to use since any errors entered into the PROM memory require that the complete memory chip be erased and reprogrammed from the beginning. In many cases there is a limited number of times that a PROM can be erased and reprogrammed before it must be replaced. To complicate matters further, most PROMs require special ultraviolet light sources for erasure. Not only is the light source damaging to the human eye, the time required to ensure complete PROM erasure can be from 30 to 45 minutes.

A caution that should be observed when using PROMs is to ensure that the proper PROM is used. Various integrated circuit manufacturers produce PROMs, and any number of manufacturers may offer a device with the same catalog number. While these "equivalent" units are the same generically, they may in fact not be totally interchangeable. A substitute PROM may be electrically different internally from the PROM recommended by the PC manufacturer. The best practice is to order and use PROMs supplied by the manufacturer of the PC equipment. The PROMs supplied by the PC manufacturer have usually been tested and verified to operate in the same environment as the PC that will house them.

CATHODE RAY TUBE PROGRAMMING TERMINALS

The CRT types of programming devices or panels are more versatile to use and offer much more capability to the user. These units consist of a keyboard for instruction and data entry, along with a black and white, or green and white, television-type display tube for the display of program data and instruction

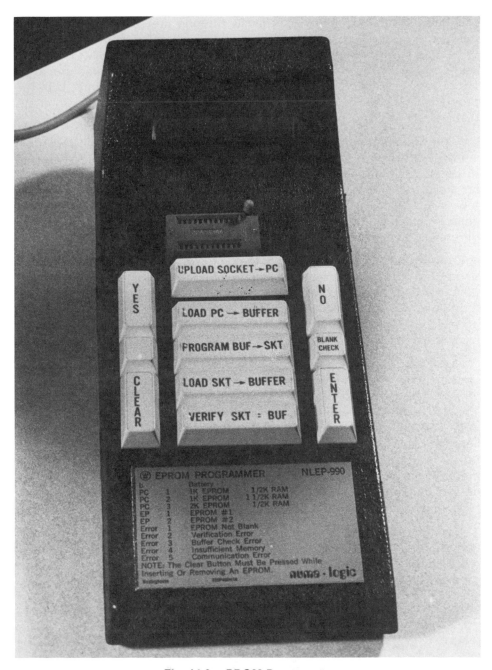

Fig. 14.2. PROM Programmer

symbology. These units are often expensive to purchase, but that expense is usually well worth the time and money saved when the full capabilities of the CRT programmer are used to program, troubleshoot, and maintain a complex control system program.

Most CRT programmers have the intelligence to offer error messages and user prompting to assist the user in his or her programming and troubleshooting efforts. Programming terminals offering user prompting are often referred to as *user-friendly* terminals. Several programmers incorporate the use of help and user prompting functions that ask the user questions, so that the user's programming efforts are easier when he or she is uncertain of the exact information that is to be supplied to the unit. The use of a printer with the programmer allows the user to obtain a hard copy of the data and instructions that are recorded in a processor's memory.

Many CRT programmers have built-in tape drives for magnetically recording a processor's memory for later use in reloading another PC system. When tape drives are not built into the unit, a separate tape drive unit is usually available for direct use with the processor. Besides the standard use as a programming tool, some CRT programmers offer a mode of operation where the unit will generate graphic displays, error messages, and system statuses for operator use in running the facility and associated equipment.

Probably the biggest drawbacks of the CRT programming terminals are their size and weight. Most terminals are awkward to carry, and the weight of some units is enough to add inches to one's arms should it be necessary to carry the terminal any distance. Even though most manufacturers advertise their units as portable, an oscilloscope cart comes in mighty handy as a permanent means to move the unit around.

Nearly all programming terminals are similar in appearance and functionality. The keyboards are the familiar touch membrane or key tape. Some incorporate full-sized typewriter keyboards with accessory keypads for special-function keys, while others have multiple overlays for operating the terminal in various modes or with several different models of PCs. To reduce the total number of keys on a terminal, some manufacturers place unlabeled keys, usually called *function keys,* directly beneath the display tube. The function of these keys can then be defined by the CRT terminal as necessary for the programming mode desired by the user. The keyboard portion of the Westinghouse Numa-Logic programming terminal is illustrated in Fig. 14.3.

The CRT programmer's display screen functions to display various information to the user. An area of the screen is allocated for the construction of the relay ladder program, with another section of the screen devoted to user prompts, system messages and error codes, or hardware operating statuses and

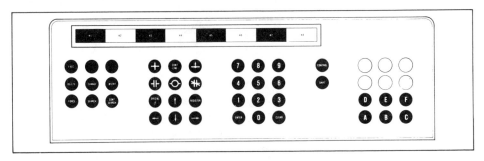

Fig. 14.3. Westinghouse Programming Terminal Keyboard

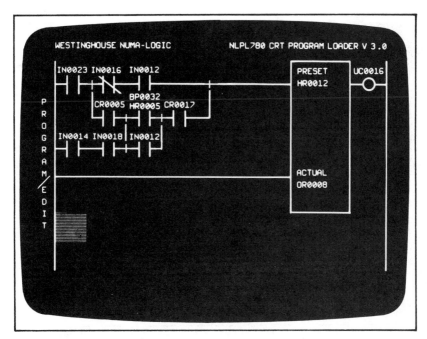

Fig. 14.4. Typical Programming Terminal Display

modes. Fig. 14.4 illustrates a relay ladder logic rung displayed on the West-inghouse Numa-Logic PC CRT.

Numerous programming terminals have multipin jacks on the rear of the unit for interfacing the terminal to a printer or a communications modem. Various user controls can be found, such as a keyswitch to prevent unauthorized programming, display screen controls for brightness and contrast, and volume controls for the loudness of the internal buzzer. A removable cover on many units permit access to small rocker switches that configure the terminal and its external jacks for communication to the processor or external devices such as a printer.

On-Line and Off-Line Programming

Most CRT terminals have several modes of operation. The *program* mode attaches the terminal to the processor and allows the user to modify the processor's memory by adding new control instructions and data, modifying existing information stored in the processor memory, or deleting previously programmed material. A *monitor* mode allows the user to examine any part of the processor's memory area, but changes are not permitted. An *executive* or *supervisory* mode allows for access to internal CRT programmer functions and operating routines. This mode may also affect various processor functions, such as starting or stopping its scan. Figs. 14.5 and 14.6 illustrate the various options available on the Westinghouse Numa-Logic programming terminal when the executive mode is placed in operation. Special function modes such as *cassette*

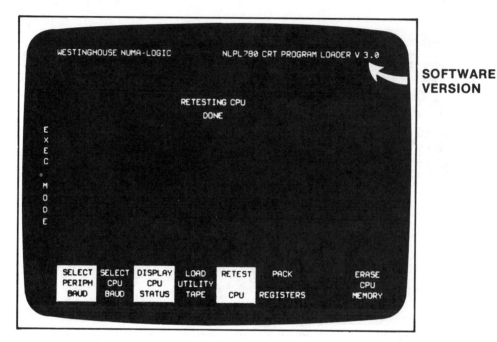

Fig. 14.5. Westinghouse Programming Terminal Executive Mode Options

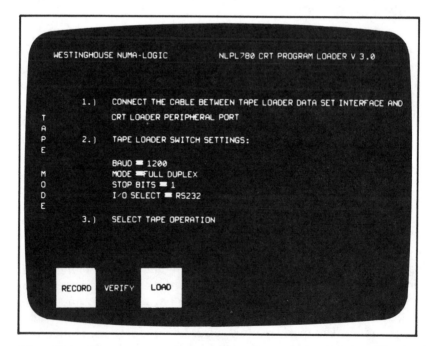

Fig. 14.6. Westinghouse Programming Terminal Tape Cassette Functions

tape and *ladder listing* may appear as dedicated operational modes, or may be part of another mode such as the supervisory or monitor mode. The programming terminal display for the Westinghouse Numa-Logic terminal is illustrated in Fig. 14.5 with the tape function mode enacted.

When programming a PC system several instruction entry modes are available depending on the manufacturer and the model of the unit. *Off-line* programming requires that the processor be stopped from scanning for the programming operation to occur. When the PC is switched into the off-line program mode, the processor scan automatically stops and the output devices are turned to their OFF state. This is the safest manner in which a PC can be programmed. Additions, changes, and deletions do not affect the operation of the system until the operator leaves the program mode. While incorrect programming may still cause control problems, the accidental entry of wrong commands or data can be corrected before there is a chance for the PC to damage expensive machinery and equipment. Often a *test* or *monitor* mode is provided to permit the processor to scan the user logic and read the input devices so that the software designer can monitor and correct any errors before the processor is switched into the run mode. In the test mode the processor indicates the status of the output devices, but they are not activated until the programmer is in the *run* mode.

An *on-line* programming mode permits the user to modify the processor's memory while the PC may be in operation. As the PC controls its equipment or process, the user can add, change, or delete control instructions and data values as desired. If the user is not attentive, any change being made could cause severe consequences to the controlled operation or process. Any modification made in the on-line programming mode is executed immediately upon the entry of the instruction to the CRT programmer. As a variation, some manufacturers allow on-line programming in two steps. The first step allows the user to modify the program as necessary. The processor continues to operate under the existing instructions, but the CRT terminal displays the various changes and also indicates how those changes operate. The user can make changes and monitor their operation to be certain they are correct before releasing them to the processor for execution. Once the user is sure that the changes and modifications are correct, he or she commands the terminal to place those changes into the processor's memory for immediate execution. The instructions present before the change are deleted, added to, or modified as required. This type of on-line programming allows the user to make a change, monitor its operation, then place it into memory when he or she is sure correct system operation will result.

In any case, on-line programming should only be done by experienced PC software personnel. They must fully understand the operation of the PC they are dealing with and the machinery being controlled. Any change should be checked and verified for accuracy and operation, and all possible sequences of equipment operation should be reviewed for any effects the new change could have. If at all possible, the change should be made off-line in order to provide a safe transition from existing programming to new programming. If the change must be done on-line, it is desirable to make the change when it can least affect the operation of the system.

When switching a processor from an off-line mode to an on-line mode of operation, extreme care should be exercised. When placed off-line, many processors reset the elapsed time and count of timers and/or counters back to either the preset or a zero condition. In addition the mechanical hardware being controlled may have been left in an intermediate state at the time the processor was stopped. Changes to the program may affect the operation of this equipment at the time of processor restart. When placing a processor back on line it is always good practice to ensure that the controlled hardware is in its "home" position, ready for initialization. Failure to develop the habit of double-checking the controlled hardware for a proper restart configuration can lead to possible equipment malfunction and damage. Returning a processor to the on-line state may involve taking the same precautions that would be taken if the unit were being programmed in the on-line mode of operation.

The Programming Terminal Screen

The CRT display screen of the CRT programming terminal varies in size and function. It may be anywhere from 5 to 12 inches in diagonal size and is always either green and white or black and white in color. The majority of the screen is allocated for building the relay ladder program instruction symbology. The amount of screen set aside for this purpose is usually in the form of a matrix. Typical row by column sizes range from 5 by 7 to 7 by 11, with an extra right-hand column position for the display of one or more output coils. Each cell of the matrix holds one instruction or command function, with multiple cells often required for more complex instructions and commands. Fig. 14.7 details the CRT screen size and functional areas for the Westinghouse Numa-Logic

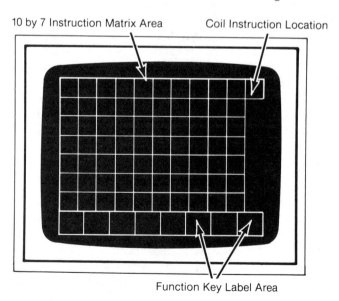

Fig. 14.7. **Westinghouse Programming Terminal CRT Screen Array**

programmer. Most programming terminals intensify those contacts that are passing power to permit the user to "see" how the circuit is operating.

Where keys are installed on a programmer without labels, the lower edge of the display screen will be used for the indication of the function of these keys. As the terminal is used to enter various instructions, the legends on the screen representing each key will change to indicate possible subcommands and subinstructions associated with the master instruction. Some of the various function key display headers for the Westinghouse Numa-Logic programming terminal are illustrated in Fig. 14.8.

Many CRT terminals have assembly areas where an instruction or command is built before it is entered into one of the matrix areas. This is done because there may be several keystrokes involved in assembling a command before it can be placed within the matrix structure. In addition to the assembly area, a part of the screen is usually set aside for error message display. Many programmers offer error messages in the form of either written displays or coded characters to aid the user in quickly determining a mistake. Other data may also be displayed in this area, either automatically or upon request. Such things as amount of memory remaining, current ladder rung identifier, programming mode, and current processor status are often important to the user, and can be displayed in this area of the programmer screen.

For many special function operations the CRT screen is designed to provide the user with instructions on how to perform a requested task. The screen may ask the user a question and provide a menu of selected answers. Once the user answers the terminal's question, the desired function will be carried out by the unit, or additional information may be requested. Often the user can request information from the terminal. In this case the information is displayed on the screen for the user until he or she clears it.

To provide versatility, a second display page may be available. The single CRT screen has an alternate display page which can be used to display additional program data or related programming information. The user develops the program on one screen area and can flip back and forth between this area and a second area that was set up to display selected program variables and information. Essentially the terminal has twice as much display area for program development and monitoring than would otherwise be available.

In order to provide usefulness to the CRT programmer when it is not being used for programming the processor, a *graphics/message* mode may be offered. Several PC systems will store operator messages and alarms as part of the processor's programming. The control program can be designed to display these messages and alarms as required on the CRT terminal. This allows the terminal to perform two functions, one being a program development tool for designing and troubleshooting a system, the second being a display device for various system messages and alarms. In addition to the messages, graphics capability may also be available. A graphic representation can be "drawn" on the CRT screen for storage in the processor's memory. The graphic can be called up by the user programming, and additional information can be added to the graphic screen regarding current machine states and conditions.

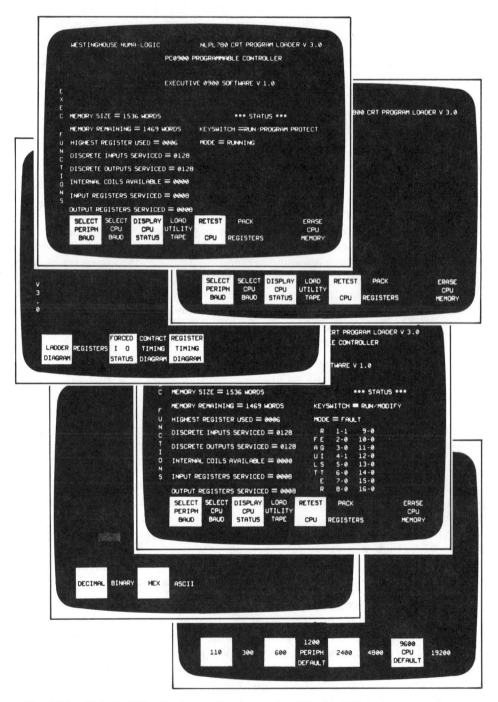

Fig. 14.8. Selected Westinghouse Programming Terminal Function Key Headers

As a security feature, many programming terminals provide keyswitches and/or access codes which must be used to place the terminal in the program mode. The use of the programming terminal keyswitch and/or access code provides the system designer with a means to *implement* some level of security against having the processor program changed by someone unfamiliar with the system and its operation. The use of a programming terminal keyswitch and/or access code may be either in addition to or in lieu of equivalent security features incorporated in the processor.

THE PROGRAMMING TERMINAL KEYBOARD

The keyboard section of a PC's CRT provides the actual input of data and commands to the processor. These keyboards are usually as diverse in appearance and functionality as the various PC models they serve. Almost all of them have certain similar keys and key functions, while the remaining keys are solely for the purpose of commanding special functions and instructions of the processor. Many units have standard typewriter keyboard layouts, while others use a design that groups similar key functions together. Some manufacturers have made their keyboards "coffee proof" by constructing them of conductive plastic sandwiches layered together to form a touch pad, while their competitors have designed CRT programmers that incorporate the standard tactile feel of a physical typewriter key in the keyboard design. As was noted earlier, unlabeled keys that have their functions defined by the CRT terminal are often found where the manufacturer desires to cut down on the size and complexity of the keyboard unit. Interchangeable plastic overlays are also used where the CRT programmer serves multiple functions. There are CRT terminals that have the keyboard unit permanently attached, units with the keyboard mounted in a hinged manner so that it can be closed up against the screen for storage and transport, and terminals where the keyboard unit is removable from the programming terminal for monitoring purposes.

STANDARD PROGRAMMING
TERMINAL KEYS AND FUNCTIONS

All PC programmer keyboards have keys that perform similar functions and operations. There are usually four keys that are used for curser control. These keys are usually engraved with arrows pointing up, down, right, and left. A numerical group of keys labeled 0 through 9 are used for inputting memory addresses, timer/counter values, and other numerical data into the processor. Where the processor accepts numerical data in the hexadecimal numbering system, a method is provided to input digits A through F. When the capability of interpreting written data is included as part of the processor's intelligence, a full ASCII typewriter keyboard is provided for the inputting of uppercase and lowercase characters. Every keyboard will generally include keys for an edit mode of operation. These keys include search, remove, insert, display, and cancel command. Since a PC could not be programmed without keys that represent the symbols of the relay ladder language, keys are allocated for each symbol or instruction that can be used with the processor. This group of keys begins with the very basic instruction symbols for normally open and normally closed contacts, coils, latches, timers, counters, and math functions, such as add, subtract, multiply, and divide. If available, various keys are added for data manipulation, bit manipulation, sequencer, and jump instructions. In most cases PCs that have many data handling and bit manipulation instructions must rely on either removable keyboard inserts for the various instruction groups, or the use of the unlabeled CRT-defined keys for the entry of these instructions and commands. It has been estimated that at least 60 separate keys would be required in addition to those listed earlier to handle the instruction sets of some of the larger PCs on the market. This quantity of special instruction keys, in addition to the quantity of standard typewriter keyboard keys, plus the PC cursor, edit, and miscellaneous control keys would compose a keyboard too awkward to use and too big to carry. In lieu of CRT-defined keys, several terminals use keys that serve multiple functions. The most used functions are engraved on the key along with the "shifted" or alternate function legend. For these terminals a shift key or special-function key is provided much the same way as a shift key is used on a typewriter to produce capitalized characters and letters.

PROGRAMMING TERMINAL USER COMMANDS
Forcing

There are several commands and functions provided by most CRT programmers that benefit the user. These functions have nothing to do with the actual instructions programmed into the processor memory, but are provided by many manufacturers to aid the user in the development and debugging of the processor software. The first of these instructions are the *enable/disable* and *force ON/force OFF* commands.

It is often helpful to override the processor's control of selected, or groups of selected, input or output points. This feature is very useful when checking out a system since it provides the capability to toggle ON and OFF an output point selectively to check the proper operation of that point. Likewise it may be necessary to operate an input point to check a group of programming instructions without going to the controlled machinery and operating the input device manually. To perform these operations the force function is often available on a CRT terminal. Upon entry of the force mode of operation, the desired I/O point or points are selected. Depending on the manufacturer of the PC hardware, the I/O point may then have to be disabled from processor control. Depression of the DISABLE key removes control of the processor from the selected I/O point. When the point is disabled, a D or a break in the coil line is usually shown on the screen to remind the user that the processor has been severed from control of the point. Through the use of the force ON/force OFF function the user can then toggle the I/O point at leisure to check for proper system response. When checking is completed, the user returns control of the I/O point to the processor via the ENABLE key.

Fig. 14.9 illustrates the concept of *input forcing*. As can be seen, the programming terminal is forcing a specified input point ON, even though the field device is actually OFF. The forcing is done jointly by the processor and the programming terminal during the update of the input image table.

The forcing of an output point occurs in a similar manner, as illustrated in Fig. 14.10. The programming terminal acts in conjunction with the processor to turn a selected output point OFF, even though the output image table indicates that the user logic is setting the point ON. The actual forcing is done by the programming terminal while the processor updates the output modules during the output update portion of the I/O scan.

Care should be exercised when disabling the I/O point to ensure that its forced state is the same as it will be immediately upon return of control to the processor. If at the time of transfer the forced state is opposite that of the unforced state, a "hiccup" could occur in the operation of the system. This could be disastrous if done to a PC I/O point that is controlling an active process or machine. If at all possible, I/O forces should *not* be done to a PC system under operation.

The user should also be aware that some models of CRT programmers remove automatically all forces and disables in effect when the terminal is switched OFF or disconnected from the processor. Also, some models of PCs allow an input point force to be in effect for only one scan. This is due to the fact

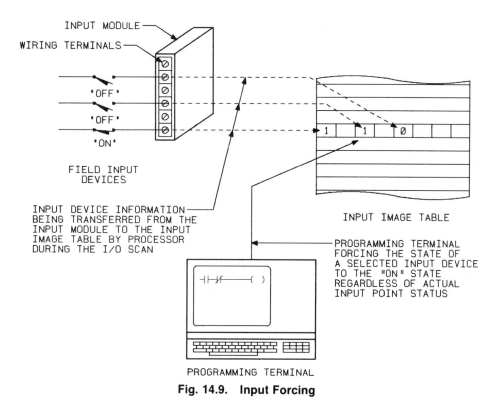

Fig. 14.9. Input Forcing

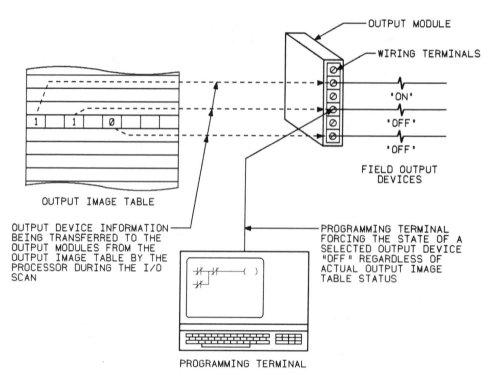

Fig. 14.10.   Output Forcing

that the processor continually updates its I/O statuses, and the force condition replaces the input point status for the particular scan it was entered on. On the next processor scan the input is corrected to its field condition.

Many programming terminals provide a *force table* or other visible means to alert the user that a force is in effect. The use of a force table or listing allows a user to know quickly the status of the force or forces in effect at any given time.

Before using a force function the user should check whether the force acts on the I/O point only, or whether it acts on the user logic as well as on the I/O point. For example, when a force operates on the processor logic as well as on the I/O points, the forcing of a single output point could produce unwanted machine operation. Consider the programming of a simple motor circuit as shown in Fig. 14.11. If the user forces the output point controlling the motor from the OFF to the ON state, the operation of the motor when the force is removed could be different, depending on how the user logic is affected by the force operation. If only the output point is controlled by the force, the processor logic would ignore the forced condition, and when the force is removed, the motor would return to the OFF state. However, if the logic is affected by the force, the normally open contact acting as the hold-in, electric latch, or seal for the motor circuit would begin passing power when the force was enacted. The motor control circuit would seal in just as if the START push button had been depressed. If the force was removed by simply returning the output point to processor control, the motor would continue to run until the stop button was depressed.

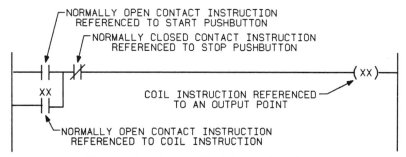

Fig. 14.11.   Seal or Electrical Latch Circuit

Search or Find

When editing or troubleshooting a processor's logic, the use of a *search* or *find* function will be extremely helpful. The programmer builds the coil, timer, counter, contact configuration, or memory location address that he or she wishes to search for. Activation of the SEARCH key locates the first use of the desired circuit element, instruction, or address in the processor memory. The circuit containing the searched element is then automatically displayed on the screen for user inspection. If desired the user can modify the searched element, or the circuit containing the searched element or instruction. Upon completion of any changes the CONTINUE SEARCH key can be depressed for location of the next successive use of the same element or instruction. Continued depression of the CONTINUE key will display each new occurrence of the desired element or instruction. Once the complete processor memory has been examined, a prompt will usually be displayed that indicates to the user that all occurrences of the element or instruction have been found and displayed.

The search function will prove extremely useful when looking for a particular reference or element. Use of the search function should be considered before doing on-line programming changes to check the use of the changed instructions in the remainder of the program. Before using the force capability of a programming terminal, a search is recommended to ensure that the forcing of an element in one logic line will not adversely affect another section of programming.

Contact Histogram

Often a PC problem may indicate a flaky I/O device or faulty programming procedures. A *contact histogram* can provide the user with the timed ON/OFF history for any specified I/O point or memory bit location. Use of the contact histogram function requires that the user input the address or identity of the desired element or instruction. Activation of the histogram function then displays the activity of the element in question. This function usually displays the time the element is OFF, followed by the time it was ON. The histogram function will continue to display each OFF and ON transition time for the specified element or instruction until interrupted by the user. Since transition times can vary from milliseconds to hours, the time is usually recorded in hours, minutes, seconds, and hundredths of a second. CRT programmers that do not offer this function can have it available if the user desires to program it into his logic.

Program Editing

A feature offered with some CRT programmers is the capability to separate a ladder rung either vertically or horizontally for the purpose of inserting a contact or a group of contacts. This process is known as PC user program *editing*. As an example to illustrate this feature, consider the circuit rung shown in Fig. 14.12. This ladder rung represents an existing circuit programmed into the processor's memory area. The row and column references do not normally appear on a CRT screen but are added here to make the discussion references to the figure more accurate. Some CRT programmers have syntax rules for placing contacts on the screen. For this example the programmer has no restraints as to placement of contact symbols. The subject of CRT programming syntax rules will be discussed later in this chapter.

The first editing change involves adding two series contacts in parallel with the series contacts in positions A1 and A2. To do this, the ladder rung must be separated horizontally between rows A and B to allow the parallel addition of the two new contact instructions. The horizontal separation generally requires the placement of the screen cursor on the contact to be moved, in this case contact B1. Once the horizontal separation command has been executed, the circuit will look like the one shown in Fig. 14.13. Notice that the contacts in row B have moved to row C, the contacts in row C were moved to row D, and the contact that was in row D now forms a new row E. Contact row B is now clear for circuit additions as necessary. Once the contact instructions for the new parallel rung have been added, the vertical short circuit that was opened up earlier in the horizontal separation will require completion. Fig. 14.14 represents the revised circuit with the two series contacts added in positions B1 and B2.

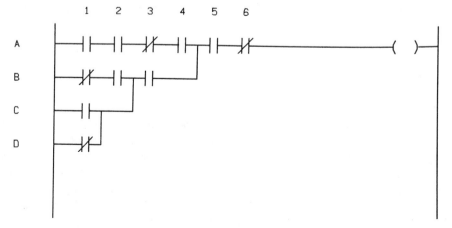

Fig. 14.12.   Ladder Rung to Be Edited

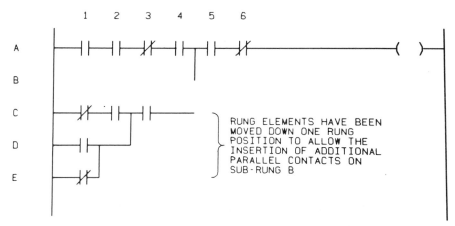

Fig. 14.13. Editing Example Step 1, Horizontal Rung Separation

The software engineer now wishes to add two more series contacts to row C after contact C3. One of the contacts could occupy screen position C4. However, the ladder rung will require expansion vertically if the second contact is to be added before the vertical connection. The vertical expansion occurs first by placing the cursor on the horizontal instruction at C4, then activating the vertical expansion command. Fig. 14.15 represents the circuit rung after the vertical expansion has occurred. Column 4 is now clear and available for contact insertion. The vertical expansion has moved the horizontal short circuit that occupied position C4 to location C5. This horizontal short circuit will require removal before the new contact can replace it. Since there are no additional contacts to be added to row A, a horizontal short circuit is placed in position A4. Fig. 14.16 represents the edited circuit with the changes made up to this point.

Two additional changes still need to be made before the editing is complete. These changes require the addition of a contact at the end of row A and another contact in row D. These additions do not require the use of the expansion

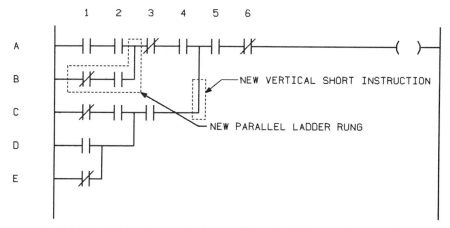

Fig. 14.14. Editing Example Step 2, New Parallel Rung Instructions

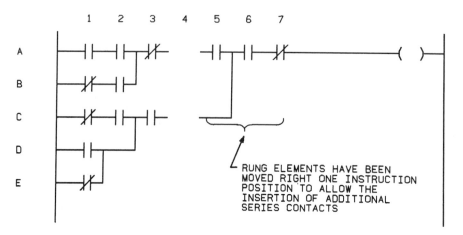

Fig. 14.15.   Editing Example Step 3, Vertical Rung Separation

edit feature and can be inserted directly into the existing ladder rung. Fig. 14.17 shows these remaining additions to the circuit.

Whenever the insertion of additional series and/or parallel contacts in a circuit is required, and the vertical or horizontal rung split editing function is used, the processor should be stopped prior to the additions. When the circuit is split apart by either of the expansion routines, the program continuity of the logic in the processor is broken. Conversely the ability to simply add a series or parallel contact to an existing ladder rung may require the contact instruction to be assembled in a series of steps. Unless the programmer has the ability to retain the existing circuit while editing is being performed, highly irregular system operation could exist. If the editing feature can only be implemented in an *off-line* programming mode, then the user should be able to perform the editing without fault.

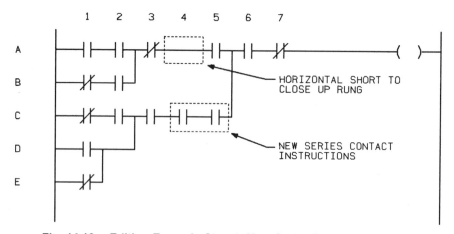

Fig. 14.16.   Editing Example Step 4, New Series Rung Instructions

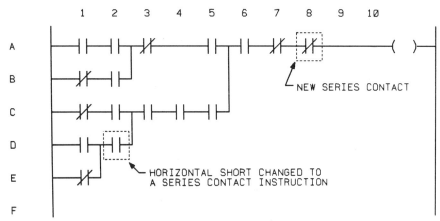

Fig. 14.17. Editing Example Step 5, Completed Ladder Rung

PROGRAMMING TERMINAL SYNTAX RULES

Many PC CRT terminals have syntax rules for the proper placement of the relay symbols on the CRT screen. Understanding these syntax rules is required if the user expects the programming terminal to assemble the ladder rung program instructions correctly as it places them in the memory of the processor. Many of these syntax rules will cause an error message to be printed on the CRT screen if the user attempts to enter a circuit in a different manner than the syntax rules will allow.

Figs. 14.18 to 14.28 represent most of the more common syntax rules for a CRT programmer. The example illustrations for the syntax rules use an arbitrary CRT screen size of 6 rungs by 10 contacts. The letter and number references shown at the top and the left-hand side of the example illustrations are for reference only and would not normally appear on the CRT screen.

Open sections of a ladder rung or short circuits around a group of circuit elements are generally not recommended. Fig. 14.18 shows a circuit with elements of the rung missing. If a contact is not required in these positions, either a duplicate contact should be inserted, or the use of a short circuit should be implemented.

If a circuit has horizontal or vertical short circuits programmed around a group of contacts, as shown in Fig. 14.19, the programmer may not be able to place the circuit properly in the processor's memory. Again the use of duplicate instructions is suggested.

Most CRT terminals require a contact programmed in the upper left position of the CRT screen for each circuit rung assembled on the screen. The top subrung of contacts must also be continuous and longer in circuit elements or instructions than any line below it. Fig. 14.20 represents a circuit without an upper left contact instruction, and how it might be corrected. The circuit shown in Fig. 14.21 shows the proper corrections for program rungs whose first subrung is shorter than those below it. Note that the first subrung of Fig. 14.21 is

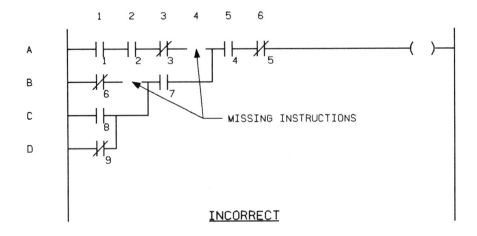

INCORRECT

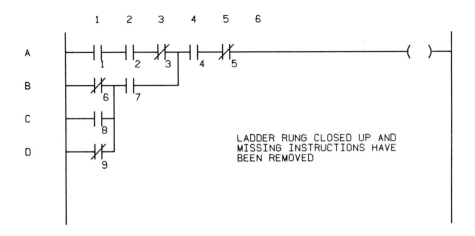

CORRECT

Fig. 14.18. Programming Terminal Syntax Rules, Missing Instructions

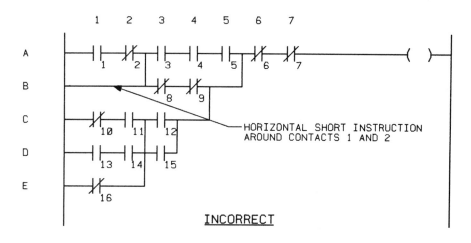

INCORRECT

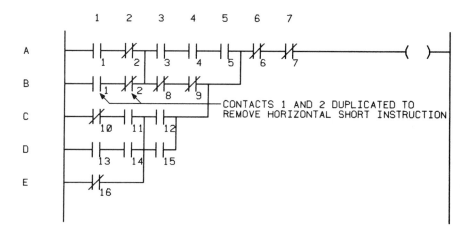

CORRECT

Fig. 14.19.   Programming Terminal Syntax Rules, Contact Short Circuits

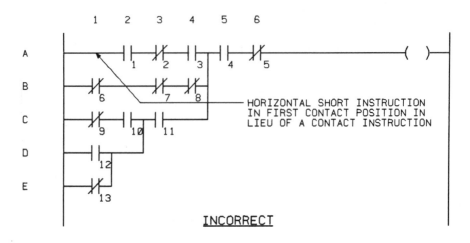

INCORRECT

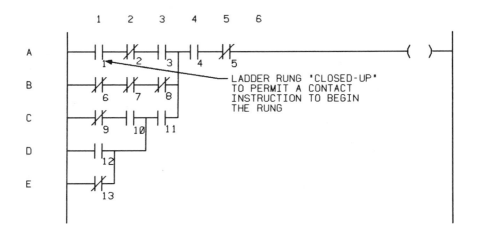

CORRECT

Fig. 14.20.   Programming Terminal Syntax Rules, Start of a Rung

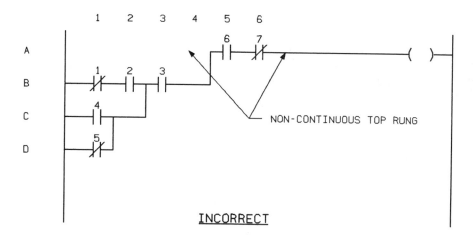

INCORRECT

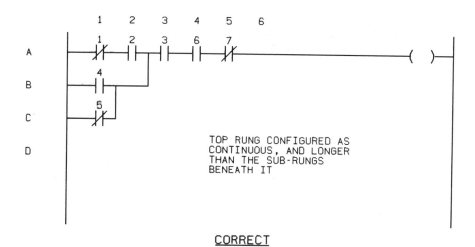

CORRECT

Fig. 14.21. Programming Terminal Syntax Rules, Rung Continuity

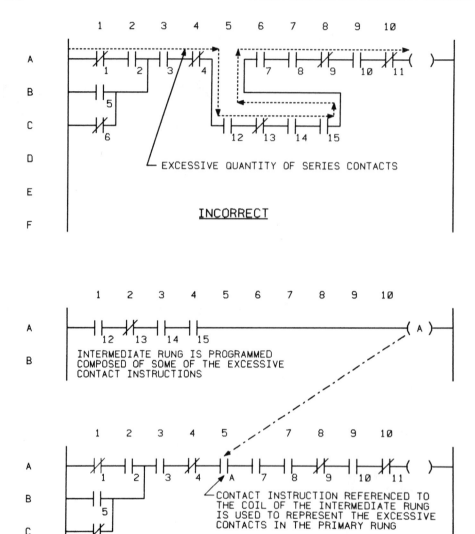

Fig. 14.22.   **Programming Terminal Syntax Rules, Excessive Series Contacts**

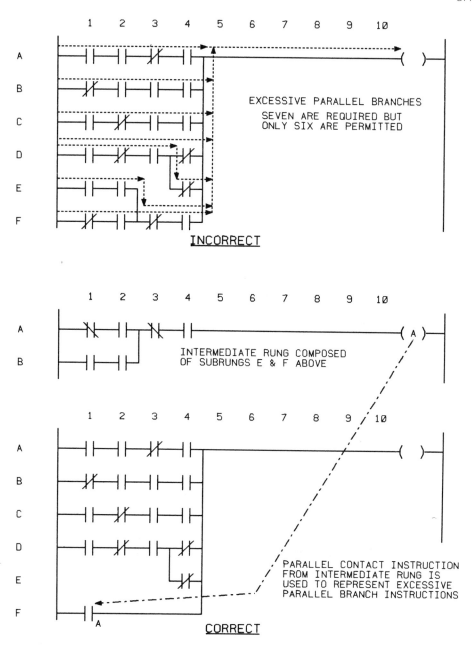

Fig. 14.23. Programming Terminal Syntax Rules, Excessive Parallel Rungs

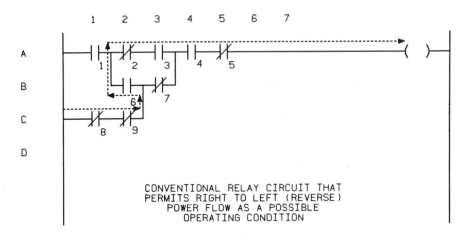

CONVENTIONAL RELAY CIRCUIT THAT
PERMITS RIGHT TO LEFT (REVERSE)
POWER FLOW AS A POSSIBLE
OPERATING CONDITION

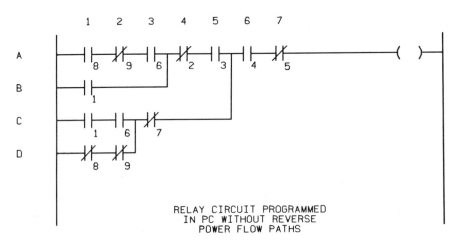

RELAY CIRCUIT PROGRAMMED
IN PC WITHOUT REVERSE
POWER FLOW PATHS

Fig. 14.24. Programming Terminal Syntax Rules, Reverse Power Flow

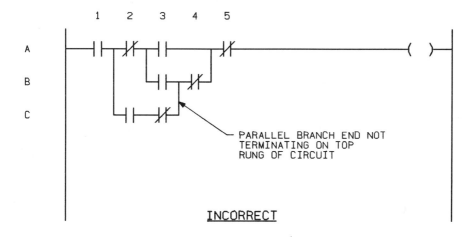

PARALLEL BRANCH END NOT
TERMINATING ON TOP
RUNG OF CIRCUIT

INCORRECT

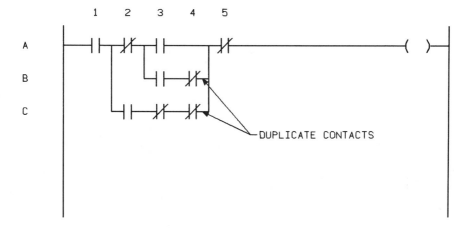

DUPLICATE CONTACTS

CORRECT

Fig. 14.25.   Programming Terminal Syntax Rules, Parallel Branch Termination

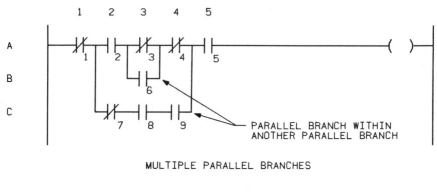

MULTIPLE PARALLEL BRANCHES

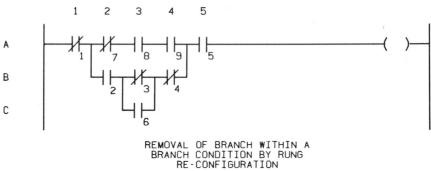

REMOVAL OF BRANCH WITHIN A
BRANCH CONDITION BY RUNG
RE-CONFIGURATION

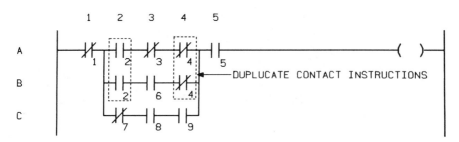

REMOVAL OF BRANCH WITHIN A
BRANCH CONDITION BY USE
OF DUPLICATE CONTACTS

Fig. 14.26.   Programming Terminal Syntax Rules, Branch within a Branch

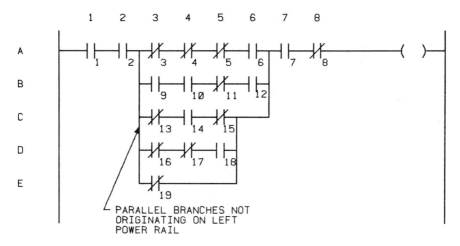

PARALLEL BRANCHES NOT
ORIGINATING ON LEFT
POWER RAIL

INCORRECT

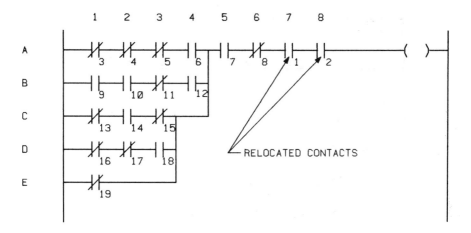

RELOCATED CONTACTS

CORRECT

Fig. 14.27. Programming Terminal Syntax Rules, Start of Parallel Branches

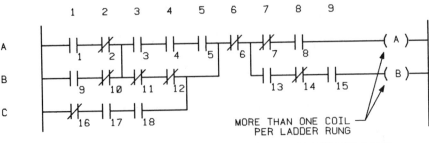

MORE THAN ONE COIL
PER LADDER RUNG

LADDER RUNG WITH MULTIPLE COIL INSTRUCTIONS

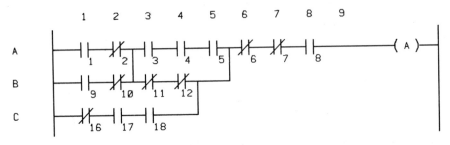

EQUIVALENT LADDER RUNG FOR COIL INSTRUCTION A

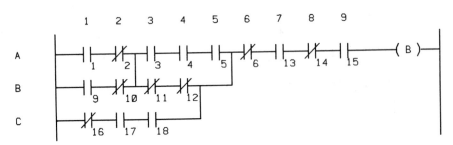

EQUIVALENT LADDER RUNG FOR COIL INSTRUCTION B

Fig. 14.28. Programming Terminal Syntax Rules, Multiple Coil Instructions

also not continuous, a condition similar to the lack of a contact instruction in the upper left-hand corner of the rung. The reconfiguration of the rung in the lower portion of the figure shows the error corrected.

All CRT programmers have a maximum display area on the screen for building a circuit rung. These boundaries should not be exceeded. Fig. 14.22 indicates an incorrect method of assembling more than the specified maximum number of contacts in any line, and how the error might be corrected. When more than the number of series instruction spaces available needs to be programmed, multiple rungs should be used, as described in Fig. 14.22.

It is also not possible to program more than the specified limit of parallel lines on a CRT screen. When additional parallel lines are required, the use of additional circuit rungs should be implemented. Fig. 14.23 shows a circuit that exceeds the CRT's limit and how to program the same circuit correctly with multiple circuit rungs.

Many PC programming terminals and processors do not permit reverse power flow through contact and branch instructions. Most terminals require that power flow in a circuit *from left to right* only. Fig. 14.24 shows a circuit that can be implemented in the design of a conventional relay control system. This circuit must be modified slightly when programmed in a PC. The proper circuit configuration for programming in a PC is also shown as part of Fig. 14.24.

Figs. 14.25 to 14.27 represent three additional parallel branch syntax rules which may need to be observed when developing a circuit on a CRT terminal. Many times a programming terminal will require all parallel circuit branches to terminate on the subrung directly above. In order to meet this requirement, it is usually necessary to insert additional contact instructions to the circuit, as shown in Fig. 14.25.

The use of parallel branches within the boundary of other parallel branches is often forbidden. Fig. 14.26 shows a ladder rung which has a parallel branch within a parallel branch, and two methods to reconfigure the circuit to eliminate the problem. Often parallel branches within parallel branches can be eliminated through the simple reconfiguration of the circuit. When the circuit does not lend itself to a simple reconfiguration, certain contact instructions may have to be duplicated to eliminate the problem. In severe cases a combination of additional contacts and circuit reorganization may be required.

Finally as indicated in Fig. 14.27, all parallel branches may be required to begin at the left-hand power rail on the programming terminal. If a circuit exists that has one or more series contacts preceding the start of a parallel branch, the circuit should be reconfigured such that the parallel branch begins with the left-hand power rail. Some programming terminals do permit multiple parallel branch start configurations on the same rung.

A final syntax rule concerns the use of multiple coils on a single rung. While most terminals permit only one coil or output-type instruction to be located on a rung, several terminals permit multiple coil or output-type instructions to occupy a ladder rung. As illustrated in Fig. 14.28, a rung that has more than one coil instruction can usually be replaced by a pair or rungs, each rung controlling one of the original coils.

The discussion on programming terminals is now complete. Due to the numerous variations of PCs and their associated programming terminals this text could not cover all of the various user and programming features available on every programming terminal. Each PC manufacturer that offers a programming terminal usually provides a unique feature or command not available from competitors. In addition those features offered by more than one manufacturer will usually not be similar in overall operation or programming format. Since detailed discussions on every possible feature and user command available on a PC programming terminal are beyond the scope of this text, the reader should consult the manufacturers of the various systems available to see whether a desired option is available or can be performed on a particular model of programmer.

15

Programmable Controller Communications

The requirement for PCs to communicate with other PCs as well as with intelligent external devices, such as computers and color graphic CRTs, spawned the development of PC-compatible communications networks. These networks, often referred to as "data highways," utilize twisted shielded-pair, coaxial, or triaxial cables to carry high-speed communications data between the various PCs and intelligent microprocessor-based devices connected to the network. While these networks may not be specifically designed for real-time control applications, they do provide an excellent medium for the sharing of data and system statuses between network devices. PC communications networks should not be confused with commercial systems, which can be adapted for use with a PC system, or control networks, which are used to communicate real-time I/O statuses between PCs.

By definition, a PC communications network is a high-speed data link designed for the passing of data and system statuses between PCs and other intelligent devices connected to the network. The communications rate of a typical data link is usually 57.6 kilobaud or higher. However, network devices may communicate with the network interface modules at data rates ranging from 300 baud to 19.6 kilobaud. Any intelligent device connected to the network can usually send and receive data from any other network device provided the data format meets the network's protocol. For communications integrity, all data networks incorporate various error-checking algorithms to ensure data integrity. Most networks can extend 15,000 cable feet from end to end. However, with the use of commercial modems these networks could extend to the planets if needed through the use of microwave and satellite communications systems.

The control network mentioned earlier has many of the same features as a PC communications network. In fact, to the casual observer the two networks might look the same. The big difference is that a control network permits limited mounts of ON/OFF discrete-type data to be passed between a limited number of network PCs. A communications network permits the same data exchange between PCs. However, the communications network can pass both numerical as well as discrete data between PCs, and in much larger quantities than is possible with a control network. The communications network gives up the ability to transfer all desired information in a single scan between all network PCs in order to accommodate the larger quantity of information being exchanged. The PC control network will transfer all information required from one PC to its network mates within a single processor scan.

MASTER-SLAVE AND PEER-TO-PEER SYSTEMS

The two basic network system formats offered by PC manufacturers are *master–slave* and *peer-to-peer*. The master–slave system illustrated in Fig. 15.1 requires the use of an intelligent device, such as a computer or a large-size PC, to manage all network communications between network devices. The programming of the master network device incorporates routines to individually address each slave device for the purpose of transmitting data between the addressed slave device and the master unit. Direct communication between slave devices is not possible. Information to be transferred between slaves must first be sent to the network master unit, which will in turn retransmit the message to the designated slave device. Master–slave systems offer the advantage of total control of network communications, but have the disadvantage of being solely dependent upon the master for all communications within the system. These types of systems often incorporate secondary backup network masters to operate the network in the event of primary master failure.

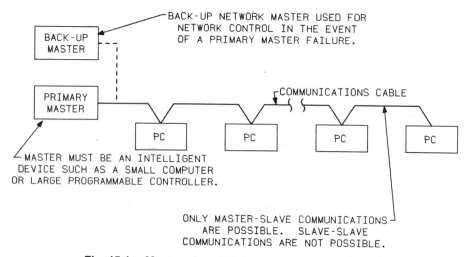

Fig. 15.1. **Master–slave PC Communications Network**

Peer-to-peer systems do not incorporate a master to control the network. Instead each network device has the ability to request use of, and then take control of, the network for the purpose of transmitting or requesting information from other network devices. This type of network communication scheme is often described as a baton or token passing system, since control of the network can be thought of as a baton that is passed from unit to unit.

The advantages and disadvantages of the peer-to-peer system are directly opposite those of the master–slave system. Since each network member has the right to use the data link, control of which devices use the network, how long a device uses the network, and the type of information passed on the network is harder to achieve. This system does have the distinct advantage of remaining in operation even with one or more units out of service. Error codes will usually be generated from active units attempting communication with faulted units, but communications can continue between active remaining units. Fig. 15.2 illustrates the configuration of a peer-to-peer communications network.

COMMUNICATIONS NETWORK COMPONENTS

Every communications network consists of two major components, the transmission cable or medium and the transmission cable interface electronics. The transmission medium is the physical method used to carry the messages between network devices. Various transmission media are depicted in Fig. 15.3. Shielded twisted-pair cable is probably the most widely used transmission medium outside of coaxial and twinaxial cable. Systems using either twisted-pair, coaxial, or twinaxial cable offer advantages in the form of ease of installation, low installed cost, easy availability, and minimum maintenance costs. Due to transmission cable losses and the types of cables available, the maximum distance a network can span is usually 10,000 to 15,000 cable feet. However, multiple networks can be interconnected to increase distances.

Local PC communications networks have incorporated baseband (passive) and broadband (cable TV) coaxial transmission techniques. Where a need has arisen to communicate over longer distances, telephone lines can often be used. Many offshore or pipeline applications incorporate the use of radio as the

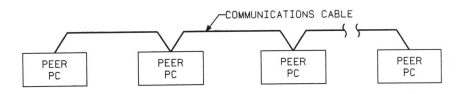

EACH PEER PC REQUESTS USE OF
THE COMMUNICATIONS NETWORK FOR
DATA TRANSMISSION TO ANOTHER
PEER UNIT AS THE NEED DICTATES

Fig. 15.2. Peer-to-peer Communications Network

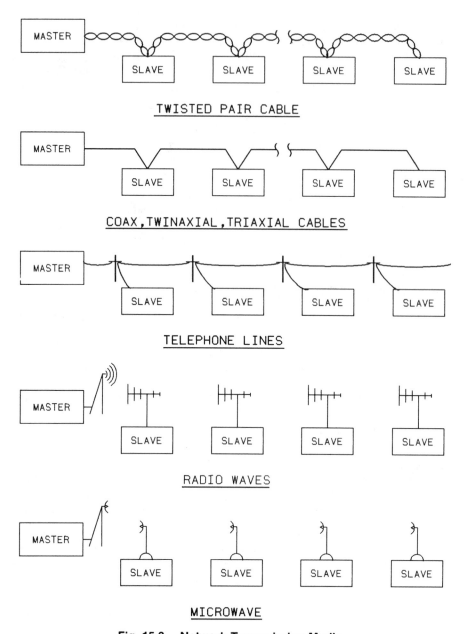

Fig. 15.3. Network Transmission Media

transmission medium. Where available, both microwave and satellite networks can be used should the application dictate the need for this type of communication medium. Finally the use of fiber-optic systems is becoming more prominent as technology in this area expands.

The interface hardware between the network communications cable and the actual PC or computer accessing the network is often referred to as a *modem* or *network adapter module*. The function of this module is to access the data network for the host PC or computer. There is usually a modem for every device connected to the network, but some systems offer the ability to connect multiple network devices to the network cable through a single modem or adapter module. Since the network must be designed to operate in harsh industrial environments, the modems communicate at frequencies specifically selected to minimize interference-induced transmission errors. Each modem is capable of interfacing with a PC or computer at data rates of 300 baud to 19.2 kilobaud. The transmission data are buffered in the modem until they can be placed on the network cable at data rates of 19.2 kilobauds or higher. The communications data are buffered in the modem so that the proper "handshaking" can occur between the network cable and the other modems on the network. Typical handshaking routines involve error checks, data cable use management, and PC/computer protocol routines.

Several physical arrangements of communications networks with current systems are possible. Daisy-chain arrangements directly tie the network device modems to the communications cable, as depicted in Fig. 15.4. Alternately some systems allow the use of "tee taps" on the main cable or trunk line, as illustrated in Fig. 15.5. This type of connection arrangement is often referred to as a "multidrop system." A modem is connected to the trunk line's tee tap via a short length (usually 500 feet maximum) of drop cable. The daisy-chain configuration might be thought of as a multidrop arrangement with "zero-length" drops from the trunk line. A third arrangement not often found is the "star" configuration of Fig. 15.6. This system has each modem wired back to a central distribution point, and is used primarily in master–slave-type applications. Finally the closed-loop system of Fig. 15.7 is also available. One advantage of the closed-loop system is that network communications are usually not lost unless the loop is severed in two locations. It should be noted that many of the available systems may be a combination of two or more of the simple arrangements listed above, or a system may have the flexibility to be configured into any one or more of the listed simple configurations, depending on the user's needs and requirements.

The number of modems that can be connected to a data communications network varies with the design of the network. Most systems currently available offer the flexibility to service either 16, 32, or 64 modems on the transmission system. This quantity of modems can be enlarged either by attaching several smaller systems together to form a larger system, or by incorporating a commercial system as part of the PC communications network.

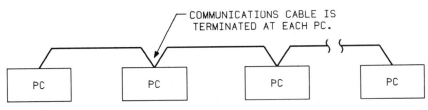

COMMUNICATIONS CABLE IS
TERMINATED AT EACH PC.

Fig. 15.4. Daisy-chain Configuration

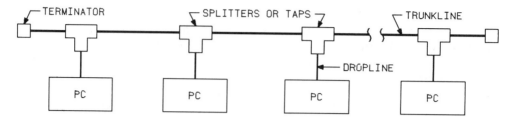

Fig. 15.5. Multi-drop Configuration

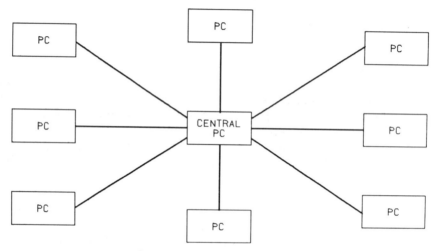

Fig. 15.6. Star Configuration

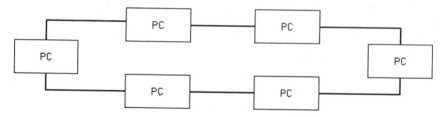

Fig. 15.7. Loop Configuration

NETWORK COMMUNICATION TRANSACTIONS

Depending on the type of network being implemented, either master–slave or peer-to-peer, several forms of network device transactions are possible. Master–slave systems usually offer either query/response or broadcast-type transactions to be implemented between devices. The peer-to-peer systems are generally limited to read/write-type message transmissions.

The query/response transaction of the master–slave system is depicted in Fig. 15.8. The communications network master unit selects the slave device to be interrogated and transmits a query message to that slave over the network. All

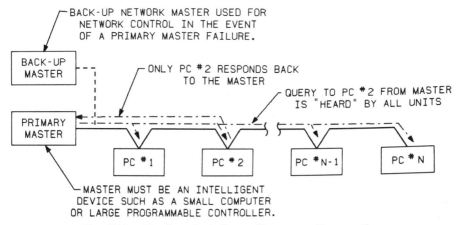

Fig. 15.8. Master–slave Query/Response Transaction

slave units on the network actually receive the message, but only the modem for the slave unit in question will decode the transmission and pass it on to its PC. The slave PC then carries out the instructions of the message and passes a response message back to the master, indicating the completion of the instruction along with any required response data.

A variation of the query/response transaction is the broadcast transaction illustrated in Fig. 15.9. This type of transaction is similar to the query/response transaction except that all slave modems decode the transmission from the master unit. Each slave PC will carry out the instructions of the message. However, none of the slaves will send a confirmation response back to the host master.

Communications networks that operate in a peer-to-peer configuration provide for the selective reading of data from one network peer to another or the selective writing of data from one peer to another. The read operation,

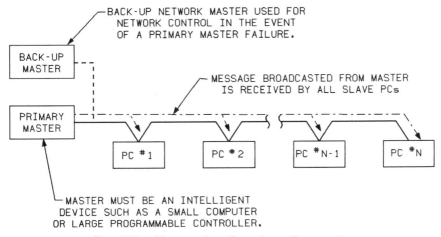

Fig. 15.9. Master–slave Broadcast Transaction

depicted in Fig. 15.10, first requires a peer PC to signal its network modem that a network transmission is required. The modem then gains access and control of the network, and once control is established, the modem can carry out the read transaction with the companion PC on the network. The write operation for the peer-to-peer system is also depicted in Fig. 15.10 and is similar to the read function. The only difference in operation is that instead of reading a portion of the selected PC's memory, the PC currently in control of the network will write new data into a portion of the selected PC's memory.

Note that two network transmissions occur as a result of a peer-to-peer read or write operation. The device requesting a read or write of information with another peer device must first gain access to the network to transmit the initial query or data. The network device being communicated with must then gain access to the network to supply or acknowledge with the proper response. This response will be either a transmission of requested read information, or a verification that the write information was received without error.

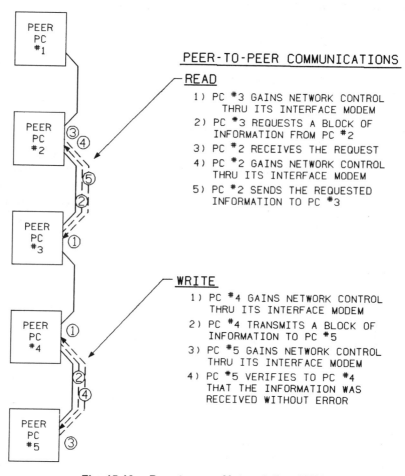

Fig. 15.10. Peer-to-peer Network Read/Write

COMMUNICATION NETWORK RESTRAINTS

The amount of data that can be transmitted in either the master–slave or the peer-to-peer configuration will depend on the design of the system. Many system designs allow for the transmission of an individual memory bit between system components as well as blocks of memory words. The upper limit on the number of memory words that can be communicated between network stations will depend on such system constraints as modem design, network communication protocol, error-check routines, and PC data-handling capabilities. The content of a message can consist of operational data, I/O statuses, program instructions or system hardware statuses.

Many systems provide the option to enable or disable certain types of communications to or from a particular station on a network. For example, peer-to-peer systems may offer both protected and unprotected read and write transactions. A system that allows protected communications is designed with additional instructions which require the specification of those memory areas that are accessible by each separate peer device on the network. Any peer attempting to access a memory area not within the specified access range for that unit will generate an error message or alarm. In the unprotected mode, a peer unit has free access to any memory location in the destination PC for read or write transactions.

As an example of the programming necessary to implement a data communications network, Fig. 15.11 represents the communications zone format for Allen-Bradley's Data Highway I. This system is a peer-to-peer communications network hosting up to 64 devices. It was chosen to illustrate the simplicity involved in programming PC communications on a data network, as well as the type of information that is required for proper system communication. The Allen-Bradley format is presented for illustration only. Each manufacturer offers his own format meeting the designs and guidelines of his system.

The first ladder rung of the Data Highway communications zone is termed the *header rung*. This rung identifies the address of the local station or PC on the data network as well as a memory address, which will contain the error code for a transmission that cannot be carried out to completion. A final piece of information encoded in this rung is a time-out code. The time-out code is an octal encoding of the number of seconds (in quarter-second increments) that the communication has to reach completion. Should the communication not reach completion in the designated time-out period, the proper error code will be set for user diagnostics.

Following the header rung is the memory access rung. It designates the memory area within the local station that can be addressed by a particular remote station. Every remote station that has access to the local station must be noted, along with the acceptable local memory area.

The actual data to be transferred are specified as part of the command rung(s) of the communications zone. One command rung is required for each transaction that is to occur, and rungs may be either a bit or word command

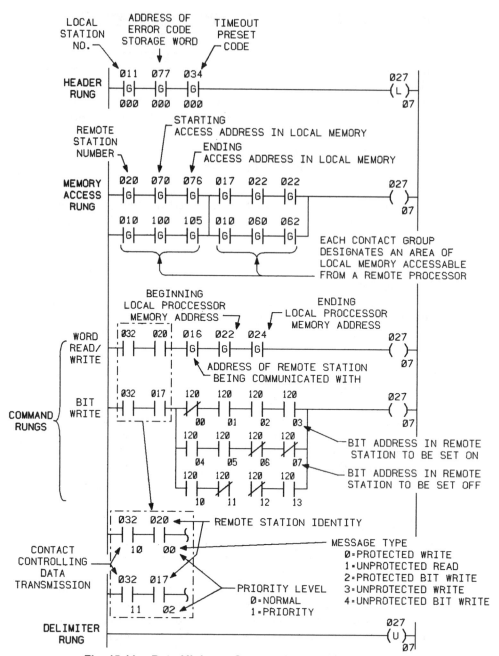

Fig. 15.11.   Data Highway Communications Zone Format

format. Bit command formats transfer individual bits of data, while word command formats are used to transfer one or more words of data memory between network devices. The significance of each instruction in the command rung is noted in the figure. The communications zone is completed with a delimiter rung. This rung signifies the completion of the communications network programming.

A communication over the network can be initiated by the ON/OFF status of an I/O device or internal coil, a timed condition, or the OFF-to-ON or ON-to-OFF transition of an I/O device or internal coil. The communication is complete when a message complete (done) status is received, or a remote or local fault status indication is activated for the transmission. This programming is necessary to monitor the done status as well as the fault statuses, and can require from two to five additional standard relay-type ladder rungs to implement.

PC REDUNDANCY

Another form of data communications network that cannot be overlooked is the data communications scheme employed with redundant PCs. While PCs are highly reliable due to their solid-state design, there is still a finite failure rate, as with any device. The PC, since its birth in the early 1970s, has grown in capability to match increasing control demands. Today larger and larger manufacturing processes and operations are relying on the PC for control. In conjunction with these larger demands are larger economic losses due to a system shutdown. An increasing need has developed to minimize the downtime of a system due to PC processor or PC-related failures.

The simplest form of PC redundancy is the stocking of spare parts. There is no need for any redundant communications network since the application usually is of a noncritical nature. Whenever there is a problem which is diagnosed as a component of the PC system, the part or parts are replaced. In many cases the critical PC system components, such as the processor and the power supply, are installed and powered with a duplicate program, in a cabinet adjacent to the operating system. In the event of on-line processor failure, maintenance personnel can manually disconnect the failed processor and substitute the backup unit. The control system is then available for restart and operation.

A manual redundant system does require that the backup processor initiate a *cold start*. The backup unit has no knowledge of past system operating history, and therefore assumes the system to be in a configuration making it available for restart. In many cases the restart of a system will require operator and often programmer interaction in order that the PC and its controlled equipment can return to normal operation in step with each other.

An improvement over cold redundancy employs additional hardware to transfer system control automatically from one PC to its backup. These systems can be employed with some success, provided that the application requirements allow for a possible interruption of system operation.

As a further enhancement, several PC manufacturers offer an optional "black box," which permits the user to pass data between the operational PC and its backup. In lieu of the black box, a PC user may wish to pass a limited amount of information, usually discrete ON/OFF data, between a pair of PCs for system backup purposes. This type of data transfer may be accomplished through I/O modules connected to each processor. A typical PC program to accomplish this type of information exchange between a pair of PCs is described in Chapter 11.

To illustrate the types of PC redundancy hardware available, several Modicon configurations will be examined. Modicon offers various levels of PC system redundancy depending on user requirements. The Modicon J340 automatically transfers selected channels of processor data between two PCs. Processor-to-processor communication in addition to automatic I/O transfer is offered with the use of a J340/J342. Total system redundancy and *hot backup* are possible using the J211/J212 redundant supervisor system. The Modicon PC redundancy systems were chosen for illustrative purposes only. The configuration, capabilities, operation, and application of any redundant PC system will be determined by the architecture of the PC system under consideration and the particular application involved.

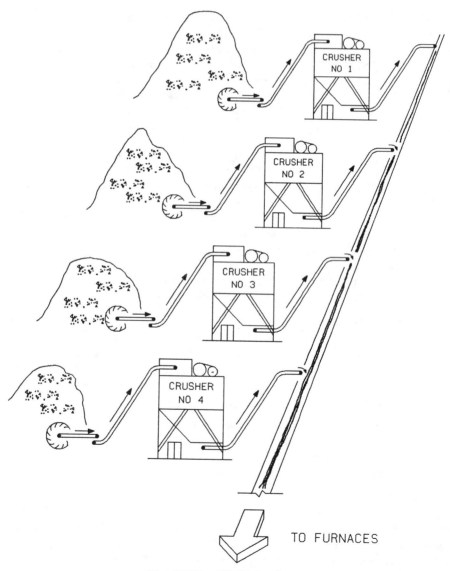

Fig. 15.12. PC Redundancy

An application that might involve the use of automated processor I/O transfer is depicted in Fig. 15.12. Many power facilities take coal and crush it before sending it to the furnaces. The coal transport conveyors and crushers are to be PC-controlled with two processors being employed. The facility can operate for a short period of time with two crushers in use. However, it has a total of four since normal operation requires three crushers with the fourth being a standby spare.

The use of a redundant PC system will be employed for the application. The supply conveyors to a crusher as well as the crusher itself will be controlled with a single I/O channel. There are four crusher systems requiring four I/O channels. The conveyor system transporting crushed coal from the crushers to the furnaces will require a single channel of I/O for control purposes. This conveyor system must have redundant PC controls since its loss for more than several minutes will require furnace shutdown.

A simple implementation of PC controls to meet the needs of the application would involve the use of an automatic I/O transfer unit similar to Modicon's J342. The first processor would control two of the crusher systems, as shown in Fig. 15.13. The second processor would be responsible for the operation of the remaining two crushing units. All system I/O would be configured for remote operation due to the long conveyor distances involved between the storage yard and the furnaces. Either processor can control the transport conveyer via the J342 automatic transfer switch. The J342 would automatically monitor the processor controlling the transport conveyor. In the event of a failure, the J342 would place the transport conveyor's control functions under the direction of the remaining PC. Both processors would contain identical programming for the transport conveyor, and the J342 transfer of I/O would be automatic with little or no system disturbance. In the event of processor failure, the facility would be operating with only two crushing units. However, the short time required for processor replacement and restart should not affect overall system performance.

Similar applications arise when there is a need to transfer operational status data from one processor to another. Consider the application depicted in Fig. 15.14, where cartons of product are conveyed to palletizers for palletizing and shipping. The cartons are transported in groups of four, which constitute a pallet, and must be handled by the conveyor system in such a manner that no breaking up of the four-group integrity occurs. Each palletizer performs the same function. However, the routing of cases to a particular palletizer is of prime importance.

Three PC processors might be used to implement the system, as depicted in Fig. 15.15. Two processors would contain the necessary programming to control two of the four palletizers. Each palletizer would be operated from its own dedicated remote I/O channel. The supply conveyor system, which transports the cartons from the manufacturing area to the palletizers, would be operated from a dedicated processor. A Modicon J340 processor-to-processor communicator could be used to transfer supply conveyor carton information automatically to either of the two palletizer processors. Since either palletizer

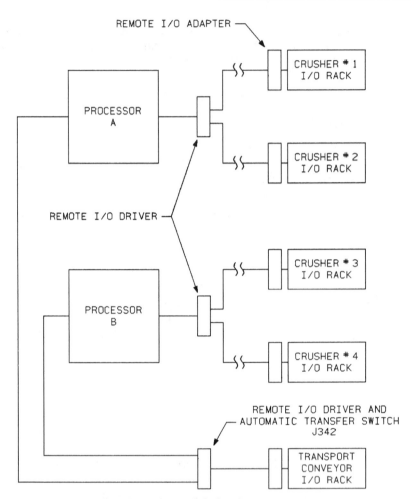

Fig. 15.13. PC Configuration

processor needs to know the current control status of the supply conveyor, the data communications capability of the J340 would be utilized. The current control status of the supply conveyor is passed from the conveyor processor to the proper palletizer processor that will be receiving a particular carton group being transported. In the event of a palletizer processor failure, the conveyor processor would have knowledge of the past supply conveyor history. Likewise if the conveyor processor were to fault and need replacement, a past history could be obtained for the replacement via the J340 from both palletizer processors. In addition, only two of the four palletizers could be lost, and the conveyor procesor would recognize the inoperability of a unit and not direct product cartons to it. Plant operation could continue, but a reduced throughput would result until the faulted unit was returned to service.

A redundant control system that affords a limited amount of data communications along with automatic I/O transfer is often referred as a *warm backup*-type system. The J340 permits the two processors to intercommunicate

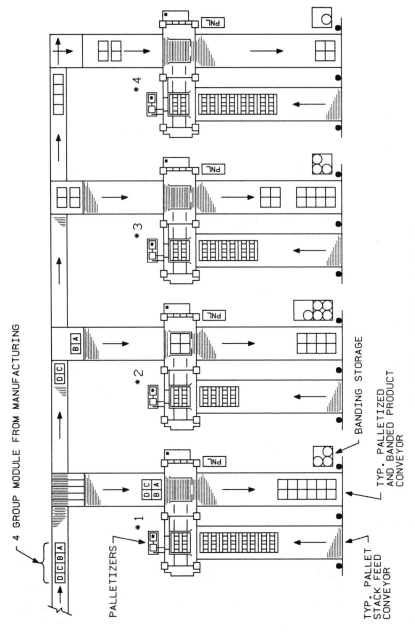

Fig. 15.14. Palletizer Application

399

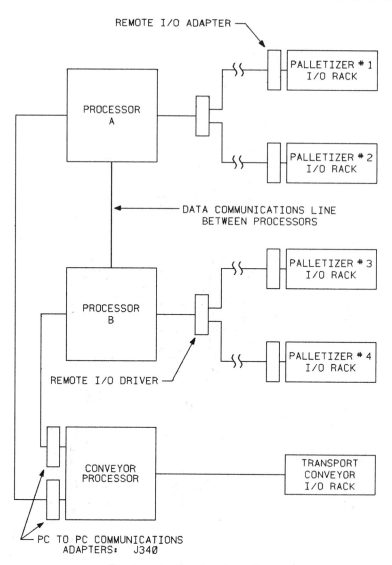

Fig. 15.15. PC Configuration

directly without the use of additional I/O hardware. Processor-to-processor communication is available, but the user must program the two processors for all data and control status exchanges. While the J340 permits the user to implement a communications scheme without the use of I/O hardware, the possibility still exists to use discrete I/O modules for processor-to-processor communications. In effect, the J340 is a small limited-capacity dedicated-application communications network, which can be employed for many PC-to-PC communications applications.

Even though it is possible to transfer data between an active processor and its running backup, there can often be large amounts of data collected and handled in the active unit. Accurately collecting and transferring the latest information and system statuses between two processors can involve extra programming for the user. In addition, there is a good possibility that information received in the standby unit is less current than the data now being executed in the controlling processor due to the scan time differences of the two machines. Also if a large quantity of data must be transferred, it may require numerous processor scans to actually complete the transmission. Should a switchover of control occur during an instant when the difference in data between the two processors is critical to system operation, unexpected and undesirable events may occur.

The only sure way to effect a transfer void of system errors and latent data is to implement a data communications scheme between processors, which can transmit all data and operational statuses from the active processor to the backup unit once every scan of both machines. To maintain absolute control, and a true *bumpless* or *hot backup* transfer, the two units should have their scans locked together as well.

One of the problems in transferring data from one processor to another involves making a decision as to exactly what to transfer and when. The majority of a PC's memory is loaded with the user logic and a small quantity of constants. This part of the memory is usually never changed unless the operation of the system is to be modified to meet newer and different system requirements. The remaining data in a processor's memory consist of coil statuses, latch states, timer and counter current values, and the results of user programmed calculations. These are the data critical to the transfer scheme. The backup unit must be updated with the most current copy of these data as soon as they are produced if a bumpless transfer is to be successful.

An additional problem of bumpless transfer lies in determining the incorrect operation or partial failure of the controlling unit. A hard failure is easy to detect, but partial failures may not surface instantly for detection. By the time a partial failure is detected and acted upon, the processor may have initiated incorrect control system operation. To complicate the problem further, the data in the operational unit may be incorrect, and a copy may have reached the backup processor prior to an operational unit's problem being detected.

The solutions to the problem can take numerous forms. One might be to have the backup processor designed to "eavesdrop" on the operational unit, keeping itself current to the actions of the first. The backup processor sees the same input statuses as the controlling processor. However, it acts on these statuses according to its programming, which is identical to that of the controlling processor. Hardware and software checks in each unit internally monitor every aspect of the system for a variation. When a variation occurs, a decision is made as to which unit is wrong, and that unit is taken off-line. The backup processor is not dependent on the operational unit since each processor arrives at independent results based on its own programmed interpretation of the I/O statuses.

A second method of bumpless transfer provides some sort of system supervisor to monitor the operation and data transfer between the two units. The system supervisor is charged with buffering the data transfer between on-line and backup units, monitoring the health of both, as well as performing hardware and software checks on each processor's solution of each user logic instruction.

The Modicon J211/J212 *fault-tolerant high-availability control scheme* is representative of a redundant PC system that is based on the supervisor concept. Two processors, as shown in Fig. 15.16, are connected for redundant operation by a J211 supervisor and a J212 I/O switchover unit. Both processors contain identical programming and data. The J211 performs a constant health check of both processors, determining their operational statuses. Should the system supervisor locate a problem developing in either of the units, the unit showing the problem will be shut down and system control will be from the healthy processor. The user logic of both units as well as the constant data are checked to ensure that they are identical. At the completion of each processor's scan the latest I/O statuses are transferred and verified between the two processors. This data transfer and verification occurs by direct memory access at data rates approaching 25 megabaud.

The supervisor is designed to bring a replacement processor back on line from cold-start conditions, as well as monitoring itself for proper operation. Should the supervisor detect a malfunction with its own operation, it is designed to take itself off-line and not affect the operation of the complete system other than a loss of the automatic redundant backup capability.

While the J211/J212 hardware combination provides bumpless transfer for the processors, the I/O system is still vulnerable to failure. No hot-backup bumpless transfer scheme can adequately address this problem. In fact the solution to this type of application can involve the implementation of complex combinations of hardware and software.

The simplest method of system backup involves the stocking of spare parts and efficient, well-trained technicians to troubleshoot, maintain, and repair the system. Whenever the costs incurred from production loss warrant the implementation of a redundant PC system, the selection of the redundant system's configuration should be made according to the requirements of the application. If limited amounts of noncritical data must be transferred between redundant processors, the use of hard-wired I/O data transfer, or the implementation of devices similar to Modicon's J340 or J342, will usually more than adequately handle the application. When downtime costs warrant the need of total on-line bumpless redundant processor backup, systems are available to handle the application requirements. The use of bumpless transfer allows a PC to operate under the most critical control system applications.

COMPUTER COMMUNICATIONS AND PROTOCOL

On occasion it becomes necessary for a PC to communicate with a peripheral device such as a modem, printer, or similar device. Several standards exist which define the signal protocol as well as the mechanical interconnections that are to

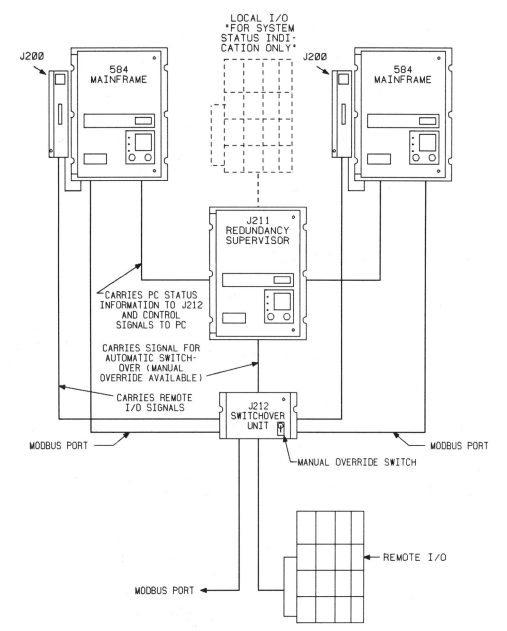

Fig. 15.16. Modicon 584 Redundancy System

be used to accomplish the device communications. The most common signal standard currently used with PCs is *RS-232C*. Three other protocols, *RS-449*, *current loop*, and *IEEE 488*, are also found in common use. The ASCII character code is the most frequently used protocol for the encoding of alphanumeric characters in a binary format. The following discussions will deal with these protocols, their specifications, and operation.

PARALLEL INTERFACES

The simplest way to communicate information between two devices is to employ a parallel signal approach, setting up one signal line between units for each individual character or piece of information to be communicated. This method of signal transmission is referred to as *parallel* and is primarily used for high data transmission rates over very short distances (usually 6 cable feet or less). As illustrated in Fig. 15.17, each parallel interface conductor serves a dedicated purpose in the transmission of information. This means that most parallel interfaces are unique and do not permit the flexibility to interconnect numerous devices. While this approach is simple, the quantity and distance limitations of the interface conductors often leave this interface unworkable in many real-world applications.

THE ASCII CHARACTER CODE

A more practical solution is to encode the data as a series of ON and OFF signals prior to transmission over a single wire. The earliest practical application of this idea is probably Samuel F. B. Morse's telegraph system. He developed a series of dots and dashes, the equivalent of today's ones and zeros, to represent the letters of the alphabet, which could be used to transmit messages between various points on the globe. The Morse code is a primitive application of a time-sequential data encoding scheme devised for the transmission of information by electromechanical means.

With the development of computers and the need to have them intercommunicate came the need to encode letters and numerals as a series of OFF and ON states. The ASCII code, which stands for American Standard Code for Information Interchange, was developed to meet this need. It represents, as a series of ones and zeros, all characters capable of being typed on a typewriter

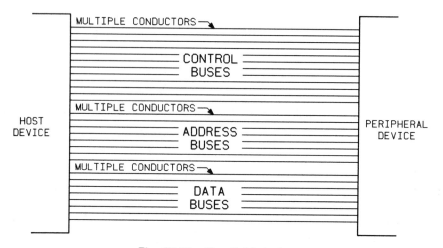

Fig. 15.17. Parallel Interface

plus a handful of special codes. The total number of elements in the ASCII character set is 128. All elements and their corresponding ASCII codes are shown in Table 15.1. Note that the 7-bit ASCII code when transmitted with a parity bit (to be discussed later in this chapter) may be referred to as 8-bit ASCII by some users.

Table 15.1. ASCII Character Codes, 0 through 127 Decimal

BINARY CODE	OCTAL CODE	DEC. CODE	HEX CODE	ASCII CHAR.	REMARKS
0000000	000	000	00	NUL	NULL; TAPE FEED; CONTROL/SHIFT/P
0000001	001	001	01	SOH SOM	START OF HEADING; START OF MESSAGE; CONTROL/A
0000010	002	002	02	STX EOA	START OF TEXT; END OF ADDRESS; CONTROL/B
0000011	003	003	03	ETX EOM	END OF TEXT; END OF MESSAGE; CONTROL/C
0000100	004	004	04	EOT	END OF TRANSMISSION (END); SHUTS OFF TWX MACHINES; CONTROL/D
0000101	005	005	05	ENQ	ENQUIRY (ENQRY); CONTROL/E
0000110	006	006	06	ACK	ACKNOWLEDGE; CONTROL/F
0000111	007	007	07	BEL	RINGS THE BELL; CONTROL/G
0001000	010	008	08	BS	BACKSPACE; FORM EFFECTOR; CONTROL/H
0001001	011	009	09	HT	HORIZONTAL TAB; CONTROL/I
0001010	012	010	0A	LF	LINE FEED OR LINE SPACE (NEW LINE); ADVANCES PAPER TO NEXT LINE; CONTROL/J
0001011	013	011	0B	VT	VERTICAL TAB (VTAB); CONTROL/K
0001100	014	012	0C	FF	FORM FEED TO TOP OF NEXT PAGE (PAGE); CONTROL/L
0001101	015	013	0D	CR	CARRIAGE RETURN TO BEGINNING OF LINE; CONTROL/M
0001110	016	014	0E	SO	SHIFT OUT; CHANGES RIBBON COLOR TO RED; CONTROL/N
0001111	017	015	0F	SI	SHIFT IN; CHANGES RIBBON COLOR TO BLACK; CONTROL/O
0010000	020	016	10	DLE	DATA LINK ESCAPE; CONTROL/P
0010001	021	017	11	DC1	DEVICE CONTROL 1; TURNS TRANSMITTER (READER) ON; CONTROL/Q (X ON)
0010010	022	018	12	DC2	DEVICE CONTROL 2; TURNS PUNCH OR AUX ON; CONTROL/R; (TAPE AUX ON)

Table 15.1 (*Continued*)

BINARY CODE	OCTAL CODE	DEC. CODE	HEX CODE	ASCII CHAR.	REMARKS
0010011	023	019	13	DC3	DEVICE CONTROL 3; TURNS TRANSMITTER (READER) OFF; CONTROL/S; (X OFF)
0010100	024	020	14	DC4	DEVICE CONTROL 4; TURNS PUNCH OR AUX OFF; CONTROL/T; (AUX OFF)
0010101	025	021	15	NAK	NEGATIVE ACKNOWLEDGE; CONTROL/U
0010110	026	022	16	SYN	SYNCHRONOUS FILE (SYNC); CONTROL/V
0010111	027	023	17	ETB	END OF TRANSMISSION BLOCK; LOGICAL END OF MEDIUM, CONTROL/W
0011000	030	024	18	CAN	CANCEL (CANCL); CONTROL/X
0011001	031	025	19	EM	END OF MEDIUM; CONTROL/Y
0011010	032	026	1A	SUB	SUBSTITUTE; CONTROL/Z
0011011	033	027	1B	ESC	ESCAPE; CONTROL/SHIFT/K
0011100	034	028	1C	FS	FILE SEPARATOR; CONTROL/SHIFT/L
0011101	035	029	1D	GS	GROUP SEPARATOR; CONTROL/SHIFT/M
0011110	036	030	1E	RS	RECORD SEPARATOR; CONTROL/SHIFT/N
0011111	037	031	1F	US	UNIT SEPARATOR; CONTROL/SHIFT/O
0100000	040	032	20	SP	SPACE
0100001	041	033	21	!	
0100010	042	034	22	"	QUOTE
0100011	043	035	23	#	POUND SIGN
0100100	044	036	24	$	DOLLAR SIGN
0100101	045	037	25	%	PERCENT SIGN
0100110	046	038	26	&	AMPERSAND
0100111	047	039	27	'	APOSTROPHE
0101000	050	040	28	(	LEFT PARENTHESIS
0101001	051	041	29	)	RIGHT PARENTHESIS
0101010	052	042	2A	*	ASTERICK OR MULTIPLY
0101011	053	043	2B	+	POSITIVE OR PLUS SIGN
0101100	054	044	2C	,	COMMA
0101101	055	045	2D	–	HYPHEN

Table 15.1 *(Continued)*

BINARY CODE	OCTAL CODE	DEC. CODE	HEX CODE	ASCII CHAR.	REMARKS
0101110	056	046	2E	.	PERIOD
0101111	057	047	2F	/	DIVIDE SIGN
0110000	060	048	30	0	
0110001	061	049	31	1	
0110010	062	050	32	2	
0110011	063	051	33	3	
0110100	064	052	34	4	
0110101	065	053	35	5	
0110110	066	054	36	6	
0110111	067	055	37	7	
0111000	070	056	38	8	
0111001	071	057	39	9	
0111010	072	058	3A	:	COLON
0111011	073	059	3B	;	SEMI-COLON
0111100	074	060	3C	<	LEFT ARROW
0111101	075	061	3D	=	EQUAL SIGN
0111110	076	062	3E	>	RIGHT ARROW
0111111	077	063	3F	?	QUESTION MARK
1000000	100	064	40	@	"AT" SIGN
1000001	101	065	41	A	
1000010	102	066	42	B	
1000011	103	067	43	C	
1000100	104	068	44	D	
1000101	105	069	45	E	
1000110	106	070	46	F	
1000111	107	071	47	G	
1001000	110	072	48	H	
1001001	111	073	49	I	
1001010	112	074	4A	J	
1001011	113	075	4B	K	

Table 15.1 *(Continued)*

BINARY CODE	OCTAL CODE	DEC. CODE	HEX CODE	ASCII CHAR.	REMARKS
1001100	114	076	4C	L	
1001101	115	077	4D	M	
1001110	116	078	4E	N	
1001111	117	079	4F	O	
1010000	120	080	50	P	
1010001	121	081	51	Q	
1010010	122	082	52	R	
1010011	123	083	53	S	
1010100	124	084	54	T	
1010101	125	085	55	U	
1010110	126	086	56	V	
1010111	127	087	57	W	
1011000	130	088	58	X	
1011001	131	089	59	Y	
1011010	132	090	5A	Z	
1011011	133	091	5B	[	OPENING BRACKET
1011100	134	092	5C	\	BACKSLASH
1011101	135	093	5D	]	CLOSING BRACKET
1011110	136	094	5E	^	CIRCUMFLEX; UP ARROW
1011111	137	095	5F	_	UNDERLINE
1100000	140	096	60	`	SINGLE OPENING QUOTE
1100001	141	097	61	a	
1100010	142	098	62	b	
1100011	143	099	63	c	
1100100	144	100	64	d	
1100101	145	101	65	e	
1100110	146	102	66	f	
1100111	147	103	67	g	
1101000	150	104	68	h	
1101001	151	105	69	i	

Table 15.1 *(Continued)*

BINARY CODE	OCTAL CODE	DEC. CODE	HEX CODE	ASCII CHAR.	REMARKS
1101010	152	106	6A	J	
1101011	153	107	6B	k	
1101100	154	108	6C	l	
1101101	155	109	6D	m	
1101110	156	110	6E	n	
1101111	157	111	6F	o	
1110000	160	112	70	p	
1110001	161	113	71	q	
1110010	162	114	72	r	
1110011	163	115	73	s	
1110100	164	116	74	t	
1110101	165	117	75	u	
1110110	166	118	76	v	
1110111	167	119	77	w	
1111000	170	120	78	x	
1111001	171	121	79	y	
1111010	172	122	7A	z	
1111011	173	123	7B	{	LEFT BRACE
1111100	174	124	7C	l	VERTICAL BAR
1111101	175	125	7D	}	RIGHT BRACE
1111110	176	126	7E	~	TILDE MAY BE ALTERNATE ESCAPE CODE
1111111	177	127	7F	DEL	DELETE, RUB OUT

NOTE: BINARY CODES 0100000 THROUGH 1111111 PRODUCE PRINTABLE CHARACTERS. ALL OTHERS ARE CONTROL (NON-PRINTABLE) CHARACTERS.

In addition to the standard numbers and letters of the ASCII code, there are 31 special codes. These codes, decimal codes 00 through 31, are referred to as *control codes*. They permit access to many of the standard special features provided with many peripheral devices. For example, hexadecimal code 0C initiates the form feed action of most printers. This code finds excellent use in printing applications where the information is to be placed on pages rather than printed continuously. Other codes, such as device control codes, permit the starting of tape or paper punch devices prior to the storage or retrieval of information. While some manufacturers do not adhere strictly to the ASCII definition of a particular control code, these codes do give a manufacturer the flexibility to incorporate special-purpose functions within a peripheral device, and they give the user access to them through the ASCII control code functions.

While the 7-bit ASCII code is used for the communication of numbers and letters, some peripherals operate from an 8-bit ASCII code. This code should not be confused with the 7-bit ASCII code with parity discussed earlier. Many peripheral devices, especially intelligent terminals and printers, incorporate alternate character sets, graphics characters, or special printing fonts. Access to these special characters, an additional 128 characters above the normal 128 7-bit ASCII characters, is achieved through the use of an 8-bit ASCII code. Peripheral devices that operate with an 8-bit ASCII code, decimal codes 128 through 255, have nonstandard assignments for the additional 128 characters. One peripheral may provide a Greek symbol for a particular code while a second device assigns the same code a graphics symbol. For those devices using the special 8-bit ASCII code, only the first 128 characters will match standard ASCII definitions.

RS-232C

Once computer-type devices began communicating using the ASCII code, additional problems developed. In order to communicate, each device had to know the communication status of the other in order to function. Just as two people alternate their conversation when talking on a phone (one usually listens while the other talks), computer- and PC-type devices need to have the same flexibility. The need then was to develop a standard that would set up guidelines for the transmission of information between computer hardware. The result of this need was the development of EIA RS-232C by the Electronic Industries Association.

The EIA RS-232C standard defines the electrical and the electro-mechanical characteristics of an interface system between two computer-type devices. In its simplest form this standard reduces the amount of interface wiring to a minimum of three wires, while a fully developed RS-232C interface can have as many as 22 parallel signal wires.

The RS-232C protocol defines *data terminal equipment* (DTE) as any device that generates or receives data, such as a printer/keyboard or a CRT terminal with keyboard. *Data communications equipment* (DCE) is any device that encodes data signals into voicelike signals for long-distance transmission, such as a modem. Since it is possible to interconnect devices such as PCs and printers without the use of a telephone modem, the full RS-232C protocol may not be required. For these applications the DTE might be a PC or other intelligent device, and the DCE could be a printer or similar "dumb" device.

In practice nine signal lines are all that is necessary for most applications. The RS-232C standard normally uses a 25-pin D-type connector as the mechanical

Table 15.2. Nine Most Common RS-232C Signals

PIN NO	RS232C CODE	SIGNAL NAME	DIRECTION	FUNCTION
1	AA	PROTECTIVE GROUND	NOT APPLICABLE	SAFETY OR POWER LINE GROUND FOR EQUIPMENT
2	BA	TRANSMITTED DATA (TXD)	TO DATA COMMUNICATIONS EQUIPMENT	TRANSMIT LINE TO DATA COMMUNICATIONS EQUIPMENT
3	BB	RECEIVED DATA (RXD)	FROM DATA COMMUNICATIONS EQUIPMENT	RECEIVING LINE FROM DATA COMMUNICATIONS EQUIPMENT
4	CA	REQUEST TO SEND (RTS)	TO DATA COMMUNICATIONS EQUIPMENT	INSTRUCTS DATA COMMUNICATIONS EQUIPMENT TO ENTER A TRANSMIT MODE OF OPERATION
5	CB	CLEAR TO SEND (CTS)	FROM DATA COMMUNICATIONS EQUIPMENT	INDICATES TO THE DATA TERMINAL THAT THE DATA COMMUNICATIONS EQUIPMENT IS READY TO TRANSMIT
6	CC	DATA SET READY (DSR)	FROM DATA COMMUNICATIONS EQUIPMENT	INDICATES TO THE DATA TERMINAL THAT THE DATA COMMUNICATIONS EQUIPMENT HAS ESTABLISHED COMMUNICA- TIONS WITH ANOTHER MODEM
7	AB	SIGNAL GROUND	NOT APPLICABLE	COMMON GROUND REFERENCE POINT FOR ALL COMMUNICATION LINES AND SIGNALS
8	CF	CARRIER DETECT	FROM DATA COMMUNICATIONS EQUIPMENT	INDICATES TO THE DATA TERMINAL THAT CARRIER SIGNALS ARE BEING RECEIVED BY THE DATA COMMUNICA- TIONS EQUIPMENT
20	CD	DATA TERMINAL READY (DTR)	TO DATA COMMUNICATIONS EQUIPMENT	INDICATES TO THE DATA COMMUNICATIONS EQUIPMENT THAT THE DATA TERMINAL IS CONNECTED, POWERED UP, AND READY

connection between the computer devices. The name and the function of the nine most used signals are shown in Table 15.2. Fig. 15.18 illustrates the proper internal wiring of a cable intended to interconnect two devices using the full RS-232C communication protocol.

When a peripheral device does not support or require the full RS-232C communication protocol, a partial RS-232C interconnection may be used. Many devices, such as displays and simple printers, do not rely on the handshaking of the RS-232C specification for operation. They are designed to interpret all signals received and either print or display them. Therefore there is little need for handshaking between the two devices. For these applications as little as three or four conductors may be used between the units. Fig. 15.19 illustrates the cabling necessary to support a partial RS-232C protocol. The cable wiring illustrated will permit two full RS-232C protocol devices to be interconnected for the transmission of information without handshaking. When this form of cable interconnection is used, the user should be aware that a host device will not be able to determine whether its peripheral device is active and available for communication. Use of this cable configuration could result in the loss of information being transmitted between the two units.

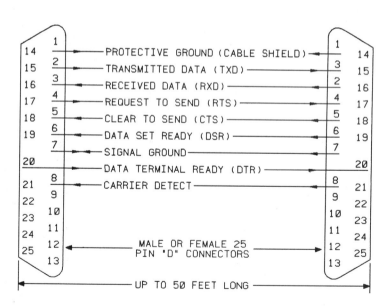

NOTES: 1. ARROWS BY SIGNAL NAME INDICATE DIRECTION OF SIGNAL.

2. SEE FIGURE 15.19 FOR FUNCTION OF EACH SIGNAL.

Fig. 15.18. Wiring Diagram of EIA Cable for Full RS-232C Protocol

CONNECT TO EIA
RS-232C PORT
ON TERMINAL OR
HOST DEVICE

CONNECT TO EIA
RS-232C PORT
ON MODEM OR
PERIPHERAL DEVICE

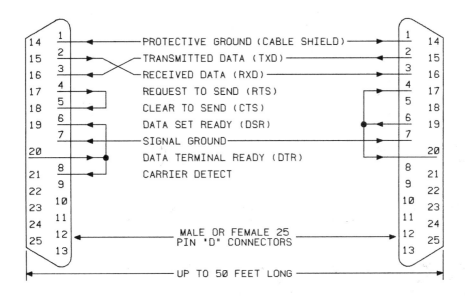

NOTES: 1. ARROWS BY SIGNAL NAME INDICATE
DIRECTION OF SIGNAL.

2. SEE FIGURE 15.19 FOR FUNCTION OF
EACH SIGNAL.

Fig. 15.19. Wiring Diagram of EIA Cable for Partial RS-232C Protocol

It should be noted that the RS-232C specification does not specify the character code to be used in the transmission of data, it only specifies the interface signal requirements. While many systems use either the 7-bit or the 8-bit ASCII code, several other codes also exist, such as a 6-bit *IBM* code, called *correspondence code.* A 5-bit *Murry and Budot* code, an 8-bit *extended binary-coded-decimal interchange code* (EBCDIC), as well as many specialty codes are also found in use to complicate the issue further. Besides addressing the RS-232C protocol, the user must know the protocol of the character code in order to operate a communications link.

In addition the mode of communication is not specified by the RS-232C protocol. As illustrated in Fig. 15.20, three communication modes are possible. *Simplex* communications involve one device, usually the DTE communicating solely to the DCE. *Half duplex* communications permit communications between a DTE and a DCE in either direction. However, communications can occur in one direction at a time only. *Full duplex* communications permit simultaneous bidirectional communications.

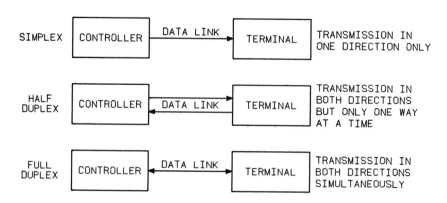

SIMPLEX - TRANSMISSION IN ONE DIRECTION ONLY WITH NO WAY OF RESPONDING. HOME RADIO RECEPTION OF A SIGNAL TRANSMITTED FROM THE RADIO STATION IS AN EXAMPLE OF SIMPLEX COMMUNICATION.

HALF DUPLEX - TRANSMISSION IN TWO DIRECTIONS, BUT ONLY ONE WAY AT A TIME. CB RADIO OPERATORS EITHER TRANSMIT OR RECEIVE, BUT CANNOT DO BOTH SIMULTANEOUSLY ON A SINGLE CHANNEL.

FULL DUPLEX - TRANSMISSION IN BOTH DIRECTIONS SIMULTANEOUSLY. A TELEPHONE LINK PERMITS COMMUNICATION IN 2 DIRECTIONS SIMULTANEOUSLY. BOTH PARTIES CAN TALK AND LISTEN AT THE SAME TIME.

Fig. 15.20.   Communications Modes

It should be noted that some RS-232C protocol users interpret half and full duplex as an indication of whether a local copy is generated of the transmitted data in conjunction with the transmitted copy. For device manufacturers who interpret the protocol in this manner, half duplex indicates that no local copy of the transmission is generated, while full duplex operation generates the local copy. Local copy is an identical duplicate, produced at the sending device (DTE), of what is being sent to the remote device (DCE).

An ASCII character code's transmission is performed in a serial manner over an RS-232C interface. Each ASCII character's series of logic ones and zeros is represented as the presence or absence of a voltage level on the RS-232C signal wire. Since the standard was developed before the use of transistor-transistor-logic (TTL) circuits, which operate on 5-volt signal levels, the RS-232C standard uses levels between −25 and +25 volts for operation. The standard also uses negative-true logic for the representation of the zero and one states, as illustrated in Fig. 15.21. A voltage level between 5 and 25 volts represents a logic zero state, and a voltage level between −5 and −25 volts indicates a logic one state.

The RS-232C control lines use the more standard system of positive logic for their operation. A positive voltage indicates a true state and a negative voltage a false state. The user should be aware of these conventions and the fact that a manufacturer employing RS-232C does not have to use any specified plus or minus voltage level for the operation of his equipment. Two pieces of

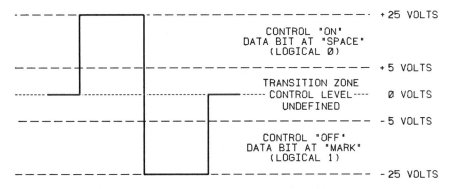

RS232C DEFINES THE LEVEL AND POLARITY OF THE SIGNALS GOING
TO AND FROM THE MODEM. A LOGICAL "1" IS REFERRED TO AS A 'MARK'
AND A LOGICAL '0" IS CALLED 'SPACE'. MARK AND SPACE ARE
TELECOMMUNICATION TERMS DATING BACK TO MORSE CODE KEY SETS.

Fig. 15.21. RS-232C Voltage Transition Levels

equipment designed to the RS-232C protocol may in fact not communicate, simply because the character codes are different or because the interface voltage levels are not exactly matched. In effect RS-232C is a nonstandard standard. Fortunately most manufacturers try to make their equipment compatible with existing hardware, but there are still those few . . .

The electrical format of the serial data being transmitted by the RS-232C standard is shown in Fig. 15.22. The transmission begins with the sending of a *start bit* to indicate that a character is being sent. This bit is followed by the *data bits* of the transmission. The data bits are the number of bits required to transmit the desired code in the character code format being used. Fig. 15.22 illustrates an example transmission for the character R in 7-bit ASCII. Note that an RS-232C transmission sends the least significant bits (LSB) first, a practice that tends to be awkward to use.

Following the data bits is a *parity* bit, which is used for error-checking the transmission. Parity is not required, but when used may be either *even* or *odd*, depending on the desires of the user. Since noise could affect the transmission or a misread could occur, the parity bit is used to confirm the correctness of the transmission. Parity is calculated by the sending unit counting the number of one state data bits transmitted. The parity bit is then set to a one or a zero so that the total number of data bit one states plus the parity bit state, is either always even (even parity) or always odd (odd parity). The receiver unit can then count the number of received data bit and parity bit ones to ensure that the correct character was received. The parity check is limited in that it can overlook certain multiple errors.

The final bits to be transmitted are the *stop bits*. These bits signal the end of the character transmission. The number of stop bits may be specified as either one, one and a half, or two. The half bit is permissible since the stop bit period actually represents a resting time for the receiver.

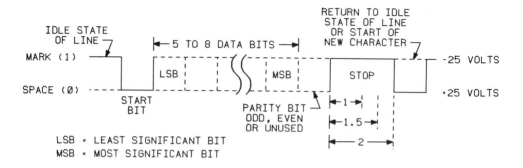

RS-232C PROTOCOL FORMAT

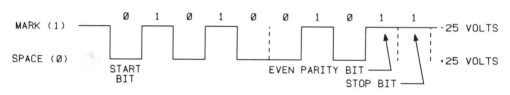

TRANSMISSION OF THE ASCII CHARACTER 'R' (1010010)

ASYNCHRONOUS OR START-STOP OPERATION

1. EACH CHARACTER IS PRECEDED BY A START BIT (OR LOGICAL ZERO), AND IS CONCLUDED WITH ONE OR MORE STOP BITS (LOGICAL ONE).

2. THE TRANSITION FROM STOP TO START TRIGGERS THE RECEIVER'S INTERNAL CLOCK. THE RECEIVER THEN ACCEPTS (AND DISCARDS) THE START BIT, ACCEPTS THE NUMBER OF BITS APPROPRIATE TO THE CODE SET, AND IS CONDITIONED BY THE STOP BIT FOR THE NEXT CHARACTER.

Fig. 15.22.   ASCII/RS-232C Character Transmission

The final consideration is the rate of bit transmission. Standard bit transmission rates are listed in Table 15.3. The transmission rate is expressed in bits per second and is commonly referred to as the *baud rate.*

So in order to use the RS-232C "standard," the user must make sure that the voltage levels for the devices, the baud rate of the transmission, the character code, the parity, the number of stop bits, and the number of handshaking lines required between the two devices are matched if the communications link is to operate correctly. For most PC systems the ASCII character code is used, the PCs recognize any voltage within the proper RS-232C range, and the 25-pin D-type connector is used on the cables. Most manufacturers provide excellent data on the handshaking lines that must be used with each type of peripheral device to be connected to the RS-232C port, as well as switches for the selection of baud rates, parity, and stop bit configuration.

RS-232C COMMUNICATIONS ERRORS

Some PCs have the capability to generate a warning or alarm message whenever there is a problem with an RS-232C communication, whether it be with a computer or a simple telephone modem. These messages are usually designed to indicate *framing, parity, overrun,* or *handshake* types of errors. Systems that have the ability to self-diagnose a problem with an RS-232C communications link will save the user untold hours of troubleshooting.

A framing error is usually an indication that the total number of bits being transmitted and received does not match between the two devices. If the parity check function is disabled, the user might check that the function is disabled at both devices. The user may also wish to check with the device manufacturer to see whether a parity bit is actually transmitted when the parity function is disabled. Some systems do not permit the parity bit to be disabled, and where this is the case, parity will have to be selected as even or odd.

A parity error indicates that either the transmission lines are picking up electromagnetic noise or the parity is set incorrectly in one of the two devices. If the parity is correct in both devices, a check of the transmission line should be made to ensure good solder connections and tight plugs and receptacles on the transmission cable. Routing the transmission cable away from sources of radio-

Table 15.3. Common Transmission Rate Uses

TRANSMISSION SPEED	USAGE
50,75	SPECIAL PURPOSE LOW SPEED APPLICATIONS
110	MECHANICAL TELETYPES AND TELEPRINTERS
134.5,150	ELECTRONIC TELETYPES AND TELEPRINTERS
200	COMMON EUROPEAN TELEPHONE DATA SPEED
300	COMMON AMERICAN TELEPHONE DATA SPEED
600	EUROPEAN HIGH SPEED TELEPHONE RATE
1200	SPECIAL PURPOSE EUROPEAN TELEPHONE RATE
1800	SPECIAL PURPOSE AMERICAN TELEPHONE RATE
2400,4800	DEDICATED LINE, SPECIAL PURPOSE RATE
9600	DIRECT CONNECTED, DEDICATED LINE RATE

MODERN COMPUTER HARDWARE IS OFTEN
SWITCH SELECTABLE TO ANY TRANSMISSION
SPEED INDICATED. THE MOST COMMONLY
USED RATES FOR PCs ARE 300, 1200,
AND 9600 BAUD.

frequency and electromagnetic interference, such as welders and arc-producing equipment, is a necessity to ensure uninterrupted communications service. Since RS-232C is a voltage-level-based communications interface, the generally recommended maximum distance the communications cable should extend is 50 cable feet. Cables that extend more than 50 feet should be shortened if at all possible, or a commercial RS-232C line driver and receiver should be employed to cover the extended distance. In all applications the RS-232C cable should be of shielded twisted-pair design. If parity has been correctly set, then a check of the number of data and stop bits is in order.

Overrun errors are caused when the rates of transmission are not matched, or where one of the two connected devices cannot process the incoming data fast enough. If the error is not caused by incorrect baud rates, the only possible solution is to select a lower baud rate for the transmission, which allows the slower device to keep up with the transmission of data.

Handshaking errors are generated whenever one of the devices can not communicate with its partner because of improper handshake signals. For example, the unit desiring to send a communication will assert (set to a one condition) its *request to send* line. If the receiving device is able to accept the message, it signals the transmitter via the *clear to send* line. If a request is made and not acknowledged within a certain amount of time, a handshake error may be generated.

There may be other specialty messages generated, and the system guides and manuals should be consulted for definitions of these miscellaneous messages, their causes, and their correction. The RS-232C interface standard is widely accepted, and once a communications link is established, it will usually be quite reliable.

CURRENT LOOP

The current loop interface is identical to the RS-232C interface except that there are no handshaking signals, and this interface is based on a current level rather than a voltage level. There are several current loop standards ranging from 20 to 60 milliamperes. The current loop standard is a four-wire standard, as shown in Fig. 15.23. The current loop uses the same transmission concepts as the four-wire RS-232C standard. There are still the start bit, five to eight data bits, a parity bit, and one, one and a half, or two stop bits. Since the current loop interface is based on a current signal, the distance the transmission cable can extend is much greater. Many PC systems offer both the current loop and RS-232C standards in the same 25-pin D-type interface connector.

IEEE 488

In order to improve on the RS-232C standard and at the same time upgrade it to be more compatible with current computer hardware, the Institute of Electrical and Electronics Engineers (IEEE) developed a general-purpose interface standard, called IEEE 488. This standard defines as many variables in an interface as possible without defining the actual uses for the interface. One of the primary goals in developing this standard was to make it a "remove-the-equipment-from-the-box-and-plug-it-in" type of standard. IEEE 488 precisely defines the connector pinouts, the actual pin functions, and the signal levels (both current and voltage) that must be used. Fig. 15.24 illustrates the major specification of the IEEE 488 interface, also known as GPIB (General Purpose Interface Bus).

There are three types of devices that can be attached to the IEEE 488 bus. A *controller* device is responsible for giving another bus device the permission to

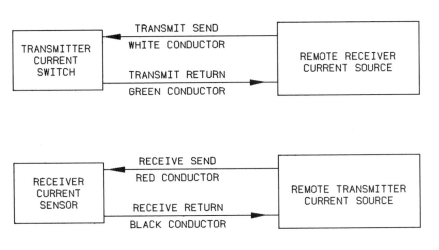

FOUR WIRE CONNECTION

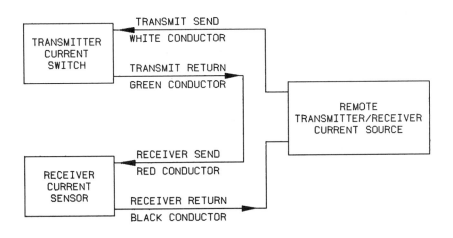

TWO WIRE CONNECTION

Fig. 15.23. Current Loop Configurations

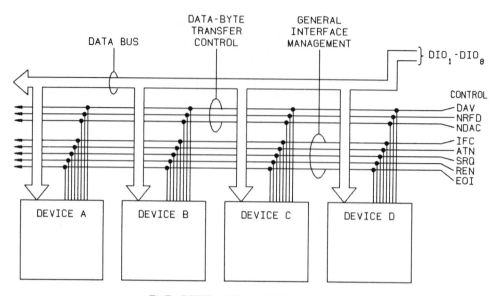

THE IEEE-488 LINES

DESIGNATION	DESCRIPTION
DIO -DIO	DATA INPUT/OUTPUT: EIGHT DATA TRANSFER LINES
ATN	ATTENTION: GAINS THE ATTENTION OF BUS DEVICES BEFORE BEGINNING A HANDSHAKE SEQUENCE
DAV	DATA VALID: TALKER NOTIFICATION THAT DATA HAS BEEN PLACED ON THE DIO LINES
EOI	END OR IDENTIFY: TALKER NOTIFICATION THAT THE DATA BYTE CURRENTLY ON THE LINES IS THE LAST ONE
IFC	INTERFACE CLEAR: ISSUED ONLY BY THE CONTROLLER TO BRING ALL ACTIVE BUS DEVICES TO A KNOWN STATE
NDAC	NOT DATA ACCEPTED: ISSUED WHEN FETCHING DATA FROM THE DIO LINES
NRFD	NOT READY FOR DATA: LISTENER IS READY TO ACCEPT DATA
REN	REMOTE ENABLE: GROUNDED TO MAINTAIN CONTROL OVER THE SYSTEM
SRQ	SERVICE REQUEST: ISSUED BY DEVICE NEEDIN SERVICE

Fig. 15.24. IEEE 488 Interface Bus

place data on the interface bus. A system can have more than one controller. However, only one can be active at a time. When more than one controller is present on the bus, a master controller has total control of all the remaining controllers.

Devices may be designated as *listeners*. These devices are permitted only to listen in on bus transmissions, not to place data on the bus. A printer might be an example of a listener device. It listens to all communications on the bus, but is only active in printing out bus data when so instructed by the bus controller.

The final type of device on the bus is a *talker*. These devices place data on the IEEE 488 bus whenever instructed to do so by the controller. It should be noted that a bus device can be a talker, a listener, and a controller as well as a combination of any two device types. For example, a computer can function as a controller as well as operate as a listener and/or a talker.

The IEEE 488 bus is actually composed of 16 signal lines, divided into three functional groups. Eight data lines form a data bus. The eight data lines permit the transmission for 7- or 8-bit ASCII data. This bus group is bidirectional in design, permitting data flow to and from a device connected to the bus. A second group of three buses is termed the data transfer control group. These three buses handle the data transmission handshaking necessary for the IEEE 488 interface to operate. The five remaining signal lines form a general interface group of buses designed to carry control signals to and operational statuses from the individual devices attached to the interface.

The best feature of the IEEE 488 interface is that it is designed to work as soon as it is plugged into the devices designed to meet the requirements of the standard. The only item not defined by the standard is the format of the data being transmitted. Data can be transmitted in ASCII provided that all devices which must use the data are designed to understand ASCII. In fact multiple

Table 15.4. EIA RS-449 Interface Signals

CIRCUIT NAME	MNEMONIC	CIRCUIT TYPE	CONTACT ASSIGNMENT	DIRECTION OF SIGNAL
SHIELD	AA	COMMON	1 (1)	—
SIGNAL GROUND	SG	COMMON	19 (5)	—
TRANSMIT COMMON	SC	COMMON	37 (9)	TO DCE
RECEIVE COMMON	RC	COMMON	20 (6)	FROM DCE
TERMINAL IN SERVICE	IS	CONTROL	28	TO DCE
INCOMING CALL	IC	CONTROL	15	FROM DCE
TERMINAL READY	TR	CONTROL	12	TO DCE
SECONDARY TERMINAL READY	STR	CONTROL	30	TO DCE
DATA MODE	DM	CONTROL	11	FROM DCE
SECONDARY DATA MODE	SDM	CONTROL	29	FROM DCE
LOCAL LOOPBACK	LL	CONTROL	10	TO DCE
REMOTE LOOPBACK	RL	CONTROL	14	TO DCE
TEST MODE	TM	CONTROL	18	FROM DCE
SELECT STANDBY	SS	CONTROL	32	TO DCE
STANDBY INDICATOR	SB	CONTROL	36	FROM DCE
PRIMARY TRANSMIT DATA	SD	DATA	4	TO DCE
PRIMARY RECEIVE DATA	RD	DATA	6	FROM DCE
SECONDARY TRANSMIT DATA	SSD	DATA	22 (3)	TO DCE
SECONDARY RECEIVE DATA	SRD	DATA	24 (4)	FROM DCE
REQUEST TO SEND	RS	CONTROL	7	TO DCE
CLEAR TO SEND	CS	CONTROL	9	FROM DCE
RECEIVER READY	RR	CONTROL	13	FROM DCE
SIGNAL QUALITY	SQ	CONTROL	33	FROM DCE
NEW SIGNAL	NS	CONTROL	34	TO DCE
SELECT FREQUENCY	SF	CONTROL	16	TO DCE
SIGNAL RATE SELECTOR	SR	CONTROL	16	TO DCE
SIGNAL RATE INDICATOR	SI	CONTROL	2	FROM DCE
SECONDARY REQUEST TO SEND	SRS	CONTROL	25 (7)	TO DCE
SECONDARY CLEAR TO SEND	SCS	CONTROL	27 (8)	FROM DCE
SECONDARY RECEIVER READY	SRR	CONTROL	31 (2)	FROM DCE
TERMINAL TIMING	TT	TIMING	17	TO DCE
TRANSMIT TIMING	ST	TIMING	5	FROM DCE
RECEIVER TIMING	RT	TIMING	8	FROM DCE
SECONDARY TERMINAL TIMING	STT	TIMING	35	TO DCE
SECONDARY TRANSMIT TIMING	SST	TIMING	23	FROM DCE
SECONDARY RECEIVER TIMING	SRT	TIMING	24	FROM DCE
NOT ASSIGNED	—		3	—
NOT ASSIGNED	—		21	—

CONTACT ASSIGNMENT IS FOR 37 PIN CONNECTOR. NUMBER
IN PARAENTHESIS IS CONTACT ASSIGNMENT FOR 9 PIN
CONNECTOR.

character formats can be used if the bus controller keeps the proper bus devices attached which understand the current character format being used.

RS-449

An enhanced version of the RS-232C communications protocol is the RS-449 protocol. It is intended to replace gradually the RS-232C standard as the interface between data terminal equipment (DTE) and data communications equipment (DCE). The RS-449 standard was developed by the Electronic Industries Association as a means to increase distances between DTE and DCE hardware, to permit higher communication rates between DTE and DCE equipment, and to accommodate the latest advances in integrated circuit design and hardware. EIA standard RS-449 actually specifies the physical interconnection between DTE and DCE hardware, while standards RS-422 and RS-423 specify the electric signal characteristics of the interface.

Table 15.4 illustrates the interface signals of RS-449. Several new circuit mnemonics have been created for the new standard, while several of the RS-232C circuits are no longer used. Ten additional circuits have been added as well as the use of an optional 9-pin connector. The 25-pin connector used with the RS-232C standard has been replaced with a 37-pin connector. In order to provide an orderly transition from the RS-232C standard to the RS-422/RS-423/RS-449 standards the new interface standard is compatible with the older RS-232C standard, and costly adapters and retrofitting of hardware should not be required. Table 15.5 provides a comparison of the RS-232C and the RS-449 interface signals.

Table 15.5. EIA RS-232C/RS-449 Interface Signal Equivalencies

RS-232-C			RS-449			
25-PIN CONNECTOR ASSIGNMENT	CIRCUIT MNEMONIC	CIRCUIT NAME	CIRCUIT NAME	CIRCUIT MNEMONIC	37-PIN CONNECTOR ASSIGNMENT	9-PIN CONNECTOR ASSIGNMENT
1	AA	PROTECTIVE GROUND	PROTECTIVE GROUND	AA	1	1
7	AB	SIGNAL GROUND	SIGNAL GROUND	SG	19	5
			SEND COMMON	SC	37	9
			RECEIVE COMMON	RC	20	6
22	CE	RING INDICATOR	TERMINAL IN SERVICE	IS	28	
20	CD	DATA TERMINAL READY	INCOMING CALL	IC	15	
6	CC	DATA SET READY	TERMINAL READY	TR	12	
			DATA MODE	DM	11	
2	BA	TRANSMITTED DATA	SEND DATA	SD	4	
3	BB	RECEIVED DATA	RECEIVE DATA	RD	6	
24	DA	TRANSMITTER SIGNAL ELEMENT TIMING (DTE SOURCE)	TERMINAL TIMING	TT	17	
15	DB	TRANSMITTER SIGNAL ELEMENT TIMING (DCE SOURCE)	SEND TIMING	ST	5	
17	DD	RECEIVER SIGNAL ELEMENT TIMING	RECEIVE TIMING	RT	8	
4	CA	REQUEST TO SEND	REQUEST TO SEND	RS	7	
5	CB	CLEAR TO SEND	CLEAR TO SEND	CS	9	
8	CF	RECEIVED LINE SIGNAL DETECTOR	RECEIVER READY	RR	13	
21	CG	SIGNAL QUALITY DETECTOR	SIGNAL QUALITY	SQ	33	
			NEW SIGNAL	NS	34	
			SELECT FREQUENCY	SF	16	
			SIGNALING RATE SELECTOR	SR	16	
23	CH	DATA SIGNAL RATE SELECTOR (DTE SOURCE)	SIGNALING RATE INDICATOR	SI	2	
23	CI	DATA SIGNAL RATE SELECTOR (DCE SOURCE)				
19	SBA	SECONDARY TRANSMITTED DATA	SECONDARY SEND DATA	SSD	22	3
16	SBB	SECONDARY RECEIVED DATA	SECONDARY RECEIVED DATA	SRD	24	4
14	SCA	SECONDARY REQUEST TO SEND	SECONDARY REQUEST TO SEND	SRS	25	7
13	SCB	SECONDARY REQUEST TO SEND	SECONDARY REQUEST TO SEND	SCS	27	8
12	SCF	SECONDARY RECEIVED LINE SIGNAL DETECTOR	SECONDARY RECEIVER READY	SRR	31	2
			LOCAL LOOPBACK	LL	10	
			REMOTE LOOPBACK	RL	14	
			TEST MODE	TM	18	
			SELECT STANDBY	SS	32	
			STANDBY INDICATOR	SB	36	

16

Programmable Controller Peripherals and Support Devices

It seems to go without saying that any microprocessor- or computer-based system always requires a host of peripheral and support hardware to enhance the operation of the finished system. For example, what started out as a single 4000-word home computer for me, has evolved into a system that is now being used to develop this text. My home computer grew from 4 K to its current 32 K memory size, moved up from color basic to extended color basic, evolved through two graphics printers (the first would have been destroyed before the final copies were printed after many editing sessions), and uses a pair of 5.25-inch floppy disk drives for information storage instead of a standard digital cassette recorder. The PC is no exception to this evolutionary phenomenon.

The first peripheral device associated with a PC was probably the programming hardware. The first PC systems required that the manufacturer be supplied with the necessary information to be programmed in the PC's memory. This was due to the special computer hardware and software programs necessary to program the early PC processors. Today every PC can be user programmed, and the only service generally supplied by the manufacturer is that of applications consultation, training, and system repair services. Since the various forms of programming devices and their use were covered in detail earlier, additional discussions are not required at this time

PC TAPE RECORDERS

Once the user was given the ability to develop his or her own PC programs, he or she also required the ability to save these programs for later use. In addition a spare processor or PC system was almost useless without a quick method to copy the user program into memory prior to use. The only easy solution was the development of a storage system that would be compatible with a PC. The development of a digital cassette recording system compatible for PC use provided PC users with a means to save and reload processor programs easily between equivalent PC models.

The selection of a cassette recording system over several other possible technologies was based on proven computer industry experience and the easy access most PC users would have to digital-grade cassettes. As a result, each PC manufacturer offers or recommends a cassette recorder that is compatible with his PC. The cassette recorder provides a means to record and later reprogram the contents of a processor memory for whatever reason the end user has in mind.

While many manufacturers custom design a cassette tape system for use with their controllers, others suggest a commercially available unit instead. One commercial manufacturer of tape loaders suitable for PC use is Electronic Processors Incorporated (EPI). Located in Englewood, Colo., EPI manufactures at least five models of cassette recorders for general use, along with many custom units designed expressly for a particular PC manufacturer. At latest count they produce tape cassette recorders compatible with a majority of PC manufacturers' PC systems. The EPI STR-LINK II digital cassette program loader is illustrated in Fig. 16.1.

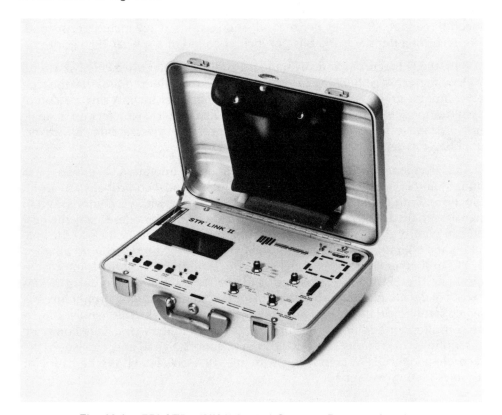

Fig. 16.1. EPI STR LINK II Digital Cassette Program Loader

Since the EPI PC tape recorder greatly resembles a standard portable audio cassette recorder with the exception of selected controls and a brushed aluminum attaché case, a detailed discussion is in order to examine why an audio cassette recorder is definitely not acceptable for PC use unless specifically indicated otherwise by the PC manufacturer. Due to the large number of PCs that are compatible with the EPI digital tape cassette recorders, these units will be examined in detail. The discussion is not intended to indicate that EPI cassette recorders are the only acceptable recorders for PC use. It is the user's responsibility to investigate the proper tape recorder to use with his or her PC system, and to ensure its proper operation and use.

EPI markets its PC cassette tape recorders under the STR-LINK name. These microprocessor-based recorders are designed to record and transmit serial data over RS-232C or 20-milliampere current loop interfaces at average data rates of up to 120 8-bit characters per second (1200 bauds) between a PC, microcomputer, or intelligent terminal and a digital tape cassette.

Data transmission baud rates can be selected between 110 and 9600 bauds, but the 1200-baud average transmission rate cannot be exceeded due to tape transport rate limitations. The data transmission can be switch selected between asynchronous half or full duplex. The error rate for the system will be less than one soft error in 1×10^7 bits and less than one hard error in 1×10^8 bits. While the standard audio cassette recorder uses a tape speed of 1.875 inches per second for both recording and playback, the EPI unit records at 4.4 inches per second, reads at 3.9 inches per second, and rewinds at 100 inches per second. The recording density is 280 bits per inch encoded in a special EPI algorithm.

Both the EPI recorder and any standard audio recorder use a Phillips cassette with write (record) inhibit tabs. However, the EPI recorder requires that the tape be computer-grade certified for 1600 flux changes per inch. A single 300-foot cassette operating at 1200 baud will store 92,000 8-bit characters on a single side. A tape can be flipped for recording on the reverse side, effectively doubling the data capacity of the tape.

The EPI tape recorder has three possible modes of operation. A series of push buttons and LED indicators can be used for total manual control of all recorder functions. Both the push buttons and the LEDs, along with PC communication signals, can be used to provide a semiautomatic control mode. Finally the host PC can control the system entirely through the communication cable for total automatic operation. To provide flexibility in adapting the recorder to a variety of PCs and other devices, 11 switches are located on circuit boards internal to the unit. The EPI manual provided with the tape cassette recorder provides excellent details as to the proper setup of the system, including proper internal switch settings for a variety of applications and PC models. This manual should be consulted for additional information. Included in the manual are complete descriptions of various RS-232C communication signal lines between the recorder and those PCs compatible with the STR-LINK, as well as schematics for troubleshooting and repair.

Operation of the EPI cassette recorder is straightforward and similar to that of PC cassette recorders offered by most PC manufacturers. The cassette recorder should be set near the processor or programming device that contains the port designated for tape functions. Often a single PC port is provided for several functions such as printers and tape recorders. In most cases the standard 25-pin D connector is used as the mechanical connection of the port.

An interface cable will be required to interconnect the recorder with the processor or programming terminal. This cable is usually available from the PC manufacturer or can be assembled by the user. The user should be warned that if he or she decides to assemble his or her own interface cables, he or she is responsible for ensuring that the proper connections are made since any damage resulting from improper cabling may void the warranty on the PC and/or the tape recorder.

Once the proper cables are in place, the tape recorder can be powered and the PC user manual consulted for operation of the PC and/or the programming terminal when performing various tape recorder operations. The user should be sure to set any control switches regarding the baud rate, the mode of operation, the parity, and the number of stop bits to the proper settings for the PC being accessed. Furthermore any switches controlling the operation of the PC port used for the tape recorder connection should be checked to verify that they are also set to the proper positions.

Making a tape of the processor's memory requires the insertion of a blank tape into the cassette recorder. The tape must be inserted in the recorder with the spool containing the most tape located on the left. The recorder provides the user with an indication that the tape is not write protected through an LED located on the control panel. Similar to an audio cassette, the digital cassette is designed with two write protect tabs at the rear edge of the tape. Whenever the tab is in place (covering the opening in the edge of the cassette), the tape cannot be written on. Care should be exercised when handling a cassette to ensure that it is protected from dust and that the tape medium itself is never contacted by a foreign object or one's fingers. To avoid accidental erasure of an existing program, no more than one program should be recorded on a single cassette side, even though most PC memory dumps use less than half of the capacity of a cassette side.

The actual steps used to record a processor's memory on tape will be a function of the particular PC being used, and how it interfaces with the tape recorder. For most PC systems the tape recorder works in a semiautomatic mode in conjunction with the PC, requiring the user to operate both the tape loader and the PC programming terminal. Normally the tape recorder is set to the record or write mode, and then the proper commands are issued to the processor to begin the recording process. When the entire content of the processor's memory has been copied to tape, the processor indicates that the taping process is complete so that the user can halt the tape recorder and rewind the tape.

Since it is possible to have an error during the transmission and writing process, it is usually a good idea to verify the just recorded tape for accuracy. Again, the user manual for the PC being used should be consulted for the proper steps to take when verifying a tape. The verify function will usually require placing the PC in a tape verify mode, followed by starting the recorder in a read mode. When the verification is complete, the user will be signaled as to a good or bad verification.

Often the processor will indicate where a miscompare exists between the processor's memory and the data on the tape. It may be necessary to repeat the write and verify operations in order to get a correct record of the processor's memory. If a verification cannot be made after several attempts, a new tape should be tried; the tape recorder heads may require cleaning and demagnetization. If several verification attempts fail, then the troubleshooting sections of the tape recorder and PC user manuals should be consulted.

It should be pointed out that the tape recorder generally records the current state of everything in a PC's memory at the time of the recording operation. Any recording made of a process or operation may contain the physical and electrical configuration of the system at that instant of recording. Depending on the PC, this means that any calculations in progress might be recorded in their current state. Any timer or counter may have its *current* elapsed time or count recorded. Any data being manipulated would be preserved on tape in its present state of completion. In some cases the current status of all coils and latches may be recorded.

While not all PCs record each of the items just listed, extreme care should be exercised to know exactly what processor memory statuses are preserved on tape and what statuses are ignored by the recording process. The main reason for the concern stems from the fact that during the read operation the entire contents of the tape is returned to the processor's memory. If an operation or process was recorded midway through its total operation, and the PC and the tape recorder are set up by the manufacturer to write every detail of memory to tape, then whenever the tape is loaded back into a processor's memory, the current state of the processor will be what is stored on the tape. If the tape was recorded with a system partway through its operation, then the program loaded from the tape will place the processor in a state where it thinks it is partway through the programmed operation. When the processor is restarted after the loading operation, the PC may attempt to operate the process or equipment from the point of the recording as well as from the actual status of the system. Timers would continue from their recorded states, latches may be incorrectly set, and data files or tables could be incorrect. The result could be damage to equipment and possibly personnel. The safest method to make a tape is to bring the operation controlled by the PC to an orderly halt, stop the processor, then make the recording. It cannot be overemphasized that it is better to know exactly what is and is not recorded than to find out later when you must dive out of the path of a flying piece of machinery.

The loading of a processor's memory from tape is usually accomplished in a manner similar to the verification of a tape. One major difference is that the processor must be halted (stopped from scanning) for the loading of a tape. This is not necessarily true for the recording operation. Halting the processor is usually accomplished via the programming terminal. Once the processor has been halted, the PC can be placed in the load tape mode, and the tape recorder set to the read mode of operation. Once the memory has been loaded, the tape can be rewound and the processor memory verified against the tape, similar to what was done during the recording operation.

BUILT-IN CASSETTE RECORDERS

While a number of PC manufacturers use a stand-alone tape cassette recorder for processor memory dumps and loads, several manufacturers have incorporated the tape loader into the programming device. Even though the marriage of the tape recorder with the programming terminal usually increases the cost of the programming terminal slightly, the benefits and conveniences that result are well worth the extra cost. Those benefits include the convenience of having the tape recorder always handy, as well as greatly simplifying the operation of making, verifying, and reading a cassette.

Several manufacturers have gone beyond simply using the tape cassette recorder portion of a programming terminal for processor memory documentation. In several instances the tape recorder is used to supply "intelligence" to the programming terminal. With PCs becoming more complex and able to handle greater and greater functions, a programming terminal designed to accommodate every possible function on a large PC would soon become too awkward to operate and handle. Instead the programming terminal is designed to be configurable to a particular operation or function. The programming necessary to accomplish each specific programming function is placed on a tape. Tapes are available for such operations as making a processor memory tape or loading a tape into the processor memory, configuring and designating the I/O allocations of an installation, configuring the data table and user memory to meet the requirements of the installation, and of course tapes are available for programming standard PC instructions as well as special functions, such as PID and ASCII messages.

The programming terminal powers up "dumb," with the exception that it has the smarts to request that a tape be loaded into the unit. Once the tape is placed in the tape drive of the programming terminal, it is automatically loaded into the memory of the terminal. Upon completion of the loading operation, the user can program the processor or carry out any other operations that the recently loaded tape permits. Should the programmer wish to perform a function not provided as part of a particular tape's capabilities, he or she resets the terminal back to the dumb mode and loads the proper tape containing those functions or instructions he or she now wishes to use. Through the use of the programming terminal tape loader the user not only benefits from being able to carry out processor memory dumps and loadings, he or she is also provided with a programming device that can be greatly simplified and made more user-friendly while still being designed to perform a multitude of PC-related functions.

HIGH-DENSITY COMPUTER-GRADE TAPE CARTRIDGE RECORDERS

Thus far the only form of tape cassette discussed was a cassette that in appearance resembles the standard audio cassette. Some manufacturers use tape recorders which use a tape system developed for the high-speed recording of a computer's memory. This form of tape, referred to as a *digital tape cartridge,* is packaged in a plastic and metal housing about 75 percent the size of the Phillips-type digital or audio cassette. These tape cartridges, and the tape drives with which they are used, have been specifically designed for the rapid storage of large amounts of data. The use of the tape cartridge is increasing since most normal digital cassettes are not capable of recording the contents of PC systems which contain user memories approaching or exceeding several tens of thousands of words. In addition the recording time for a 300-foot digital cassette can reach 15 to 20 minutes per side. Pictured in Fig. 16.2 is the EPI STR-LINK III digital minicartridge program loader. This unit includes many enhancements over the STR-LINK II cassette unit and is designed to be a direct replacement.

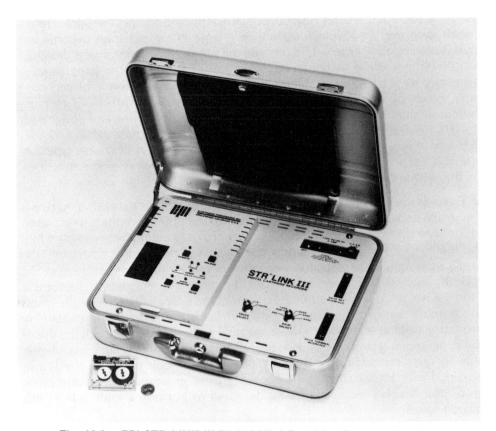

Fig. 16.2. EPI STR LINK III Digital Mini Cartridge Program Loader

FLOPPY DISK PC RECORDING SCHEMES

In lieu of either the digital tape cassette or the cartridge, several systems incorporate a computer-grade floppy disk as part of the programming terminal. The function and operation of a system using the floppy disk is the same as that of the cassette or cartridge discussed earlier. An advantage of the floppy disk is its ability to access any stored data within several milliseconds. Any cassette or cartridge must be serially read to locate the desired data before it can be accessed, whereas the floppy disk permits random access to programs and information that has been stored. Probably the biggest disadvantage to the floppy disk is the care that must be given the floppy disk unit, especially concerning vibration and impacts. The alignment of a floppy disk's read/write heads is critical to the success of the system. Rough handling can upset this balance and require servicing of the floppy disk drive. Also floppy disks themselves must be kept away from any dust and/or contamination, often making them unacceptable for industrial plant floor use.

PRINTERS

Most PC users find the convenience of a printer second to that of the tape loader. Many programming terminals incorporate an RS-232C port at the rear of the terminal for hard-copy printouts of the processor's memory in ladder diagram format. The Texas Instruments Silent 700 and Omni 800 series printers are probably most often recommended for use with a PC in obtaining a ladder listing. The Silent 700 series printers use 8.5-inch wide thermal paper for printing, while the Omni 800 series printers offer adjustable paper widths to 15 inches and use a ribbon for printing. Both series of printers are available with built-in RS-232C ports and full typewriter keyboards. In lieu of the Texas Instruments printers, practically any printer will work provided it meets the interface requirements of the programming terminal.

When selecting a printer for use with a PC other than the unit recommended by the manufacturer, there are several items to check to ensure compatibility. First the size of paper and the number of columns to be printed should be checked. Many PCs produce a ladder listing on an 80-column printer. However, if cross references are printed with each ladder rung, a 132-column printer may be required.

The *line feed/carriage return* operation of the programming terminal and the printer must be compatible. Many printers are designed to perform both a carriage return (return to the left-hand side of the paper from the right-hand side) and a line feed (index from the current line to the next sequential line to be printed) every time a return command is issued by the programming terminal. Other printers require separate line feed and carriage return commands instead of a return command to start the next line. Obviously a programming terminal issuing a return only will not work with a printer requiring both a carriage return and a line feed.

The easiest way to check for compatibility is to review the ASCII codes for both devices. A carriage return is a HEX 0D, while a line feed is a HEX 0A. If the programming terminal issues just a HEX 0D, then the printer must interpret the signal as both a HEX 0D and a HEX 0A. Should the terminal issue both the HEX 0D and the HEX 0A signals, the printer must also use both the HEX 0D and the HEX 0A signals. If the printer interprets the HEX 0D as both a HEX 0D and an HEX 0A, a blank line may appear between lines since the printer performs a carriage return and line feed when it receives the HEX 0D and then another line feed when it receives the HEX 0A signal.

When checking the ASCII codes for line feeds and carriage returns, a check on form feed operation might be helpful. While the execution of a form feed command will not inhibit a printout, correct form feed operation will give a more readable finished document. Form feed is the indexing of the paper from its current position to the top of the next page. While printers using roll-type paper do not have to rely on a form feed, printers using fan-folded paper (paper with perforations for tearing) should use a form feed to prevent printing across a perforation. Many manufacturers have designed the programming terminal to issue consecutive carriage returns (and/or linefeeds) to the printer in lieu of the form feed command. If the programming terminal issues a form feed command, it is best to use a printer with form feed capability.

One final comment before leaving the discussion on printer form feed control involves whether the printer provides an automatic form feed during the printing process. Many printers are termed "intelligent printers" since they have internal microprocessors to control the various options available on the unit. Some of the options available on an intelligent printer include automatic form feed control options where the printer controls the indexing from one page to another internally rather than using a signal from the device controlling the unit. Unless the internal printer form feed or the PC programming terminal form feed signal is deactivated, large amounts of printer paper may be used due to the two units, each issuing form feed commands at incorrect page locations.

The communications link between the programming terminal and the printer must be correct if the printer is to operate. Most programming terminals use the RS-232C protocol and the ASCII character set for printing operations. Chapter 15 discusses both the RS-232C communications protocol and the ASCII character code. A minimum of three wires must be connected between the terminal and the printer. If a cable is not available from the PC manufacturer for a particular printer, it can usually be made up by any good electronics technician. Fig. 16.3 illustrates the proper pin connections for a typical programming terminal to printer cable. Assuming the standard 25-pin D connector, pin 1 of either the printer or the programming terminal should be connected to the interconnecting cable shield. The shield should never be connected at both ends of the cable, since a ground loop may be created and cause interference. Pin 2 of the programming terminal connector (transmitted data) should be connected to pin 3 of the printer connector (received data). Finally pin 7 of the programming terminal and printer connectors (signal ground) should be connected together. It may also be necessary to jumper pins 4 (request to send) and 5 (clear to send) together, as well as jumpering

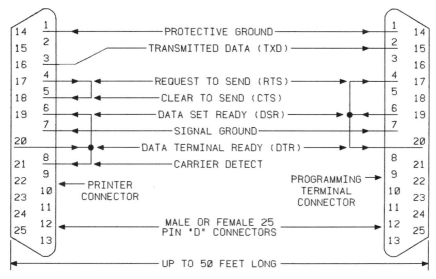

Fig. 16.3. Typical Programming Terminal to Printer Cable

pins 6 (data set ready) and 20 (data terminal ready) together at the programming terminal connector in order to transmit data to the printer. Likewise it may be necessary to jumper pins 4 (request to send) and 5 (clear to send) together, as well as jumpering pins 6 (data set ready), 8 (carrier detect), and 20 (data terminal ready) together at the printer connector for the printer to accept and print incoming data. The result of these jumpers is to "trick" the programming terminal and printer into thinking they are always connected and ready.

It should be noted that the above cable will usually work for any printer. However, if the printer is buffered, or the baud rate between the programming

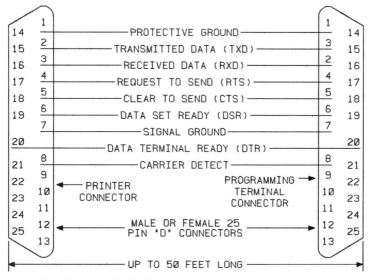

Fig. 16.4. Full RS-232C Programming Terminal Cable

terminal and the printer is faster than the actual printing speed of the printer, characters or garbage may be printed. The solution to this problem is to reduce the data transfer rate to a level acceptable to the printer. Most printers can keep up with 300-baud transmission rates, while units offering bidirectional printing will usually operate at rates as high as 1200 baud. Printers incorporating buffers (memories to store data temporarily until they can actually be printed) cannot fully benefit from the cable connections of Fig. 16.3 when the buffer is used and/or the transmission rate is faster than the printer speed. The buffer allows the sending device to dump quickly a small amount of data to the printer, then continue with other tasks, instead of having to operate at the much slower speed of the printer for the duration of the print operation. For these applications a cable wired according to Fig. 16.4 should be used.

PROGRAMMING TERMINAL LADDER LISTING

Often a PC program is developed on paper before it is actually entered into the processor. Since the entry of a program is similar to the typing of a prewritten page of text, and thereby subject to input errors, the use of a printed ladder listing is often handy in checking for mistakes. Also, during troubleshooting a quick reference to a particular rung of logic can be easier with a ladder diagram printout than scrolling through the processor logic with the programming terminal. For these needs many PC manufacturers have enhanced their programming terminals to print out the content of a processor's memory exactly as it is displayed on the programming terminal CRT screen.

The actual printing out of a ladder listing is generally straightforward. Once the printer has been connected and is found operational, the printing process can begin. Usually most PCs permit printing of a ladder listing while the processor is operating. Care should be exercised to ensure that both the programming terminal and the printer are connected to the same filtered power source as the processor. This ensures that the possibility of noise entering the PC system through the printer connections will be minimized. The programming terminal should be placed in the mode required for ladder listing, as indicated by the user's manual. Often the ladder lister routine of a programming terminal permits the user to place a header at the top of each printed page. Included in this heading should be the date, processor identification, or serial number and a brief description to identify the printout. Many ladder listers offer a user the options of printing out the complete processor memory, including the I/O image table and data memory, or just a portion of the user program. In addition many listers permit a user to select whether or not a cross reference listing is printed. A cross reference listing is handy in that it tells what inputs and outputs have been used, how many times, and where, as well as what data table locations and internal storage locations (coils) have been used, how many times, and where. Fig. 16.5 illustrates a typical PC programming terminal printout.

The printing of a processor's memory immediately after the entering of a new program can be especially helpful in the locating of program entry errors. The

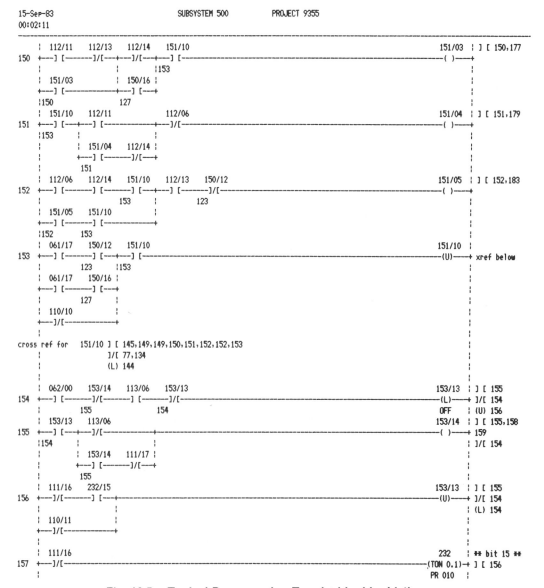

Fig. 16.5.   Typical Programming Terminal Ladder Listing

printout can be checked against the handwritten programming sheets for errors. These errors can be corrected in the editing mode prior to the actual operation of the system, saving many initial startup headaches.

PROGRAM MONITORING ACCESSORIES

Many industrial control applications require that an operator manually change the preset value of a timer, counter, or similar function while a process or machine is functioning. In effect, a means must be available to fine-tune a control system to meet varying operating conditions. Relay-based control systems often incorporate timers and/or counters as part of the operator stations, which can be changed as needs dictate. Due to the solid-state computer-based design of a PC, limited means often exist to change or monitor a program variable while the system is in operation.

The first and most obvious system monitoring device is the programming terminal. This solution poses several possible problems, should it be implemented. First, many programming terminals allow access to *all* parts of a program. Usually it is only a handfull of values in the data memory that must be changed during system operation. In fact, the changing of actual control logic while a system operates is never recommended. Second, designing a programming terminal into an operator station creates problems due to the size and configuration of most terminals. Finally the programming terminal is designed for programming a PC, and not for operating a control system. While the terminal may last for a short time under production operation, sooner or later repairs may become necessary. These disadvantages are in addition to the increased costs of using a dedicated programming terminal as an operating station device, and possible cable distance limitations between the terminal and the processor.

An alternate method for operator modification of operating data in a PC's data memory is the use of special I/O devices. For example, a set of selector switches, thumbwheel switches, and digital displays can be wired to the PC for the purpose of inputting and monitoring data changes. The processor is programmed to examine the position of the selector switch in order to determine what variable is to be changed in the data memory. The thumbwheel switches are used to set the desired value, while the digital display indicates the current value in the location being addressed. A keyswitch or push button is wired to the processor to operate as a load signal to the system. Again, disadvantages include the costs involved in purchasing the I/O modules and associated field hardware, as well as the additional memory and programming time required to implement the scheme.

Realizing this need, many PC manufacturers have developed a peripheral device commonly referred to as a *monitor/access panel.* The monitor/access panel permits the operator or PC user to monitor various control system parameters as well as to modify selected variables and system operating conditions without having access to the actual control system logic. In general these peripheral devices include a membrane keypad, a keyswitch for device function selection, and a digital display. Monitor/access panels are designed for either portable use or permanent panel mounting.

Square D manufactures a monitor/access panel for use with its SY/MAX family of PCs. This panel is illustrated in Fig. 16.6. Typical of many loader/ monitor panels offered by various PC manufacturers, the SY/MAX loader/ monitor includes a 26-key membrane-type keypad, a 16-digit alphanumeric fluorescent digital display, and a three-position mode selection keyswitch. In the monitor mode a user has access to the status of any I/O point, internal storage coil location, or any data memory storage location. Operation of the

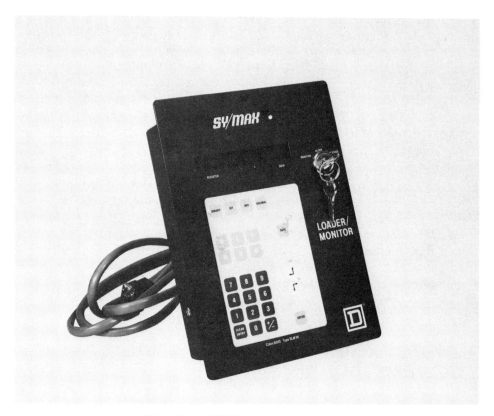

Fig. 16.6. SY/MAX Loader/Monitor

loader-monitor in the monitor mode simply requires the user to enter the desired address to be monitored. The loader/monitor automatically displays the contents of the desired address. Accessed data can only be modified when the loader/monitor is in the alter mode. Once the desired address is entered into the loader/monitor, the DATA key is depressed. This signals the monitor that the current displayed contents of the addressed location are about to be changed. The operator is now free to modify the current contents of the addressed location to meet his or her needs. The modified data are not entered into the processor memory until the ENTER key is depressed on the loader/monitor.

The SY/MAX loader/monitor provides the ability to display the contents of a specified data memory location in several formats. Selection of the decimal display format provides a decimal numbering system representation of the specified location contents. Depressing the HEX key places the loader/monitor in a hexadecimal display format. Should the user wish to view the one and zero status of each individual bit of an addressed location, the BINARY key can be selected to place the loader/monitor in the binary display format. Selection of the bit format permits the user to display the ON/OFF status of a single I/O point or bit location in a designated data memory location.

A final operational mode of the SY/MAX loader/monitor is the message mode. While in the message mode, the loader/monitor provides the PC with a means to output alarms and report data to operating personnel regarding current control system status. The desired messages are programmed as part of the user logic using the *message instruction* available in the SY/MAX programming terminal instruction set. Typical alarm messages might include "Motor P-38 OFF" or "High temp-oven 4." Reports could include information relating to part counts, downtime, or even current product codes.

The SY/MAX loader/monitor provides the user with several hardware error messages as well. Error messages include illegal address or data indications when the data being entered are out of range (outside the −32,768 to +32,767 SY/MAX numerical range) or the addressed location is invalid. A fenced error message appears whenever a data change is attempted on a "protected" (fenced) memory location. The SY/MAX controllers permit the user to specify certain data memory locations as protected from alteration by the loader/monitor panel. Error messages are also provided to alert the user that he or she is attempting to modify an external input location or a forced output location.

REMOTE PROGRAMMING TERMINAL MODEMS

One of the advantages of using a PC for some applications is its remote I/O capability. Many installations rely on remote I/O to drastically cut the costs of wiring each field I/O device back to the main control enclosure. While the use of remote I/O can offer definite cost incentives, the PC user may be faced with a problem when operating a programming terminal with a PC system employing remote I/O. Most programming terminals must be connected to the processor by a short multiconductor manufacturer-supplied cable. This means for remote I/O locations, where the I/O devices are not installed within sight of the processor, that the user will not be able to watch directly how the control system is performing. A PC user may desire to work with a coworker and an intercom or radio in order to monitor a PC-based control system properly as changes are implemented. For most installations the need to rely on additional help can be very cumbersome, definitely costly, and may even be somewhat dangerous.

The problem can be solved by the use of a special modem, similar to a standard telephone modem, which permits the programming terminal to be located at various remote I/O installations. The function of the modem is to accept digital data from either the processor or the programming terminal, transform them into a frequency-modulated signal, and pass this signal down a transmission cable where it can be reconverted back to a digital signal by another modem. The modem employed must be specifically designed to handle the unique signals that pass between a processor and a programming terminal, and at the same time to insulate those signals from a noisy industrial environment.

In order to provide remote programming terminal capability for remote I/O locations, many PC manufacturers offer a stand-alone industrial PC modem as a special PC peripheral. An example of a multifunction industrial PC modem is Modicon's J478 stand-alone modem. While this modem is designed for use with Modicon's Modbus data communications network, it is also designed to permit remote programming terminal connection to the processor as well as voice communications between modems.

The J478 modem is designed to transmit processor/programming terminal digital data over a maximum of 15,000 cable feet of four-conductor shielded twisted-pair cable. It will asynchronously convert digital data into a frequency-modulated signal at transmission rates of 50 to 19,200 bits per second (baud). A maximum of 33 modems (32 remote I/O locations plus the processor location) can be connected to the transmission cable. As an industrial modem, the J478 offers certain features not available with most commercial modems. These features include ultrahigh noise immunity and the ability to operate at temperatures between 32 and 140°F with humidity levels reaching 95 percent noncondensing.

The operation of the J478 modem is similar to that of many commercial modems, with the exception that the Modicon unit has been specially enhanced for industrial environments. Fig. 16.7 illustrates the operation of the J478 modem. Digital data sent from the processor to the master modem in the form of ones and zeros are used to frequency-modulate (FM) a signal being transmitted between modems. The method used to modulate the signal is called *frequency-shift keying* (FSK). Frequency-shift keying involves sending a constant frequency signal, called the *carrier signal,* from one modem to another. The carrier signal is then shifted between two predetermined frequencies. One frequency is used to represent a zero digital state, while a second frequency is used to represent a digital one state. The J478 uses 50,000 hertz to represent a zero state and 80,000 hertz to represent a one state on the transmission line. The receiving modem transforms the two received frequencies back to the proper digital states and passes this digital information onto the receiving device, a programming terminal remotely located from the processor.

It should be noted that the Modicon J478 modems are designed to send and receive digital data. They allow communication between a processor and the programming terminal in either direction, but in only one direction at a time (half duplex). The mode and form of communications between two modems, as well as the FSK encoding of the carrier signal between two modems described previously, are very simplistic descriptions of a highly complex data communications routine. As an industrial modem, the J478, in conjunction with a processor and a programming terminal, uses a series of error detection schemes to ensure the proper transmission of information. These error-checking routines may include cyclic redundancy checks (CRC), longitudinal redundancy checks (LRC), or vertical redundancy checks (VRC or parity).

Use of a modem similar to the J478 allows the PC user to be in direct contact with the process or machinery being controlled, even though the PC processor

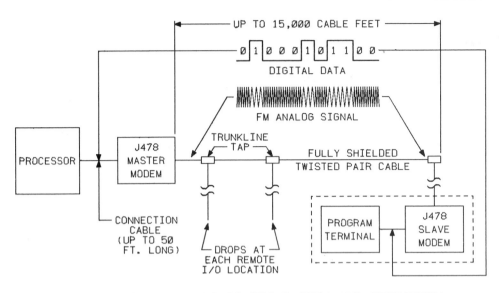

TYPICAL MODICON J478 CONFIGURATION AND OPERATION

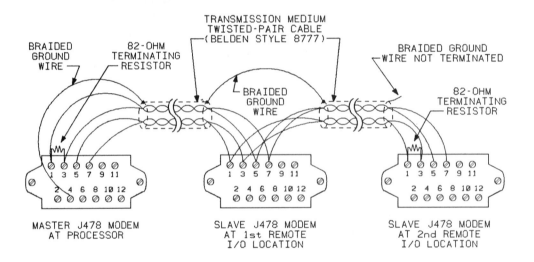

PROPER MODICON J478 MODEM WIRING

Fig. 16.7. Modicon J478 Modems Used to Provide Remote Programming Terminal Capability

may be located some distance away. Installation of the modem requires the installation of a dedicated processor/programming terminal communications line in conjunction with the remote I/O communications line. In most cases these two lines can be run in the same conduit or raceway. A modem is required at the processor as well as a second unit for the programming terminal. Many PC users prefer to install dedicated modems at each remote I/O location for use with the programming terminal in lieu of the single modem attached to the programming terminal. It should be noted that when an industrial modem, such as the J478, is used for processor to remote programming terminal functions, only one terminal can be active at a time. If the remote programming terminal communications link permits more than one terminal to be connected at a time, the additional terminals must be operated in a slave mode as monitors only.

It may be desirable in some facilities to provide modems at every processor at the site, along with modems at every remote I/O location at the site. All of the modems would be interconnected by a central communications line used to provide site-wide processor/programming terminal communications. It would then be possible to communicate to any processor at the site from any remote location having a programming terminal available. This type of installation can provide a great deal of flexibility in quickly accessing any processor operating at the site for programming and/or system monitoring. It should be noted that the installation just described requires the PCs used to be individually addressable via special access codes, and that use of a modem for remote programming terminal operation should not be confused with the use of modems for PC data communications networks.

TELEPHONE MODEMS

Another type of modem often sold by a PC manufacturer or recommended for PC use by a manufacturer is the commercial telephone modem. It differs from the industrial modem previously described in several respects. First the telephone modem uses the telephone lines for data transmission in lieu of a dedicated communications line. Second, the data transmitted between a pair of telephone modems may not receive the same rigorous error detection checks that are performed on communications between industrial modems.

Telephone modems are often used to monitor and troubleshoot remotely a PC-based control system. One application cited by a PC manufacturer involves the use of multiple PCs on a pipeline in Texas. The pipeline stretches for many miles and uses PCs at strategic points to control the flow of fluid in the pipeline. Each processor is connected to a telephone modem for monitoring purposes. The maintenance mechanics for the pipeline have mobile telephones installed in their vans along with 120-volt ac alternators and programming terminals. Whenever a problem shuts down the pipeline, the mechanics can "call" the various PCs along the pipeline and check their operational statuses. Often the problem can be diagnosed in the van on the way to the problem area, and an alternate mode of pipeline operation placed in effect until the mechanic arrives to correct the malfunction. Other telephone modem uses include similar applications where it is advantageous to monitor a control system for one reason or another from a distant location over standard telephone lines.

Like its industrial counterpart, the telephone modem also uses frequency-shift keying to encode digital data for transmission. However, it must use lower transmission frequencies and baud rates for operation due to the design of the commercial telephone system. A telephone modem will employ frequencies between 300 and 3300 hertz since that is the optimum bandwidth of the telephone system, as illustrated in Fig. 16.8. Two pairs of frequencies, one pair

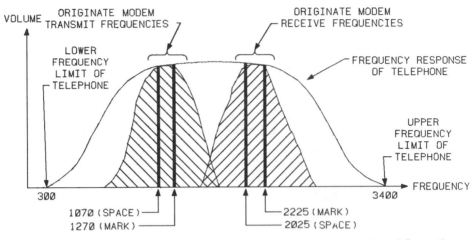

Fig. 16.8. 300 Baud Full-duplex Modem, Frequency Assignment and Operation

centered at 1170 hertz and the other centered at 2125 hertz, are used to both transmit and receive data between a pair of modems. The binary-formatted digital data from a modem's host device are converted to a frequency-modulated signal, as shown in Fig. 16.9, for transmission over the phone lines to a remote modem. Likewise the incoming frequency-modulated signal is converted back to binary-formatted digital data by the modem prior to being sent to the modem's host device. Internally a telephone modem includes both high-pass and low-pass frequency filters, a modulator, and a demodulator to encode and decode properly the data being passed through the unit. Fig. 16.10 illustrates in block form the filters, modulators, and demodulators for a pair of modems, and the signal paths and frequencies between them.

Most modems will transmit over standard telephone lines at either 300 or 1200 baud. However, special dedicated line modems may employ transmission

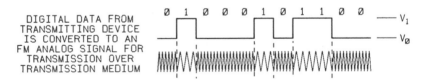

MODULATING DIGITAL DATA TO FM ANALOG SIGNALS

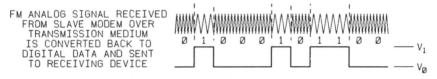

DEMODULATING FM ANALOG SIGNALS

Fig. 16.9. Frequency-Modulated Modem Signal

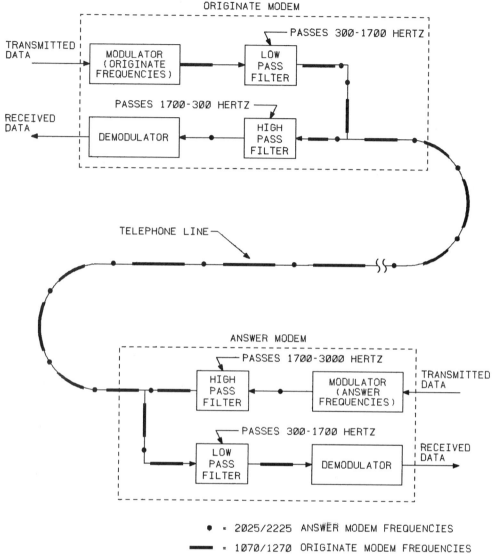

Fig. 16.10. 300 Baud Full-duplex Modem Frequency Assignment and Operation

rates as high as 9600 baud. Finally most commercial telephone modems are designed to receive and transmit data to a device via RS-232C protocol instead of a dedicated processor/programming terminal protocol.

A telephone modem can be connected to the telephone line in either of two methods. Modems employing an acoustic coupler, Fig. 16.11, require the insertion of the telephone handset in a special receptacle for connection to the telephone lines. The modem employs an audio amplifier and a speaker to transmit the modulated audio carrier frequency to the telephone handset receiver for signal transmission to a distant modem. A microphone is used by the distant modem to pick up the audio frequency signals that are received over

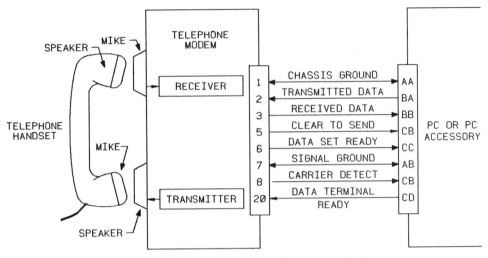

Fig. 16.11. Acoustic Coupled Telephone Modem

the telephone handset from the originating modem. In lieu of an acoustic coupler, many modems are designed to be directly connected to the telephone line by a Y-type modular telephone jack, as shown in Fig. 16.12. Once the telephone has been used to dial up the receiving modem, the handset is either placed in the acoustic coupler or simply hung up in the case of the direct connect modem.

Connection of the telephone modem to a PC or other computer-type device is usually accomplished through a 25-pin D-type connector. This connector provides interfacing for standard RS-232C signals between the modem and the host hardware connected to the modem. While only four signal lines are actually necessary to connect the modem to a host device, most modem manufacturers will recommend the full use of all RS-232C signal lines. The cable

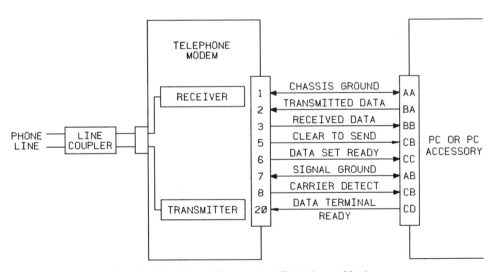

Fig. 16.12. Direct Connected Telephone Modem

wiring of Fig. 16.2 can also be used for the connection of a modem to a PC's RS-232C port. Again note that pins 2 and 3 have been switched in the cable wiring.

Telephone modems are often referred to as answer only or answer/originate. The difference lies in a modem's ability to answer automatically an incoming call and then self-connect its peripheral device to the remote calling modem via the telephone line. For example, most modems connected to computers are of the answer/originate type. A user wishing to connect to the computer dials the phone number of a modem connected to a port on the computer. When the computer's modem detects that the phone is "ringing," it automatically answers and immediately begins transmitting a return carrier signal over the telephone line. When the calling user hears the steady pitched carrier tone in the telephone handset, he or she connects the local modem to the telephone line by either inserting the handset in the acoustic coupler or switching his or her direct connect modem to "data." In short, the answer/originate modem has the capability to originate a carrier tone. Any modem that does not have the ability to originate a carrier signal is referred to as an answer-only modem. All answer-only modems must receive a carrier tone from an answer/originate modem before they will connect their peripheral device to the telephone line. Note that any answer/originate modem has the capability to operate as a simple answer-only modem, provided that it connects to a second modem configured for originate operation.

In addition to providing the proper cables and signal connections between the modem and its host device, the user must also check for proper data transmission protocol between the modem and its peripheral device. The modem will need to be configured for either half or full duplex operation. If the RS-232C protocol is used, the proper number of stop bits must be selected, as well as the use of either even, odd, or no parity. The number of data bits to be transmitted must be determined as well as the coding format to be used, such as ASCII. Many modems offer auto time-out and telephone line disconnect options for user selection as well as the ability to operate at selected baud rates. Chapter 15 discusses the various transmission protocols in common use, along with the parameters of each.

As a word of recommendation, it is usually best to get a telephone modem connection initially established with hardware that can be located in the same room. For example, if a PC processor installation is to have an answer/originate modem connected to the processor, and answer-only modems at each remote programming terminal location, initial system communications checks should be attempted with all necessary hardware in the same room. This will require having a processor, programming terminal, two telephone modems, and two different telephone services all available in the same local area. The direct-connect answer/originate modem can be connected to the processor with the proper cables as well as to one of the telephone lines. Likewise the programming terminal can be connected to its modem. Once the communications protocols have been selected and set at all devices, the communications link can be tested. The processor modem can be called from the remaining phone. When the carrier tone is heard in the handset, the programming terminal modem can be connected to the line. If all options have been selected correctly, it should be possible to use the programming terminal in the same modes of operation that are possible with a directly connected programming terminal configuration.

Should problems arise it might first be advantageous to slow the rate of communications over the phone lines (the baud rate) to the slowest rate possible. Many times the telephone equipment will operate better at slower transmission rates. If problems still exist, it will be necessary to recheck all option settings as well as to consult the "In Case of Problems" section of both the PC and the modem user manuals. Sometimes the selection of an alternate modem option can clear a problem. If the user is unable to locate the cause of a particular problem, it may be necessary to contact the PC manufacturer's applications engineer and/or the modem manufacturer. Once the telephone connections have been thoroughly checked out to user satisfaction, the hardware can be placed in actual operation with reasonable assurance that the first time the telephone link is required it will operate.

SIMULATION PANELS

For many PC applications it is desirable, and often necessary, to verify the proper operation of the PC programming prior to actual use. This necessity creates the need to emulate the operation of a PC's field I/O devices. One approach is to fabricate a *simulator panel* consisting of rows of small indicator lights and toggle switches. Each indicator or switch is connected to a PC I/O point and labeled with the function of that I/O point. The processor user logic can be verified by watching the statuses of the output device indicators as toggle switches are flipped to emulate the operation of the field input devices. In this manner the correct operation of the user logic can be verified.

The construction of a simulator panel and wiring it to the I/O modules can become costly and time consuming. Recognizing these drawbacks, several PC manufacturers have developed simulator I/O modules for use with their PC systems. These I/O modules permit the PC user to temporarily remove the I/O modules designated for use with the field I/O devices and substitute in their place the proper simulator I/O module. Several types of simulator I/O modules are generally available. One type is the input simulator I/O module. These I/O modules contain toggle switches mounted as part of the module in lieu of field wiring terminals. One toggle switch is provided for each input point available on the module. Another simulator I/O module type is the output simulator I/O module. This module provides one indicator for each output point normally available on an output module. Fig. 16.13 illustrates a typical input and output simulator often available as a PC accessory for user program verification prior to actual program operation with real-world I/O devices. No external power supplies are usually required to operate a simulator module, and sufficient space is provided to label the function of a particular indicator or switch. Once the simulation has been completed, the simulation modules can be removed and the regular I/O modules reinserted.

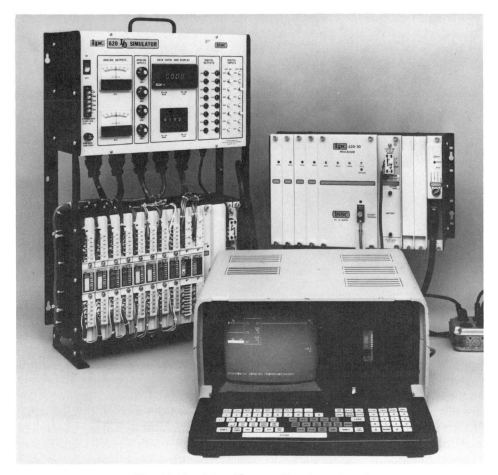

Fig. 16.13. Input/Output Simulator Panel

COMMERCIAL COMPUTER-BASED SIMULATORS

Often simulation exercises which make use of simulator I/O modules or panels cannot provide all necessary data concerning the operation of a PC-based control system. Since a series of switches must be activated, and many indicators must be monitored during the simulation process, it is difficult to get the "big picture" of exactly how a control system is functioning. Usually the person performing the simulation can only concentrate on one event at a time, thus other concurrent problems which develop may be overlooked. In addition it often becomes costly to perform "what if" exercises due to the limited flexibility in the hardware used to perform a simulation. Several PC users, realizing the drawbacks of using switches and indicators, have developed computer-based simulators to assist them in their simulation exercises. An example of a commercially available computer-based PC simulator is the HEI real-time computer simulator illustrated in Fig. 16.14.

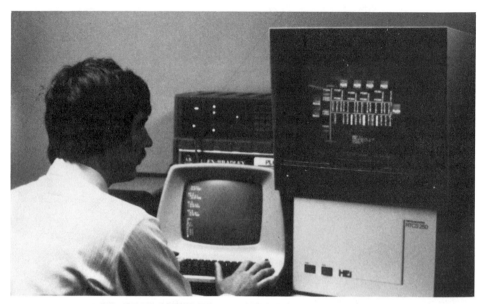

Fig. 16.14. The HEI Real Time Computer Simulator

The HEI simulator incorporates multiple 8086 microprocessors, a color CRT, and the user's PC to provide a real-time simulation of a PC-based control system. The simulation is performed with the HEI simulator connected directly to the actual PC that will be used for the application. All I/O racks and modules are disconnected from the PC since the simulator will act as the I/O for the PC. If a conveyer system is being simulated, the PC operates as if it were actually controlling the conveyer system, but what the processor thinks are signals from real-world input devices, are actually signals from the HEI simulator computer. Likewise what the PC thinks is a real-world output device, is really a signal to the simulator. A graphic display is presented on the simulator CRT, which not only shows the conveyor configuration, but the current status of every I/O point at each location along the conveyor. Simulated product "loads" can be placed on the conveyor, and their movement through the conveyor system can be observed. The proper operation of a transfer can be watched in detail as well as the correct accumulation and release of loads from various conveyor zones.

The HEI simulator permits the PC user to halt a simulation at any time for closer examination of system conditions. A scenario mode of operation permits the user to develop special simulation conditions or tests which he or she wishes to examine further. The use of a single-step mode permits the stepping through of a critical PC control algorithm. Any number of "what if" conditions can be generated to fully test the reaction of the PC control logic to abnormal conditions. A proposed automated system can be simulated under existing operating conditions, then resimulated with future modifications and capacity increases to ensure future system expandability. The ability to observe visually the operation of an automated system provides plant production and operating personnel with a way to locate and correct system bottlenecks and ineffi-ciencies. As a monitoring device, the HEI simulator provides a means to display visually the current status of a system, making overall system monitoring and troubleshooting easier to perform in critical applications.

Since the HEI simulator actually emulates the operation of the I/O devices for a PC-based conveyor control system, the simulator must be programmed with the physical configuration of the system as well as the operation and location of the I/O devices themselves. The simulation computer reads output statuses sent from the PC, and based on its programming of the configuration and operation of the system being examined, it provides signals that are read as input device statuses by the PC. The PC executes the same program that will be used to control the physical plant hardware, except that the physical hardware is the HEI computer simulator.

Programming of any system and its configuration and operation is done by a "fill in the blanks" type of tabular program entry. First HEI-supplied tables are filled out for each system element. A typical table will contain data concerning device types, lengths, locations, and speeds, as well as I/O device locations and functions. The operation of any special system equipment or hardware is detailed in specific tables. Once the system has been fully defined, the tables are edited and stored in a host computer system. The function of the host computer is to compile the tables into data blocks for use by the simulation computer. Any business-quality computer operating under Microsoft's MS-DOS operating system, or the IBM personal computer, may serve as the host computer. Once the tables have been compiled by the host computer for simulator use, they are loaded into the simulator's nonvolatile memory in preparation for simulation. After the PC has been connected to the simulator through the processor programming terminal port, the simulation can commence. The simulator accepts output commands directly from the PC, responds with a proper PC input response according to the conveyer's operation, and graphically displays every system condition and I/O status requested. As entities move through the automated system, they are visually displayed. Limit switches and photocells visually react to the flow of product, indicating the input conditions being supplied to the PC, as illustrated in Fig. 16.15. Product "flows" according to motor and solenoid operating statuses. When the simulation is complete the PC software engineer can be reasonably sure that the PC-based control system will perform as required with a minimum of system startup headaches.

COMMERCIAL COLOR GRAPHICS TERMINALS

Pictures, symbols, and drawings are used every day as the most efficient means to communicate an idea or thought between two or more people. The control industry is no exception to this observation. One often encounters large graphic displays of lights and buttons in a production facility, which confirms the often heard adage that a picture is worth a thousand words. With processes and operations becoming both more complex and more computer/microprocessor oriented, the use of pictures and displays is greatly increasing. Multiple-color monitors driven by large computer systems are replacing the conventional graphic display panel since better, quicker, more concise data can

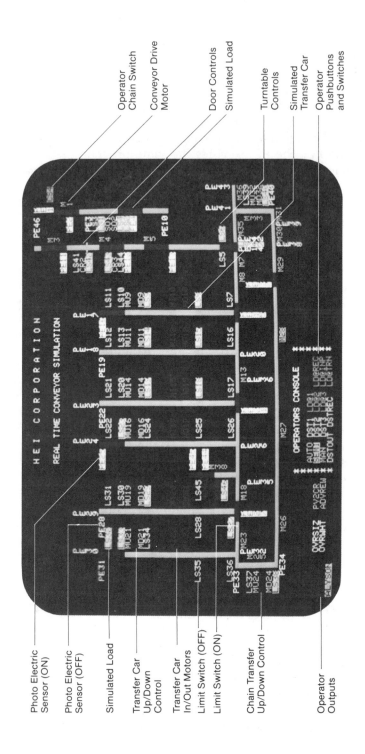

Fig. 16.15. Typical HEI Simulator Graphic Display

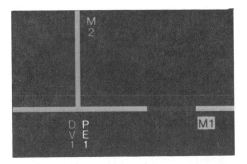

Package moves down mainline conveyor powered by motor (M1).

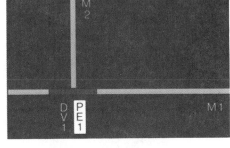

Package blocks photo electric sensor (PE1).

Mainline conveyor halted.

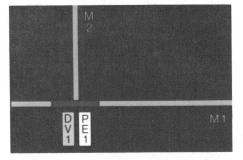

Diverter (DV1) activated to transfer package to spur conveyor.

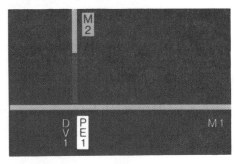

Diverter turned off.

Spur conveyor motor (M2) turned on.

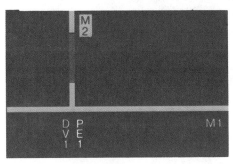

Package moves down spur conveyor.

Photo electric sensor cleared.

be displayed to an operator by such monitors. For their many benefits, these computer-driven monitors are expensive to purchase and implement, except for the very largest control systems. The use of a PC to support a color graphics terminal has changed this picture, allowing the use of a color monitor on many smaller, less costly installations.

One such manufacturer of PC-compatible color graphics terminals is Industrial Data Terminals (IDT). The IDT Classimate system, pictured in Fig. 16.16, is a hardware and software package designed to allow easy graphics

Fig. 16.16. IDT Classimate Color Graphics System

construction and communication with many models of PCs. Both graphic and text information may be stored for later recall by a PC. IDT terminals offer eight foreground and eight background colors, plus a blink function on a standard 19 inch color monitor. Full ASCII characters in either 5 × 7 or 10 × 14 dot character matrixes are immediately available to the user. A dot addressable 512 × 512 dot (pixel) matrix is provided for user graphics generation. Each terminal is packaged in an industrially rugged enclosure, allowing it to be used in factory environments.

Four basic stages are required to develop graphics and link them to the PC. First, graphic symbols must be created, from which entire pictures may be assembled. Classimate terminals are supplied to the PC user with many standard symbols, including pipe sections, valves, pumps, etc. An easy to use menu-driven method is also provided for the user to develop a library of custom symbols. Symbols are not restricted to character cell boundaries, and may be hundreds of pixels in size.

The second step involves the actual generation of customized display pages. A graphics page can be thought of as a single sheet of paper containing the final picture, which had been assembled as creation and editing routines "mapped" previously constructed symbols onto the screen. Each finished picture is then stored in bubble memory within the system. The use of bubble memory allows any graphic page to be instantly recalled to the screen for display. A 10-megabyte Winchester disk drive is provided as a memory resource for the system.

During the third step the dynamic elements of each graphics page are identified and referenced to the proper data locations within the host PC. For example, during actual PC operation a valve symbol may need to change colors and/or blink to indicate a change of state. Bar graphs may need to change in height in response to new PC data. Even messages and alarms may need to be displayed on the screen or sent to a printer in response to current PC information. Through a menu-prompted process a data file is created in the Classimate which links each dynamic picture element to the required PC data memory address. The data files, which have been created with the use of the Winchester disk drive, are transferred into bubble memory for quick access.

Finally after the development process is complete, the Classimate system can be connected to the host PC. Since the system's bubble memory provides nonvolatile information storage, the Winchester drive is not needed for on-line operation. Depending upon the requirements of the application and the design of the graphic displays, operators can monitor and modify setpoints and other process limits. Motors can be monitored and controlled by a display that is interactive with both the PC controlling the motor and the operator controlling the overall system. Status and alarm messages, shift reports, and so on, may be displayed on the screen or sent to a printer. Displays typical of those that can be created with the Classimate system are illustrated in Fig. 16.17. The design and use of a graphic display system is practically limitless, with the only real constraints being the memory storage area of the host PC and the user's imagination.

PC MANUFACTURER-SUPPLIED COLOR GRAPHICS SYSTEMS

As an alternate to the use of a generalized color graphics package, many PC manufacturers have developed a color graphics system specifically tailored for their particular PCs. Eagle Signal's graphic display package (GRPHPK) is one example of a color graphics software package designed for use with a particular PC system. The GRPHPK software package, operating with a commercially available color terminal as illustrated in Fig. 16.18, supplies all necessary

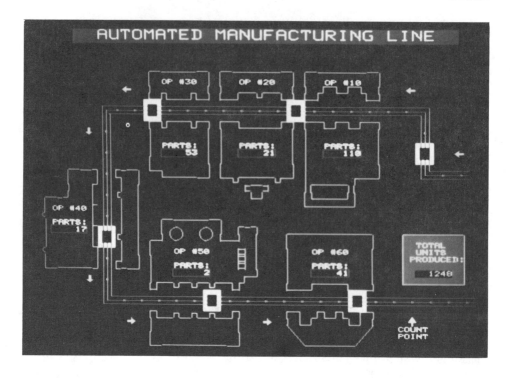

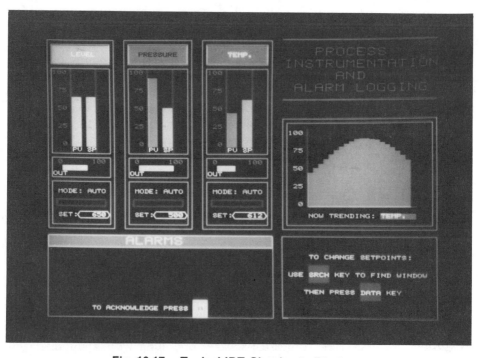

Fig. 16.17.   Typical IDT Classimate Displays

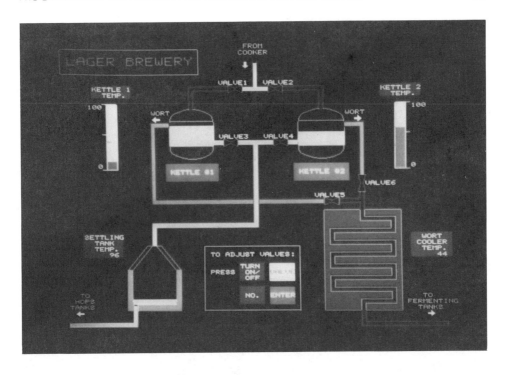

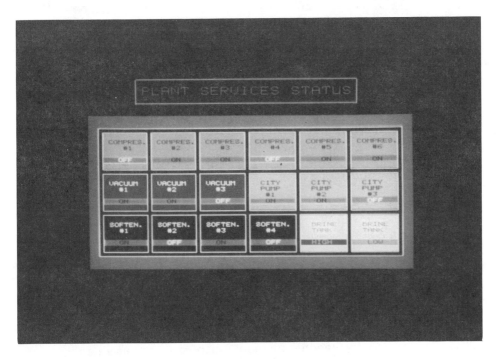

Fig. 16.17. *(Continued)*

Fig. 16.18. Eagle Signal's GRPHPK Software Package Operating a Color Terminal

intelligent color computer and EPTAK 700 PC software required to provide analog control, alarm displays, historical and real-time trending, dynamic graphics, and data logging for process-type applications which rely on the Eagle EPTAK 700 PC for operation.

Any system alarm can be displayed on the CRT and logged on a printer automatically along with the time the alarm occurs, the alarm number, and a description of the alarm. Both analog and digital alarms may be displayed and logged as desired. Analog displays are provided for open-loop inputs, open-loop outputs, and closed-loop controllers. In addition the ability to incorporate auto loop backup and bumpless transfer is provided along with setpoint ramping. These types of software/hardware packages can provide the PC user with an economical alternative to an expensive computer-based control system. They reduce the amount of user programming and interfacing required, and are fully supported and documented by the PC manufacturers offering them.

PC MANUFACTURER-SUPPLIED PROGRAM DOCUMENTATION

One problem faced by every PC user is how to document a PC program. Drawings are required for I/O wiring and system configurations in order to purchase, fabricate, and install a PC-based control system. But the actual processor program is often developed on the backs of napkins and envelopes. By the time the system is operational, changes to the initial program have made these "programmer's notes" useless. The actual documentation of the final PC program is often left to the maintenance personnel who must troubleshoot the system.

As a first step solution to this problem most PC manufacturers design the programming terminal with the capability to print out the contents of a processor's memory. This allows the user to generate a hard-copy document of exactly what is in the processor's memory, including cross-reference listings where contacts are used for each and every I/O point or coil. This document, along with a moderate amount of patience and determination, will permit a maintenance technician to possibly maintain the system. Fig. 16.5 illustrates a typical PC programming terminal ladder listing.

Realizing the inadequacies of this form of program documentation, many PC manufacturers offer a special-purpose documentation package for use with their PC systems. These systems provide the ability to generate a mnemonic descriptor for each contact, I/O point, coil, timer, and so on. Once these mnemonic descriptors have been generated, they are entered into a host computer as part of the manufacturer's PC documentation package. The host computer is connected to the PC, and the documentation software package is run. The resulting printout of the processor memory contains mnemonic descriptors in conjunction with the standard PC references for each processor instruction, as illustrated in Fig. 16.19. This form of PC program documentation usually eliminates much of the confusion generated when trying to understand the operation of an existing PC system.

COMMERCIAL PC PROGRAM DOCUMENTATION SYSTEMS

While the level of documentation obtained with a PC manufacturer's documentation package is often quite good, these packages can still have some drawbacks. First, in order to run one of these packages, a mainframe computer may be required, such as a Digital Equipment Corporation PDP-11. Many smaller companies using PCs cannot afford the investment of a large computer, or the continued expense to operate one. If they can, the one they own may not be compatible with the PC manufacturer's documentation package software. Second, many original equipment manufacturers (OEMs) offer a variety of PCs with their hardware in order that their equipment can be compatible with the standard PC equipment of their various customers' facilities. The purchase of

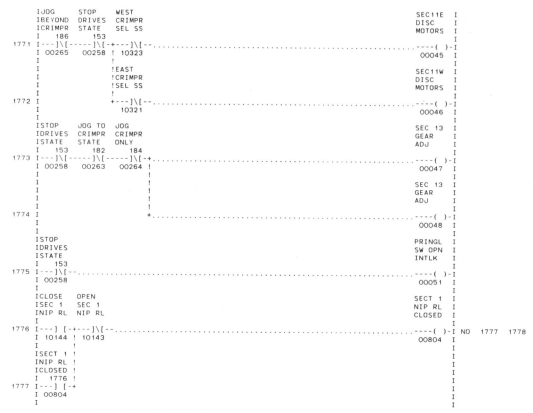

Fig. 16.19. Typical PC Manufacturer Supplied Program Documentation

two, three, or even four specialized PC documentation packages and associated support hardware can get quite expensive. The same problem exists for control system fabricators and PC system design and engineering concerns to an even greater extent since they must deal with various PC users, each employing a different make or model unit. Finally, many of the documentation systems offered by various PC manufacturers require the processor to be connected to the host computer in order for the documentation package to operate. Realizing these drawbacks, at least several companies have developed a commercially available PC documentation package which is self-contained and capable of being used with several PC systems.

One such offering is the XYCOM ladder diagram translator package (LDT) illustrated in Fig. 16.20. The XYCOM LDT is a relatively inexpensive documentation system consisting of a self-contained microcomputer, disk drive, and printer. The microcomputer is part of a keyboard/CRT terminal, and all necessary software for system operation is supplied on 8-inch floppy disks. The LDT system can be connected directly to a PC, a PC data communications network, or a PC tape cassette/cartridge loader through an RS-232C port located at the rear of the LDT terminal.

Fig. 16.20. Xycom Ladder Diagram Translator Package

Once the PC program source has been connected to the terminal, the program is loaded into the LDT system memory. The documentation system user then enters whatever mnemonic descriptions and ladder rung commentary necessary to define adequately the operation of the PC program. Once the commentary and mnemonic data are permanently stored on floppy disk for future reference and use, the actual documented ladder listing is produced.

The LDT has the capability to produce a variety of PC documentation reports, as shown in Fig. 16.21. A user may request the printing of the PC ladder diagram or any part thereof. A typical ladder diagram report is illustrated in Fig. 16.22. A summary report of every accessible coil within the PC can be produced indicating whether the coil is used as an internal storage location, as either a used or an unused output point, or just available for program use. Undefined and/or unreferenced program symbols can be displayed for user interpretation and action. A reporting of how and where every PC instruction is utilized is possible, as well as a complete listing of the contents of a processor's data memory. Specialized reports include processor configuration printouts, ASCII message listings, sequencer instruction data, PID loop instruction data, as well as various mnemonic and commentary listings sorted alphabetically, and by PC rung number or address. Portions of various XYCOM ladder diagram reports are illustrated in Fig. 16.23.

INTELLIGENT ALPHANUMERIC DISPLAYS

Many PC applications can be enhanced with the addition of an alphanumeric display. A simple single-line display of several dozen characters can often replace a much larger panel of many indicators. In fact, with proper design, a display unit can help to greatly reduce the number of push buttons and indicators required for a large control system. Implementation of an alphanumeric display with a PC requires PC programming to generate the proper message at the right time, along with storing each message in the PC memory. In

```
27-Sep-83                           SUBSYSTEM 101          PROJECT 9355
00:07:20                      Xycom 4820 LDT  Release 4.3   Allen-Bradley
-----------------------------------------------------------------------------
```

```
      P R O G R A M   I N F O R M A T I O N
      -----------------------------------

        PC program name .......... : SS101
        Description .............. : SUBSYSTEM 101 MAY PLANT PROJECT 9355
        Date/time stored on disk .. : 27-Sep-83 at 00:03:19
        PC type .................. : Allen-Bradley PLC-2 series
        Program size ............. : 1039
        Data table size .......... : 3968

      L I S T I N G   I N F O R M A T I O N
      -------------------------------------

        Time at which listing finished ......... : 02:23:10
        Ladder listing uses .................... : rung numbers
        Cross reference data ................... : shown to right
        Closest control for each element ....... : shown
        Maximum number of elements per line .... : 11
        Maximum cross references listed per rung : 0
        Implied cross references ............... : included
        One rung per page ...................... : no
        Lines per page ......................... : 60
        Data table listing format .............. : default

      R E P O R T   T A B L E   O F   C O N T E N T S
      -----------------------------------------------

        Ladder diagram ......................... : 1
        Address usage report ................... : not printed
        Unreferenced description report ........ : not printed
        Undefined description report ........... : not printed
        Full cross reference report ............ : not printed
        Data table listing ..................... : not printed
        Data handling report ................... : not printed

      L A D D E R   L I S T I N G   S Y M B O L   K E Y
      -------------------------------------------------


        !27 character                                 ! output
        !contact                                      ! coil
        !description                                  ! description
        ! aaaaa         aaaaa                 aaaaa    !
  bbbb  +---]/[---------...] [...-------------------( )---+ optional
        !ccc                                          ! cross
        !                                             ! reference


        aaaaa = Contact address
        bbbb  = Line or rung number
        ccc   = Closest line or rung number where this
                contact is an output coil
        ...   = Marks a contact that is a real I/O address

      A D D R E S S   U S A G E   R E P O R T   S Y M B O L   K E Y
      -------------------------------------------------------------

        .   = Address not used in program and not assigned as real I/O
        S   = Address used in program but not assigned as real I/O
        0   = Address not used in program but is assigned as real I/O
        *   = Address used in program and assigned as real I/O
```

Fig. 16.21. Xycom Reports Listing

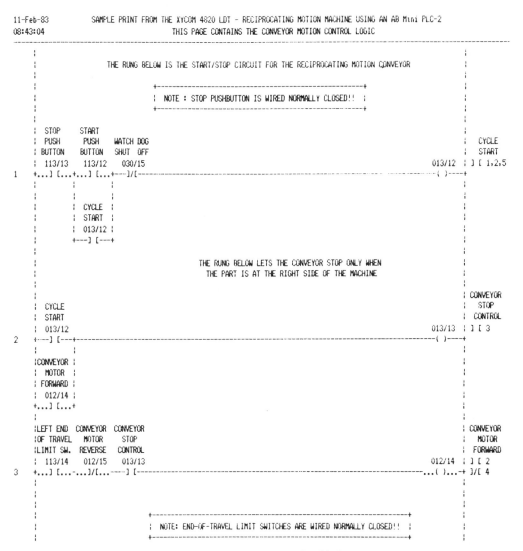

Fig. 16.22. Xycom Ladder Listing

addition, PC to display hardware interfaces will be required to connect a display electrically to the PC being used. For single applications the costs of display hardware, PC to display interfacing, and programming and debugging time may not justify the use of a display system.

One way to reduce the costs of implementing a PC display is to use a commercially available system. A typical example of a display system designed for use with a PC is Comptrol's COMP 7010 and COMP 7210 family of intelligent display devices. Each of the Comptrol display systems shown in Fig. 16.24 provides the user with a great deal of flexibility without a great deal of PC programming required. The Comptrol intelligent displays communicate to standard 5-volt dc TTL-type discrete PC output modules, or serial RS-232C,

11-Feb-83   SAMPLE PRINT FROM THE XYCOM 4820 LDT - RECIPROCATING MOTION MACHINE USING AN AB Mini PLC-2
08:43:04   Data Table Listing

Address	Contents in Hexadecimal															
	-00-	-01-	-02-	-03-	-04-	-05-	-06-	-07-	-10-	-11-	-12-	-13-	-14-	-15-	-16-	-17-
000	0000	0000	0000	0000	0000	0000	0000	0000	0000	0000	0000	0000	0000	0000	0000	0000
020	0000	0000	0000	0000	0000	0000	0000	0000	0000	0000	0000	0000	0000	0000	0000	0000
040	0000	0000	0000	0000	0000	0000	0000	0000	0000	0000	0000	0000	0000	0000	0000	0000
060	0000	0000	0000	0000	0000	0000	0000	0000	0000	0000	0000	0000	0000	0000	0000	0000
100	0000	0000	0000	0000	0000	0000	0000	0000	0000	0000	0000	0000	0000	0000	0000	0000
120	0000	0000	0000	0000	0000	0000	0000	0000	0999	0000	0000	0000	0000	0000	0000	0000
140	0000	0000	0000	0000	0000	0000	0000	0000	0000	0000	0000	0000	0000	0000	0000	0000
160	0000	0000	0000	0000	0000	0000	0000	0000	0000	0000	0000	0000	0000	0000	0000	0000
200	0000	0000	0000	0000	0000	0000	0000	0000	0000	0000	0000	0000	0000	0000	0000	0000
220	0000	0000	0000	0000	0000	0000	0000	0000	0000	0000	0000	0000	0000	0000	0000	0000
240	0000	0000	0000	0000	0000	0000	0000	0000	0000	0000	0000	0000	0000	0000	0000	0000
260	0000	0000	0000	0000	0000	0000	0000	0000	0000	0000	0000	0000	0000	0000	0000	0000
300	0000	0000	0000	0000	0000	0000	0000	0000	0000	0000	0000	0000	0000	0000	0000	0000
320	0000	0000	0000	0000	0000	0000	0000	0000	0000	0000	0000	0000	0000	0000	0000	0000
340	0000	0000	0000	0000	0000	0000	0000	0000	0000	0000	0000	0000	0000	0000	0000	0000
360	0000	0000	0000	0000	0000	0000	0000	0000	0000	0000	0000	0000	0000	0000	0000	0000

11-Feb-83   SAMPLE PRINT FROM THE XYCOM 4820 LDT - RECIPROCATING MOTION MACHINE USING A TI PM550
08:58:21   Mnemonic Listing

Address	Contents		Description		
0000	STR	X2	STOP	PUSH	BUTTON
0001	STR	X1	START	PUSH	BUTTON
0002	OR	CR1		CYCLE	START
0003	AND STR				
0004	AND NOT	CR3		WATCH DOG SHUT OFF	
0005	OUT	CR1		CYCLE	START
0006	STR	CR1		CYCLE	START
0007	OR	Y23	CONVEYOR	MOTOR	FORWARD
0008	OUT	CR2	CONVEYOR	START	CONTROL
0009	STR	X3	LEFT END	OF TRAVEL LIMIT SW.	
0010	AND NOT	Y24	CONVEYOR	MOTOR	REVERSE
0011	AND	CR2	CONVEYOR	START	CONTROL
0012	OUT	Y23	CONVEYOR	MOTOR	FORWARD
0013	STR	X4	RIGHT END	OF TRAVEL LIMIT SW.	
0014	AND NOT	Y23	CONVEYOR	MOTOR	FORWARD
0015	OUT	Y24	CONVEYOR	MOTOR	REVERSE
0016	STR	CR1		CYCLE	START
0017	STR	X2	STOP	PUSH	BUTTON
0018	TMR				
0019	preset	C100			
0020	current	V100	CYCLE	WATCH DOG	TIMER
0021	OUT	CR3		WATCH DOG SHUT OFF	
0022	STR NOT	X255		DUMMY	CONTACT
0023	MOVE				
0024	source	V100	CYCLE	WATCH DOG	TIMER
0025	destination	A0	CYCLE	TIMER	DISPLAY

Fig. 16.23. Sample of Typical Xycom Documentation Reports

11-Feb-83   SAMPLE PRINT FROM THE XYCOM 4820 CRT - RECIPROCATING MOTION MACHINE USING AN AB Mini PLC-2
08:43:04   Full Cross Reference Report

Address	PC operation and references
012/14	CONVEYOR MOTOR FORWARD
	] [2
	]/[4
	( ) 3
012/15	CONVEYOR MOTOR REVERSE
	]/[3
	( ) 4
013/12	CYCLE START
	] [1,2,5
	( ) 1
013/13	CONVEYOR STOP CONTROL
	] [3
	( ) 2
030/15	WATCH DOG SHUT OFF
	]/[1
113/12	START PUSH BUTTON
	] [1
113/13	STOP PUSH BUTTON
	] [1
	]/[6
113/14	LEFT END OF TRAVEL LIMIT SW.
	] [3
113/15	RIGHT END OF TRAVEL LIMIT SW.
	] [4
011	CYCLE TIMER DISPLAY
	(PUT) 7
030	CYCLE WATCH DOG TIMER
	(RTO 1.0) 5
	[G] 7
	(RTR) 6
130	
	(RTO 1.0) 5
	(RTR) 6

Fig. 16.23. *(Continued)*

RS-243, RS-422, or current-loop PC interfaces. A maximum of 112 messages can be stored in each 8085A microprocessor-based display unit. In addition to display lengths of 8 and 20.5 inch-high characters, a printer interface option is available for hard-copy printouts of each display message.

Setup and operation of a Comptrol display is quite simple and straight-forward. The self-powered unit is mounted and connected to a suitable 120-volt

Fig. 16.24.   Comptrol's Comp 7220 Display (left) and Comp 7010 Display (right)

ac power source. Each message is programmed through a software-encoded keyboard (available from Comptrol) or an ASCII keyboard using RS-232C serial interfacing, both capable of being connected directly to a special port provided on the display. Along with each message there are special codes to tell the display processor whether the message should be scrolled, flashing, or blocked (displayed as word groups rather than scrolled word groups). Once programmed with messages, the unit is ready for connection to a PC. A total of 2045 characters are available, with any given message capable of being a maximum of 245 characters in length.

Even though there are serial interface options available for the various displays Comptrol manufactures, the standard parallel interface is probably the easiest to use for most applications. This interface uses a special terminal strip mounted on a snap track to connect the display to a PC's discrete output module. Eight signal leads and ground must be installed between a display's interface terminal strip and the PC. If use of the display's internal 5-volt dc power supply is not desired for signal interfacing, a correctly specified user power supply may be substituted. The PC output module may be either a TTL, a reed relay, or a similar low-voltage discrete output module.

Activation and display of a message is initiated by simply sending the 7-bit message number, in binary format, to the display. An eighth bit, which is transmitted with the 7-bit message number, acts as a *strobe* or message-ready signal. The display decodes the incoming number, then displays the desired message or messages until instructed to stop by the PC. In addition, an operating mode is offered whereby variables or data can be imbedded into a message that has been previously programmed with blanks or "subs." The PC programming is reduced to that necessary to detect a condition for which a message has been designed, and then to identify the message by number. The message number is output to the display along with the strobe signal to inform the display that a new message exists. Since the message number is output in a binary format, the PC should provide a convenient means to output numerical data encoded in a binary format. This will usually involve the use of a BCD-to-binary conversion instruction prior to transmission of data to the display for those PCs handling numerical data in a BCD format within the processor. For those PCs not having the ability to output numerical data in a binary format, the user's programming will be increased since the relay ladder program will have to generate the proper binary number signals in a relay ladder format. A one-shot contact instruction is recommended as the strobe bit. Also, the time between successive strobes should not be faster than 10 to 20 milliseconds, allowing for adequate setup and settling time of the data bits.

PART
V

User
Reference

17

Selection and Purchase of a Programmable Controller

The actual purchasing of a PC and the support hardware for a control system will involve writing a specification detailing the requirements of the intended application. The specification is then distributed to manufacturers and their representatives to obtain pricing for evaluation of the most cost effective method to implement the desired control scheme. This chapter will provide many basic ideas and considerations that can be used for the development of a specification relating to the purchase of a PC, a PC enclosure, and a complete PC-based control system. The statements listed are of a general nature. Each statement is designed to be adaptable to many specific control system applications. Those persons responsible for the development of control system specifications might wish to use the items listed in each of the example specifications as a basis from which to develop their own control system specifications.

This chapter contains three separate reference specifications. The first deals with the purchase of a PC and the PC-related hardware and accessories for a control system. Every PC manufacturer can provide a "prewritten" PC specification to a potential customer. These manufacturer's specs are usually written in such a way that they eliminate many of the competitors' products if they are not carefully examined and edited. The items listed in this text have been compiled from several PC manufacturers' specifications and depersonalized in order to allow as many systems as possible to be quoted. The specification can be "personalized" by the user to suit a particular need.

Many PC system designers require that the PC, its support hardware, and other control system devices be assembled into enclosures and cabinets prior to installation. The second specification notes many of the fabrication and assembly considerations which may be important during the construction of cabinets, consoles, and enclosures.

The final specification addresses the considerations and details of employing an outside source to develop the PC program for the intended application. Many companies prefer to have an experienced PC software supplier develop the programming necessary for their control applications. The third specification details many of the procedures and application considerations connected with the development of PC software by a third party.

Many statements in the specifications require additional information for completion. For example, the statement "electrical supply voltage to the programmable controller shall be (115 volts ac) plus or minus (10) percent" contains two fields that must be completed. A specification topic will indicate a field requiring user input by parentheses (. . .). The field will contain an example or a description of the data required to complete the statement. It shall be noted that many statements listed in this chapter will not be applicable to all installations. These statements shall be modified or deleted as necessary in order that the finished specification is tailored to suit a particular application.

Prior to listing the three specifications of this chapter, a quick review of specification writing is in order. Any purchase order between a buyer and a seller is considered a legal contract in most states. If the purchase agreement or order references or includes a detailed specification for the materials and/or hardware being supplied and/or purchased, that specification is legally binding on both the buyer and the seller. If a dispute shall arise, the legal interpretation of the specification will be the deciding factor as to which party wins or loses. It is of paramount importance that the specifications, as well as the purchase agreement, be concise and well written.

Correct grammatical construction and punctuation as well as exacting detail are of highest priority in the development of a specification. In many respects the seller develops his attitude toward the buyer when he reads the specification. A sloppy specification can cause a seller to "relax" his usual good-quality and production requirements in the misconception that the buyer "is not really interested." In many cases the specification is a buyer's first contact with a potential vendor.

The purchase order or agreement between a buyer and a seller contains the business elements of a transaction. It specifies the terms of payment and shipping, pricing, scheduling, and similar details. It is also the actual specification, which contains the engineering and related technical details, describing the hardware or service that is being purchased. With each party's legal and contractual responsibilities being detailed in the specification portion of a purchase agreement, the specification must be written without expressions and phrases that are not legally enforceable. Listed below are 14 common specification pitfalls:

1. Use of the term "shall"

The use of terms such as "will," "should," or "may" generally indicates a nonmandatory requirement. The term "shall" indicates a binding requirement and shall be used in lieu of will, should, and may.

2. Use of the term "suitably"

Suitably is a nondescriptive term. For example, "suitably attached" could mean anything from the use of baling wire to welded brackets. Always provide specific details if the item is worth mentioning in a specification.

3. Use of the term "design objective"

This term implies that part or all of the requirement may not be practicable to obtain, and that the seller shall "do the best he can." Always specify requirements that can be achieved and measured against a known, detailed standard.

4. Use of the term "shall be capable"

Sentences containing the term "shall be capable" usually detail a requirement secondary to the topic of the specification. "Shall be capable" really does not matter—something is or is not.

5. Use of the terms "intent" or "intended"

Use of the terms "INTENT" or "INTENDED" is difficult to enforce since only the requirements of a contract can be enforced and not a contract's "intent."

6. Use of the term "consistent with"

Consistency is a relative term and implies no definite standards or requirements. If something can be "consistent with," it shall really "meet the requirements of" a specific code or standard.

7. Use of the term "in accordance with"

Another relative term that implies no definite standard or requirement. If something shall be "in accordance with," it can really "meet the requirements of" the specific code, standard, or requirement.

8. Use of the term "suitable"

"Suitable" is a generality open to interpretation and therefore not legally enforceable.

9. Use of the term "or equal to"

Equality is another relative term since it leaves the determination of whether two items are equivalent to both the buyer and the seller's judgments. If the term "or equal to" must be used in a specification, the rules, tests, and requirements for equivalency must be specified, and both parties shall agree in writing to the equivalent prior to use.

10. Use of terms such as "long life, industrially rated, heavy duty," etc.

These terms are often used to specify the performance requirement of a device, system, or piece of hardware. If specific performance requirements exist, they shall be detailed along with the tests used to determine the particular performance requirements.

11. Symbols

Avoid symbols such as 55'–6¼". Detail the units as 55 feet, 6¼ inches. Avoid trade names, colloquialisms, and slang substitutions for definable, well-known terms or definitions.

12. Quantity details

State all quantities in the units of measure, being sure to separate a quantity from a size, such as "bore 2 5/32" holes." Instead use "Bore 2 holes, each 5/32 inch in diameter." Be aware of various units of measure, such as

Tons: metric, long, or short

Weight: troy, avoirdupois, or apothecaries'

Gallons: imperial or U.S.

13. Size

Specify any size in an exact dimension or capacity with proper units. Be sure to include where and how a particular measurement is to be obtained.

14. Thickness

Specify all thicknesses as a physical dimension instead of a "gauge."

SPECIFICATION FOR THE PURCHASE OF A PROGRAMMABLE CONTROLLER

General

This specification covers the technical, mechanical, and electrical requirements of a programmable controller which shall be used to (Supply a single sentence to indicate a broad application. Example: "control the pumping of petroleum products in a tank farm.").

The PC hardware supplied as part of this specification shall be currently available as vendor catalog items. Equipment which might be under development or in the commercialization stage (shall/shall not) be considered for this specification. Those items that are not current catalog items must be broken out and listed separately as part of the quote.

The PC equipment furnished as part of this specification shall meet or exceed the following specifications: (List all JIC, NEMA, NEC, Military, and specialized company specifications which shall apply to the selection and installation of the proposed PC hardware.)

All equipment supplied as part of this specification shall operate continuously under the environmental conditions listed below.

> Temperature (0 to 60 degrees celsius).
> Humidity (5 to 95) percent, noncondensing.

The electric power supply to the processor and the remote power supplies shall be:

> *Processor*—Line voltage: (120 volts ac) plus or minus (10) percent, at (60) hertz.
> *Remote power supplies*—Line voltage: (120 volts ac) plus or minus (10) percent, at (60) hertz.

All power supplies shall be equipped with fuses or circuit breakers for internal short-circuit and overload protection. Manufacturer shall provide a listing of the supply line current requirements so that purchaser can adequately size his circuit feeders and power distribution system. Any isolation and/or line conditioning transformers which shall be required or shall be recommended by the manufacturer for this application shall be noted as part of the quotation.

All PC assemblies, subassemblies, circuit cards, cables, and other related hardware items shall be permanently identified with the manufacturer's part or catalog number. All processor and support equipment firmware identifications and revision levels shall be clearly indicated.

All PC hardware supplied as part of this specification shall be manufactured to those recognized industry standards listed below, and regularly sold for the purposes or uses stated in this specification. All components shall be fabricated to withstand, without premature failure, all normal use and handling. (List any industry standards which shall be applicable as well as any performance specifications that shall be required.)

The general purchaser's arrangement and location of processor(s) and I/O devices is shown on (List all drawings and sketches.). The I/O structure for this application shall require the I/O hardware to be configured for (local, remote, both local and remote) operation. The wiring distances shown on the drawings and sketches represent the distances between PC components in cable feet.

The engineering contact for this specification is:

Name:
Phone:
Address:

Input/Output Hardware

The number of I/O devices by location is detailed on (list all applicable drawings, sketches, and tables). Vendor shall supply sufficient quantities and types of I/O modules to handle the I/O requirements of this specification. A detailed listing of all I/O hardware by quantity and catalog number shall be provided with the quotation. Any support PC equipment required to implement the proposed system, which is to be supplied by the purchaser, is to be noted as part of the quotation.

It shall not be necessary to remove the field wiring to an I/O module during maintenance or replacement. All purchaser's field wiring shall terminate on terminal blocks with a (600-volt ac) minimum rating. Field device terminal blocks shall be designed to facilitate the quick removal of the I/O module from the terminal block and its associated field wiring. All field device terminal blocks shall be designed to accommodate a minimum of (2), size (14) AWG wires.

Diagnostic indicators shall be provided on all I/O points. Each discrete I/O point shall include at least one indicator designed to display the status of the field device connected to it. A blown fuse indicator shall be provided on output modules. A blown fuse indicator is required for each (output point, group of eight output points) minimum. Analog and special-purpose I/O modules shall include diagnostic indicators to indicate the proper operation of the module.

Discrete input points shall be guaranteed to be ON if 75 percent or more of the I/O card's working voltage is present. A discrete input point shall be guaranteed to be OFF if 25 percent or less of the I/O module's working voltage is present.

All I/O points shall provide a minimum of 1500 volts rms electrical isolation between the field device I/O circuitry and PC logic voltage levels contained on the I/O module. Isolation is to be provided by optoelectronics, transformer magnetics, or read relays.

All I/O points shall have provisions for being identified with purchaser's (seven) place alphanumeric description. Identification strips shall be provided adjacent to the I/O module terminals, as well as the I/O module indicator lights, for the purpose of labeling I/O device functions, wire numbers, and I/O point identifications. The identification strips shall be easily relocated from one I/O module to another in the event of module replacement or I/O point relocations, deletions, and additions.

Each I/O card shall be clearly identified as to function and type. Different colors shall be required for the I/O module labels to differentiate between various voltage levels and module types.

Each I/O module shall include a terminal strip connection label which indicates the arrangement and function of each terminal of the module. This label shall not be visible unless the I/O module is removed from its mounting hardware.

All I/O modules shall be keyable in their mounting hardware. I/O module keying devices shall not be easily removed. No two electrically different I/O modules shall have the same keying identity, only identical modules shall be keyed identically.

All I/O hardware shall automatically shut down if there is no communication between the location and the processor in any contiguous 250-millisecond time period. The processor shall generate an alarm in the event of a communications fault with an I/O location. All I/O points shall be switched OFF at the I/O location in the event of a communications fault with that location.

All I/O modules shall be securely fastened to the hardware which mounts them by screws or hold-down clips to prevent accidental movement due to vibration.

Each I/O module shall be designed to minimize damage in the event of improper wiring of the module. The accidental application of a voltage level higher than the design level of the module shall not damage any PC system hardware other than the module receiving the improper voltage level. The incorrect wiring of an I/O module to a field device shall not damage any other PC hardware except the involved module.

In addition to the I/O requirements detailed as part of this specification, an additional spare I/O capacity shall be provided for each I/O location. Sufficient spare I/O modules shall be included as follows:

    Installed spare discrete input points: (10) %
    Installed spare analog input points: (5) %
    Installed spare discrete output points: (10) %
    Installed spare analog output points: (5) %
    Installed mounting space for additional discrete input modules: (15) %
    Installed mounting space for additional analog input modules: (10) %
    Installed mounting space for additional discrete output modules: (15) %
    Installed mounting space for additional analog output modules: (10) %

All output points shall be sized to match purchaser's loads as follows:

DEVICE FULL-LOAD CURRENT VOLTAGE INRUSH CURRENT
Indicators
Solenoid valves
Motor starters
Relays

All input points shall be designed to interface with input devices having solid-state output electronics such as transistors or triacs. Each PC input point shall be designed to sense an OFF input state when connected to a solid-state input device having a (5) milliampere or less leakage current in the OFF state. Any input module not meeting the above specification shall require vendor-provided RC networks as part of the field wiring circuitry. The vendor is to indicate, as part of his quotation, those input modules that do not meet the above specification.

No output point shall have a voltage drop of more than (2) volts at the maximum rated current of the output point.

Central Processor Unit (CPU or Processor)

The central processor unit shall provide the following instructions as part of its total instruction set:

Relay   Latches
Timers   Counters
Addition Subtraction
Equality   Inequality

(Provide additional specialty instructions such as data handling, bit manipulation, subroutines, jumps, and so on, as required for the application.)

The memory capacity of the processor shall be (state memory size in 1024-word increments, such as 4096) thousand 16-bit words. For this specification, one 16-bit word shall be equivalent to the amount of memory required to store a single relay-type contact or coil instruction. The memory shall be capable of storing the following amounts of user program logic and data:

User logic (State number of words.).
Data storage (State number of words.).

The storage of I/O and internal processor data shall be in addition to the above requirements.

The memory system used shall be (CMOS, Core, PROM). It shall be easily expanded in the field to the maximum capacity of the system by changing a memory circuit card or by the addition of memory ICs. Memory expansion shall be performed by a manufacturer's service technician or a trained purchaser's representative.

The processor, all I/O devices, and any support system hardware shall shut down orderly in the event of loss of processor scanning, loss of processor logic power, or the detection of a memory or operating error or fault. Where the possibility exists for a processor to perform illegal operations due to improper user programming techniques, the processor shall initiate a "soft fault" or "running error." Typical illegal programming instructions might include jumps to nonexistent instructions or infinite program loops. A soft fault or running error shall shut down the processor and its related I/O equipment. All faults of this nature shall be diagnosed with the programming device.

Where battery backup is required to support the memory system during limited power outages, visual as well as internal program monitoring of the battery status shall be provided. The status indicator and program monitoring functions shall be designed to indicate poor battery conditions up to (seven) days prior to battery failure. The batteries shall be replaceable without interruption of the processor or any memory functions. The type of battery backup system to be employed shall not be (nickel-cadmium, lithium, lead-acid—list those not acceptable, if any).

The processor shall incorporate a key-lock switch to prevent unauthorized program alteration. Programming of the processor shall only be possible in (the off-line, the on-line, both the off-line and on-line) mode of system operation. In the off-line programming mode all output devices are to be deactivated. The processor shall actively execute the user programming and control the OFF/ON statuses of all outputs during the on-line programming mode.

All programming and monitoring hardware shall not affect the operation of the processor when it is attached or disconnected and the processor is in normal operation.

The statuses of any math functions, timer/counter current and preset values, system constants, relay latch statuses, and other stored data values shall be retained through a power outage.

All user programming and data shall be retained without loss for a minimum of (30) days with power disconnected from the processor.

Programming Device

Processor programming shall be performed with the use of a cathode ray tube (CRT) based programming terminal designed for the purpose. The programmer shall be designed for industrial use as well as for ease of portability.

The programming terminal shall alert the user to programming errors, such as illegal references, improper instruction construction, and so on. The terminal shall provide continuous status indication of the processor's mode of operation. In addition the programming device shall be able to monitor any data stored in the processor as well as the operation of the user programming.

A rung of relay ladder instructions shall not require reprogramming in the event of a change, addition, or deletion. Additional instruction rungs shall be added anywhere in the program without difficulty. Any ladder diagram rung, or groups of consecutive rungs, shall be deleted without difficulty.

The programming terminal shall have two modes of operation. The first mode permits the addition, changing, and removal of instructions contained in the processor's memory. This mode shall be termed the program mode for off-line operations, and the edit mode for on-line operations. A second mode shall permit only the monitoring of the processor's memory for the purpose of troubleshooting and monitoring the user's program.

Where a PC processor supports remote I/O, it shall be possible to access the processor with the programming terminal at each remote I/O location. The processor to remote I/O communications shall be separate of the programming terminal to processor communications in order that the operating status of the remote I/O hardware be kept independent of the terminal operation. The processor shall communicate with only one programming terminal at a time.

The programming device shall include an RS-232C port for the connection of a printer. The programming terminal shall provide a hard-copy printout of the processor's data memory in ladder diagram format similar to the CRT display format. In addition the programmer shall include as part of the printout the contents of all data stored in the processor's memory, as well as a cross-reference listing of each location where an input, output, or internal storage reference is used.

The ability to quickly search the user program for a particular instruction and/or reference shall be included as part of the programming terminal's capabilities.

The programming device shall override, on user command, the status of any input or output point through the use of special programming terminal instructions or functions. Any input or output point which is overridden shall be indicated on the terminal display. Overridden I/O addresses (shall/shall not) remain active when the programming terminal is disconnected or power is interrupted to the processor.

Training, Documentation, and Warranties

A listing of all training courses and their curricula, time, date, and cost shall be provided as part of the quote. These classes shall be held at the manufacturer's training location and shall include classroom instruction as well as hands-on programming experience. Costs of training at the purchaser's facility shall also be included as part of the quotation. For training classes at the purchaser's facility, a listing of required purchaser supplied equipment and training aids shall be included.

The successful vendor shall provide (3) sets of installation, assembly, and programming instruction manuals with equipment delivery.

The seller shall warrant to the purchaser that the equipment delivered with reference to this specification complies with this specification unless specifically noted in writing at time of seller order acceptance.

The seller shall warrant the equipment as to defects in material and workmanship for a period of one year from the time it is received by the purchaser. Seller shall include a copy of his warranty with the quotation.

The warranty specified by this specification shall be exclusive, and in lieu of all other warranties whether written, implied, orally presented, or statutory. No warranties of merchantability or fitness for purpose shall be inferred or apply.

Vendor shall be capable of providing a service representative and replacement hardware at the purchaser's facility within twenty-four hours of the purchaser's request for such assistance and/or hardware. Vendor is to provide the cost of this service at time of quotation. All requests for hardware assistance and/or replacement during the time the warranty is in effect shall be the financial responsibility of the seller. Requests made for equipment out of warranty or for programming assistance shall be the financial responsibility of the purchaser.

Information to be Provided in Quotation

The following dates shall be provided: (List all dates pertinent to the purchase of the PC hardware.).

Provide a listing by catalog number and description of all hardware, software, accessories, and so on, being provided.

Provide a listing of shipping schedules and method of shipment.

Provide any alternate quotations that the seller and/or buyer deems necessary or applicable.

List any exceptions to this specification.

CONTROL SYSTEM HARDWARE SPECIFICATION

General Conditions

Vendor shall provide a completely assembled and tested control system as described by this specification. All assembled equipment shall be packaged to facilitate shipment and minimize installation.

Vendor shall supply all necessary equipment shown on control system drawings unless otherwise noted on the drawings or in writing from the purchaser: (List all drawings, sketches, tables, and their current revision levels that shall apply.).

All equipment shall be assembled and installed to meet (JIC, NEMA, ANSI, NEC, OSHA, others as necessary) standards as listed below: (Include a listing of all specific standards which shall apply and those sections that shall have particular importance.).

Environmental and operational conditions for the final system installation location are as follows:

Class (3), Division (1) area

Ambient temperature range: (0 to 60) degrees Celsius.

Ambient humidity range: (5 to 95) percent, noncondensing.

System power supplies: (List all power sources, their frequency, number of phases, current capacity, fault capacity, and voltage levels.).

System operation: (24) hours per day, (5) days per week, (52) weeks per year continuously.

Engineering contact for this specification shall be as follows.

Sean P. McHugh
1234 USA Avenue
This City, That State 98765
(999) 123-4567

All equipment specifications and catalog numbers shall be adhered to. Substitution of alternate hardware shall be done only with the written approval of the purchaser.

No equipment shall be fabricated without the written release of the purchaser. Order placement does not constitute a fabrication release. Pricing for all changes and revisions shall be required prior to fabrication release of the change or revision.

All labels shall be fastened with the use of (nuts and bolts, epoxy glues, pop rivets, screws) unless otherwise noted. All system devices and hardware must be labeled as shown on drawings using (lamacoid, metal, tape, punched) labels.

Vendor shall meet all shipping dates. Items that shall cause a delay in system shipment shall be called to the attention of the purchaser as soon as possible.

The following equipment shall be provided to the vendor by the purchaser: (List all purchaser supplied equipment, its catalog number, and when it shall be shipped to the vendor.).

All prices shall be firm as of vendor order acceptance. Price changes resulting from purchaser system changes and revisions shall be agreed upon prior to vendor fabrication of the proposed change or revision. Purchaser shall not be responsible for vendor cost increases due to labor or material, unless prior written agreement is obtained from the purchaser.

Wiring Practices

All wiring to be (THHN, THWN, MTW) insulation rated for (300, 600) volts minimum, unless otherwise noted on drawings. Wire size shall be (#14 AWG) stranded copper for control and (#10 AWG) stranded copper for power buses, unless otherwise noted on drawings. All flexible cords to be TYPE 'SO'. Logic wiring (48 volts maximum) shall be (#22 AWG) stranded copper, unless otherwise noted on drawings. All wiring runs shall be continuous with no splices permitted.

All wiring shall be installed such that the removal of an electric device can be made without the removal or relocation of the wiring bundles and harnesses adjacent to the device being removed.

Where wiring crosses the hinged edge of a panel, stranded wiring shall be provided. In addition the wiring shall be securely bundled and wrapped with plastic spiral wrap for abrasion protection. The wiring bundle shall be securely fastened in a flexible loop to both surfaces supporting the hinge. Plastic self-stick fasteners shall not be used for this purpose. A stud-mounted plastic cable clamp is preferred.

All enclosure, cabinet, panel, or operator station wiring shall be run as secured wax twine laced bundles, or in plastic wiring ducts with covers.

All wiring shall be run in a professional manner consistent with industry standards for installation and workmanship. No wiring shall be permitted to contact sharp edges, rough surfaces, burrs, or pinch points. Insulated bushings shall be used wherever wiring passes through a partition or barrier. Where wiring enters conduit, square duct, or other type of metal raceway systems, insulating-type bushings and connectors shall be used to prevent possible mechanical injury.

All wiring shall be labeled with the wiring identification shown on schematics and electrical drawings. Wire labels shall be the nonmetallic sleeve type or hot-stamped into the insulation. Tape-type labels and clip-on wire identifiers (are/are not) acceptable.

All fuses, circuit breakers, disconnects, and switching-type terminal blocks shall be labeled according to the electrical drawings.

Ground continuity between all enclosures, panels, back panels, enclosure doors, hinged panels, and operator stations shall be provided. All grounding conductors shall be bare braided copper or green insulated stranded copper. All terminals provided for the termination of shielded cable shields shall be tied to the enclosure ground system. All grounds shall be tied to a central location and not "daisy-chained" together. Where the grounding conductor is attached to a painted surface, the paint shall be removed in the area of the connection. Star-type washers shall be used on all bolted ground connections.

Where openings larger than (12) square inches are provided for the entry of cables into an enclosure, cable lashing bars shall be provided. Cable lashing bars shall be constructed of half-inch-diameter rod on three-inch centers.

Wire coding according to function shall be adhered to. The following color code shall be used: (List all wire colors to be used and the electrical function for that color. For example, neutral wiring: white.).

All wiring attached to nonclamp-type terminals or terminal blocks shall be attached to the termination point using the proper sized crimp-on wire lug. Wiring that must be attached to devices having wire pigtails shall require the use of (crimp connectors, wire nuts, wire tap connectors). Where wiring must be soldered to a component or other device, 60/40 rosin core electric solder shall be used. A mechanically sound joint shall be provided with a soldering iron of sufficient power to eliminate any cold solder joints while not damaging the wire insulation with excessive heat. All flux shall be removed from the joint with a flux remover designed for the purpose.

All terminal blocks shall be labeled according to electrical drawings. No more than (2) wires shall be permitted under any terminal block screw.

Fabrication Practices

All enclosures, panels, cabinets, and operator stations shall be NEMA (12) construction.

All removable covers which are larger than (12 by 28 inches) shall be sectionalized. All covers shall be gasketed.

All sharp edges, burrs, and rough edges which might cause personnel injury or wiring abrasion shall be removed prior to painting. All welds shall be ground smooth. All dents, creases, nicks, and gouges shall be filled with weld and sanded smooth. Automotive-type body fillers shall not be used.

All enclosures, cabinets, and panels over (400) pounds, or larger than (5) feet in any dimension, shall be equipped with lifting bars for handling as well as braced for fork truck lifting. Lifting eyes (are/are not) permitted.

Minimum enclosure metal thickness is ($\frac{1}{8}$) inch. All doors, as well as enclosure and panel sides, shall be braced to inhibit buckling, sag, or other movement.

All enclosures and cabinets shall have hydraulic heavy-duty storm door closers, interior fluorescent lighting, and a duplex utility receptacle.

All enclosure interiors shall be painted (white). All enclosure exteriors shall be painted (storm blue). All enclosures shall be primed (specify type or specification) prior to painting. Specification (list any paint specifications) shall be followed when painting all enclosures.

All conduits and metal raceways shall have their cutting burrs removed prior to installation.

Witness Tests and Inspections

All hardware must be inspected prior to shipment. Vendor shall notify purchaser in writing (3) days prior to equipment availability for inspection. Vendor shall have completed all electrical and mechanical testing prior to purchaser inspection. Shipment of equipment without notification of inspection shall subject vendor to all field charges relating to any corrections that shall be required in order to bring equipment up to specification.

Purchaser shall provide a detailed description of the acceptance tests he shall perform at time of inspection. All acceptance tests are to be completed by the vendor prior to inspection. Acceptance tests shall include pressurization of pneumatic systems, powering of electric systems, and operation of all components of the system. Any PC programming required for the acceptance tests shall be provided by the (purchaser, vendor). Any additional hardware which shall be required by the purchaser for the acceptance tests and inspections shall be provided by the (purchaser, vendor).

Acceptance of the equipment does not relieve the vendor of his responsibility to meet the system specifications.

All equipment shall be crated for shipment using (pallets, crates, shipping containers). To provide maximum protection from moisture, all equipment shall be wrapped with heavy plastic (state thickness), and sealed with cloth tape. All delicate instruments shall be secured, and any components that might jar loose or break from their mounts shall be braced. All plastic and glass bezels, indicator dials, door cutouts, and instrument covers shall be covered with heavy cardboard to prevent accidental breakage or damage.

Purchaser shall determine method and route of shipment to his facility.

Information to Be Provided in Quotation

The following dates shall be provided: (List all dates pertinent to the purchase of the PC hardware.).

Provide a listing by catalog number and description of all hardware, software, accessories, and so on, being provided.

Provide a listing of shipping schedules and method of shipment.

(Provide any alternate quotations which the seller and/or buyer deems necessary or applicable.)

List any exceptions to this specification.

SYSTEM PURCHASE SPECIFICATION: HARDWARE AND/OR SOFTWARE DESIGN

General Requirements

Vendor shall provide all drawings, equipment fabrication and installation specifications, PC software, operation manuals and instructions, and maintenance and troubleshooting manuals required for the purchaser to fabricate, install, start up, and maintain the control system and integral hardware as detailed by this specification and its attachments.

All drawings are to be provided on (purchaser/vendor) supplied blank tracings.

A listing of the purchaser's standard suppliers and the devices supplied by that supplier or manufacturer is given in (Provide listing as an attachment.). All equipment specified for use in the purchaser's facility shall be from suppliers listed in attachment. Design vendor is responsible for providing alternate sources for system equipment not listed in attachment, provided the alternate source can supply material equal in quality and function. Purchaser is to approve all alternate substitutions.

Design vendor and purchaser shall jointly develop a schedule of project timing. This schedule shall detail the following.

> System functional descriptions due from purchaser.
> Begin drawing development.
> Approval of drawings submitted to purchaser for review.
> System controls reviews.
> Start of construction.
> Start of system checkout.
> Startup of system hardware and equipment.

System Design Considerations

Purchaser shall provide all necessary sketches, drawings, logic flow sheets, written and verbal descriptions necessary to describe fully the function, implementation, and operation of the control system and its associated components and assemblies.

Design vendor shall be available for review meetings (one) day per week at purchaser's facility. Purchaser shall provide personnel capable of answering any phone questions raised by the design vendor.

Design vendor shall be responsible for the development of a written document detailing the total operation of the system. This document shall include a general system overview as well as a step-by-step detailing of the operation of the system. A thorough description of the PC logic by ladder diagram rungs shall be included as part of this description. The sequence for checking out the installed system shall be included detailing all checks and tests to be made as well as any expected results. The routine maintenance procedures as well as the methods of troubleshooting the system shall be fully detailed. The format of this document, as well its approval, shall be the responsibility of the purchaser.

Drawings and Documentation

Design vendor shall provide all arrangement drawings showing locations, mounting details and dimensions, materials of construction, and all equipment and component identifications for the control system.

A bill of material shall be provided for all equipment required to implement the control system. Complete ordering information shall be provided, including source and catalog number for each item listed in the bill of material.

All electrical schematics and connection diagrams for each panel, cabinet, enclosure, and so on, shall be provided. These drawings shall include wiring identifications, descriptions of component function and operation, as well as any installation and operation notes that shall be necessary.

System legend and key drawings are to be provided. A total system key plan is required as well as key drawings for the purpose of detailing sections of the total control system. A key drawing shall be included to cross reference all other system drawings as to function and/or use. A master legend drawing shall be developed defining all symbols, abbreviations, and any special nomenclatures used on any system drawings.

All drawings prepared by design vendor shall be according to purchaser's drawing standards and methods.

All written descriptions shall be provided on 8½ by 11-inch three-hole-punched typewritten sheets. Design vendor shall use word processing equipment for development of all written material. However, printout must be via letter-quality printers. Page formats and layouts shall be subject to purchaser approval. All final documents shall be printed on heavy bond paper. Correction tape to cover typing errors shall not be used on final documents.

Design vendor shall be responsible for updating all drawings and system documents to "as-built" status within (3) weeks after startup. Purchaser shall provide copies of all system drawings and documents with corrections and changes noted within (1) week after startup.

Information to be Provided in Quotation

The following dates shall be provided: (List all dates pertinent to the purchase of the PC hardware.).

Provide a listing by catalog number and description of all hardware, software, accessories, and so on, being provided.

Provide a listing of shipping schedules and method of shipment.

(Provide any alternate quotations which the seller and/or buyer deems necessary or applicable.)

List any exceptions to this specification.

18

Assembly, Installation, and Maintenance

$\mathbf{T}$he time and effort spent on the evaluation of various PCs, along with the time and effort spent in the development and simulation of the PC program itself, can be worthless if the system is installed incorrectly and not maintained periodically. All PC manufacturers offer guidelines which provide in-depth explanations of exactly how their PC hardware should be assembled, installed, and maintained to provide years of trouble-free service. While many of the recommendations' do's and don'ts are similar for all manufacturers, each PC system will undoubtedly have requirements not generally found for other competitive systems. These differences arise from the basic design criteria that were used when the manufacturer first laid out a system. Mounting clearances, wiring lengths, sizes, and routing can differ as well as hardware arrangement restrictions, to note several of the more obvious system differences. More subtle items might include the use of lithium or alkaline batteries in the processor, grounding requirements, and field device cabling.

Many chapters in this text have noted the various differences between PCs. Where applicable, the discussion has included do's and don'ts for the installation and operation of a PC-based system. This chapter will include a reminder on these items, as well as discussions on many new ones.

In order to provide a coherent discussion on the assembly, installation, and maintenance of a PC system, it will be assumed that the PC user has received the PC hardware from the manufacturer and is in the process of installing it within the control enclosures. No particular manufacturer's system will be considered, and only the typical items of concern will be discussed. Since every PC system is different, the user must supplement the statements in this chapter with a detailed reading and understanding of the assembly, installation, and maintenance information provided by the manufacturer of the system.

The order in which each item is presented does not reflect the exact order of installation which must be followed. For example, the installation of hardware in the control enclosure requires drawings detailing component layout. These drawings can be completed before the PC hardware arrives and not immediately prior to the actual component installation.

DELIVERY OF THE PC HARDWARE

Many PC users instruct the PC vendor to deliver all of the equipment by a specified date. This often causes problems for the manufacturer at the production facilities. Many PC systems are installed in a step-by-step method, especially the larger systems which may have design, fabrication, installation, and checkout occurring over many months. When the user knows that the system will be installed in phases, he or she can often instruct the PC manufacturer to supply hardware on a firm schedule. Most manufacturers will be happy to ship equipment in phases while still maintaining prices fixed as quoted at time of order placement and not at time of shipment. The foresight to request phased delivery with prices held to those at time of order placement will permit uniform production schedules by the PC manufacturer, with the added benefit that the manufacturer will be able to meet any shipment dates quoted. As a side benefit for the user, storage space is saved since it is the manufacturer who will hold the hardware until it is needed.

When the PC hardware arrives at the user's facility, it should be immediately checked for shipping damages and shortages. Any problems should be directed to the manufacturer and/or shipper. Many PC items, such as processors, programming terminals, power supplies, and other support items, contain serial numbers. These numbers should be verified against the shipping lists, and all equipment serial numbers, dates of arrival, model numbers, and the user's purchase numbers should be filed away in a safe location. Many manufacturers use this information, especially the serial numbers, for documentation of problems and return authorizations. All manufacturers issue design change orders (DCOs) to their field service and applications personnel when a change is made to a particular PC model or hardware component. Accurate records will speed solution to problems and difficulties encountered during the installation and operation of a PC system.

The shipping container or crate should be carefully checked for any loose parts or paperwork. Stray parts may indicate internal hardware damage and should be reported. Many manufacturers include data and instruction sheets in the shipping containers. These sheets should be checked against the user's current sheets to assure that the latest information has been provided. Many times updated data sheets will arrive with the hardware before the updated material is received by the end user through the manufacturer's sales and applications staff. The updated information may be of prime importance in the installation of the system. Often manufacturers issue special instructions not appearing in other standard publications. These special instructions may note a change in the hardware, necessitating a change in installation or use from earlier similar hardware.

Several manufacturers have developed ingenious methods to place "secret compartments" of accessories within the foam packing material. The user should always probe all packing materials for hidden parts and accessories. Foam-type packing materials should be compressed by hand to ensure that no hidden storage compartments exist. The "peanut"-type packing materials should be dumped slowly into another receptacle to make sure that nothing is buried. Careful unpacking of the shipping containers and boxes will eliminate the time and paperwork required to have a manufacturer ship a missing part.

INITIAL ENCLOSURE LAYOUT

Once the hardware is received and determined free of shipping damages and/or shortages, the user can begin initial assembly of the individual components into subassemblies. Initial assembly should begin with the identification of the hardware needed for a subassembly. This hardware should be located and placed together. Depending on the system, a processor may require an external power supply, power supply cables, memory boards, remote I/O adapters and interfaces, batteries, miscellaneous cables, jumpers, connectors, and other assorted odds and ends. Each I/O rack should be placed with its allotment of I/O cards, interconnect cables, connectors, terminators, jumpers, power supplies, I/O module terminal strips, mounting screws, and labels. Once the PC hardware has been laid out according to use, it can be prepared for installation in the control enclosures.

Preparation of PC hardware for enclosure installation usually involves assembly and setup of the individual I/O racks along with setup of the processor. Drawings will have been prepared by this time to indicate I/O module locations within an I/O rack, as well as the address of each individual I/O module. If the PC system provides for keying of the individual I/O modules, these keying bands can be inserted in the proper locations of the I/O rack. Small switches, called *dual inline package* (DIP) switches, should be adjusted to the proper settings to provide correct I/O module addressing. If a remote I/O power supply is required as part of the I/O rack assembly, the manufacturer's instructions for this installation can be followed at this time. Remote I/O adapter modules may require installation in the I/O rack prior to rack installation. The I/O modules can be labeled with their proper I/O point addresses as well as with any mnemonic descriptions added where necessary.

The majority of processors requiring battery backup will not be shipped with the batteries installed. The batteries should be checked to ensure that they are fresh, and then dated with the installation date. The manufacturer's recommendations on battery replacement should be consulted, and the date when the batteries require changing should be noted on a calendar as a future reminder. If the processor memory is shipped separate from the processor, the detailed memory installation instructions should be followed. Any switches or jumpers that require attention should be taken care of at this time.

Once the PC hardware has been prepared, the support hardware for the control system and the PC can be prepared. Practically all installations require terminal blocks, fuse blocks, external relays, and operator devices such as pilot lights and push buttons to be installed in conjunction with the PC hardware. The terminal blocks required for the installation can be assembled as required in preparation for installation. Push-button switches can be assembled with the proper contact modules, operators, and legends according to the electrical schematics. Relays used for emergency stop or special control can be assembled and made ready for installation. The proper overload devices can be installed in motor starter overload relays.

Once the individual components are ready for installation in the control enclosures, a rough enclosure assembly can take place. This operation involves placement of the individual system components in relative relationship to one another according to the component layout drawings. At this time the backplate of the enclosure can be removed and placed horizontally so that the various control and PC devices can be placed on it. Special attention should be paid to the manufacturer's recommendations concerning spacing of various PC components with respect to proper separation from high-level voltage sources. Sufficient clearances to permit convection cooling of the PC hardware should be ensured, in conformity with the manufacturer's recommendations. The user must bear in mind that many of the manufacturer supplied cables are of specific length. It may be advantageous to place them temporarily in position to ensure sufficient cable length for later hookup. Many processors have receptacles for the connection of programming terminals and other peripherals. It is advisable to connect the devices to these receptacles to ensure proper plug and receptacle clearances, especially when the enclosure door is closed. PC components that are designed to "hinge open" for access to internal components should be tested to ensure that no obstructions exist which might later hinder servicing.

ENCLOSURE FABRICATION

Once the user is sure that the arrangement of PC hardware and control components is correct and meets his or her needs, the process of mounting each individual piece can begin. Before removal of the components from the backplate, each piece needs to be marked for placement. Many PC manufacturers provide templates for their PC components which mount in an enclosure. These templates can be taped to the mounting surface, and the necessary mounting holes drilled and tapped as specified.

Once the necessary mounting holes have been drilled, actual installation of the components begins. Under some conditions it may be desirable to mount the backplate in the enclosure prior to the actual component mounting. Many PC assemblers prefer to mount and wire as many of the control system components on the backplate as possible, before the backplate is mounted in an enclosure. No matter what the preference, all PC components and support hardware should be mounted prior to system wiring.

Once all components have been mounted, component identification should be done according to the electrical schematics and cabinet arrangement drawings. All terminal blocks installed in the enclosure need to be labeled as well as all fuse blocks. Disconnect switches which interlock with the panel door need to be checked and adjusted as required. Any PC component having access doors or hinged parts needs to be rechecked for proper clearances and free access.

Once the user is satisfied with the physical mounting of components, the enclosure wiring can begin. Usually the easiest place to start is with the manufacturer supplied cables. In most cases it is simply a matter of plugging each cable to the proper plug and/or jack of the PC hardware. These cables should be secured as necessary to prevent them from falling loose in the cabinet. As each individual wire is installed in the enclosure, the electrical schematic should be high-lighted with a marker to indicate that the wire has been installed. All cabinet wiring should have a nonmetallic wire identification band on each end for later ease in checking, troubleshooting, and maintenance. Proper attention must be paid to the manufacturer's requirements and specifications concerning the installation and routing of wiring within a PC enclosure. In order to provide a neat appearance, all wiring should be bundled and properly fastened within the enclosure.

After completion of the wiring, special care should be taken to inspect for loose bits of wire and/or metal which may have fallen into the enclosure components during assembly and installation. It might be helpful to use an air jet to dislodge any loose materials before vacuuming the enclosure. Any remaining labels or identification should be applied, and all screws, nuts, and similar fasteners should be checked for tightness. Once the user is satisfied with the enclosure assembly, actual checkout of the internal hardware and components can begin.

INITIAL SYSTEM CHECKOUT

Prior to powering up the enclosure in order to check the operation of the internal electric systems, a short-circuit test should be performed on all raw power sources used inside the enclosure. This test consists of connecting an ohm meter across each pair of line and neutral input power terminals to the enclosure with no external power sources connected. While referring to the electrical schematics on the system, each fuse and/or circuit breaker in the circuit under examination is closed. The readings on the ohm meter should indicate some finite resistance and not a short circuit. Should a short circuit be indicated, its cause should be found and corrected. This operation should be continued for all supply circuits in the system to avoid possible problems when actual power is applied. It may be advantageous to trace out more complex circuits for incorrect wiring and short circuits.

At this time sources of power can be connected to the enclosure as required, paying particular attention to the voltage, current, phase, and frequency requirements specified in the schematics. The power sources should employ overload and short-circuit protection fuses or circuit breakers that are sized to handle the loads of the enclosure under test conditions, and not the field devices which may be connected when the unit is finally installed.

Many engineers install a power-rated contractor in the test supply circuit for emergency power cutoff purposes. In the area of testing, an overhead pull-cord system is installed with cords running parallel on 6-foot centers. Should a problem develop during any test phase, one of the cords can be pulled to deenergize instantaneously all power sources to the system under test. In spite of the extra expense, this type of precaution can more than pay for itself the first time sparks begin to fly, or a checkout team member occasionally connects with a live circuit.

Actual checkout of the control system components begins with the power-up of each individual power circuit. It is perhaps best to start with the general-purpose lighting and power circuit, followed by the PC processor and power supply circuits. The last circuits to be checked would be the various PC input and output circuits. The exact order, and the precautions taken during the power-up and checkout of the individual circuits, must be determined by the engineers and designers of the system. This procedure should avoid any damage to the PC hardware and the devices associated with the control system.

When the PC processor is first energized, a thorough check should be performed to ensure that it is totally operational. Often a processor will display a processor fault indication, or the fact that it is not scanning. Every PC manufacturer includes detailed instructions covering the initial startup of the processor. These instructions usually call for the connection of the programming device to the processor, along with a detailed set of instructions for activating the processor.

Before the I/O circuits are powered, the processor's memory should be cleared, and all real outputs for the system should be checked to ensure that the coil instructions are OFF. This operation will ensure that the subsequent testing of the I/O devices will only affect the input or output circuit being tested, and should not cause operation of any other circuit unless they are interconnected intentionally.

Checkout of each input and output circuit can be done quickly and effectively with the use of a programming device, such as a CRT programmer. Since most of the I/O devices will not be connected to the enclosure, it will be necessary to simulate an input condition in order check the operation of the input circuitry, input module, and I/O system. Each input should be powered and the programming terminal referenced to the address of the input under test in order to check the actual operation of the circuit.

A safe method to energize each input circuit involves the use of a standard light bulb connected in series with the hot test probe. Should the hot test probe accidentally touch the enclosure or a neutral or ground terminal in the enclosure, the bulb will light up instead of causing a short circuit and fireworks. Since PC input points are high-impedance loads, there should be sufficient voltage to activate the input point while only dimly lighting the lamp. If a short circuit exists, the light will shine brightly, indicating the short-circuited connection. Any input point that is activated should be so indicated on the programming terminal. Should any input point not operate correctly, the cause should be found and corrected.

Testing of the output circuits will require use of the forcing feature available with most PC systems. An indicator lamp is connected to the output point being tested. The output point is then addressed on the programming terminal and toggled ON and OFF via the force function. If the lamp operates according to the commands of the programmer, the circuit is operational. Should the circuit fail to operate correctly, the cause must be found and corrected. If the PC system does not include forcing capability, a rung of logic must be programmed for the output under test. An input point wired to a push-button switch may be used as the control (through programming) for the output.

Once all circuits for the system have been checked and determined operational, the enclosure can be readied for shipment to the plant or end user. All test power sources must be disconnected, the enclosure circuit breakers switched off, and any fuses installed for the testing of the enclosure removed. Any finishing touches can be completed, and the enclosure(s) can be palletized for shipment. It is recommended that the enclosure(s) be wrapped in plastic to seal out any moisture. Several companies produce small clear plastic impact-detection devices which can be applied to the enclosure to detect sudden impacts during shipping and handling. These devices consist of ball bearings spring-loaded against internal stops on the sides. There are usually two or more spring-loaded bearings in a single unit, each spring/bearing combination oriented on a different axis to detect impacts from multiple directions. The units come calibrated for different impact forces, and the manufacturer of the device usually supplies data sheets recommending the proper calibrations to use for various enclosures. These devices are one-time devices and cannot be reused once tripped. Any enclosure arriving at a facility with one of the impact indicators tripped should be thoroughly inspected for damage prior to installation. Moisture absorbtion packets should be placed inside each enclosure and cabinet to prevent rust and oxidation from forming.

If the enclosure is to be stored for a short time prior to actual installation, it should be protected from moisture and extremes in temperature. This may mean that a small heating device may need to be installed in the enclosure if it is to be stored in an unheated area subject to condensation. A small screw-in heating element placed in a porcelain lamp socket will usually work well. The socket should be firmly attached to a strong magnet and placed in the bottom of the enclosure. Care should be taken to ensure that the heat rising from the unit does not damage any wiring insulation or other enclosure components. The addition of a small thermostat to control the heater will save energy and ensure that the enclosure does not become excessively hot.

CHECKOUT OF THE SYSTEM AT THE PLANT

Installation and checkout of a PC-based control system at a plant requires the implementation of a well thought out step-by-step procedure. While the actual hardware installation of the control system will vary from application to application, the steps followed during checkout remain basically similar. Any system being installed will usually have experienced electricians, mechanics, and related craftspeople performing the actual installation according to the specifications, drawings, and details of the job. The actual checkout and startup of the system should involve the guidance of the designers and engineers who developed it. This ensures that all components and hardware have been installed correctly and according to the drawings and specifications.

Unless the system designers and engineers jointly develop a procedure for the checkout phase, items may be overlooked which could cause damage to the controlled equipment or operating personnel. While the individual steps of a checkout and startup procedure will vary between systems, the general format and sequence of items checked will usually remain the same.

Once the installation phases of a project are complete, the installed equipment must be checked for proper mechanical and physical installation. This first step in the checkout of a system can usually be completed in a short period of time. At this point in the checkout a "punch list" should be started to keep track of each item that needs attention, whose responsibility the corrective action falls on, and when the item was finally corrected and reinspected as satisfactory. While this inspection does not uncover faulty equipment operation, it will usually uncover a quantity of missing fasteners, loose or missing guards, covers, sight gauges, empty oil reservoirs, and other minor items.

Once all parties associated with the design of the system are satisfied that the installation is correct, phase 2 of the checkout can commence. This phase involves the checking of all supply circuits for proper wiring. With the supply circuits disconnected at their source, each circuit should be checked for short circuits and incorrect wiring terminations and/or practices. Once all power circuits have been verified as installed correctly, they can be reconnected in preparation for power-up of the electrical hardware.

The first component to be energized will most likely be the PC processor. Checkout of the processor will usually follow the same procedural steps that were implemented when the unit was initially checked out before shipment to the site. No other power sources except those necessary for the processor and remote power supplies should be energized until the processor is fully operational with a cleared memory. It is strongly recommended that the processor's memory be cleared for the initial checkout steps to prevent the processor from reacting to test signal input conditions.

Phase 3 of the checkout involves energizing each input device circuit and checking for the proper operation of each connected input device. As was required when the system was checked in the fabrication facility, the CRT programmer will be used to ensure the correct operation of the input device from the physical device to the PC processor. Each input device must be activated and monitored on the CRT programmer for correct operation. The electrical schematics should be reviewed for any special device operation notes or instructions. Each input device should be adjusted for correct operation if at all possible. The I/O card indicators can be used in conjunction with the programming device to troubleshoot and correct any faulty input devices and/or wiring.

Phase 4 of the checkout will energize all output devices to ensure correct operation. Again, each output power circuit will need to be energized and the programming terminal used to toggle the output point ON and OFF. Special attention should be paid to dual operation type output devices, such as reversing motor starters and solenoid valves, to ensure correct device functions. Output devices which cause motion should first be checked with the source of motive force (air, hydraulic, motor power, etc.) locked OFF. This practice will ensure that the control device is operating properly before actual motion is permitted. Once the control device is proven operational, the source of motive force can be applied to check out further the output device's operation. Motors can be checked for proper rotation, solenoids can be checked for proper piping and flow control, and the mechanical hardware operated by the control system can be checked for correct operation according to any checkout and adjustment procedures prepared by the mechanical designers and engineers.

Note that it may be necessary to program the processor temporarily with a test routine to facilitate the operational checkout of the mechanical hardware. Before completing the output device and mechanical checkout, the input devices should be rechecked for proper operation and adjustment. All equipment operated by the PC system should be fine-tuned as much as possible prior to the actual loading and operation of the PC program.

The final phase of checkout will involve operation of the system under the control of the PC, using simulated loads and materials where possible. The processor program should be loaded and verified. Following a highly detailed procedure written by the control system engineers and designers, in conjunction with the facility operating personnel, the system should be operated. Special attention should be paid to ensure not only that the system operates correctly, but also that it operates and fails correctly when an abnormal condition or operation occurs.

While many people place a lot of emphasis on the proper operation of a system under normal operating conditions, the real test of an experienced PC systems programmer lies in his or her ability to anticipate any normal system fault as well as the most unusual system faults, and to program the PC to react to them correctly. In short, the programmer must be able to program the PC not only for correct normal operation, but also for correct incorrect operations. It is much better to create system faults at this time with no product present, than to generate a disaster later when the plant management is watching. It cannot be emphasized too strongly that this is the time to test the system thoroughly, and not later under production operation. There will always be enough unexpected problems to keep one busy during actual production without the need to solve those that could have been prevented by a thorough checkout.

Once the system designers, engineers, and associated personnel are satisfied with the operation of the system and its components, the actual product can be introduced and the control system brought on line. Exactly how this is done will be determined solely by the application. The process of bringing the new installation on line will usually mean starting out slowly at first. Once the system is operating, it can then be fine-tuned and sped up to production-rate operation. There will undoubtedly be problems which require immediate attention. These will have to be solved as they occur. Once the system is on line and operational for several days, the control system personnel can be reasonably sure that the system will operate correctly and that the PC programming is correct. The remaining job will be to complete the documentation for the system and train any plant operating personnel on the operation of the control system.

MAINTENANCE

All PCs are designed to minimize the amount of time spent in trouble-shooting problems associated with the controller itself or a portion of the control system. Usually no special test equipment is required outside of a volt meter and standard screw drivers, pliers, and possibly a wrench or two. Every PC provides status and diagnostic indicators to aid in the isolation and diagnosis of a fault, either in the PC system itself or in the user's hardware. While these indicators may not directly identify the source of a problem, they do allow the user to narrow down quickly the many possibilities that can exist.

Probably the biggest deterrent to system faults is a proper preventive maintenance program for the PC and the control system. The periodic inspection of terminal strip connections, processor batteries, plugs, sockets, and similar electrical hardware during periods of reduced system use will uncover items that, if left unattended, could cause later system faults. All field I/O devices should receive a periodic inspection to ensure that they are properly adjusted and not being damaged. Many control systems operate processes that must be shut down for short periods due to product changes. Maintenance items can often be logged for attention during these short shut-down periods.

Every PC is composed of electric circuit boards that are vulnerable to dust and dirt. While most manufacturers place these boards inside a metal or plastic enclosure or module, the boards still require some attention. All electronic equipment operates best in a clean environment, and the PC is no exception. While the PC is designed to tolerate a much higher degree of contamination than a computer, excessive buildups of dirt and dust can be detrimental. The control enclosure should be kept clean, and any dust that has built up should be removed. Cabinets containing fans and blowers for air circulation purposes should have those devices, and the screens and filters associated with them, cleaned periodically. Ensuring that the enclosure door is kept closed will prevent the rapid buildup of these contaminants. Plant environments that are fogged with oily mists are especially prone to rapid dust and dirt buildups, and PCs installed in these areas should be cleaned more frequently. Every PC manufacturer issues guidelines on the care of the system. These guidelines should be understood and a maintenance program developed for proper maintenance of the PC-based control system.

SPARE PARTS

It is surprising to see the number of businesses that spend a large portion of their profits on expansion and facility upgrading, but neglect to commit a small portion of that money to spare parts and training. They spend hundreds of thousands of dollars for machinery and controls, then elect not to spend several additional percent of that investment to train their maintenance personnel properly, and stock spare parts.

Every manufacturer provides a recommended list of spare parts, whether it be a machine-tool or a PC manufacturer. No matter how quick and experienced the people are in maintaining a system, they cannot repair a problem without the proper spare parts. Even in spite of the recommendations of the manufacturer, many companies lose thousands of dollars in production time due to the lack of availability of a replacement PC component worth only several hundred dollars.

Many managers insist that their engineers incorporate an on-line backup PC in their control systems to take over automatically in the event the primary PC faults. However, they do not approve the purchase of a third spare processor, as they are living under the false security of the backup unit. Since a faulted primary processor may need to be exchanged for another unit, what does the plant manager do if an accident or a component failure faults the backup processor before the primary unit can be returned to service? This sounds like a remote possibility, but it has happened numerous times in the past. The real clinker is to have the primary processor fail, only to find that the backup unit has been bad for some time. There has been no routine inspection of the system, a failure alarm was not installed, and the faulted backup unit went undetected.

An extremely efficient use of spare parts is training of the personnel who must maintain the system. Instead of storing the spares for a PC system in their boxes in some dark storage area, they may be installed in an enclosure or simulator for use by the plant maintenance personnel. They can connect lights, switches, thumbwheels, and displays to the I/O modules. Many PC manufacturers provide video cassettes and programmed learning courses on such topics as PC programming, PC installation, and PC maintenance. Scheduling maintenance personnel as well as the engineers and designers inexperienced with the operation of a PC not only educates these people, but also helps them become more effective in their jobs. The experienced PC users on the site should also make use of this equipment periodically in order to keep their knowledge of the plant's PCs sharp.

As a side use for the spare equipment, alternate programs can be developed and simulated to improve the efficiency of existing operating programs. This equipment would also lend itself to the troubleshooting of existing programs. An indirect benefit of having spare PC hardware available lies in the fact that the spare parts are not lying on the shelf but are actually being used. Management spends money for items that serve multiple purposes, and the maintenance department is assured that the spare parts on hand are operational in case of actual need.

TROUBLESHOOTING

In the event of a PC fault, a careful systematic approach should be employed when troubleshooting the system to resolve a malfunction. The objective of the troubleshooting should be to *solve* a problem and not to *fix* the problem. While the operations of solving a problem versus fixing it sound the same, they are actually quite different. The act of solving a problem involves proper diagnosis of the cause, correction of the cause, and then repair or replacement of any damaged or faulty components. Any fault can be "fixed" by simply replacing the damaged or malfunctioning parts with operational ones. This type of temporary fix eventually leads to more involved failures, resulting in extended downtimes and increased service costs.

As an example of the difference between fixing and solving a problem, consider a motor that faults periodically due to the overload devices tripping. The fix for the problem is simply to reset the overloads or replace them with a higher rated unit. The solution to the problem involves examination of the motor and its mechanical load for an actual overload condition that occurs occasionally. The overload devices should be examined to see whether they were improperly selected for the application, or whether they are in fact bad. In order to keep the facility in production the problem may require temporary fixing, but unless the problem is actually solved, it will never correct itself.

When attempting to solve a problem associated with a PC directly, the manufacturer's guidelines should be consulted. Many manufacturers publish these guidelines in the form of *troubleshooting flowcharts*. They usually offer a step-by-step procedure for examining the fault indicators of the processor to determine the nature of the problem. Once the problem is located, additional flowcharts are usually provided along with detailed instructions on how to correct the problem.

The PC user should be aware that most control system problems will concern the field I/O devices and the hardware they operate. When a problem occurs, a common tendency is to examine the processor program for the cause of the malfunction. While this operation is usually the fastest way to locate a problem, many PC users then change the program to correct the problem. There is usually no need to change a PC program that has been in operation for many months or even years. The malfunction is probably caused by a faulty field device, and it is this device that must be located and repaired. The only programming that *may* be necessary is that needed to "trap" the problem.

19

Selected Programming Applications and Routines

INTRODUCTION

The programming applications and routines listed in this chapter have been combined from various sources. The routines should give the reader a flavor for some of the ways a particular application can be solved, as well as a feel for the power of the relay ladder programming language.

The programs illustrated in Figs. 19.1 to 19.15 are not necessarily the shortest, fastest, best, or only way to perform the required programming. The reader is invited to explore other methods, as well as build upon those cited in this text. The routines are listed in a generic nature, and as such will require the reader to "convert" them into the actual commands and instructions of the PC he or she is using. The conversion of these programs, as well as of those listed elsewhere in the text and in the manuals of various PC manufacturers, should provide excellent exercises for the reader to become acquainted with a particular PC model or models.

OPERATION :

1) SOME PC PROGRAMS, OR SYSTEMS CONTROLLED BY A PC, MAY REQUIRE
   CERTAIN CONDITIONS OR ACTIONS TO TAKE PLACE, OR BE SET OR CLEARED,
   WHEN A PROCESSOR IS STARTED. THE EXAMPLE LADDER RUNG SHOULD BE
   PROGRAMMED AS THE LAST RUNG OF THE USER PROGRAM.

2) ANY ACTION OR EVENT THAT REQUIRES AN OPERATION ON PROCESSOR
   POWER-UP, SHOULD BE INITIATED BY A NORMALLY CLOSED CONTACT
   INSTRUCTION REFERENCED TO COIL INSTRUCTION 'A'.

3) COIL INSTRUCTION "A" WILL REMAIN OFF FOR THE FIRST PROCESSOR
   SCAN. AT THE END OF THE FIRST PROCESSOR SCAN, COIL INSTRUCTION
   "A" WILL BECOME ENERGIZED AND REMAIN ENERGIZED UNTIL THE NEXT
   TIME THE PROCESSOR LOSES POWER.

4) COIL INSTRUCTION "A" MUST NOT BE REFERENCED TO A PROCESSOR
   MEMORY LOCATION THAT RETAINS ITS STATE THRU THE POWER CYCLING
   OF THE PROCESSOR.

Fig. 19.1. First Scan Indicator

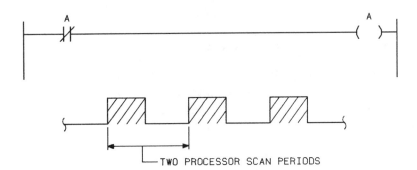

TWO PROCESSOR SCAN PERIODS

OPERATION :

1) ON ODD PROCESSOR SCANS, COIL INSTRUCTION 'A" IS ENERGIZED.
   THIS CAUSES THE COIL INSTRUCTION TO BECOME DE-ENERGIZED ON EVEN
   PROCESSOR SCANS. THE TIME REQUIRED TO MAKE A COMPLETE ON-OFF
   CYCLE IS TWO PROCESSOR SCANS.

2) NOTE THAT SINCE MOST PROCESSOR'S SCAN TIMES VARY DUE TO THE
   AMOUNT AND COMPLEXITY OF USER LOGIC WHICH MUST BE SOLVED DURING
   ANY GIVEN SCAN, THEREFORE THE PERIOD OF THE OSCILLATOR WILL VARY
   OVER SOME FINITE AMOUNT OF TIME.

Fig. 19.2. Scan Oscillator

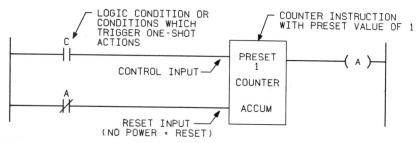

OPERATION :

1) WHEN LOGIC CONDITIONS TRANSITION FROM OFF TO ON, A ONE-SHOT ACTION IS INITIALIZED AND COUNTER COUNTS FROM 'Ø' TO '1'.

2) SINCE COUNTER ACCUMULATED AND PRESET VALUES EQUAL '1', COIL INSTRUCTION 'A' IS ENERGIZED.

3) ON NEXT PROCESSOR SCAN, COUNTER IS RESET BACK TO 'Ø', AND COIL INSTRUCTION 'A' IS DE-ENERGIZED, BECAUSE A NORMALLY CLOSED CONTACT INSTRUCTION REFERENCED TO COIL INSTRUCTION 'A' HAS BEEN PROGRAMMED AS THE COUNTER RESET INPUT CONDITION.

4) SINCE THE COUNTER CONTROL INPUT MUST CYCLE OFF TO ON FOR THE COUNTER TO INCREMENT, COUNTER AWAITS NEXT OFF-TO-ON CYCLE OF TRIGGERING LOGIC CONDITIONS TO INITIATE ANOTHER ONE-SHOT ACTION.

5) ALL ONE-SHOT CONTACT INSTRUCTIONS SHOULD BE REFERENCED TO COIL INSTRUCTION 'A'.

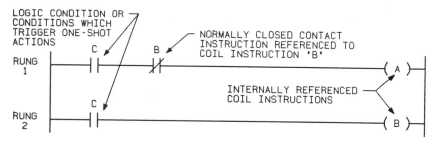

OPERATION :

1) WHEN LOGIC CONDITIONS ARE TRUE FOR ONE-SHOT ACTION TO OCCUR, COIL INSTRUCTION 'A' IS ENERGIZED.

2) AS LONG AS LOGIC CONDITIONS FOR ONE-SHOT ACTION ARE TRUE, COIL INSTRUCTION 'B' IS ENERGIZED.

3) COIL INSTRUCTION 'A' IS ENERGIZED FOR THE SCAN IN WHICH THE TRIGGER LOGIC CONDITIONS ARE TRUE. NORMALLY CLOSED CONTACT INSTRUCTION 'B' IN RUNG 1 INSURES COIL INSTRUCTION 'A' REMAINS DE-ENERGIZED FOR ALL REMAINING SCANS THAT THE TRIGGER CONDITION REMAINS TRUE.

4) ALL ONE-SHOT CONTACT INSTRUCTIONS SHOULD BE REFERENCED TO COIL INSTRUCTION 'A'.

Fig. 19.3. One-Shot or Transitional Contact Circuits

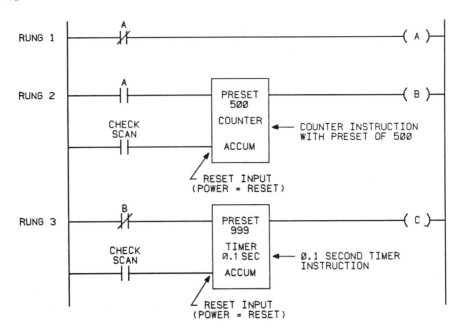

OPERATION :

1) RUNG 1 FORMS A FREE-RUNNING OSCILLATOR COMPLETING A CYCLE EVERY TWO PROCESSOR SCANS.

2) RUNG 2 COUNTS THE NUMBER OF OSCILLATOR CYCLES. EVERY TWO PROCESSOR SCANS EQUALS ONE COUNT ON THE COUNTER.

3) COIL INSTRUCTION 'B' OF RUNG 2 BECOMES ENERGIZED WHEN 500 COUNTS ARE ACCUMULATED ON COUNTER, INDICATING 1000 PROCESSOR SCANS HAVE OCCURRED.

4) RUNG 3 TABULATES THE AMOUNT OF TIME REQUIRED TO ACCUMULATE 1000 PROCESSOR SCANS. THE TIMER INSTRUCTION IS PROGRAMMED WITH A TIME BASE OF 0.1 SECONDS AND IS INHIBITED FROM TIMING WHEN THE COUNTER INSTRUCTION OF RUNG 2 RECORDS 500 COUNTS OR 1000 PROCESSOR SCANS.

5) A 'CHECK SCAN' CONTACT INSTRUCTION IS USED TO RESET BOTH THE COUNTER AND TIMER TO ZERO TO INITIATE   CALCULATION OF THE AVERAGE PROCESSOR SCAN TIME.

6) THE ACCUMULATED VALUE OF THE TIMER INSTRUCTION IN RUNG 3 INDICATES THE AVERAGE SCAN TIME OF THE PROCESSOR IN MILLISECONDS.

Fig. 19.4.   Scan Counter

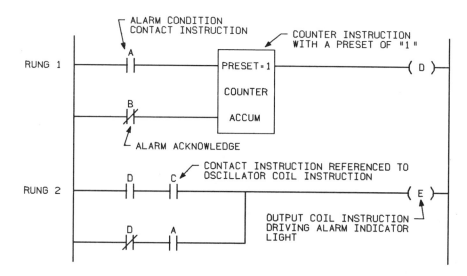

CONTACT FUNCTIONS :

1) 'A' IS ALARM POINT OR CONDITION INPUT

2) 'B' IS ALARM ACKNOWLEDGE INPUT

3) 'C' IS FLASHER CONTACT FROM ONE SECOND ON/
ONE SECOND OFF TIMED OSCILLATOR

4) 'D' IS INTERNALLY REFERENCED COIL INSTRUCTION

5) COIL INSTRUCTION "E" OPERATES ALARM INDICATOR

OPERATION :

1) COUNTER INSTRUCTION IN RUNG 1 HAS A PRESET OF '1'. COIL
INSTRUCTION "D" IS ENERGIZED WHENEVER ALARM CONDITION OCCURS
AND HAS NOT BEEN ACKNOWLEDGED EVEN IF ALARM CONDITION CLEARS
IN MEAN TIME.

2) RUNG 2 OPERATES ALARM INDICATOR IN FLASHING MODE THRU NORMALLY
OPEN CONTACT INSTRUCTIONS "D" AND 'C'.

3) RUNG 2 OPERATES ALARM INDICATOR IN STEADY "ON" MODE WHEN ALARM
CONDITION STILL EXISTS BUT HAS BEEN ACKNOWLEDGED, THRU NORMALLY
CLOSED CONTACT INSTRUCTION 'D' AND NORMALLY OPEN CONTACT
INSTRUCTION "A".

Fig. 19.5.   Alarm Point Monitor

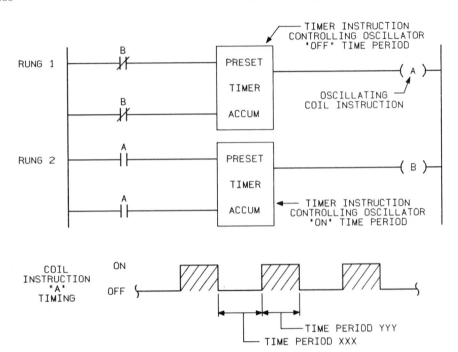

OPERATION :

1) RUNG 1'S TIMER INSTRUCTION TIMES FOR THE AMOUNT OF TIME PRESET. THIS TIME PERIOD REPRESENTS THE 'OFF' TIME FOR A COMPLETE OSCILLATOR CYCLE.

2) WHEN RUNG 1'S TIMER INSTRUCTION COMPLETES ITS TIMING CYCLE, RUNG 2'S TIMER IS ACTIVATED TO TIME FOR ITS PRESET TIME THRU THE NORMALLY OPEN CONTACT INSTRUCTION "A'.

3) RUNG 2'S TIMER PRESET VALUE REPRESENTS THE 'ON' TIME FOR A COMPLETE OSCILLATOR CYCLE. WHEN RUNG 2'S TIMER COMPLETES ITS TIMING CYCLE, COIL INSTRUCTION 'B' IS ACTIVATED.

4) WHEN COIL INSTRUCTION "B' IS ENERGIZED, THE TIMER INSTRUCTION OF RUNG 1 IS RESET. THE RESETTING OF THE TIMER IN RUNG 1 RESETS THE TIMER OF RUNG 2 SO THE CYCLE CAN REPEAT.

5) COIL INSTRUCTION "A' OSCILLATES AT THE RATE SET BY THE PRESET VALUES OF RUNG 1 AND RUNG 2'S TIMERS.

Fig. 19.6. Timing Oscillator

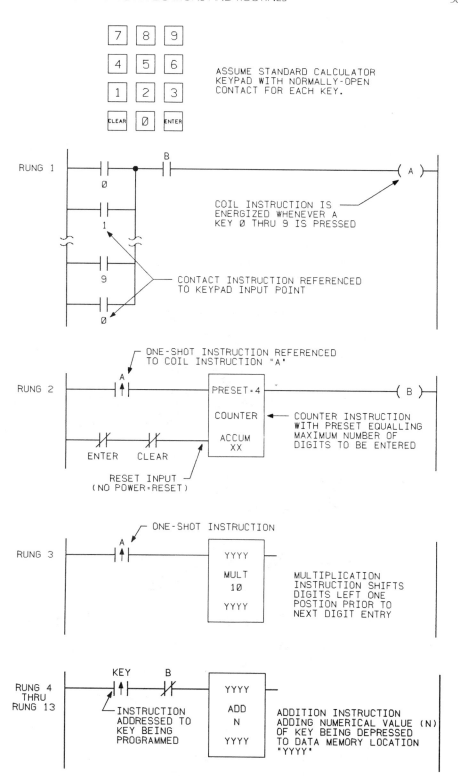

Fig. 19.7. Numeric Keypad Entry Program

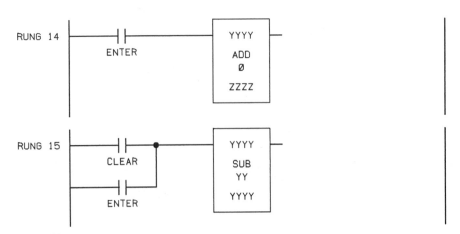

OPERATION :

1) RUNG 1 DETECTS WHENEVER A NUMERICAL KEY HAS BEEN DEPRESSED.

2) RUNG 2 KEEPS TRACK OF THE NUMBER OF KEY DEPRESSIONS. THE PRESET
   VALUE SELECTED FOR THE COUNTER DETERMINES THE MAXIMUM NUMBER OF
   DIGITS WHICH CAN BE ENTERED.

3) THE NUMBER BEING ENTERED WILL BE ASSEMBLED IN DATA MEMORY LOCATION
   "YYYY". WHEN THE 'ENTER' KEY IS DEPRESSED, THE INPUTTED NUMBER
   WILL BE TRANSFERRED FROM DATA MEMORY LOCATION 'YYYY' TO "ZZZZ".

4) RUNG 3 MULTIPLIES THE NUMBER CURRENTLY IN DATA MEMORY LOCATION
   "YYYY" BY TEN, EFFECTIVELY MOVING ITS DIGITS ONE PLACE TO THE
   LEFT, PRIOR TO THE ENTRY OF A NEW DIGIT (UNITS DIGIT).

5) RUNGS 4 THRU 13 ENTER THE NUMERICAL VALUE OF THE DEPRESSED KEY
   INTO DATA MEMORY LOCATION 'YYYY'. THE ADDITION INSTRUCTION IS
   PROGRAMMED TO ADD THE NUMERICAL VALUE OF THE DEPRESSED KEY, TO
   THE CURRENT VALUE ALREADY IN LOCATION 'YYYY'. THE ADDITION
   INSTRUCTION IS TRIGGERED BY A ONE-SHOT INSTRUCTION REFERENCED TO
   THE KEY REPRESENTING THE NUMBER BEING ADDED. THE NORMALLY-CLOSED
   CONTACT INSTRUCTION REFERENCED TO COIL INSTRUCTION 'B' INHIBITS
   ADDITIONAL DIGITS FROM BEING ENTERED ONCE THE LIMIT SET BY THE
   COUNTER INSTRUCTION IN RUNG 2 HAS BEEN REACHED.

6) RUNG 14 MOVES THE ENTERED NUMBER FROM DATA MEMORY LOCATION "YYYY'
   TO 'ZZZZ' WHEN THE "ENTER" KEY IS DEPRESSED.

7) RUNG 15 CLEARS DATA MEMORY LOCATION 'YYYY' WHENEVER THE 'ENTER' OR
   "CLEAR" KEYS ARE DEPRESSED.

8) THE PROGRAMMING ORDER SHOWN IS CRITICAL FOR THE CIRCUIT'S OPERATION.

9) IF A MATRIX CONTACT TYPE KEYPAD IS USED, THE PROGRAM INSTRUCTIONS
   OF RUNGS 1, AND 4 THRU 13, MUST BE CHANGED ACCORDINGLY.

10) DATA MEMORY LOCATION 'YYYY" MAY BE ASSIGNED AS AN OUTPUT REGISTER
    OR WORD SO THAT A NUMERIC DISPLAY MAY BE USED TO VIEW THE NUMBER
    AS IT IS ENTERED.

11) IF SCROLLING OF DIGITS IS DESIRED, I.E. CONTINUOUS SHIFTING TO
    THE LEFT WITHOUT A MAXIMUM DIGIT COUNT PRIOR TO ENTRY, THE LOGIC
    OF RUNG 2 CAN BE DELETED.

Fig. 19.7. *(Continued)*

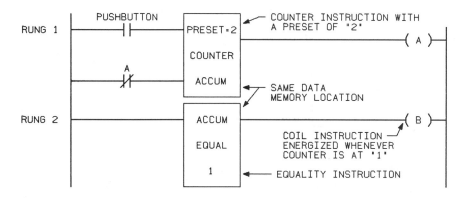

OPERATION :

1) COUNTER INSTRUCTION OF RUNG 1 COUNTS EACH PUSHBUTTON OPERATIONS. THE COUNTER RESETS TO ZERO VIA COIL AND NORMALLY CLOSED INSTRUCTIONS REFERENCED AS 'A'.

2) EQUALITY INSTRUCTION OF RUNG 2 ACTIVATES COIL INSTRUCTION 'B' WHENEVER COUNTER ACCUMULATED VALUE EQUALS '1'.

3) ON FIRST DEPRESSION OF PUSHBUTTON, COUNTER COUNTS TO '1' AND COIL INSTRUCTION 'B' IS ENERGIZED.

4) ON SECOND DEPRESSION OF PUSHBUTTON, COUNTER COUNTS TO '2' AND IS IMMEDIATELY RESET BACK TO '0'. COIL INSTRUCTION 'B' IS DE-ENERGIZED.

Fig. 19.8.   Alternate-Action Pushbutton

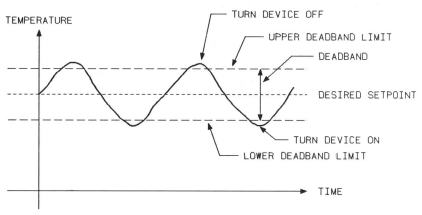

DESIGNATE THREE DATA MEMORY STORAGE LOCATIONS:

1) R1 = MEMORY LOCATION WHICH IS UPDATED WITH CURRENT ANALOG INPUT VALUE
2) R2 = DESIRED SYSTEM SETPOINT
3) R3 = DEADBAND VALUE; PERCENT ABOVE OR BELOW SETPOINT WHERE NO CONTROL ACTION OCCURS
4) R4/R5 = MEMORY LOCATION PAIR STORING RESULT OF SETPOINT DEADBAND MULTIPLICATION
5) R6/R7 = MEMORY LOCATION PAIR STORING DEADBAND DEVIATION ABOVE/BELOW SETPOINT
6) 68 = UPPER DEADBAND LIMIT
7) R9 = LOWER DEADBAND LIMIT

Fig. 19.9.   ON/OFF Analog Control

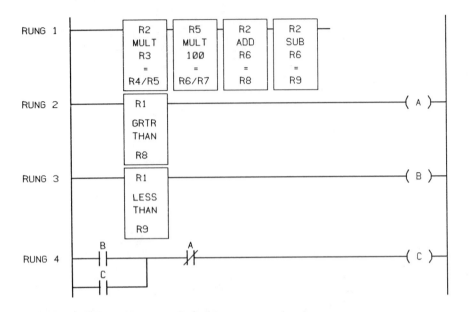

OPERATION:

1) RUNG 1 PERFORMS TWO MULTIPLICATIONS, AN ADDITION, AND A SUBTRACTION TO CALCULATE THE UPPER AND LOWER DEADBAND LIMITS ABOUT THE SETPOINT

2) RUNG 2 ENERGIZES COIL INSTRUCTION 'A" WHENEVER THE INPUT EXCEEDS THE UPPER DEADBAND LIMIT

3) RUNG 3 ENERGIZES COIL INSTRUCTION 'B' WHENEVER THE INPUT IS LESS THAN THE LOWER DEADBAND LIMIT

4) RUNG 4 TOGGLES ON AND OFF KEEPING THE ANALOG INPUT WITHIN THE DEADBAND LIMITS

5) THE CALCULATIONS ASSUME FOUR DIGIT MEMORY STORAGE LOCATIONS. FOR THREE DIGIT MEMORY LOCATIONS, CHANGE THE SECOND MULTIPLICATION OPERATION MULTIPLIER FROM 100 TO 10.

Fig. 19.9. *(Continued)*

```
        [AAAA][BBBB]
      + [CCCC][DDDD]         [----] = DATA MEMORY
      ──────────────────            STORAGE LOCATION
      [EEEE][FFFF][GGGG]
```

OPERATION :

   IN LIEU OF ACTUAL PROGRAM INSTRUCTIONS, A WRITTEN DESCRIPTION
   WILL BE PROVIDED.

   RUNG 1 :
      CLEAR LOCATIONS [EEEE], [FFFF], AND [GGGG].

   RUNG 2 :
      ADD LOCATION [BBBB] AND LOCATION [DDDD], AND STORE RESULT IN
      LOCATION [GGGG].

   RUNG 3 :
      ADD LOCATION [AAAA] AND LOCATION [CCCC], AND STORE RESULT IN
      LOCATION [FFFF].

   RUNG 4 :
      IF AN OVERFLOW OR CARRY IS GENERATED AS A RESULT OF THE ADDITION
      OPERATION IN RUNG 2, ADD ONE (1) TO LOCATION [FFFF].

   RUNG 5 :
      IF AN OVERFLOW OR CARRY IS GENERATED AS A RESULT OF THE ADDITION
      OPERATION IN EITHER RUNG 3 OR RUNG 4, ADD ONE (1) TO LOCATION
      [EEEE].

Fig. 19.10.   Double-Precision Addition

```
        [AAAA][BBBB]
      − [CCCC][DDDD]         [----] = DATA MEMORY
      ──────────────             STORAGE LOCATION
        [EEEE][FFFF]
```

OPERATION :

   IN LIEU OF ACTUAL PROGRAM INSTRUCTIONS, A WRITTEN DESCRIPTION
   WILL BE PROVIDED. FOR NEGATIVE DIFFERENCES, ROUTINE AUTOMATICALLY
   BORROWS FROM THE MILLIONS DIGIT FOR THREE DIGIT STORAGE LOCATIONS,
   AND FROM THE 100 MILLION DIGIT FOR FOUR DIGIT STORAGE LOCATIONS.

   RUNG 1 :
      SUBTRACT LOCATION [CCCC] FROM 999 FOR THREE DIGIT STORAGE
      LOCATIONS AND FROM 9999 FOR FOUR DIGIT STORAGE LOCATIONS. STORE
      THE RESULT IN LOCATION [EEEE].

   RUNG 2 :
      ADD LOCATION [EEEE] TO LOCATION [AAAA] AND STORE THE RESULT IN
      LOCATION [EEEE].

   RUNG 3 :
      SUBTRACT LOCATION [DDDD] FROM 999 FOR THREE DIGIT STORAGE
      LOCATIONS AND FROM 9999 FOR FOUR DIGIT STORAGE LOCATIONS. STORE
      THE RESULT IN LOCATION [FFFF].

   RUNG 4 :
      ADD LOCATION [FFFF] TO LOCATION [BBBB] AND STORE THE RESULT IN
      LOCATION [FFFF].

   RUNG 5 :
      ADD ONE (1) TO LOCATION [FFFF].

   RUNG 6 :
      ADD ONE (1) TO LOCATION [EEEE] IF THE ADDITION OPERATION OF
      EITHER RUNG 4 OR RUNG 5 GENERATES A CARRY OR OVERFLOW CONDITION.

Fig. 19.11.   Double-Precision Subtraction

```
            [AAAA][BBBB]
          × [CCCC][DDDD]           [----] = DATA MEMORY
    ———————————————————                     STORAGE LOCATION
    [EEEE][FFFF][GGGG][HHHH]
```

OPERATION :

    IN LIEU OF ACTUAL PROGRAM INSTRUCTIONS, A WRITTEN DESCRIPTION
WILL BE PROVIDED.   DATA MEMORY STORAGE LOCATIONS [1111] THRU [8888]
WILL BE USED FOR TEMPORARY STORAGE.

RUNG 1 :

    MULTIPLY LOCATION [BBBB] BY LOCATION [DDDD] AND STORE THE
RESULT IN DOUBLE REGISTERS [3333] AND [4444].

RUNG 2 :

    MULTIPLY LOCATION [AAAA] BY LOCATION [DDDD] AND STORE THE
RESULT IN DOUBLE REGISTERS [5555] AND [6666].

RUNG 3 :

    MULTIPLY LOCATION [AAAA] BY LOCATION [CCCC] AND STORE THE
RESULT IN DOUBLE REGISTERS [1111] AND [2222].

RUNG 4 :

    MULTIPLY LOCATION [BBBB] BY LOCATION [CCCC] AND STORE THE
RESULT IN DOUBLE REGISTERS [7777] AND [8888].

RUNG 5 :

    ADD LOCATION [2222] TO LOCATION [5555] AND STORE THE RESULT
IN LOCATION [2222].

RUNG 6 :

    IF CARRY OR OVERFLOW IS GENERATED AS A RESULT OF THE ADDITION
OPERATION IN RUNG 5, ADD ONE (1) TO THE CONTENTS OF LOCATION
[1111].

RUNG 7 :

    ADD LOCATION [2222] TO LOCATION [7777] AND STORE THE RESULT
IN LOCATION [FFFF].

RUNG 8 :

    IF CARRY OR OVERFLOW IS GENERATED AS A RESULT OF THE ADDITION
OPERATION IN RUNG 7, ADD ONE (1) TO THE CONTENTS OF LOCATION
[1111].

RUNG 9 :

    ADD LOCATION [3333] TO LOCATION [6666] AND STORE THE RESULT
IN LOCATION [3333].

RUNG 10 :

    ADD LOCATION [3333] TO LOCATION [8888] AND STORE THE RESULT
IN LOCATION [GGGG].

RUNG 11 :

    IF CARRY OR OVERFLOW IS GENERATED AS A RESULT OF THE ADDITION
OPERATION IN RUNG 9, ADD ONE (1) TO THE CONTENTS OF LOCATION
[FFFF].

RUNG 12 :

    IF CARRY OR OVERFLOW IS GENERATED AS A RESULT OF THE ADDITION
OPERATION IN RUNG 10, ADD ONE (1) TO THE CONTENTS OF LOCATION
[FFFF].

RUNG 13 :

    IF CARRY OR OVERFLOW IS GENERATED AS A RESULT OF THE ADDITION
OPERATION IN EITHER RUNG 11 OR RUNG 12, ADD ONE (1) TO THE
CONTENTS OF LOCATION [1111].

RUNG 14 :

    MOVE CONTENTS OF LOCATION [1111] TO LOCATION [EEEE].

RUNG 15 :

    MOVE CONTENTS OF LOCATION [4444] TO LOCATION [HHHH].

Fig. 19.12.   Double-Precision Multiplication

```
      [AAAA].[BBBB]
    + [CCCC].[DDDD]        [----] = DATA MEMORY
      ————————————                  STORAGE LOCATION
      [EEEE].[FFFF]
```

OPERATION :

    IN LIEU OF ACTUAL PROGRAM INSTRUCTIONS, A WRITTEN DESCRIPTION
WILL BE PROVIDED.

    RUNG 1 :

        ADD LOCATION [AAAA] TO LOCATION [CCCC] AND STORE RESULT IN
LOCATION [EEEE].

    RUNG 2 :

        ADD LOCATION [BBBB] TO LOCATION [DDDD] AND STORE RESULT IN
LOCATION [FFFF].

    RUNG 3 :

        IF AN OVERFLOW OR CARRY IS GENERATED AS A RESULT OF THE ADDITION
OPERATION IN RUNG 2, ADD ONE (1) TO LOCATION [EEEE] AND STORE
RESULT IN LOCATION [EEEE].

    RUNG 4 :

        IF AN OVERFLOW OR CARRY IS GENERATED AS A RESULT OF THE ADDITION
IN EITHER RUNG 1 OR RUNG 3, THE FINAL RESULT IS GREATER THAN
THE NUMERICAL CAPACITY OF THE PROCESSOR'S STORAGE REGISTERS.
IF THIS OCCURS, THE ANSWER IS 1[EEEE].[FFFF].   IF AN OVERFLOW OR
CARRY IS NOT GENERATED AS A RESULT OF THE ADDITIONS IN EITHER
RUNG 1 OR RUNG 3, THE ANSWER IS [EEEE].[FFFF].

Fig. 19.13.   Decimal Addition Routine

```
      [AAAA].[BBBB]
    − [CCCC].[DDDD]        [----] = DATA MEMORY
      ————————————                  STORAGE LOCATION
      [EEEE].[FFFF]
```

OPERATION :

    IN LIEU OF ACTUAL PROGRAM INSTRUCTIONS, A WRITTEN DESCRIPTION
WILL BE PROVIDED.

    RUNG 1 :

        CHECK THAT LOCATION [AAAA] IS GREATER THAN LOCATION [CCCC].
IF TRUE, PROCEED WITH RUNG 2.   IF FALSE, EXCHANGE CONTENTS OF
LOCATIONS [AAAA].[BBBB] AND [CCCC].[DDDD], AND RESULT WILL
BE NEGATIVE.

    RUNG 2 :

        SUBTRACT LOCATION [CCCC] FROM LOCATION [AAAA] AND STORE RESULT
IN LOCATION [EEEE].

    RUNG 3 :

        SUBTRACT LOCATION [DDDD] FROM LOCATION [BBBB] AND STORE RESULT
IN LOCATION [FFFF].

    RUNG 4 :

        IF AN UNDERRUN OR BORROW IS GENERATED AS A RESULT OF THE
SUBTRACTION OF RUNG 3, LOCATION [EEEE] MUST BE REDUCED BY 1
WITH AN ADDITIONAL SUBTRACTION INSTRUCTION.

    RUNG 5 :

        THE RESULT OF THE SUBTRACTION APPEARS IN LOCATIONS [EEEE].[FFFF].
A COIL INSTRUCTION CAN BE USED TO INDICATE A NEGATIVE ANSWER.

Fig. 19.14.   Decimal Subtraction Routine

```
                    [AAAA].[BBBB]
                  ×[CCCC].[DDDD]          [----] = DATA MEMORY
          [EEEE][FFFF].[GGGG][HHHH]                STORAGE LOCATION
```

OPERATION :

    IN LIEU OF ACTUAL PROGRAM INSTRUCTIONS, A WRITTEN DESCRIPTION
WILL BE PROVIDED. DATA MEMORY STORAGE LOCATION [1111] THRU [8888]
WILL BE USED FOR TEMPORARY STORAGE.

    RUNG 1 :

        MULTIPLY LOCATION [BBBB] BY LOCATION [DDDD] AND STORE THE
        RESULT IN DOUBLE REGISTERS [3333] AND [4444].

    RUNG 2 :

        MULTIPLY LOCATION [AAAA] BY LOCATION [DDDD] AND STORE THE
        RESULT IN DOUBLE REGISTERS [5555] AND [6666].

    RUNG 3 :

        MULTIPLY LOCATION [AAAA] BY LOCATION [CCCC] AND STORE THE
        RESULT IN DOUBLE REGISTERS [1111] AND [2222].

    RUNG 4 :

        MULTIPLY LOCATION [BBBB] BY LOCATION [CCCC] AND STORE THE
        RESULT IN DOUBLE REGISTERS [7777] AND [8888].

    RUNG 5 :

        ADD LOCATION [2222] TO LOCATION [5555] AND STORE THE RESULT
        IN LOCATION [2222].

    RUNG 6 :

        IF CARRY OR OVERFLOW IS GENERATED AS A RESULT OF THE ADDITION
        OPERATION IN RUNG 5, ADD ONE (1) TO THE CONTENTS OF LOCATION
        [1111].

    RUNG 7 :

        ADD LOCATION [2222] TO LOCATION [7777] AND STORE THE RESULT
        IN LOCATION [FFFF].

    RUNG 8 :

        IF CARRY OR OVERFLOW IS GENERATED AS A RESULT OF THE ADDITION
        OPERATION IN RUNG 7, ADD ONE (1) TO THE CONTENTS OF LOCATION
        [1111].

    RUNG 9 :

        ADD LOCATION [3333] TO LOCATION [6666] AND STORE THE RESULT
        IN LOCATION [3333].

    RUNG 10 :

        ADD LOCATION [3333] TO LOCATION [8888] AND STORE THE RESULT
        IN LOCATION [GGGG].

    RUNG 11 :

        IF CARRY OR OVERFLOW IS GENERATED AS A RESULT OF THE ADDITION
        OPERATION IN RUNG 9, ADD ONE (1) TO THE CONTENTS OF LOCATION
        [FFFF].

    RUNG 12 :

        IF CARRY OR OVERFLOW IS GENERATED AS A RESULT OF THE ADDITION
        OPERATION IN RUNG 10, ADD ONE (1) TO THE CONTECTS OF LOCATION
        [FFFF].

    RUNG 13 :

        IF CARRY OR OVERFLOW IS GENERATED AS A RESULT OF THE ADDITION
        OPERATION IN EITHER RUNG 11 OR RUNG 12, ADD ONE (1) TO THE
        CONTENTS OF LOCATION [1111].

    RUNG 14 :

        MOVE CONTENTS OF LOCATION [1111] TO LOCATION [EEEE].

    RUNG 15 :

        MOVE CONTENTS OF LOCATION [4444] TO LOCATION [HHHH].

Fig. 19.15.   Decimal Multiplication Routine

20

Glossary of Common Programmable Controller Terms

INTRODUCTION

Probably one of the hardest tasks in learning about a new technical subject is becoming familiar with the jargon used by those who are already experienced with the subject matter. The glossary of terms presented in this chapter is provided as a means to assist the reader in understanding PC terminology. It is a generic glossary and does not contain specific references to any manufacturer's products or product lines. Various computer terms have been included since the PC can be considered a special-purpose computer, and since it is often called upon to interface with a computer or computer-based hardware.

The glossary has been composed of entries from many PC-related sources, including PC manufacturers' dictionaries and glossaries as well as the author's experiences in applying PCs. The terms listed in this glossary have had their definitions generalized in order to provide and maintain simplicity. Many PC manufacturers and applications personnel use similar terms to describe their products and hardware. Where a single term may have multiple meanings, each of the meanings applicable to PCs has been included. While the glossary cannot contain every possible term related to PCs, it does provide an excellent starting point for the reader.

Access Time   The amount of time between a request for data and the instant that data are received.

Accumulator   A memory or internal microprocessor location used for the storage of data and addresses.

AC Input   Term applied to a module or card of electronic circuitry that converts and isolates ac signals from user field devices prior to transmission to the processor.

Acoustic Coupler   A device that converts electric signals into audio signals for transmission over telephone lines via a conventional telephone handset.

AC Output   Term applied to a module or card of electronic circuitry that converts low-power processor signals into ac voltage and current levels for use by the user field devices.

Acquisition   The function of obtaining information or data from another source.

Active   A diagnostic indicator used to signify the proper operation, electrically and/or of a programming nature, of a particular PC component or section of hardware.

Active Elements   Those elements of a circuit that have signal power gain, such as transistors, silicon-controlled rectifiers, and triacs.

Adapter   A device used to effect operative compatibility between two or more different parts of a single system or between dissimilar systems.

Adapter Module   1) An electronic module or card that converts serial communications data from the processor into parallel communications data for use by a remote I/O system; 2) Any electric or electronic circuit that interconnects two distant systems by providing a signal conversion from one system to the other.

Add   A PC instruction that provides for the mathematical addition of numerical data.

Address   A code that indicates the location of data to be used by a program, or the location of additional program instructions.

Algorithm   A special set of rules or instructions that perform a specific operation.

Alphanumeric   A character set that contains both numerical and alphabetic characters and possibly other special-purpose symbols as well.

Alternating Current (ac)   Current flow, which is characterized by a reversal of charge flow, usually on a periodic basis.

ALU   See *arithmetic and logic unit*.

Ambient Temperature   The temperature within a given room or enclosure.

American Wire Gauge (AWG)   A standard system of designating the mechanical size (diameter) of an electric conductor. The numerically larger the wire gauge, the smaller the physical diameter of the conductor.

Ampacity   The current-carrying capacity of an electric conductor. Ampacity is expressed in amperes.

Ampere   A current flow that is equivalent to a charge of 1 coulomb (6.24 $\times$ 10^{18} electrons) past a given cross section of conductor in a time period of 1 second is defined as 1 ampere of current flow.

Analog   A value that varies continuously with time over some given range.

Analog Input   Term applied to an electronic module or card that derives a scaled number based on an analog input signal from a user field device proportional to some maximum fixed level.

Analog Output   Term applied to an electronic module or card that converts a scaled digital number supplied by the processor into a voltage or current level proportional to some maximum fixed level.

Analog-to-Digital Conversion A hardware and/or software process that converts a scaled analog signal or quantity into a scaled digital signal or quantity.

AND (Logic) A logic operation between one or more memory bits whereby all bits must be a one or true for the result to be a one or true.

ANSI Abbreviation for American National Standards Institute.

Arithmetic and Logic Unit (ALU) Performs various mathematical computations and/or logic operations on the contents of one or more processor internal data storage locations or registers.

Array See *matrix*.

ASCII Abbreviation for American National Standard Code for Information Interchange.

ASCII Code An 8-bit (7 bits plus parity) code that represents all characters of a standard typewriter keyboard, both uppercase and lowercase, as well as a group of special characters that are used for control purposes.

Assembly Language A machine-oriented language that uses mnemonic symbols to represent one or more discrete machine language codes.

Asynchronous Shift Register A shift register that is not clocked but operates to the requirements of the input data or system needs.

Asynchronous System A system that operates on an as-needed basis rather than on a definite time period-related basis.

Auxiliary Power Supply A power supply that is not associated with the processor. Auxiliary power supplies are usually required to supply logic power to I/O racks and other PC support hardware, and are often referred to as remote power supplies.

AWG Abbreviation for American Wire Gauge.

B&S Gauge Brown and Sharpe wire gauge is the same as American wire gauge.

Backplane An electronic circuit board usually containing sockets and copper foil buses used to pass signals between other circuit boards plugged into the sockets.

Base (I/O Base) A plastic and/or metal assembly that supports the I/O modules and provides a means of supplying power and signals to each I/O module or card.

BASIC A computer language using brief English-language statements to instruct a computer or microprocessor.

Battery Indicator A diagnostic aid that provides a visual indication to the user, and/or an internal processor software indication, that the memory power-fail support battery is in need of replacement.

Baud An indication of the rate of transmission of data and information between two devices.

Baud Rate Data transmission rate expressed in bits per second.

BCD See *binary-coded decimal*.

BCD-to-Binary Conversion A software or hardware routine that converts numerical data stored in a BCD format to an equivalent binary format.

BCD-to-Decimal Conversion A software or hardware routine that converts numerical data stored in a BCD format to an equivalent decimal format.

Bidirectional Data communications between two devices in either direction.

Binary   A number system using 2 as a base, or radix, rather than 10. The binary number system requires only two digits, zero (0) and one (1), to express any alphanumeric quantity desired by the user.

Binary-Coded Decimal (BCD)   A system of numbering that expresses each individual decimal digit (0 through 9) of a number as a series of 4-bit binary notations. BCD is often referred to as 8421 code.

Binary-to-BCD Conversion   A hardware or software routine that converts numerical data stored in a binary format to an equivalent BCD format.

Binary Word   A contiguous group of binary digits (ones and zeros) that have a meaning due to their assigned position or natural binary numerical value.

Bit   An acronym for *binary digit*. The bit is the smallest unit of information in the binary numbering system. It represents a decision between one of two possible and equally likely values or states. It is often used to represent an OFF or ON state as well as a true or false condition.

Bit Clear   An instruction that sets a specified bit location in a matrix to the zero, OFF, or false state.

Bit Manipulation Instructions   A family of PC instructions which exchange, alter, move, or otherwise modify the individual bits of single or groups of processor data memory words.

Bit Pick   Set *bit sense*.

Bit Rate   The rate at which a bit or a group of bits is acted upon or passes a given point in a communications link.

Bit Reset   See *bit clear*.

Bit Sense   An instruction that examines a specified bit in a matrix for either a one, ON, or true state, or a zero, OFF, or false state.

Bit Set   An instruction that sets a specified bit location in a matrix to the one, ON, or true state.

Bit Storage   A single location in a processor data memory which may be used for the ON/OFF, one/zero, or true/false storage of a processor result. Bit storage locations are usually not available for output to field devices directly; rather they are commonly used for internal user program status storage.

Block Diagram   A method of representing the major functional subdivisions, conditions, or operations of an overall system, function, or operation.

Block Transfer   1) An instruction that copies the contents of one or more contiguous data memory words to a second contiguous data memory location; 2) An instruction that transfers data between an intelligent I/O module or card and specified processor data memory locations.

Boolean Algebra   A mathematical shorthand notation that expresses logic functions, such as AND, OR, EXCLUSIVE OR, NAND, NOR, and NOT.

Branch   1) A parallel logic path of relay ladder instructions; 2) The point in a program where a decision must be made as to which one of several paths is to be taken.

Branch Instructions   A family of PC instructions which effect a change in the flow of a program. Examples of branch instructions include jump, jump to subroutine, and skip.

Breakdown Voltage   1) The ac or dc voltage level at which an insulator or dielectric permits a disruptive discharge to occur either through or over the surface; 2) The voltage point at which there is a dramatic change in the voltage/current characteristics of a semiconductor device.

Buffer 1) A circuit placed between other circuit elements to prevent electrical interactions or provide additional power or drive capability; 2) A temporary data storage area used whenever it is necessary to verify or gather data prior to processing.

Bumpless Transfer 1) The transference of control from one PC processor to its backup unit without any interruption of I/O signals or control during the actual control transfer. Also called dipless transfer or hot backup; 2) The transference of PID control from auto to manual or vice versa.

Bus 1) One or more electric conductors that collectively carry information and data from at least one source device to one or more destination devices; 2) A conductor that carries power to numerous load devices.

Byte A single contiguous group of bits which are processed as a single parallel entity. The number of bits that compose a byte can vary. However, most PC systems have standardized on 8 bits as equal to 1 byte.

Cabinet A steel box with hinged doors used to house electric equipment.

Cable 1) A group of conductors bound by an outer protective covering; 2) A conductor composed of one or more uninsulated strands of wire.

Carrier A continuous frequency that can be modulated with a signal for purposes of data transmission.

Cascading 1) A technique used when programming timers and counters to extend the timing or counting range beyond what would normally be available. This technique involves the driving of one timer or counter instruction from the output of another similar instruction; 2) Term applied to a PID loop that has input from one or more support loops.

Cassette Recorder A specially designed audio cassette tape recorder that is used to document the contents of a PC's memory on a cassette of magnetic tape. The resulting tape is used as a record of the contents of the PC. The cassette recorder can also be used to reload or verify the contents of a PC's memory with the information that is stored on a tape.

Cassette Tape (Computer Grade) 300 feet of ⅛-inch computer-grade magnetic tape enclosed in a protective metal or plastic housing similar in appearance to a standard Phillips-type audio cassette. Computer-grade magnetic tape is certified as to continuous oxide coating.

Central Processing Unit (CPU) The part of a computer-based system that contains the main memory, arithmetic and logic unit, and other special-purpose electronic circuitry. The CPU is the electronic circuitry that is responsible for the interpretation and execution of user instructions.

Channel See *I/O channel*.

Character 1) A symbol of a larger group of similar symbols used to represent information on a display device. The letters of the alphabet and the decimal numbers are examples of characters used to convey information; 2) A group of memory bits that, when considered as a group, represent a symbol.

Chassis 1) A housing or framework that is used to hold assemblies. When the chassis is filled with one or more assemblies, it is often referred to as a rack; 2) The housing or framework that is used to mount I/O modules.

Chattering The prolonged undesirable cyclic opening and closing of an electrical contact.

Checkout   The process of verifying the wiring and installation of a control system and its controlled hardware prior to initial operation.

Chip   A single substrate which has active and passive electronic components mounted on it. It is manufactured by semiconductor technology and must be mounted in a plastic or ceramic package with leads prior to use.

Circuit Card   A glass-epoxy card with copper foils and electronic components.

Clear   1) An instruction or a sequence of instructions which removes all current information from a PC memory; 2) An instruction or operation which returns the accumulated or current value of a counter or timer to either its preset value or zero.

Clock   An electronic circuit that generates extremely accurate electric pulses used to achieve synchronization in a digital computer system.

Clock Rate   The rate at which information is transferred from one device to another. For parallel transmissions the clock rate is expressed in words per second and for serial transmissions the clock rate is expressed in bits per second (baud).

CMOS   An acronym for complementary metal oxide semiconductor. CMOS-based logic offers low power consumption and high-speed operation.

Coaxial Cable   A transmission line that is constructed such that an outer conductor forms a cylinder around a central conductor. An insulating dielectric separates the inner and outer conductors, and the complete assembly is enclosed in a protective outer sheath. Coaxial cables are not susceptible to external electric and magnetic fields and generate no electric or magnetic fields of their own.

Code   A system of communications that uses arbitrary groups of symbols to represent information or instructions. Codes are usually employed for brevity or security.

Coding   1) A method of representing characters within a computer system; 2) The conversion of instructions or signals into a form understandable by a computer-based system.

Cold Backup   A PC backup scheme that requires a total shutdown and restart of the controlled process or machinery in order to transfer control between two PC processors.

Color Code   A system of colors developed to identify the electrical characteristics and/or function of various materials and devices.

Color CRT   A specially designed television used with computer-based devices to display information and data graphically in numerous color combinations.

Communications Network   A communications system consisting of adapter modules and cabling used to transfer data and system statuses between network PCs and computers.

Compare   An instruction that compares the contents of two designated PC data memory locations for equality or inequality.

Compatibility   The ability of one unit or device to replace another similar unit or device with little or no degradation of performance or use.

Compiled Language   Any computer language (BASIC, Fortran, etc.) that requires the user program (instructions) to be converted (using a compiler) from the symbology of the language into machine language prior to the running of the program.

Compiler   A special computer program that translates the instructions of a high-level language into machine language.

Computer   Any electronic device which is able to accept information, manipulate it according to a set of preprogrammed instructions, and supply the results of the manipulation. Both computers and PCs may or may not execute their instructions in the same order every time, depending on external conditions.

Computer-Grade Magnetic Tape   A specially produced ⅛-inch magnetic tape used to record digital signals as contrasted to the standard audio signals of a tape recorder. Computer-grade tape must be certified to record and playback 1600 FCI (flux changes per inch). It must also be certified not to have any breaks (dropouts) in the magnetic oxide coating.

Computer Interface   An electronic circuit designed to communicate instructions and data between a PC and a computer or another computer-based intelligent device.

Conditional Zones   A user program area in a PC that is executed or not executed due to the status of a zone-conditioning rung.

Conducting   See *passing power.*

Conductor   Single- or multiple-wire strands used for the conveyance of electric current. A conductor may or may not be insulated.

Conduit   A solid or flexible metal and/or plastic tube used to run insulated electric conductors.

Consecutive Message Listing   The sequential listing of messages and system statuses in a predetermined orderly fashion. The messages are listed in the order received, with no priority given to type or function.

Constant   A predetermined alphanumeric data value that is stored in a PC for later use in the program.

Contact   The current-carrying part of an electric relay or switch that engages to permit power flow and disengages to interrupt power flow to a load device.

Contact Arc   The electrostatic discharge that occurs whenever a contact engages or disengages.

Contact Bounce   The uncontrollable making and breaking of a contact during the initial engaging or disengaging of the contact.

Contact Histogram   An instruction sequence that monitors a designated memory bit or I/O point for a change of state. A listing is generated by the instruction sequence that displays how quickly the monitored point is changing state.

Contactor   A special-purpose relay that is designed to establish and interrupt the power flow of high-current electric circuits.

Contact Symbology Diagram   See *relay ladder diagram.*

Control   1) A relay or a PC-based unit which operates an industrial process or machine; 2) The operation of causing a machine or process to behave in a predetermined manner; 3) A device used to manipulate the manner in which a control system reacts such as a potentiometer.

Control Block   The electronic circuitry of the CPU which decodes the processor instructions and then manipulates the other portions of the CPU in the proper manner to execute the current instruction.

Control Character   A special-purpose character used to initiate a special function or control input operation on a computer terminal or peripheral device.

Control Network   A system of adapter modules and cabling used to transmit real-time logic states between network PCs.

Control Relay   A relay used to control the operation of an event or a sequence of events.

Convert   A family of PC instructions which provide for the conversion of PC data between numerical bases or between an intelligent I/O device and the processor.

Core Memory   A type of memory system that employs ferrite cores to store information. Core memory operates by magnetizing the ferrite core in one direction to represent a one, ON, or true state, and in the opposite direction to represent a zero, OFF, or false state. This form of memory is nonvolatile.

Counter   1) An electromechanical device in relay-based control systems that counts numbers of events for the purpose of controlling other devices based on the current number of counts recorded; 2) A PC instruction that performs the functions of its electromechanical counterpart.

CPU   See *central processing unit.*

CRC   See *cyclic redundancy check.*

Cross Coupling   The unwanted transfer of energy from one signal line to an adjacent signal line.

Cross Reference   A reference listing of the locations in a user program where a particular I/O point, data memory location, or instruction is used and the form in which it is used.

Crosstalk   The unwanted transfer of energy or noise from one circuit, termed the disturbing circuit, to another adjacent circuit, termed the disturbed circuit.

CRT Pointer   See *cursor.*

CRT Terminal   A portable enclosure containing a cathode ray tube (CRT), a special-purpose keyboard, and a microprocessor which is used to program a PC. The CRT terminal converts user commands and instructions into a binary machine code that is understandable by the PC processor, and it displays a selected portion of the user ladder diagram program.

CSA   Abbreviation for Canadian Standards Association, which is the Canadian equivalent of Underwriters Laboratories.

Cumulative Lost Time   See *downtime.*

Current   The rate of electrical electron movement, measured in amperes. The symbol for current is *I*.

Current-Carrying Capacity   The maximum amount of current a conductor can carry without heating beyond a predetermined safe limit.

Cursor   A CRT position indicator used to indicate the current point of data entry. A PC cursor will usually appear as a reverse-video block, and may flash at a slow rate as well.

Cursored Rung   The ladder diagram rung of the CRT terminal display which currently contains the cursor. The cursored rung is the only ladder rung that can be affected by a program change.

Cycle   1) An OFF/ON repeated application of power; 2) A regular repeated sequence of operations or instructions; 3) The part of an alternating waveform that begins from a zero reference point, reaches a maximum value, returns to the zero reference, then reaches a negative maximum, followed by a return to the zero reference point.

Cyclic Redundancy Check   An error detection scheme that generates a check character from the remainder of a division of serialized data bits when divided by a predetermined binary number.

Daisy Chain   A communications link that has the link elements connected in such a manner that the signal passes from one unit to the next in a serial fashion.

Damped PID Control   A PID loop that has been desensitized to rapid control changes due to rapidly varying input data.

Data   Information encoded in a digital form, which is stored in an assigned address of data memory for later use by the processor.

Data Access Panel   A calculator-like keypad and display used to monitor and often to modify the contents of selected data memory locations within a PC.

Data Bus   A bus that is dedicated to the transmission of data.

Data Communications Network   A communications network operating in half or full duplex which carries serial data between multiple stations connected to the network. A PC data communications network permits multiple PCs, computers, data terminals, and other intelligent devices to communicate with each other via the network. It eliminates the necessity of having dedicateed wiring between would-be network station devices.

Data Field   A group of data words or elements.

Data File   A group of data memory words which are acted upon as a group rather than singularly.

Data-Handling Module   A special-purpose processor module or accessory devoted to the manipulation of data within the processor. Data-handling modules often generate messages and reports as well as perform data file manipulation and mathematical operations.

Data Link   The equipment that makes up a data communications network.

Data Manipulation   The process of exchanging, altering, or moving data within a PC or between PCs.

Data Manipulation Instructions   A classification of processor instructions that alter, exchange, move, or otherwise modify data memory words. Data manipulation functions include move, FIFO, search, and block instructions.

Data Memory   The portion of a PC's memory that is set aside for data storage. This portion of processor memory contains the user program constants, timer and counter preset and accumulated (current) values, math instruction data, as well as any special-purpose user program alphanumeric information. Some PCs include the I/O image table as part of the data memory.

Data Processor A separate processor dedicated to data manipulation functions, which works in conjunction with a PC's main processor to improve overall system efficiency and throughput.

Data Register A single word of data memory. Also referred to as a data word.

Data Set See *modem*.

Data Table 1) An alternate term for data memory; 2) An alternate term for data file.

Data Terminal Any peripheral device that can load, monitor, program, or save the contents of a PC's memory. These devices include tape loaders, printers, paper tape punches and readers, PC programming terminals, and other like data terminal hardware.

Data Transfer 1) The process of exchanging data between two devices over a communications link; 2) A more common name for data manipulation instructions.

Data Word A single word of data memory, also referred to as a data register.

DC Input Term applied to a module or card of electronic circuitry that converts and isolates dc signals from field devices prior to transmission to the processor.

DC Output Term applied to a module or card of electronic circuitry that converts low-power processor signals into dc voltage and current levels for use by the user field devices.

Dead Band The small range of PID input signal variation that occurs when the PID loop is operating at its setpoint, where no output control is achieved due to the small error being generated by the PID calculation.

Debug The act of searching for and eliminating sources of program or machine errors. Sources of errors include wiring mistakes, hardware misalignment or operation, software syntax errors, and incorrect software algorithms.

Decay Time The time required for a signal's trailing edge to fall from 90 percent of its final value to 10 percent of its final value.

Decimal A numbering system with a radix of 10 using the symbols 0, 1, 2, 3, 4, 5, 6, 7, 8, 9.

Decimal-to-BCD Conversion Term applied to a software or hardware routine that converts numerical data stored in a decimal format to an equivalent BCD format.

Decrement The act of reducing the contents of a storage location or value in varying increments.

Dedicated The act of limiting a piece of hardware and/or software to performing a specific function or operation even though the unit may have the ability to provide other functions or operations.

Deenergize 1) The physical removal of power to a circuit or device in order to deactivate it; 2) The act of setting the OFF, false, or zero state of a PC relay ladder diagram output device or instruction.

Delimiter A character used to separate or organize data into fields.

Demodulator A hardware device or circuit used to detect and decode specific signals while rejecting any other signals or noise that may be present.

Derivative Control See *proportional, integral, derivative*.

Destination Register, Table, Word, File The location where information is to be stored upon completion of an operation.

Deviation   The difference between desired and actual levels of a signal or quantity.

Diagnostic Indicator   A hardware element, such as a lamp, or a memory location used to indicate the functional status of a device or system.

Diagnostic Program   A program used to test the integrity of a system. Diagnostic programs cycle a system through its routines, or simulate real-time operating conditions in order to uncover flaws and weaknesses in the system prior to use.

Dielectric   A nonconductive insulating medium.

Digital   The use of scaled numbers to express a specific variable. A PC uses the binary number system to store and manipulate information pertaining to hardware statuses (OFF or ON), numerical quantities, and control system data.

Digital-to-Analog Conversion   Term applied to a hardware or software routine that converts a scaled digital signal or data into an equivalent scaled analog signal.

Dipless Transfer   See *bumpless transfer.*

Direct Current (dc)   An electric current that flows in only one direction.

Direct Memory Access (DMA)   The ability to access a portion of the system memory directly with a peripheral device or a microprocessor.

Direct Memory Addressing   The ability to directly reference a memory location in the user program for the purpose of data manipulation.

Discrete   Term used to indicate an item which is separate and distinct, such as the elements of a circuit, characters, memory words, and so on.

Discrete I/O   A group of input and/or output modules that operate with ON/OFF signals as contrasted to analog modules that operate with continuously variable signals.

Disk Storage   A storage means that utilizes circular magnetic disks rather than tape for the storage of information and data.

Disruptive Discharge   The sudden burst of high current flow through a dielectric exposed to intense electrostatic stress.

Distributed Control System   A control scheme relying on two or more PCs and/or computers for complete system operation. Each PC or computer controls a portion of the system and may rely on communications with one or more other other units for operation.

Distributed I/O   An I/O layout that places I/O hardware at various strategic locations rather than at a single central location.

Divide   A PC instruction which performs a numerical division of one number by another.

Documentation   A collection of both hardware and software reference materials including system drawings, diagrams, memory maps, program listings, maintenance procedures, operating instructions, troubleshooting guides, etc.

Double Precision   The storing and handling of data in two consecutive data memory words to obtain an increase in magnitude and accuracy.

Downtime   The length of time that a piece of equipment or system is out of operation due to mechanical, electrical, or electronic malfunction. The lack of equipment or system as well as the absence of an operator may or may not be included in the downtime figure.

Drain Wire An insulated wire placed in continuous contact with the shield foil of shielded cable for the purpose of making shield connections to terminal blocks and connectors.

Driver A circuit attached to the output of a system or device for the purpose of increasing the power or current-handling capability of the system or device.

Drop Line The signal cable connecting a device to the distribution cable.

Drum Controller A control device that uses either cams or pins on the surface of a drum to operate numerous contacts.

Dry Contact A pair of contacts which are not immersed in a liquid such as oil and do not rely on a mercury film for operation.

Duct A protective metal wireway with a removable cover.

Dump The operation of producing either a hard-copy printout or a magnetic tape recording of the contents of a PC's memory.

Duplex A communications scheme that supports simultaneous two-way transmissions over the communications link. See *full duplex* and *half duplex*.

Dwell An axis positioning term that describes the period of time where constant velocity occurs.

Dynamic Memory A memory system that must be constantly refreshed in order to retain information. Dynamic memories are generally faster and consume less power than static memories.

E The symbol for voltage.

Edit The act of modifing a PC program to eliminate mistakes and/or simplify or change system operation.

EEROM See *electrically erasable read-only memory*.

EIA Abbreviation for Electronic Industries Association, an organization that produces and distributes standards relating to electronic hardware and systems.

EIA Interface An interface port designed to the requirements of a particular EIA specifications such as RS-232C.

Electric Latch See *seal*.

Electrically Erasable Read-Only Memory A type of read-only memory which can be erased electrically and reprogrammed. The erasing and reprogramming require more time than a standard Random Access Memory.

Electromagnetic Interference A phenomen responsible for noise in electric circuits.

Element A single instruction of a relay ladder diagram program.

Emergency Stop Relay A relay used to inhibit all electric power to a control system in the event of an emergency or other event requiring that the controlled hardware be brought to an immediate halt.

EMI See *electromagnetic interference*.

Enable To permit a particular function or operation to occur under natural or preprogrammed conditions.

Enclosure A steel box with a removable cover or hinged door, used to house electric equipment.

Encoder An optical, electromechanical, or electronic device used to monitor the circular motion of a device. An absolute encoder provides a specific code for each angular position of a shift or device. An incremental encoder transmits a specific quantity of pulses for each revolution of the shaft or device.

End of Scan   A PC instruction that instructs the processor not to continue scanning the user memory, but to return to the beginning of user memory or to begin its housekeeping routines prior to the start of another program scan.

Energize   1) The physical application of power to a circuit or device in order to activate it; 2) the act of setting the ON, true, or one state of a PC relay ladder diagram output device or instruction.

EPROM   See *erasable programmable read-only memory.*

Erasable Programmable Read-Only Memory   A generic term applied to a family of memory devices which can be erased and reprogrammed a limited number of times.

Even Parity   The setting of the parity bit such that the number of data bits plus the parity bit equal an even number.

Examine OFF   Term used by some PC manufacturers to refer to a normally closed contact instruction in a relay ladder program. The examine OFF instruction passes power whenever the referenced I/O or internal storage location is in an OFF, false, or zero state.

Examine ON   Term used by some PC manufacturers to refer to a normally open contact instruction in a relay ladder program. The examine ON instruction passes power whenever the referenced I/O or internal storage location is in an ON, true, or one state.

Exclusive OR   A logic operation between memory bits whereby the result is zero or false only if all bits being compared are either false and/or zero, or true and/or one.

Execute   The performance of a specific operation or a selected group of operations.

Execution Time   The time required to process a specific instruction or group of instructions.

Executive   A machine language program responsible for interpreting user program instructions and executing them, as well as handling all PC housekeeping and operating requirements, such as I/O communications, timer/counter incrementing, and system health checks.

Extended Instruction Set   A group of PC instructions not standardly available. Extended instruction sets usually include special data and bit manipulation instructions as well as other special-purpose algorithms not offered in base price models of PCs.

False   As related to PC instructions, an OFF or zero state.

Fast-Response Input Module   An input module not incorporating any contact bounce suppression electronics for the purpose of providing minimal input signal delay to the processor.

Fault   1) A wiring defect that permits the unintentional grounding, break in the path, or short-circuiting of an electric system; 2) An electronic defect that permits the improper operation of a semiconductor or electronic device or circuit; 3) The malfunction of an electronic system from a normal operation.

Fault Current The amount of current that flows in a faulted electric or electronic circuit or device.

Fault Indicator A diagnostic aid that provides a visual indication and/or an internal processor software indication that a fault is present in the system.

Fault Zone A section of user program that alters the operation of a system, or the portion of a system, whenever a fault condition is sensed.

FCI See *flux changes per inch.*

Feed An axis positioning term used to indicate a period of constant velocity.

Feedback The return signal to a control system or device indicating the response of the controlled hardware to the control system's commands.

Feedforward PID Control A PID loop that uses supplementary system operational data to enhance the control action of the PID loop. For example, a temperature control loop on a heated vessel might incorporate the incoming flow and temperature of the material being heated to supplement the PID controls of the vessel's temperature control loop.

Fence Codes Special PC instructions which delimit a specific block of user programming. Fence codes permit specified sections of the user program to be skipped or treated in a predetermined manner under system operating conditions.

FIFO An acronym for a *first-in, first-out* type of data storage queue. The FIFO queue is similar to a cup dispenser in that data placed in the queue are stored in the sequence entered, and later discharged in the same order as previously entered.

File A contiguous group of data memory words in a PC usually operated on as a single unit by a PC program.

Filter A specially designed resistor, inductor, capacitor, and/or surge suppressor network used to suppress electric noise.

Firmware A user-transparent software program placed in read-only memory (ROM) and used to operate a system or microprocessor. The executive programming of a PC is an example of firmware.

Flag Bit A processor memory bit used to signify that a specific condition or action is or has occurred. Flag bits may be designed into a PC system for status indication (example: battery low indicator, blown fuse indicator), or they may be designated through user programming.

Floating Point A number expressed as a bounded number (mantissa) and a scale factor (exponent) indicated by a power of a number base.

Flowchart/Flow Diagram A graphic representation of the logic steps of a computer or microprocessor program. A software program is developed by writing down the successive instructions that cause a computer or PC to perform the operations defined by the flowchart.

Flux Changes per Inch (FCI) The number of magnetic polarity reversals possible with computer-grade magnetic tape. Most computer-grade tapes must be capable of 1600 FCI.

Foil A thin metal tape wrapped immediately under the outer insulation jacket of a cable to minimize the effects of EMI and RFI interference. The tape forms a continuous tubular capacitor surrounding the wires of the cable. This foil is often referred to as the shield of a cable. The foil should be grounded at only one end of the cable by means of a drain wire.

Force Instruction   A PC programming terminal function that allows the user to override the operation of a specified input or output point. The processor responds to the forced status of the I/O point rather than to the actual field status of the I/O module. Forces may or may not remain in effect when the programming terminal is removed from the processor.

Force OFF   A PC programming terminal function that deenergizes a selected I/O point regardless of the actual field status of the selected point.

Force ON   A PC programming terminal function that energizes a selected I/O point regardless of the actual field status of the selected point.

Free Instruction   A user program instruction placed on a programming terminal's screen for monitoring purposes only. The free instruction is not part of the active user program but does respond to status changes due to control system operation.

Frequency   The number of periodic recurrences of an event within a specific time period. Frequency is expressed in units of cycles per second or hertz.

Frequency-Shift Keying (FSK)   A means of data transmission that utilizes frequency modulation to achieve transmission. Most standard telephone modems use FSK as the method of transmission, selecting one audio frequency to represent a digital one and a second audio frequency to represent a digital zero.

Full Duplex (FDX)   A mode of data communication that permits data to be transmitted and received simultaneously between two devices.

Gain   The increase ratio of an input signal to the output control signal of a PID loop.

Get Byte Instruction   A PC instruction that fetches the contents of either the upper or the lower byte of a specified data memory word.

Get Instruction   A PC instruction that fetches the contents of a specified data memory word. Get instructions are often used to fetch data prior to programming mathematical and data manipulation instructions.

Gray Code   A binary coding scheme that allows only 1 bit in the data word to change state at each increment of the code sequence.

Ground   A metallic connection with the earth to establish ground potential.

Ground Conductor   A conductor that normally carries no power flow, but is used to connect various electrical surfaces to ground.

Grounded Conductor   A current-carrying conductor intentionally connected to the ground, such as the neutral conductor of 120-volt ac single-phase circuits.

Grounding Conductor   See ground *conductor*.

Ground Potential   Zero voltage potential with respect to the ground.

Half Duplex (HDX)   A mode of data communication that permits data to be transmitted and received between two devices, but not simultaneously.

Handshaking   A colloquial term describing the initial transmission and reception of predetermined characters used by communicating devices to establish contact with each other.

Hard Copy   Any printed document produced by a computer- or microprocessor-based system.

Hardware   Any physical, mechanical, magnetic, electrical, electromechanical, or electronic devices that are part of a system.

Hard-Wired   The physical interconnection of electric and electronic components with wire.

Hertz   Unit of measure for frequency in terms of cycles per second.

Hexadecimal   A number system having a base of 16. This numbering system requires 16 elements for representation, and thus uses the decimal digits zero (0) through nine (9) and the first six letters of the alphabet, A through F.

Hierarchical Control   A control scheme incorporating various levels of control responsibility and/or equipment.

Hierarchy   The listing or classification of items according to a predetermined set of rules or requirements.

High-Level Language   A programming language based on the application requirements of the system rather than the processor's machine language.

Histogram   A graphic representation of the frequency at which an event occurs.

Host   1) A master computer or PC that monitors and/or controls the operation of other computers and PCs; 2) An intelligent device which controls the operation of a dumb device. A PC processor is a host device to a cassette tape loader.

Hot Backup   See *bumpless transfer.*

Housing   1) The plastic or metal covering of a circuit card or electronic device; 2) A plastic and/or metal assembly that supports the I/O modules and provides a means of supplying power and signals to each I/O module or card.

I   Symbol used to designate the current in amperes flowing in an electric circuit.

IEEE   Abbreviation for Institute of Electrical and Electronics Engineers, an organization that produces and distributes standards and other latest technology data relating to electronic and electric systems.

Image Table   See *input image table* and/or *output image table.*

Immediate Input Instruction   A PC instruction that temporarily halts the user program scan so that the processor can update the input image table with the current status of one or more user-specified input points.

Immediate I/O Instruction   See *immediate input instruction* and/or *immediate output instruction.*

Immediate Output Instruction   A PC instruction that temporarily halts the user program scan so that the current status of one or more user-specified output points can be updated to current output image table statuses by the processor.

Impedance   The total resistive and inductive opposition that an electric circuit or device offers to a varying current at a specified frequency. Impedance is measured in ohms and is denoted by the symbol *Z.*

Inclusive OR (OR)   A logic operation between memory bits whereby the result is one or true whenever one or more of the bits being compared is a one or true.

Increment   The act of increasing the contents of a storage location or value in varying amounts.

Indirect Memory Addressing   An addressing scheme whereby the addressed location contains the address of the data rather than the actual data itself.

Inductance   A circuit property that opposes any current change. Inductance is measured in henrys and is represented by the letter H.

Infrared   The spectral region beyond red from 1×10^4 to 1×10^6 angstroms.

Initialize   The process of setting all system switches, addresses, data, and hardware to a predetermined state, ready for system startup and/or operation.

Input Device   Hardware devices used to provide data and status information to a computer or PC. Input devices may be either discrete or analog in nature.

Input Image Table   A portion of a processor's data memory reserved for the storage of input device statuses. A one, ON, or true field device condition is represented by a one state in the input image table storage location reserved for that particular input point.

Input Impedance   The effective impedance "seen looking into" the input terminals of an electric or electronic device.

Input Instruction   1) Any instruction or instructions that bring data from an input module into the processor; 2) The instruction or instructions that determine the operation of an output instruction on PC relay ladder rung.

Input Register/Input Word   A particular word in a processor's input image table used to store numerical data being input from a field input device.

Insert   A PC programming terminal function that permits the insertion of an instruction between other previously programmed instructions.

Instantaneous Contact   A normally open and/or normally closed PC contact instruction, which is actuated immediately whenever a timer receives an input signal.

Instruction   A command that initiates a prescribed set of operations within a computer or PC. Instructions control the movement of data, the performance of logic and mathematical functions, control of peripheral devices, and initiation of succeeding instructions in a computer or PC.

Instruction Set   A list of all instructions, with a short description of their operation, which are available for use with a particular computer or PC.

Insulator   A material with a high resistance to the flow of electricity due to the close bonding of the material's electrons in the atom.

Integral Control   See *proportional, integral, derivative control.*

Intelligent I/O   Input and/or output modules which utilize a special-purpose microprocessor built into the module to precondition the flow of I/O data between the processor and the field devices. Intelligent I/O modules reduce the amount of data handling and manipulation a PC processor must perform prior to the use of input data or the transmission of output data. Many intelligent I/O modules have the ability to operate in a semiautomatic or manual mode should communications with the processor be interrupted.

Interface   The common boundary between adjacent components, circuits, or systems, which allows for the transmission of data, information, power levels, or modes of operation between one and another. The terms buffer, handshake, and adapter are often used to refer to an interface.

Interfacing   The interconnecting of a computer or PC with its field devices through the use of cables and/or special-purpose hardware.

Interference   An EMI and/or RFI disturbance or emission which causes, or could cause, the malfunction or degradation of an electric or electronic system.

Interlock   The use of mechanical, electric, or software devices and/or routines to govern the succeeding operations of a machine or process.

Internal Coil Instruction   A relay coil instruction used for internal storage or buffering of an ON/OFF logic state. An internal coil instruction differs from an output coil instruction in that the ON/OFF status of the internal coil is not passed to the I/O hardware for control of a field device.

Interpretive Language   Any computer language that executes the language symbols and instructions directly without being processed by a compiler. Interpretive language programs are easier to run than compiled language programs but are also slower in execution since the actual conversion from the language's symbols and instructions is done as each instruction is executed.

Interrupt   A break in the normal flow of a computer program by either an internal or an external signal to handle a sudden request to perform a particular operation or function. Upon completion of the necessary operations or functions, program execution returns to continue at the point where it was interrupted. An interrupt should not be confused with program skips, jumps, or subroutines that modify the normal flow of a program.

I/O   Abbreviation for input/output.

I/O Base   See *Base.*

I/O Card   See *I/O Module.*

I/O Channel   A prespecified number of input and output points or a mix of input and output points serviced as a group by a PC processor. The number of I/O points permissible in an I/O channel is determined by the capacity of the system power supplies, the physical size of the I/O modules and associated I/O hardware, and the communications scheme employed between the I/O modules and the processor.

I/O Chassis   See *Chassis.*

I/O Housing   See *Housing.*

I/O Module   An electronic assembly containing the necessary interface and isolation electronics for interfacing user field devices to the processor.

I/O Node   A smaller portion of an I/O channel. Many PCs allow a single channel of I/O to be subdivided into smaller stand-alone I/O units for increased flexibility.

IOP Board   Term commonly applied to the circuit board of a PC processor that performs the communications between the processor I/O image tables and I/O modules.

I/O Rack   See *Rack.*

I/O Scan   The portion of a PC's scan dedicated to the reading of input conditions and the transmission of output statuses.

I/O Scan Time   The amount of time required for the processor to perform an I/O scan.

Isolated I/O   An I/O module having each I/O point on the module completely electrically isolated from the other points on the same module in such a manner that each individual point requires a dedicated supply and return wire for operation.

Jacket   The protective outer covering of a cable.

Jump   An instruction that permits the bypassing of selected portions of the user program. Jump instructions are conditional whenever their operation is determined by a set of preconditions, and unconditional whenever they are executed to occur every time programmed. Many jump instructions permit reverse as well as forward jumps to occur, and therefore should not be confused with the skip instruction which only permits the bypassing of selected portions of the user program in the forward direction.

Jumper   1) A short length of wire used to bypass selected portions of an electric circuit; 2) The short length of wire used to connect various electric devices to a central point.

K   A letter used to denote memory size. One K (1 K) of memory is equivalent to 2 to the tenth power (2^{10}) or 1024 words of memory.

k   A letter used to denote kilo or one thousand.

Keying   A mechanical interlocking scheme used on circuit boards and I/O modules to ensure that they cannot be placed in the wrong location of a chassis or rack. A plastic clip is usually placed between specified contacts of a socket on the chassis or rack backplane such that it aligns with a slot in the edge of the proper mating circuit card or I/O module. If an incorrect card or module is inserted, the plastic clip, sometimes called a keying band, does not match the slot in the card or the module edge, and therefore the unit cannot be firmly seated in the chassis or rack.

Key Lock   A key-operated switch used on PC processors and/or programming terminals to inhibit unauthorized access to the processor memory or selected PC programming terminal functions and operations.

Label   1) A PC instruction that assigns an alphanumeric designation to a particular location in a program. This location is used as the target of a jump, skip, or jump to subroutine instruction; 2) An alphanumeric identifier assigned to an enclosure, device, component, or other unit for identification purposes.

Ladder Diagram   A control industry method of representing relay-based logic control. The diagram resembles a ladder in that the vertical supports of the ladder appear as power feed and return buses, and the horizontal rungs of the ladder appear as series and/or parallel connections of relay contact symbols feeding a relay-type coil, timer, or counter.

Ladder Diagram Program   A PC program written in relay ladder format and symbology.

Language   A set of symbols and rules, which form a code that distinctly expresses a singular meaning, function, or operation. A language is used to convey information between people and machines.

Large-Scale Integration   Any monolithic integrated circuit which consists of 100 or more logic gates or logic gate equivalent circuits.

Latching Relay   An electromechanical relay designed such that the contacts of the device lock in either an energized or a deenergized state until reset either electrically or mechanically.

Latch Instruction   One-half of a PC instruction pair (the second instruction of the pair being the unlatch instruction) which emulates the latching action of a latching relay. The latch instruction energizes a specified output point or internal coil until deenergized by a corresponding unlatch instruction. The output point or internal coil remains energized regardless of whether or not the latch instruction is energized. The output point or internal coil retains its ON or OFF state through the power cycling of the processor.

Lay   The distance required along a cable or stranded conductor for one cable conductor or conductor strand to make a full revolution about the central axis of the cable or conductor. The direction of the lay, either right- or left-handed, is defined as the direction of rotation of the conductors or strands looking away from the observer.

LCD   See *liquid crystal display.*

Leakage   The small amount of current that flows in a semiconductor device when in the OFF state.

Least Significant Bit   The rightmost bit of a word or byte of memory.

Least Significant Digit   The rightmost digit of a number or the lowest place value in a number.

LED   See *light-emitting diode.*

LED Display   A display device incorporating light-emitting diodes to form the segments of the displayed characters and numbers.

LIFO   An acronym for a *last-in, first-out* type of data storage queue. The LIFO queue stacks incoming data in a file in the order received. When the data are called from the queue, the latest data received are removed first.

Light-Emitting Diode   A semiconductor *PN*-type junction that emits light when biased in the forward direction.

Limit Switch   An industrially designed switch containing normally open and/or normally closed mechanical or solid-state contacts used to sense the location or position of a mechanical device through the movement of an arm or plunger associated with the switch unit.

Line   The incoming power feed to a system.

Line Driver   An electronic device designed to transmit digital information over a dedicated communications link from one digital device to another. Line driver/receivers are often referred to as modems.

Line Loss   The sum total of the various forms of energy losses in a transmission line.

Line Printer   A printer designed to print complete lines at a time rather than single characters at a time.

Line Receiver   An electronic device designed to receive the data transmitted over a communications link from one digital device to another. Line driver/receivers are often referred to as modems.

Liquid Crystal Display   A display device using reflected light from liquid crystals to form the segments of the displayed characters and numbers.

Load   1) The power consumed by a machine or circuit while operating; 2) A device that consumes power and converts it to a desired form or function; 3) The act of filling a memory system with data and/or program instructions.

Load Resistor A resistor connected in parallel with a load device so that sufficient power can be consumed by the combination to correctly operate a solid-state switch controlling the power flow to the load device and resistor.

Local I/O I/O hardware which is located within 75 to 100 cable feet of a processor.

Local Power Supply The power supply used to provide power to the processor and a limited amount of local I/O modules.

Logic A mathematical approach using simple functions to solve a complex problem. The three basic symbols of logic are AND, OR, and NOT. These three symbols, when combined according to Boolean algebra formats, are analogous to addition and subtraction.

Logic Diagram A diagram of logic symbols and their interconnections.

Logic Levels The voltage magnitude associated with signals which represent a zero, OFF, or false state and a one, ON, or true state.

Logic Power One or more dc power sources supplying power to the logic components of an electronic system or device.

Longitudinal Redundancy Check (LRC) An error-checking technique similar to parity (VRC), where a check character (parity character) is transmitted after a specified number of data characters. Each check character bit is a parity bit calculated from the corresponding bit locations of the transmitted characters.

Loop 1) A computer or PC software routine which is repeated until a predetermined set of results is obtained; 2) The control system path that is followed by a command signal, which directs the actions of a control device, and the return path to the control system by the feedback signal from the controlled device or system.

Low Battery Indicator See *Battery Indicator.*

LSB See *Least Significant Bit.*

LSD See *Least Significant Digit.*

LSI See *Large-Scale Integration.*

Machine Language The simplest of all languages and the only language actually understood by a microprocessor. Machine language is based on a set of binary codes which represent instructions to the microprocessor. All microprocessor instructions must be converted to a series of machine language statements prior to execution.

Magnetic Tape A plastic tape coated with ferric oxide or a similar magnetic oxide, used to store information. Magnetic tapes designed for PC or computer use must be certified for a minimum number of flux changes per inch and have no oxide voids.

Magnetics 1) A branch of science that deals with magnetic properties; 2) Electromechanical devices that rely on magnetic forces for operation, such as relays.

Main Power Supply The power supply associated with the processor and a limited amount of local I/O modules. The terms local and main power supply are often used interchangeably.

Malfunction The incorrect functioning of any device.

Manual 1) A text used as documentation; 2) To operate by hand.

Mark   The presence or occurrence of a communications link signal. The Mark represents the closed condition, power flow, an ON signal, or a binary one state.

Mask   A programming technique using logic instructions to inhibit certain memory bits from affecting the outcome of an operation.

Master   Any device that provides instruction and/or control of one or more slave devices.

Master Control Relay   A PC instruction that performs the equivalent function of an emergency stop relay. Due to the failure nature of certain solid-state devices used in all PCs, the master control relay should not be used as the sole source of emergency stop protection. The physical, hard-wired, mechanical emergency stop relay should still be employed where a risk to life, property, and/or limb is present.

Master-Slave   A system of two or more interconnected devices where one of the devices is responsible for the overall operation of the rest.

Matrix   1) A two-dimensional arrangement of elements; 2) One or more data memory words whose individual bits are used to store ON/OFF, one/zero statuses.

Maximum Operating Temperature   The maximum safe temperature at which a system or device can be operated continuously.

MCR Instruction   See *Master Control Relay.*

Mean Time Between Failures   The limit of the ratio of a device's operating time to the number of failures observed as the latter reaches infinity.

Mean Time to Failure   The operating time of a device divided by the total number of failures within that same time period.

Mean Time to Repair   The total accumulated maintenance time for a device divided by the total number of times maintenance had to be performed on the device.

Mechanical Drum Programmer   See *Drum Controller*

Medium-Scale Integration (MSI)   A monolithic integrated circuit which contains between 10 and 100 logic gates or logic gate equivalent circuits.

Megahertz   One million cycles per second or one million hertz.

Megohm   One million ohms of resistance.

Memory   The equipment and components used to store binary-coded computer information for later retrieval and use. "Memory" is usually located adjacent to the microprocessor it supports, whereas "storage" often refers to disk drives, magnetic tape, or punched cards stored external to the system.

Memory Dump   See *Dump.*

Memory Load   See *Load.*

Memory Module   The modular element of a PC processor containing the actual PC memory and memory support electronics.

Memory Protect   A hardware and/or software capability that prevents unauthorized changes to be made to the user program or processor data memory. Some PCs permit user-selected areas of the data memory to be externally altered while the processor is operating.

Message   A predetermined set of characters used to convey information.

Metal-Oxide Semiconductor (MOS)   A semiconductor device that utilizes a metal-oxide film as the insulator between conducting channels.

Metal-Oxide Varistor (MOV) A variable resistor used for spike and transient protection on solid-state circuits. MOVs are metal-oxide semiconductors with a nearly infinite resistance below the device's operating point and very low resistance above their operating point.

Micro PC A very small PC controller usually having less than 64 I/O points and housed in a "shoe-box"-sized package.

Microprocessor An electronic computer implemented on an LSI circuit chip. The microprocessor integrated circuit contains all the arithmetic, logic, control, memory, and I/O functions necessary to be classified as a computer.

Migration The movement of some metals from one location to another. The migrating effect is caused by a plating action set up whenever a metal is exposed to electric current flow and moderate moisture levels.

Military Connector A round multicontact connector designed for very rough handling as well as for extremely harsh environments. Military connectors are often designed to be water- and dusttight when made up.

Military Standards A group of standards that define the operating parameters, construction and fabrication details, as well as life expectancy of devices used by the military.

Milli A prefix indicating one thousandth of a base unit.

Mini Programmable Controller A very small PC usually offering less than 128 I/O points and 4K of memory.

Mnemonic An abbreviation given to a particular function or device. Often mnemonics are acronyms that can be formed from the combination of letters and parts of words used to describe the function or device fully.

Mode A term used to refer to the selected operating method such as automatic, manual, test, program, diagnostic, etc.

Modem An acronym for *modulator/demodulator*, a device that employs frequency-shift keying (FSK) for the transmission of data over a telephone line. A modem converts a two-level binary signal to a two-frequency audio signal and vice versa.

Module An element of a packaging scheme that offers regularity in size and general appearance. A module of electronic hardware is usually designed to "plug in" to a larger rack, a cage, or a larger module. The use of modules in the design and fabrication of a system permits quick replacement of faulted subassemblies as well as flexibility in configuration.

Module Group Two or more modules which as a group perform a specific function or operation, or are thought of as a single unit.

Monitor 1) Any display device incorporating a cathode ray tube (CRT) as the primary display medium; 2) The act of listening to or observing the operation of a system or device.

MOS See *Metal-Oxide Semiconductor*.

Most Significant Bit The leftmost bit of a word or byte of memory.

Most Significant Digit The leftmost digit of a number, or the highest place value in a number.

Motherboard See *backplane*.

Motor Control Center (MCC) A modular enclosure housing power buses which feed multiple motor starters also housed within the enclosure. Each motor starter is contained in its own modular compartment along with a control transformer, short-circuit protection fuses or circuit breakers, and a disconnect switch.

Motor Overload See *Overload Relay.*

Motor Starter A specially designed heavy-duty power-type relay used to control the power flow to a motor. A motor starter will usually incorporate some form of overload protection for the motor it serves.

MOV See *Metal-Oxide Varistor.*

Move Instructions A family of instructions which move data from one location to another within a PC. Various forms of move instructions are offered, including instructions to copy data from one file or table to another, to copy the contents of a single word or register to a specific indexed point in a destination file or table, and to move the contents of an indexed location in a file or table to a destination word or register location.

MSB See *Most Significant Bit.*

MSD See *Most Significant Digit.*

MSI See *Medium-Scale Integration.*

MTBF See *Mean Time Between Failures.*

MTTF See *Mean Time to Failure.*

MTTR See *Mean Time to Repair.*

Multiconductor Cable A cable consisting of two or more insulated conductors.

Multimeter A portable test instrument used to measure electrical properties such as resistance, ac and dc voltage, and ac and dc current.

Multiple-Rung Display The capability of a programming terminal to display two or more ladder diagram rungs at one time. The display is "windowed" in that only a contiguous section of user program relay ladder rungs can be displayed at one time.

Multiplex A process of transmitting more than one signal at a time over a single communication link. Two methods are commonly used, the first being a frequency sharing of channel bandwidth much like runners competing on a track. Each lane is similar to a frequency band of the link where the runner (communication of data) operates within an assigned lane (bandwidth). The second method time-shares multiple signals on the link, each signal taking its turn, using the link, like pole vaulters using the same bar to jump over.

Multiply A PC instruction that provides for the mathematical multiplication of two numbers.

Nand A logic operation between memory bits or signal lines whereby all bits or signals must be a zero or false for the result to be a one or true.

Nano A prefix meaning one-billionth, or ten to the minus nineth power.

National Electrical Code (NEC) A set of regulations developed by the National Fire Protection Association which govern the construction and installation of electric wiring and electric devices. The NEC is recognized by many governmental bodies, and compliance is mandatory in much of the United States.

National Electrical Manufacturers Association An organization of electric device and product manufacturers. NEMA issues standards relating to the design and construction of electric devices and products.

Natural Binary See *Binary.*

NEC   See *National Electrical Code.*

NEMA   See *National Electrical Manufacturers Association.*

NEMA Standards   A group of standard practices developed by the National Electrical Manufacturers Association.

NEMA 12   A standard developed by the National Electrical Manufacturers Association defining the construction requirements for an electrical enclosure that is dust- and oiltight.

Nesting   The practice of placing a software routine or block of data within a larger software routine or block of data. Software subroutines that call other subroutines are an example of nesting.

Network   1) A combination of electric and/or electronic elements that perform a particular function or task; 2) An organization of devices that are interconnected for the purpose of intercommunication; 3) A group of one or more relay ladder language rungs that are displayed as a single unit on a programming terminal screen. The term *page* is also used to refer to a programming terminal network.

Node   1) See *I/O node;* 2) A particular position on a programming terminal screen where an instruction is located.

Noise   The presence of undesirable electric voltages and/or currents which disrupt the normal operation of a system. Noise is generated from numerous sources and can enter an electronic device through signal leads, power sources, and adjacent metal parts.

Noise Filter   An electronic filter network used to reduce and/or eliminate any noise that may be present on the leads to an electric or electronic device.

Noise Immunity   A measure of insensitivity of an electronic system to noise.

Noise Spike   A short burst of electric noise with more magnitude than the background noise level.

Noise Suppressor   See *Noise Filter.*

Nominal   The average value of a measurable quantity halfway between a maximum and a minimum limit.

Nonretentive Instruction   Any PC instruction that must be continuously controlled for operation. For example, nonretentive relay coils return to their OFF state when power flow is lost, as contrasted to retentive relay coils, which remain in their current state until intentionally switched to the opposite state.

Nonvolatile Memory   A memory structure that retains its information whenever power is removed without the aid of a backup power source.

NOR   A logic operation performed between memory bits or signals whereby the result is zero or false whenever one or more of the bits or signals being compared is a one or true state.

Normally Closed   A designation given to a pair of relay contacts that are mated, or closed, whenever the relay coil is not energized. Normally closed contacts pass power whenever the relay coil is deenergized.

Normally Closed Instruction   A PC instruction that operates in a manner similar to the normally closed contact pole of an industrial control relay.

Normally High   A device that provides a high voltage level output (sourcing) in the rest or OFF state.

Normally Low   A device that provides a low voltage level output (sinking) in the rest or OFF state.

Normally Open   A designation given to a pair of relay contacts that are not mated, or closed, when the relay coil is not energized. Normally open contacts pass power whenever the relay coil is energized.

Normally Open Instruction   A PC instruction that operates similar to the normally open contact pole of an industrial control relay.

NOT   A logic function whereby an input state is negated. A one or true input state is changed to a zero or false resultant state and vice versa.

Notation   The process of representing information as a series of predetermined characters, marks, signs, or figures.

Numerical Control   A control scheme based on highly accurate computer-generated control commands used primarily in the machine tool industry.

Octal   A numbering system having 8 as its base or radix. The octal numbering system uses the first eight decimal digits, 0 through 7 for representation.

Odd Parity   The setting of the parity bit such that the number of data bits plus the parity bit equal an odd number.

Off-Delay Timer   1) An electromechanical relay that has contacts which change state a predetermined time period after power is removed from its coil. Upon reenergization of the coil, the contacts return to their shelf state immediately; 2) A PC instruction that emulates the operation of the electromechanical off-delay relay.

Off-Line   Any equipment not actively operating or connected to a system. Off-line equipment may be running or operational, but not performing its intended function.

Off-Line Editing   See *Off-line programming.*

Off-Line Programming   1) A method of PC programming and/or editing where the operation of the processor is stopped and all output devices are switched off. Off-line programming is the safest manner to develop or edit a PC program since the entry of instructions does not affect operating hardware until the program can be verified for accuracy of entry; 2) The capability of a programming terminal to store a program for active edit or development without the need of a processor. The program is usually stored on magnetic tape for later off-line programming sessions, or loaded into the processor for operational checks.

Ohm   The unit of resistance. It is defined as the resistance of a mercury column 106.300 centimeters long, at zero degrees Celsius, and weighing 14.4521 grams. An ohm of resistance is equivalent to a potential difference of 1 volt across a conductor when 1 ampere of current is flowing.

Ohmmeter   A test instrument used to measure resistance.

Ohm's Law   The resistance R in a circuit is equal to the total voltage E applied to the circuit, divided by the current I flowing in the circuit.

On-Delay Timer   1) An electromechanical relay that has contacts which change state a predetermined time period after the coil is energized. The contacts return to their shelf state immediately upon deenergization of the coil; 2) A PC instruction that emulates the operation of the electromechanical on-delay timer.

One-Shot   1) A circuit that produces an output signal for a predetermined time period whenever an input signal is received; 2) A PC instruction that passes power for one processor scan whenever the I/O or internal location to which the instruction is referenced receives power.

On Line   Any operating equipment that is serving its intended function.

On-Line Data Change   The ability of a programming terminal or data access panel to accept and store changes to the contents of selected data memory locations while the processor is solving and executing the user program.

On-Line Editing   See *On-line Programming.*

On-Line Operation   The status of a PC when it is actively solving and executing a user program.

On-Line Programming   The ability of a processor and programming terminal to jointly make user directed additions, deletions, or changes to a user program while the processor is actively solving and executing the commands of the existing user program. Extreme care should be exercised when performing on-line programming to ensure that erroneous system operation does not result. Whenever possible it is usually much safer to make programming changes off-line.

Operand   The quantity upon which a mathematical operation is performed.

Optical Isolator   An electronic device used to isolate electrically and mechanically two electric and/or electronic systems or circuits through the use of optoelectronics.

Optional Features/Instructions   Additional or special-purpose capabilities available in a PC at additional cost to the purchaser.

Optoelectronics   The technology of generating and detecting light with the aid of semiconductor and electronic devices.

Optoisolator   See *Optical Isolator.*

OR (Logic)   See *Exclusive OR* and/or *Inclusive OR.*

OSHA   Abbreviation for Occupational Safety and Health Act, which is a set of governmental rules for employers to follow to guard against job-related employee injuries and illnesses.

Output   1) The current or voltage delivered by a device or circuit; 2) Information transferred from the internal storage locations in a computer or PC to external hardware or storage mediums.

Output Coil Instruction   A PC relay coil instruction that is addressed to a specified output point and used to control the ON/OFF condition of a field output device. During the I/O scan the processor reads the output image table's ON/OFF status for each output coil and either activates or deactivates the addressed output point accordingly.

Output Device(s)   Any physical device that receives data from a computer, PC, or control system. Output devices may be either discrete or analog in nature.

Output Image Table   A portion of a processor's data memory reserved for the storage of output device statuses. A one, ON, or true state in an output image table storage location is used to switch on the corresponding output point.

Output Impedance   The impedance at the output terminals of an electronic device which has had its load disconnected and all power switched OFF.

Output Instruction   The term applied to any PC instruction that is capable of controlling the discrete or analog status of an output device connected to the PC.

Output Register/Output Word   A particular word in a processor's output image table where numerical data are placed for transmission to a field output device.

Overflow   A condition that occurs whenever a data storage location used for a mathematical operation is insufficient to hold the result.

Overload   A load greater than a component or system is designed to handle. An overload condition is characterized by excessive heating, signal distortion, or erroneous system operation.

Overload Relay   A special-purpose relay designed such that its contacts transfer whenever its coil current exceeds a predetermined value. Overload relays are used with electric motors to prevent motor burnout due to mechanical overload.

Override   The ability to switch the control of a device or system from a predetermined sequence of operation to a secondary or manual sequence of operation.

Page   A segment of one or more PC program rungs which are displayed together as a unit on the programming terminal. A programming terminal page may also be referred to as a network.

Parallel Circuit   The ability for an electric current to use two or more paths to reach the load.

Parallel Instruction   A PC instruction used to begin and/or end a parallel branch of instructions being programmed on a programming terminal.

Parallel Operation   The processing of all the digits of a word simultaneously.

Parallel Output   The outputting of data or information from more than one output port simultaneously.

Parallel-to-Serial Conversion   The conversion of binary data contained on parallel buses or wires to binary serial data in a single bus or wire.

Parity   The addition of a noninformation parity bit to a data transmission to detect transmission errors. Parity, often referred to as a vertical redundancy check, is not always reliable for the detection of multiple transmission errors.

Parity Bit   A noninformation bit added to a group of serially transmitted data bits such that the total number of bits in the group is either even or odd.

Passing Power   A term applied to any normally open or normally closed contact which has its contact surfaces mated such that power can flow through the contact assembly. A PC programming terminal displays a normally open or normally closed contact as intensified on the CRT screen when it is passing power.

PC   Abbreviation for programmable controller.

Peak   The greatest positive or negative instantaneous value of a quantity.

Peer-to-Peer   A system of two or more interconnected devices where each device is independently responsible for its actions.

Peripheral Equipment   Any device that is used in conjunction with a PC but is not directly part of a PC, such as a tape loader, printer, etc.

Photocell   An industrial sensor using a beam of light as the means to sense the presence of an object. The photocell emits a narrow beam of incandescent or infrared light to a detector for sensing. Whenever the light beam is interrupted, the photocell signals the condition with the opening or closing of a switch contact.

Photoisolator   See *Optoisolator*.

PID   See *Proportional, Integral, Derivative Control*.

Pilot Device   Any control system sensor or device that indicates that a specific condition is or is not present to either the control system or the operator. The pilot device indicates whether a condition is or is not present and does not directly perform the control function itself.

Pin   A terminal on a connector, socket, plug, or circuit board edge connector.

Plated Wire Memory   A memory system containing wires plated with magnetic materials which can be magnetized in either of two magnetic orientations for the representation of zeros and ones.

Pointer   1) A data memory word used to store the current reference point of an instruction. Any PC instruction that moves or manipulates data often uses a pointer as an index to indicate the current location of operation in the data file or table; 2) A programming terminal CRT position indicator used to indicate the current point of data entry. A PC cursor will usually appear as a reverse-video block (black lettering on a white background) and may flash at a slow rate as well.

Polarity   1) The directional indication of electrical flow in a circuit; 2) The indication of charge as either positive or negative, or the indication of a magnetic pole as either north or south.

Polarized   To configure two devices such that they can only mate in the proper orientation and with the proper partner.

Polling   The routine checking of a device to ensure its correct operation or the routine checking of a terminal or memory location for a specific condition.

Port   A connector or terminal strip used to access a system or circuit. Generally ports are used for the connection of peripheral equipment.

Potential   See *Voltage*.

Potential Difference   The voltage difference between two points in a circuit.

Power Supply   A device used to convert an ac or dc voltage of specific value to one or more dc voltages of a specified value and current capacity. The power supplies designed for use with PCs convert 120 or 240 volts ac to the dc voltages necessary to operate the processor and I/O hardware.

Precondition   A condition or set of conditions that must be met prior to the occurrence of an operation or function.

Preset   A predetermined value placed in a data memory location and used as the limit for a timer or counter instruction.

Printed Circuit Board   A glass-epoxy card with copper foils for electric conductors and electronic components.

Priority   An order of relative importance based on some predetermined value or condition.

Privileged Logic/Privileged Memory   See *Protected Logic/Protected Memory*.

Process   A continuous manufacturing operation.

Process Control   The automatic control of a continuous process or operation. Many process control applications require the continuous comparison and manipulation of large amounts of numerical data.

Processor   A microprocessor-based device that reads input statuses, solves user programming, and controls output devices.

Process Variable   The signal used for control of a PID loop.

Program   A sequence of user-entered processor instructions which when executed provide control of a machine or process.

Programmable Controller   NEMA standard ICS3-1978, Part ICS3-304: "A digitally operating electronic apparatus which uses a programmable memory for the internal storage of instructions for implementing specific functions such as logic, sequencing, timing, counting, and arithmetic to control, through digital or analog input/output modules, various types of machines or processes. A digital computer which is used to perform the functions of a programmable controller is considered to be within this scope. Excluded are drum and similar mechanical type sequencing controllers."

Programmable Read-Only Memory (PROM)   A form of read-only memory (ROM) that is not programmed when fabricated but can be user programmed prior to use. Some PROMs can be erased and reprogrammed a limited number of times, and are referred to as erasable read-only memory (EPROM).

Programming Terminal   A device used to insert, monitor, and edit a PC program. A programming terminal contains a CRT screen for displaying the user program as well as offering printer, tape, and other user functions.

Program Panel   A programming device without a CRT screen and offering less features than a programming terminal.

Program Scan   The solving of all instructions entered into a PC's memory in a designated order from the first to the last instruction.

Program Scan Time   The amount of time required to complete a fixed number of program scans divided by the number of program scans within that time period.

PROM   See *Programmable Read-Only Memory.*

Proportional, Integral, Derivative Control (PID)   A closed-loop process control scheme whereby an output device is controlled by the solution of a mathematical equation containing a proportional, integral, and derivative function. Solution of the proportional function provides output control in direct ratio to the input signal variation. Integral function solution affects the output signal by the summation of input signals up to the present time. The derivative function varies the output signal according to the rate at which the input signal varies within a fixed time period. PID control involves the reading of analog input signals and the control of analog output devices rather than ON/OFF-type discrete I/O devices.

Protected Logic/Protected Memory   A processor feature that permits a user specified portion of the user program and/or data memory to be "protected" from change once entered. PCs offering the protected logic feature often require the manufacturer to enter the user logic that is to be protected or require that the user program the protected logic first following a special set of rules.

Protocol   A predetermined set of conventions specifying the format and timing of message transmissions between two or more communicating devices.

Proximity Switch   An input device utilizing a magnetic field or light source to sense the presence or absence of an object. Magnetic proximity switches generate a magnetic field which can be affected by a metallic object. The magnetic field is sensed by an oscillator which changes frequency as the magnetic field is deformed by the presence of the metallic object. The frequency shift of the oscillator is then used to close the contacts of the device. The photoelectric sensor emits light which is reflected off the object to be sensed. Light reflected back to the device is sensed and used to control the contacts of the device.

Pull Instruction   A PC instruction that fetches the contents of a specified data memory word. Pull instructions are often used to fetch data prior to programming mathematical and data manipulation instructions.

Pulse   A short change in the value of a voltage or current level. A pulse has a definite rise and fall time and a finite duration.

Push Instruction   A PC instruction that places data retrieved buy a pull instruction in a specified data memory location specified as part of the push instruction.

Put Instruction   A PC instruction that places the data retrieved by a get instruction in a specified data memory location specified as part of the put instruction.

Raceway   Any plastic or metal channel or pipe designed to hold wires, cables, or bus conductors.

Rack   1) A housing or framework that is used to hold assemblies; 2) A plastic and/or metal assembly that supports I/O modules and provides a means of supplying power and signals to each I/O module or card.

Rack Fault   Any one of a number of malfunctions occurring with an I/O system. Common rack faults include blown fuses on the I/O modules, loss of communication between I/O modules and processor, and loss of power to the I/O system.

Radio-Frequency Interference (RFI)   Any energy generated at frequencies in the radio transmission band that can affect the operation of electronic equipment.

Radix   The total number of distinct symbols used to represent a number system.

RAM   See *Random Access Memory*.

Ramp   An axis positioning term used to describe a period of time of acceleration or deaceleration.

Random Access Memory   A memory system that permits the random accessing of any storage location for the purpose of either storing (writing) or retrieving (reading) information. RAM systems allow the data to be retrieved and stored at speeds independent of the storage locations being accessed.

Reactance   The opposition that a capacitor (capacitive reactance) and/or resistor (resistive reactance) offer to the flow of alternating current. Reactance is symbolized by X and measured in ohms.

Read 1) The accessing of information from a memory system or data storage device; 2) The gathering of information from an input device or devices or a peripheral device.

Read-Only Memory (ROM) A permanent memory structure where data are placed in the memory's storage locations at time of fabrication, or by the user at a speed much slower than it will be read. Information entered in a ROM is usually never changed once entered.

Readout 1) A device used to convey alphanumeric information from a control system. Most readouts employ fluorescent, light-emitting diode, or liquid crystal display devices; 2) The displaying of information for user purposes.

Read-Write Memory Any memory structure that permits the storing (writing) and retrieving (reading) of information. The reading of information from the memory may destroy the contents of the location being accessed. To prevent information loss during the read cycle, the information may be written back to the same location as part of the read cycle.

Read-Write Time See *Access Time*.

Real Estate A slang term used to describe the area required for a device or component in an enclosure or on a circuit board.

Real Time Any operation that takes place in the actual time required to complete the physical events.

Real-Time Operation 1) Any operation performed with a computer or PC that exactly duplicates the time required to complete the physical event; 2) Any system that responds to the real-time requirements of the physical operation in time to influence the outcome.

Record 1) A grouping of related characters or data stored in memory or on a storage device; 2) The operation of transferring data from a PC to a permanent storage medium.

Recovery Time The amount of time required for a device to return to normal operation after an interruption.

Rectifier A solid-state device that converts alternating current to pulsed direct current. A capacitor added to the output side of a rectifier reduces the amount of pulsing to a steady-state value.

Redundancy A hardware configuration that has the critical components installed in duplicate. Each redundant component provides backup to an identical operating mate. Should the operating component fail, the backup takes over. This transfer from operating unit to backup unit can be manually initiated (cold backup requiring operator assistance), semiautomatically initiated (warm backup requiring little or no operator assistance), or fully automatic (hot backup requiring no operator assistance).

Reed Relay A relay employing reed switches as the method of controlling the power flow in a circuit.

Reed Switch A switching device utilizing two magnetic strips encapsulated in an inert gas filled glass envelope. Each magnetic strip overlaps the other and contains a contact surface for mating. When subjected to a magnetic field from a coil or permanent magnet, the magnetic strips are attracted to each other permitting power to flow from one side of the switch to the other.

Refresh 1) The act of updating a system to the latest available information; 2) The process of periodically reading and rewriting the contents of a dynamic memory in order that the data are not lost. Dynamic memories require refreshing due to the fact that they store information by the placement of electric charges. The level of the charge bleeds down and must be constantly replenished if the memory is to retain its data.

Register   A data memory word used to store information for later use by the user program.

Regulation   The ratio of maximum and minimum levels of a quantity expressed as a percent. The output voltage regulation of a power supply, for example, indicates how well the supply can maintain a constant output voltage level with variations in supply voltage and load current.

Relative Addressing   An addressing scheme that indicates a desired address in terms of another quantity or function.

Relative Humidity   The ratio of the amount of water vapor in the air to the amount required to saturate the air at a given temperature, expressed as a percent.

Relay   An electromechanical device having contacts that can be opened and/or closed by an electric coil. A relay permits the controlling of power flow in one circuit through the actions of another circuit.

Relay Contacts   The contacts of a relay which are either opened and/or closed according to the condition of the relay coil. Relay contacts are designated as either normally open or normally closed in design.

Relay Instruction   See *Internal Coil Instruction; Normally Closed Instruction; Normally Open Instruction;* and *Output Coil Instruction.*

Relay Ladder Diagram   See *Ladder Diagram.*

Relay Ladder Program   See *Ladder Diagram Program.*

Remote I/O   Any I/O system that permits communications between the processor and I/O hardware over a coaxial or twin axial cable. Remote I/O systems permit the placement of I/O hardware any distance from the processor, from immediately adjacent to a maximum of 20,000 cable feet away.

Remote Mode Selection   The ability to select the operating mode of the processor (run, test, program, diagnostic, etc.) from a location remote to the processor using either the programming terminal or a special remote mode selection panel.

Remote Power Supply   A power supply not associated with the processor (local or main power supply). Remote power supplies are used to supply power to an I/O rack, chassis, or base that is part of a remote I/O system. Remote power supplies do not power the I/O devices themselves. The remote power supply is often referred to as an auxiliary power supply.

Report   Any information or data that are output on a printer, CRT screen, or display, regarding the operational status or condition of a system.

Report Generation   The printing or displaying of a report.

Reserved Memory   A portion of memory that is reserved for special processor-related functions and operations.

Reset   1) The act of returning a storage location or device to its initial predetermined condition or status; 2) The integral action of a PID equation.

Resistance   A property of electric conductors that produces a voltage drop and/or a heating effect with current flow.

Resistance Drop   The voltage drop that occurs between two points on a conductor when current is flowing due to the resistance of the wire between those two points.

Resistivity   A measure of the ability of a material to resist the flow of current.

Resistor   A device placed in an electrical circuit that introduces a predetermined amount of resistance to the circuit.

Resolution   A measure or indication of the smallest increment of change that can be determined in the measurement of a variable.

Response Time   The amount of time required for a device to react to a change in its input signal, or a request.

Retentive Instruction   Any PC instruction that does not need to be continuously controlled for operation. Retentive instructions must be initiated to perform their function. Loss of power to the instruction does not halt execution or operation of the instruction. The instruction halts operation only when physically complete or stopped by a specific command.

Retentive Output Instruction   A relay coil instruction that retains its ON/OFF status during power cycling of the processor, provided that the instruction remains "sealed" on power-up.

Retentive Timer   An electromechanical relay that accumulates time whenever the device receives power, and maintains the current time should power be removed from the device. Once the device accumulates time equal to its preset value, the contacts of the device change state. Loss of power to the device after reaching its preset value does not affect the state of the contacts. The retentive timer must be intentionally reset with a separate signal in order for the accumulated time to be reset and for the contacts of the device to return to their shelf state.

Retentive Timer Instruction   A PC instruction that emulates the timing operation of the electromechanical retentive timer.

Retentive Timer Reset Instruction   A PC instruction that emulates the reset operation of the electromechanical retentive timer.

Retroreflective   Employing the use of a special reflector to return a source of light back upon itself.

Retroreflective Photocell   An industrial sensor using a beam of light as the means to sense the presence of objects. The photocell emits a narrow beam of incandescent or infrared light to a retroreflector, which returns the beam back to the unit for sensing. Whenever the light beam is interrupted, the photocell signals the condition with the opening or closing of a switch contact.

Reverse Video   A CRT display characterized by black alphanumeric characters on a white background as contrasted to the standard display of white alphanumeric characters on a black background.

Revision   An alphanumeric designation given to any hardware, firmware, or software portions of a system to indicate the latest changes that have been incorporated for that part of the system.

RFI   See *Radio Frequency Interference.*

Rise Time   The amount of time required for a signal to rise from 10 percent of its final value to 90 percent of its final value.

ROM   See *Read-Only Memory.*

Root Mean Square (RMS)   The effective alternating voltage or current value that produces the same amount of heating effect as is produced with a direct current of equal value. RMS is the square root of the average of the squares of values taken of an alternating current or voltage. The RMS value for 60-hertz ac power used in the United States is 0.707 times the peak value.

Rotate Instructions   A family of PC instructions that perform the rotating of matrix bits or the contents of contiguous data memory words. Rotate instructions may be referred to as shift instructions. However, rotate instructions recycle the outshifted data back to the beginning of the matrix instead of discarding them like the shift instructions.

Routine   A series of instructions that perform a specific function or task.

RS-232C   A standard specifying both the electrical and mechanical characteristics of an interface between data terminal equipment (DTE) and data communications equipment (DCE) developed by the Electronic Industries Association.

Run   The single continuous execution of a program by a PC.

Run-In   The initial operation of a system immediately after it has been checked out and before it is placed in full production operation.

Rung   A group of PC instructions that control a single PC output instruction. A group of rungs constitute a ladder diagram program.

Sample Time   The time between successive readings of input signals to a PID loop.

Scan   The sequential examining and solving of each user instruction by a PC processor from the first instruction to the last.

Scanner Module   A circuit board or module contained within a PC processor which is responsible for updating the input image table with the current system statuses, as well the output modules or cards with the current output image table conditions.

Scan Time   The time required to scan and solve a PC program completely, update all I/O statuses, perform communications with a programming terminal and/or data communications network, and perform any internal processor housekeeping routines.

Schematic   A diagram of graphic symbols representing electric and/or electronic devices representing the electrical scheme of a circuit.

Scratchpad Memory   A section of processor memory that is not user accessible and is used solely by the processor for temporary storage of subtotals, logic equation results, or other operating system and/or user logic results.

Screen   The surface of a cathode ray tube (CRT) upon which a visible pattern is produced.

Scroll   A programming terminal feature that permits the user to index sequentially through each ladder diagram program rung. Most PCs permit the scrolling of the user ladder diagram program from either top to bottom or bottom to top, and on either a one rung at a time or a multiple rung at a time basis.

Seal   A programming technique whereby once a coil instruction is energized, a normally open contact associated with the coil (called the seal contact) is used to maintain power to the coil even though the conditions which initially energized the coil may no longer be present. The coil instruction is deenergized by a contact or multiple contact instructions programmed in series with the contact instruction acting as the seal contact. A seal is often referred to as an electric latch.

Search   A programming terminal feature that permits the user to perform a systematic examination of the entire user program to identify each occurrence of a particular instruction or data memory location for purposes of program editing or verification.

Segment   See *Page*.

Self-Diagnostic   The ability for equipment and/or software to monitor itself continuously for improper operation and to take the necessary steps to correct the problem or shut down, as well as to indicate the nature of the problem that exists.

Semantics   The relationship between a symbol or a group of symbols and their meaning.

Semiconductor   A material whose resistivity is inbetween that of a conductor and an insulator. The material also exhibits the ability to have its resistivity varied by light, electric and magnetic fields, or pressures and vacuums.

Sensitivity   The ratio of a device's response with respect to the magnitude of the quantity that can be measured by the device.

Sensor   A device used to gather information by the conversion of a physical occurrence to an electric signal or signals.

Sequencer   A mechanical, electric, or electronic device that can be programmed by setting cams, pins, or switches such that a predetermined set of events occurs repeatedly.

Serial Operation   The time-sequential transmission of data over a single data channel or wire as contrasted to parallel operations involving the simultaneous transmission of data over multiple data channels or wires. Serial transmissions are slower than parallel transmissions for equivalent clock speeds.

Series Circuit   The connecting of circuit components end to end to form a continuous chain of elements such that equal current flows through each component.

Series Instructions   The logic ANDing of instructions such that each instruction of the sequence must be true or passing power if the result or output device is to be true or energized.

Series Operation   The processing of all the bits of a word one after another.

Series-to-Parallel Conversion   The conversion of binary serial data contained on a single bus or wire to binary data contained on multiple parallel buses or wires.

Setpoint   The desired operating point of a system or operation.

Settling Time   The amount of time required for an analog I/O module to settle within its specified error band following a specified change in an operating condition or status.

Shelf Conditions   The conditions under which a device or material should be stored for long periods of time.

Shelf Life   The length of time that a device or material can be stored under specified conditions and still remain usable.

Shelf State   The characteristics of a device during storage. Is used to describe the characteristics of an installed device when it is in its home or rest state.

Shield   See *Foil*.

Shielded Cable   A single-conductor or multiconductor cable that contains one or more shields intended to reduce the effects of RFI and/or EMI on the cable conductors.

Shielding   The practice of placing a shield either over or under the insulation jacket of a cable in order to reduce the effects of the electric field generated by the cable when current flows through the conductors, or to reduce the effect of outside RFI and/or EMI on the conductors of the cable.

Shield Wire   See *Drain Wire*.

Shift Instructions   A family of PC instructions that perform the shifting of matrix bits or the contents of contiguous data memory words. Shift instructions may be referred to as rotate instructions. However, shift instructions do not recycle the outshifted data back to the beginning of the matrix as the rotate instructions do.

Shift Register   A digital storage circuit consisting of relays or electronic latches used to shift binary information from one position to the next upon receipt of a clock pulse. Information can be shifted to either the right or the left and at more than one position at a time if additional circuitry is present. New information is shifted into the circuit at the first relay or latch, and shifted from the circuit at the last relay or latch. All shift registers are clocked either asynchronously or synchronously.

Shock   A sudden impact applied to an object.

Shock Mount   A resilient support used to isolate shock between two structural members.

Shock Test   A special-purpose test used to determine the ability of a device to withstand shock and/or vibration.

Short Circuit   An undesirable path of very low resistance in a circuit between two points.

Short Circuit Protection   Any fuse, circuit breaker, or electronic hardware used to protect a circuit or device from severe overcurrent conditions or short circuits.

Shrink Tubing   A nonmetallic tubing slid over a wire splice to provide an insulation jacket. Shrink tubing is fabricated such that when heated it shrinks to the contour or the wire or splice. Shrink tubing may be applied to cables and multiple conductors to provide mechanical protection as well.

Signal   Any visible, audible, or similar action that is intended to convey information.

Signal Ground   The ground circuit for any low-level signal especially susceptible to noise.

Signal-to-Noise Ratio   The magnitude ratio of a signal to its background noise. This ratio is usually expressed in decibels.

Sign Bit   A memory bit used to indicate the sign of a number. In general if the sign bit is a zero, the number is positive, and if the sign bit is a one, the number is negative.

Significant Bits   Any bit that contributes to the significance of the data being conveyed. The quantity of significant bits begins with the bit contributing the least to the data (least significant bit) and ends with the bit contributing the most to the number (most significant bit).

Significant Digits   Any digit that contributes to the significance of the number being conveyed. The quantity of significant digits begins with the digit contributing the least to the number (least significant digit) and ends with the digit contributing the most to the number (most significant digit).

Simplex   A communications scheme that permits data to be transferred in one direction only, between two devices.

Simulation   The testing of a device or system to ensure correct operation under various conditions.

Simulator   Any computer, instrumentation, bank of switches and/or lights, or test equipment used to perform simulation on a device or system.

Single-Line Diagram   A form of electrical diagram whereby a single line is used to represent one or more conductors between devices and components. The purpose of a single-line diagram is to convey the overall interconnection of devices and components in a system rather than the discrete wiring necessary for the system.

Single Phasing   A condition occurring in three-phase systems whereby one of the phases is lost. If left undetected, single phasing can damage electric equipment subjected to the condition.

Sinking   The internal electrical configuration of an electronic device operating on direct current (dc). A device is referred to as sinking when the current flow is into its signal terminals. For example, a sinking input module has the current flow from the field input device (input device connected to positive) to the input module (input module connected to negative).

Skip   An instruction that permits the bypassing of selected portions of the user program. Skip instructions are conditional whenever the operation is determined by a set of preconditions, and unconditional whenever they are programmed to occur every time executed. Skip instructions can only instruct the processor to pass over selected portions of the user program after the skip instruction.

Slave   Any device that requires and/or receives instruction from another device, usually referred to as the master. A slave may often be capable of short-time operation without direction from the master.

Sleeving   A tough tubular insulation placed over a wire or cable to provide protection against physical damage.

Slow Death   The long-term reduction in the performance of an integrated circuit or transistor due to the collection of ions in the semiconductor material of the device.

Small-Scale Integration   A monolithic integrated circuit which contains less than ten logic gates or logic gate equivalent circuits.

Software   Any program, subroutine, operating system, assembler, or compiler used with a PC as contrasted to the physical equipment itself (hardware).

Solenoid   A bobbin of wire containing a movable steel core which when energized causes the core to move. The operation of the solenoid is used to move a small mechanical part a short distance of usually 1 inch or less.

Solenoid Valve   An electrically operated valve used to control the ON/OFF flow of liquids and gases.

Solid Conductor/Wire   A conductor composed of a single strand of wire.

Solidly Grounded   Any electric system or device that is directly grounded and contains minimal ground circuit impedance.

Solid State   Any circuit or component that uses semiconductors or semiconductor technology for operation.

Solid-State Contact   An electronic equivalent to a mechanical-type electric contact. Any electronic device incorporating a transistor, a silicon-controlled rectifier (SCR), or a triac to control the ON/OFF flow of electric power is often referred to as containing a solid-state contact or switch.

Solid-State Device   Any semiconductor-based element that controls the flow of current without the use of a moving part, heated electrode, or vacuum gap.

Source Register, Table, Word, File   The location of information to be retrieved from for an instruction or operation.

Sourcing   The internal electrical configuration of an electronic device operating on direct current. A device is referred to as sourcing when the current flow is out of its signal terminals. For example, a sourcing input module has the current flow from the input module (input module connected to positive) to the field input device (input device connected to negative).

SPACE   The absence of a communications link signal. SPACE represents the open condition, lack of power flow, an OFF signal, or a binary zero state.

Spike   See *Noise Spike.*

Splice   An in-line connection between two conductors.

SSI   See *Small-Scale Integration.*

Standard Features   Any feature offered with a device at its base price.

Standby   Any duplicate piece of hardware or system component available for immediate use should the primary hardware or component become inoperative.

Startup   The time period between the completion of equipment or system installation and the moment when the equipment or system is in full operation. The startup period can be broken down into a checkout period and run-in period.

State   1) The current condition of a circuit or device; 2) The zero or one status of a signal or storage location.

Static Memory   A memory system that does not require constant refreshing in order to maintain information. A static memory system retains its information as long as power is present. Loss of power to the system results in total memory loss.

Station   Any device connected to a data communications or control network for the purpose of transmitting and/or receiving data.

Status Area   A portion of memory set aside for the storage of processor and system statuses.

Step   The next sequential unit of a contiguous group of elements, functions, or operations.

Stepping Motor   A bidirectional motor designed to rotate in predetermined angular increments rather than continuous rotation.

Stepping Relay   A relay containing multiple sets of contacts designed such that different sets or combinations are activated each time the relay is energized. The number of relay steps determines the number of times the relay can be energized before it must be reset back to the first set of contacts. Each step of the relay may have various arrangements of normally open and normally closed contacts.

Stepping Switch See *Stepping Relay.*

Storage See *Memory.*

Strand One of the wires or a group of wires that make up a larger stranded conductor.

Stranded Conductor/Wire A conductor composed of a group of uninsulated wires or a combination of groups of uninsulated wires. Stranded wire offers much greater flexibility and resistance to breaking than solid conductor wire.

Stranding Effect The property of stranded conductors to exhibit greater resistance than a solid conductor of equal length because of the effective increased length due to the spiral wrapping of the individual wires making up the stranded conductor.

Subroutine A portion of a larger user program, which may be "called" and executed any number of times during a single scan of the PC.

Subtract A PC instruction that performs the mathematical subtraction of one number from another.

Suppression The reduction or elimination of undesirable noise, EMI, and/or RFI by the use of shielding, filtering, grounding, or the relocation of components.

Surge Any sudden current or voltage change in a circuit.

Surge Suppressor A semiconductor device that exhibits a quick reduction in its resistance whenever the voltage across the device exceeds a certain value. Surge suppressors are often used to limit the magnitude of a noise spike on the power lines to electronic equipment.

Switch A mechanical or solid-state device used to close or open a path of electric current flow as well as redirect it from one circuit to another.

Switch Gear Any electric hardware used in the generation, transmission, and distribution of electric power.

Symbol A character or graphic used to represent a piece of hardware, component, or device.

Synchronous Shift Register A shift register that is clocked such that data are shifted one position per clock period.

Synchronous System A system that operates on a constant-time-interval rather than an as-needed basis.

System An assembly of components, devices, and/or hardware linked together in such a manner as to form an organized unit capable of performing a specific function or operation.

System Reliability A stated probability that a system will perform its intended function or operation under a specified set of conditions.

Table A contiguous group of data memory words of specified length.

Tap A branch off a main trunk cable or conductor.

Tape A strip of magnetic material, paper, or metal foil used to store information on.

Tape Cassette   A magazine or holder designed to hold a specified length of tape. The tape cassettes used for PC data recording resemble the standard audio cassette in appearance. The latter should not be substituted for a digital cassette since digital cassettes are designed to different specifications.

Tape Recorder   A peripheral device used to record the binary data of a PC's memory onto a digital tape cassette for storage. The tape recorder is similar in appearance to that of the standard audio cassette with several major exceptions. The PC tape recorder is housed in an industrial package, designed to automatically operate with a PC, and has a frequency response designed for optimum digital signal response rather than a flat audio response.

Temperature Derating   The lowering of the voltage, current, or power rating of a device or system whenever it is used above a specified temperature.

Temperature Rise   The difference between a component or device's final temperature and its initial temperature.

Terminal   1) A connection point for one or more wires to a circuit board, module, or second grouping of wires; 2) An electronic device used to gain access to a PC or computer.

Terminal Address   The alphanumeric address assigned to a particular I/O point.

Terminal Box   An enclosure containing terminal blocks used for the termination of wires.

Terminator   A load device used at the end of a transmission cable. A terminator is designed to have the same impedance as the line it terminates in order to eliminate reflected waves within the cable.

Test Mode   An operating mode that permits checking out of a system or device without actual production operation of the system.

Thermal Cutout   See *Overload Relay.*

Thermocouple   A temperature-measuring device that utilizes two dissimilar metals for temperature measurement. As the junction of the two dissimilar metals is heated, a proportional voltage difference is generated which can be measured.

Three Phase   An electric power distribution and utilization scheme requiring three power wires, each wire differing in phase angle by 120 electrical degrees from the other two.

Throughput   A measure of a system's overall efficiency in performing its intended function or operation.

Thumbwheel Switch   A rotating numerical switch used to input numerical information into a PC. A thumbwheel switch will encode the 10 decimal digits in a binary, binary-coded-decimal (BCD), or similar digital format prior to input to the PC.

Timed Contact   A normally open and/or normally closed contact that is actuated at the end of a timer's time delay period.

Time-Delay Relay   See *Off-Delay Timer; On-Delay Timer;* and *Retentive Timer.*

Time-Out   A predetermined interval of time allotted for a specific operation to be completed before the operation is automatically terminated. Timeouts are often used in communications links to guarantee the completion of a transmission.

Timer   See *Off-Delay Timer; On-Delay Timer;* and *Retentive Timer.*

Toggle   To cycle ON and OFF repeatedly.

Toggle Switch   A two- or three-position snap switch designed for panel mounting. The toggle switch is operated by an extended lever and operates one or more normally open and/or normally closed contacts.

Tolerance   The permissible deviation of a quantity from the desired value.

Transducer   A device used to convert mechanical, sonic, light, radioactive, or vibratory signals to electric signals.

Transfer Line   A manufacturing system composed of individual stations performing dedicated operations. The manufactured assembly must sequentially pass from one station to the next.

Transformer   An electric device that converts the electric energy of a circuit into one or more circuits of different voltage and current ratings.

Transformer Coupling   The use of a transformer to isolate two circuits from each other electrically. Early PCs employed transformer coupling as part of each I/O point to isolate the field wiring circuits from the processor circuits.

Transient   An undesirable momentary surge or spike in a signal or power line due to rapid voltage and/or current changes within the system.

Transistor   A three-terminal active semiconductor device composed of silicon or germanium, which is capable of switching or amplifying an electric current.

Transitional Contact Instruction   A PC contact instruction that passes power whenever the coil referenced by the instruction changes state. Two types of transitional contact instructions are available, the off-transitional and the on-transitional contact instructions. PCs that do not offer a transitional or one-shot instruction may be programmed by using a self-resetting counter instruction set to a preset of one.

Translator   A special-purpose computer program designed to examine the memory contents of a PC and provide a documented printout of those contents with mnemonics and cross-referencing.

Transmission Line   One or more electric conductors used to convey information between various locations.

Triac   A five-layer semiconductor device used to control or switch alternating current.

Triaxial Cable   A special three-conductor coaxial cable.

Trickle Charge   A very slow charge rate applied to a battery to compensate for internal losses. A trickle charge is able to restore a battery to a fully charged condition after small discharges and to keep the battery in a fully charged condition for future use.

Trip Relay   A relay designed to trip a circuit breaker or contactor into the open state even though the circuit breaker or contactor is attempting closure by an external source or a control source.

True   As related to PC instructions, an ON or one state.

Truncation   The dropping of one or more digits from a number without affecting any of the remaining digits. For example, 12345.678 would become 12345.6 when truncated to six digits, but it becomes 12345.7 when rounded to six digits.

Truth Table A matrix-type listing of the output state for a logic circuit according to each possible combination of input states.

Twinaxial Cable A single twisted pair of conductors with a single outer shield.

Twisted Pair Two insulated wires twisted about each other.

UART See *Universal Asynchronous Receiver Transmitter.*

UL See *Underwriters Laboratories.*

Ultrasonics The technology of generating and detecting sounds above the frequency of 20,000 hertz.

Ultraviolet-Erasable PROM A PROM that can be erased by exposure to strong ultraviolet light. These types of PROMs cannot be erased in part, and must be completely erased prior to reprogramming. They have a limited number of times they may be erased and reprogrammed.

Underload Relay A special-purpose relay designed such that its contacts transfer whenever its coil current drops below a predetermined value.

Underwriters Laboratories An independent laboratory that tests equipment to ensure its safe operation under normal conditions.

Universal Asynchronous Receiver Transmitter (UART) A device designed to interface a group of parallel data buses to a serial data bus. This device is commonly used to convert processor data to serial data for transmission over a communications link.

Unlatch Instruction One-half of a PC instruction pair which emulates the unlatching action of a latching relay. The unlatch instruction deenergizes a specified output point or internal coil until reenergized by a latch instruction. The output point or internal coil remains deenergized regardless of whether or not the unlatch instruction is energized. The output point or internal coil retains its ON or OFF state through the power cycling of the processor.

UV EPROM See *UltraViolet-Erasable Programmable Read-Only Memory.*

V The symbol for voltage.

VA See *Volt-Ampere.*

Variable A quantity that can be changed, measured, or altered by a PC processor as the user program is executed.

Variac An autotransformer designed such that the output voltage from the device is continuously variable from zero to 117 percent of the input voltage value.

Vertical Redundancy Check See *Parity.*

Volatile Memory A memory structure that loses its information whenever power is removed. Volatile memories require a battery backup to ensure memory retention during power outages.

Volt A unit of electromotive force. The volt is defined as the potential required to make a current of 1 ampere flow through a resistance of 1 ohm.

Voltage The electric pressure or force that causes electric current to flow through a conductor.

Voltage Drop The difference in voltage level at two points in a circuit. Voltage drop is due to the electric losses of the conductors and components located between the two measured points of the circuit.

Voltage Rating The maximum applied working voltage level that a device or component can withstand continuously without damage or breaking down.

Volt-Ampere A unit of apparent power in an ac circuit. The volt-ampere rating of a circuit or device is the working voltage of the circuit or device multiplied by the current required by the circuit or device. Any difference in phase between the voltage and current readings is not taken into consideration.

W The symbol for watt.

Warm BackUp A redundancy technique that requires the on-line PC or device to inform constantly its redundant backup as to current operating conditions and statuses. In the event of on-line device failure the backup unit may or may not be able to proceed exactly from where the on-line unit left off.

Watchdog Timer A special-purpose timer designed to monitor the scan of a PC. The PC must generate a cyclic ON and OFF signal with each processor scan for proper operation of the watchdog timer. The timer is preset to monitor both the ON and OFF duration of the signal being generated by the PC. Should the PC stop scanning, the output signal would remain in either an ON or an OFF state. This condition would be picked up by the watchdog timer and used to emergency stop the system being controlled by the PC. Watchdog timers are available for external connection to a PC as well as being offered on some models as an internal processor safety shutdown circuit.

Watt A unit of electric power equivalent to a joule per second or the amount of power expended when 1 ampere of current flows through a resistance of 1 ohm.

Weighted Value The numerical value assigned to a bit or digit as a function of its position within a complete word or number.

Wetware Term used in reference to the human mind. Wetware is the intelligence that creates hardware, firmware, and software and is responsible for its correct operation and maintenance.

Wire A solid or stranded conductor and its associated insulation jacket.

Wire Gauge See *American Wire Gauge*.

Word A group of contiguous bits treated as a unit.

Word Length The total number of bits that comprise a word. Most PCs use either 8 or 16 bits to form a word.

Write 1) The transferring or recording of information from a memory system to a storage device; 2) The outputting of information to an output device or devices or to a peripheral device.

21

Who's Who in Programmable Controllers

INTRODUCTION

The surveys included as part of this chapter are designed to give the reader a flavor for the overall capabilities of various PCs available in the marketplace. The survey was sent to all of the PC manufacturers listed in the Appendix. The results of those who responded are included as part of this chapter. Black-and-white photographs were requested, and have been included if they were provided.

The reader is warned to be cautious of the information listed by each manufacturer. Surveys are always open to interpretation, on both the part of the people filling them out and those who are reviewing them. The information contained in each survey should be reviewed with the respective manufacturer before it is taken for gospel. The information is unedited by the author, and questions regarding a particular survey entry should be directed to the PC manufacturer involved.

The reluctance of a particular manufacturer to respond should not be taken as an indication that he is not interested in selling hardware. Legal and time restraints may have preempted a manufacturer from responding within the time period allotted. This survey is similar to those often published in the magazines listed in the Appendix, and these periodicals can be consulted for similar information. It will be the actual day-to-day contacts and working with a manufacturer that will provide the potential PC user with information and insight on the true abilities of a manufacturer.

ADATEK, INC.		SYSTEM I	SYSTEM II	SYSTEM III

PROCESSOR

YEAR INTRODUCED		1982	1982	1983
MEMORY SIZE TOTAL		9K	9K	9K
MEMORY INCREMENTS		2K	2K	2K
MEMORY TYPES		1K RAM,2K EPROM 6K EITHER	1K RAM,2K EPROM 6K EITHER	1K RAM,4K EPROM 4K EITHER
VOLTAGE		120VAC	120VAC	120VAC
FREQUENCY		60	60	60
HUMIDITY	MIN			0°C
	MAX			50°C
TEMPERATURE	MIN	0°C	0°C	0°C
	MAX	50°C	50°C	50°C
USER MEMORY	MIN	2K	2K	2K
	MAX	6K	6K	4K
DATA MEMORY	MIN	2K	2K	2K
	MAX	6K	6K	4K
USER CONFIGURABLE		YES	YES	YES
WORD SIZE		8 BITS	8 BITS	8 BITS
WORDS/CONTACT - COIL		N/A	N/A	N/A
WORDS/TIMER - COUNTER		N/A	N/A	N/A
SCAN TIME - USER LOGIC (SEE NOTE 1)		VARIABLE	VARIABLE	VARIABLE
I/O SCAN TIME		2 MS	2 MS	2 MS

SUPPORT

	SYSTEM I	SYSTEM II	SYSTEM III
FACTORY APPLICATIONS ENGINEERS	YES	YES	YES
DOCUMENTATION SERVICE	YES	YES	YES
SERVICE CENTER (24 HRS.)	FACTORY	FACTORY	FACTORY
MAINTENANCE TRAINING	YES	YES	YES
APPLICATION TRAINING	YES	YES	YES
WARRANTY TERMS	2 YEARS	2 YEARS	2 YEARS
DISTRIBUTION	DISTRIBUTORS	DISTRIBUTORS	DISTRIBUTORS
MAINTENANCE CENTERS	FACTORY	FACTORY	FACTORY

NOTES :
  1) ONLY ACTIVE STATE MODULES ARE SCANNED

ADATEK, INC.	SYSTEM I	SYSTEM II	SYSTEM III

DISCRETE I/O MODULES

AC VOLTAGES/CURRENT			
DC VOLTAGES/CURRENT		USE ANY OPTO 22	
ISOLATED AC VOLTAGES/CURRENT		QUAD MODULE	
ISOLATED DC VOLTAGES/CURRENT			
CONTINUOUS OUTPUT RATING			
COMMON POINTS/MODULE	2	2	2
OUTPUT FUSING CONFIGURATION	1 FUSE/COMMON	1 FUSE/COMMON	1 FUSE/COMMON
INDICATORS — I/O POINTS	LOGIC	LOGIC	LOGIC
INDICATORS — FUSING	YES	YES	YES
ISOLATION			

ANALOG I/O MODULES (SEE NOTE 2)

VOLTAGES: SINGLE			0-5,0-10
VOLTAGES: DIFFERENTIAL			0-5,0-10
CURRENT: SINGLE			4-20MA
CURRENT: DIFFERENTIAL			4-20MA
SPECIAL PURPOSE INPUT			THERMOCOUPLE
SPECIAL PURPOSE OUTPUT			
INDICATORS:			RACK STATUS FAULT

SPECIAL I/O MODULES

SIMULATION	YES	YES	YES
ASCII			
AXIS POSITIONING	NO	NO	YES (SEE NOTE 3)

I/O NODE DATA

TOTAL NODES/CHANNEL	1	10	10
TOTAL DISCRETE POINTS	24	24	24
MAXIMUM POINTS ASSIGNABLE — INPUT	24	24	24
MAXIMUM POINTS ASSIGNABLE — OUTPUT	24	24	24
MAXIMUM ANALOG ASSIGNABLE — INPUT	N/A	N/A	N/A
MAXIMUM ANALOG ASSIGNABLE — OUTPUT	N/A	N/A	N/A
DISTANCE FROM DROP	50 FT	50 FT	50 FT

NOTES :
2) SINGLE CHANNEL, 8 PER RACK, ANY MIX.
3) SUPPORTS CLOSED LOOP SERVO FROM ABSOLUTE CONTROLS, INC.

ADATEK, INC.	SYSTEM I	SYSTEM II	SYSTEM III

I/O STATISTICS

TOTAL DISCRETE POINTS AVAILABLE		48	240	240
MAXIMUM POINTS ASSIGNABLE	INPUT	48	240	240
	OUTPUT	48	240	240
NUMBER OF CHANNELS		2	2	2
NUMBER OF NODES/CHANNEL		1	10	10
MODULE I/O POINT DENSITY		4	4	4
ISOLATED MODULE I/O DENSITY		ALL I/O ISOLATED	ALL I/O ISOLATED	ALL I/O ISOLATED
TOTAL ANALOG CHANNELS AVAILABLE		0	0	48
MAXIMUM CHANNELS ASSIGNABLE	INPUT			48
	OUTPUT			48
I/O MAPPING		USER DEFINED	USER DEFINED	USER DEFINED
INTERRUPTS		1	1	1
I/O LOGIC UPDATING	ASYNC			
	SYNC	YES	YES	YES
ERROR CHECKING METHOD		IMAGE REFRESH	IMAGE REFRESH	IMAGE REFRESH
MODULES PER CHANNEL		6	6	6
APPLICABLE PROCESSORS		ALL	ALL	ALL
YEAR OF INTRODUCTION		1982	1982	1983
I/O SYSTEM STRUCTURE		I/O POINT	RACK CHANNEL	RACK CHANNEL

CONTROL NETWORK

TYPE			HOST - PC
NUMBER OF PC'S			32
MAXIMUM LENGTH			1500 FT
TYPE OF CABLE			TWISTED PAIR
OPERATION W/PROCESSOR			ALL
DISCRETE SIGNALS TRANSFERRED			
DATA SIGNALS TRANSFERRED			8 BITS
COMMUNICATION SPEED			19.2K ASYNC 56K SYNC

ADATEK, INC.	SYSTEM I	SYSTEM II	SYSTEM III

INSTRUCTION SET

	SYSTEM I	SYSTEM II	SYSTEM III
CONTACTS			
TRANSITIONAL			
LATCHES			
RETENTIVE COILS			
SKIP/JUMP			
SUBROUTINE	YES	YES	YES
TIMERS	YES	YES	YES
UP COUNTER	YES	YES	YES
DOWN COUNTER	YES	YES	YES
ADD/SUBTRACT	INTEGER BASIC ±32767	INTEGER BASIC ±32767	INTEGER BASIC ±32767
MULTIPLY/DIVIDE	INTEGER BASIC ±32767	INTEGER BASIC ±32767	INTEGER BASIC ±32767
>,=,<			
FILE TRANSFER			
LOGICAL			
BIT ROTATE/SHIFT			
FIFO/LIFO			
SEARCH			
DIAGNOSTIC			
SEQUENCER			
PID			
ASCII			
AXIS POSITIONING	NO	NO	NO
MACHINE LANGUAGE	YES	YES	YES
BOOLEAN ALGEBRA	YES	YES	YES
HIGH LEVEL LANGUAGE (SEE NOTE 4)	YES	YES	YES

NOTES :
   4) TINY BASIC AND ADATEK'S PROPRIETARY CONTROL LANGUAGE, PSM.

ADATEK, INC.		SYSTEM I	SYSTEM II	SYSTEM III

I/O CHANNEL DATA

TOTAL DISCRETE POINTS		24	240	240
MAXIMUM POINTS ASSIGNABLE	INPUT	24	240	240
	OUTPUT	24	240	240
MAXIMUM ANALOG ASSIGNABLE	INPUT			48
	OUTPUT			48
MAXIMUM LOCAL CHANNELS		48	240	240
MAXIMUM REMOTE CHANNELS		N/A	N/A	N/A
LOCAL CHANNEL DISTANCE		50 FT	50 FT	50 FT
REMOTE CHANNEL DISTANCE		N/A	N/A	N/A
TYPE LOCAL CHANNEL CABLE		PARALLEL	PARALLEL	PARALLEL
TYPE REMOTE CHANNEL CABLE		N/A	N/A	N/A
LOCAL CHANNEL COMMUNICATION		PARALLEL	PARALLEL	PARALLEL
REMOTE CHANNEL COMMUNICATION		N/A	N/A	N/A
LOCAL CHANNEL DATA RATE		100K	100K	100K
REMOTE CHANNEL DATA RATE		N/A	N/A	N/A
FAULT MONITORING				

DATA NETWORK

TYPE			HOST - PC
NUMBER OF NODES			32
MAXIMUM LENGTH			15000 FT
TYPE OF CABLE			TWISTED PAIR
COMMUNICATION SPEED			19.2K ASYNC 56K SYNC
COMPUTER INTERFACES			
SPECIAL INTERFACES			RS232, MODEM

PROGRAMMING TERMINAL

USES ANY DUMB TERMINAL

ALLEN-BRADLEY		PLC-2 MINI	PLC-2/15	PLC-4

PROCESSOR

YEAR INTRODUCED		1979	1981	1983
MEMORY SIZE TOTAL		1K	2K	640 TO 4K
MEMORY INCREMENTS		1/2K	1/2K	640 WORDS
MEMORY TYPES		RAM	RAM & EPROM	RAM & EPROM
VOLTAGE		120V,220VAC	120V,220VAC	120,220VAC,24VDC
FREQUENCY		47-63HZ	47-63HZ	47-63HZ
HUMIDITY	MIN	5%	5%	5%
	MAX	95%	95%	95%
TEMPERATURE	MIN	0°C	0°C	0°C
	MAX	60°C	60°C	60°C
USER MEMORY	MIN	128	128	N/A
	MAX	1K	1920	512
DATA MEMORY	MIN	48	48	N/A
	MAX	128	1920	128
USER CONFIGURABLE		YES	YES	NO
WORD SIZE		16 BIT	16 BIT	16 BIT
WORDS/CONTACT - COIL		1	1	1
WORDS/TIMER - COUNTER		2	2	1
SCAN TIME - USER LOGIC		22 MSEC	18 MSEC/K	15 MSEC
I/O SCAN TIME		1 MSEC	2 MSEC	10 MSEC
COLOR GRAPHICS		YES	YES	YES
REPORT GENERATION		YES	YES	YES

SUPPORT

	PLC-2 MINI	PLC-2/15	PLC-4
FACTORY APPLICATIONS ENGINEERS	YES	YES	YES
DOCUMENTATION SERVICE	YES	YES	YES
SERVICE CENTER (24 HRS.)	YES	YES	YES
MAINTENANCE TRAINING	YES	YES	YES
APPLICATION TRAINING	YES	YES	YES
WARRANTY TERMS	1 YEAR	1 YEAR	1 YEAR
DISTRIBUTION	YES	YES	YES
MAINTENANCE CENTERS	YES	YES	YES

ALLEN-BRADLEY	PLC-2 MINI	PLC-2/15	PLC-4

INSTRUCTION SET

	PLC-2 MINI	PLC-2/15	PLC-4
CONTACTS	YES	YES	YES
TRANSITIONAL	NO	NO	NO
LATCHES	YES	YES	YES
RETENTIVE COILS	ES	YES	YES
SKIP/JUMP	NO	YES	NO
SUBROUTINE	NO	YES	NO
TIMERS	40	1000	32
UP COUNTER	40	1000	32
DOWN COUNTER	40	1000	32
ADD/SUBTRACT	YES	YES	NO
MULTIPLY/DIVIDE	YES	YES	NO
>,=,<	YES	YES	YES
FILE TRANSFER	NO	YES	NO
LOGICAL	NO	YES	NO
BIT ROTATE/SHIFT	NO	NO	NO
FIFO/LIFO	NO	NO	NO
SEARCH	NO	NO	NO
DIAGNOSTIC	NO	NO	NO
SEQUENCER	NO	YES	YES
PID	YES	YES	NO
ASCII	YES	YES	NO
AXIS POSITIONING	YES	YES	NO
MACHINE LANGUAGE	YES	NO	NO
BOOLEAN ALGEBRA	NO	NO	NO
HIGH LEVEL LANGUAGE	NO	NO	NO

ALLEN-BRADLEY		PLC-2 MINI	PLC-2/15	PLC-4

I/O STATISTICS

		PLC-2 MINI	PLC-2/15	PLC-4
TOTAL DISCRETE POINTS AVAILABLE		128	128	32 TO 256
MAXIMUM POINTS ASSIGNABLE	INPUT	128	128	20 TO 160
	OUTPUT	128	128	12 TO 96
NUMBER OF CHANNELS		1	1	N/A
NUMBER OF NODES/CHANNEL		1	1	N/A
MODULE I/O POINT DENSITY		8	8	N/A
ISOLATED MODULE I/O DENSITY		6	6	N/A
TOTAL ANALOG CHANNELS AVAILABLE		64	64	N/A
MAXIMUM CHANNELS ASSIGNABLE	INPUT	64	64	
	OUTPUT	32	32	N/A
I/O MAPPING		YES	YES	YES
INTERRUPTS				NO
I/O LOGIC UPDATING	ASYNC			
	SYNC	YES	YES	YES
ERROR CHECKING METHOD		PROPRIETARY	PROPRIETARY	PROPRIETARY
MODULES PER CHANNEL		16	16	N/A
APPLICABLE PROCESSORS		ALL	ALL	/A
YEAR OF INTRODUCTION		1979	1981	1983
I/O SYSTEM STRUCTURE		RACK/MODULE	RACK/MODULE	FIXED

CONTROL NETWORK

	PLC-2 MINI	PLC-2/15	PLC-4
TYPE	N/A	N/A	RING
NUMBER OF PC'S	N/A	N/A	8/CHANNEL
MAXIMUM LENGTH	N/A	N/A	4000
TYPE OF CABLE	N/A	N/A	TWINAX
OPERATION W/PROCESSOR	N/A	N/A	
DISCRETE SIGNALS TRANSFERRED	N/A	N/A	64
DATA SIGNALS TRANSFERRED	N/A	N/A	
COMMUNICATION SPEED	N/A	N/A	100MS

ALLEN-BRADLEY		PLC-2 MINI	PLC-2/I5	PLC-4

DISCRETE I/O MODULES

		PLC-2 MINI	PLC-2/I5	PLC-4
AC VOLTAGES/CURRENT		24,120,220	24,120,220	120,220
DC VOLTAGES/CURRENT		5-30,5,12,24,48	5-30,5,12,24,48	12-24VAC
ISOLATED AC VOLTAGES/CURRENT		120-220	120-220	N/A
ISOLATED DC VOLTAGES/CURRENT		CONTACT	CONTACT	N/A
CONTINUOUS OUTPUT RATING		1.5A	1.5A	2A
COMMON POINTS/MODULE		8	8	N/A
OUTPUT FUSING CONFIGURATION		8/8	8/8	1
INDICATORS	I/O POINTS	YES	YES	YES
	FUSING	YES	YES	YES
ISOLATION		1500V	1500V	1500V

ANALOG I/O MODULES

	PLC-2 MINI	PLC-2/I5	PLC-4
VOLTAGES: SINGLE	1-5V,0-10V, 0-5V,-10-+10V	1-5V,0-10V, 0-5V,-10-+10V	NONE
VOLTAGES: DIFFERENTIAL	1-5V,0-10V, 0-5V,-10-+10V	1-5V,0-10V, 0-5V,-10-+10V	NO
CURRENT: SINGLE	0-20MA,4-20MA, -20-+20MA	0-20MA,4-20MA, -20-+20MA	NO
CURRENT: DIFFERENTIAL	0-20MA,4-20MA, -20-+20MA	0-20MA,4-20MA, -20-+20MA	NO
SPECIAL PURPOSE INPUT	THERMOCOUPLE	THERMOCOUPLE	NO
SPECIAL PURPOSE OUTPUT PID	PID	PID	NO
INDICATORS:	YES	YES	NO

SPECIAL I/O MODULES

	PLC-2 MINI	PLC-2/I5	PLC-4
SIMULATION	NO	NO	YES
ASCII	YES	YES	NO
AXIS POSITIONING	YES	YES	NO

I/O NODE DATA

		PLC-2 MINI	PLC-2/I5	PLC-4
TOTAL NODES/CHANNEL		1	1	1
TOTAL DISCRETE POINTS		128	128	32
MAXIMUM POINTS ASSIGNABLE	INPUT	128	128	20
	OUTPUT	128	128	12
MAXIMUM ANALOG ASSIGNABLE	INPUT	64	64	N/A
	OUTPUT	32	32	N/A
DISTANCE FROM DROP		N/A	N/A	N/A

ALLEN-BRADLEY		PLC-2 MINI	PLC-2/15	PLC-4

I/O CHANNEL DATA

TOTAL DISCRETE POINTS		128	128	32
MAXIMUM POINTS ASSIGNABLE	INPUT	128	128	20
	OUTPUT	128	128	12
MAXIMUM ANALOG ASSIGNABLE	INPUT	64	64	NONE
	OUTPUT	32	32	NONE
MAXIMUM LOCAL CHANNELS		1	1	1
MAXIMUM REMOTE CHANNELS		0	0	0
LOCAL CHANNEL DISTANCE		N/A	N/A	N/A
REMOTE CHANNEL DISTANCE		N/A	N/A	N/A
TYPE LOCAL CHANNEL CABLE		N/A	N/A	N/A
TYPE REMOTE CHANNEL CABLE		N/A	N/A	N/A
LOCAL CHANNEL COMMUNICATION		N/A	N/A	N/A
REMOTE CHANNEL COMMUNICATION		N/A	N/A	N/A
LOCAL CHANNEL DATA RATE		N/A	N/A	N/A
REMOTE CHANNEL DATA RATE		N/A	N/A	N/A
FAULT MONITORING		YES	YES	YES

DATA NETWORK

	PLC-2 MINI	PLC-2/15	PLC-4
TYPE	HDLC	HDLC	HDLC
NUMBER OF NODES	64	64	64
MAXIMUM LENGTH	10,000	10,000	10,000
TYPE OF CABLE	BELDEN 9463	BELDEN 9463	BELDEN 9463
COMMUNICATION SPEED	57.6K	57.6K	57.6K
COMPUTER INTERFACES	YES	YES	YES
SPECIAL INTERFACES	YES	YES	YES

ALLEN-BRADLEY	PLC-2 MINI	PLC-2/I5	PLC-4

PROGRAMMING TERMINAL

	PLC-2 MINI	PLC-2/I5	PLC-4
SCREEN SIZE (DIAGONAL)	9"	9"	LIQUID CRYSTAL
WEIGHT	35 LBS.	35 LBS.	3.1 LBS
SIZE (W X H X D)	14.5"X10"X23"	14.5"X10"X23"	8.2"X11"X2.8"
POWER REQUIREMENTS	120/220	120/220	NONE
DISPLAY SIZE (CONTACT × ROWS)	12X7	12X7	1
GRAPHICS	YES	YES	NO
VIDEO OUT	YES	YES	NO
LADDER LISTER	YES	YES	NO
OFF-LINE PROGRAM	NO	NO	NO
REMOTE OPERATION	NO	NO	NO
SCREEN MNEMONICS	NO	NO	NO
BUILT-IN TAPE	NO	NO	NO
KEYBOARD	YES	YES	YES
HANDHELD	YES	YES	YES
DOUBLES AS A COMPUTER TERMINAL	YES	YES	NO

ALLEN-BRADLEY		PLC-3	PLC-2/20	PLC-2/30

PROCESSOR

YEAR INTRODUCED		1981	1980	1982
MEMORY SIZE TOTAL		96 TO 512K	8K	16K
MEMORY INCREMENTS		16,32,64K	2,4,8K	2,4,8,16K
MEMORY TYPES		CORE,CMOS RAM EDC RAM	CORE,CMOS RAM	CORE,CMOS RAM
VOLTAGE		120,220	120,220,240	120,220,240
FREQUENCY		47-63	47-63	47-63
HUMIDITY	MIN	5%	5%	5%
	MAX	95%	95%	95%
TEMPERATURE	MIN	0°C	0°C	0°C
	MAX	60°C	60°C	60°C
USER MEMORY	MIN	0	128	128
	MAX	512K	8K	16K
DATA MEMORY	MIN	0	128	128
	MAX	512K	8K	8K
USER CONFIGURABLE		YES	YES	YES
WORD SIZE		16 BITS	16 BITS	16 BITS
WORDS/CONTACT - COIL		1	1	1
WORDS/TIMER - COUNTER		3	2	2
SCAN TIME - USER LOGIC		2.5 MS/K	5 MS/K	5 MS/K
I/O SCAN TIME		6 MS	.5 TO 7 MS	.5 TO 7 MS

SUPPORT

	PLC-3	PLC-2/20	PLC-2/30
FACTORY APPLICATIONS ENGINEERS	YES	YES	YES
DOCUMENTATION SERVICE	YES	YES	YES
SERVICE CENTER (24 HRS.)	YES	YES	YES
MAINTENANCE TRAINING	YES	YES	YES
APPLICATION TRAINING	YES	YES	YES
WARRANTY TERMS	1 YEAR	1 YEAR	1 YEAR
DISTRIBUTION	YES	YES	YES
MAINTENANCE CENTERS	YES	YES	YES

ALLEN-BRADLEY	PLC-3	PLC-2/20	PLC-2/30

INSTRUCTION SET

	PLC-3	PLC-2/20	PLC-2/30
CONTACTS	YES	YES	YES
TRANSITIONAL	YES	NO	YES
LATCHES	YES	YES	YES
RETENTIVE COILS	YES	YES	YES
SKIP/JUMP	YES	NO	YES
SUBROUTINE	YES	NO	YES
TIMERS	10,000	4000	4000
UP COUNTER	10,000	4000	4000
DOWN COUNTER	10,000	4000	4000
ADD/SUBTRACT	YES	YES	YES
MULTIPLY/DIVIDE	YES	YES	YES
>,=,<	YES	YES	YES
FILE TRANSFER	YES	NO	YES
LOGICAL	YES	NO	YES
BIT ROTATE/SHIFT	YES	NO	YES
FIFO/LIFO	YES	NO	YES
SEARCH	YES	NO	YES
DIAGNOSTIC	YES	NO	ES
SEQUENCER	YES	NO	YES
PID	YES	YES	YES
ASCII	YES	YES	YES
AXIS POSITIONING	YES	YES	YES
MACHINE LANGUAGE	NO	NO	NO
BOOLEAN ALGEBRA	NO	NO	NO
HIGH LEVEL LANGUAGE	YES	NO	NO
FLOATING POINT & INTEGER MATH	YES	NO	NO

ALLEN-BRADLEY		PLC-3	PLC-2/20	PLC-2/30

I/O STATISTICS

TOTAL DISCRETE POINTS AVAILABLE		8192	1792	1792
MAXIMUM POINTS ASSIGNABLE	INPUT	4096	896	896
	OUTPUT	4096	896	896
NUMBER OF CHANNELS		4 TO 60	2	2
NUMBER OF NODES/CHANNEL		16	14	14
MODULE I/O POINT DENSITY		8	8	8
ISOLATED MODULE I/O DENSITY		6	6	6
TOTAL ANALOG CHANNELS AVAILABLE		3968	448	448
MAXIMUM CHANNELS ASSIGNABLE	INPUT	2048	448	448
	OUTPUT	1920	168	168
I/O MAPPING		PROGRAMMABLE	FIXED	FIXED
INTERRUPTS		YES	NO	NO
I/O LOGIC UPDATING	ASYNC	YES		
	SYNC		YES	YES
ERROR CHECKING METHOD		CRC PLUS PROPRIETARY	CRC PLUS PROPRIETARY	CRC PLUS PROPRIETARY
MODULES PER CHANNEL		256	112	112
APPLICABLE PROCESSORS		ALL	ALL	ALL
YEAR OF INTRODUCTION		1981	1980	1983
I/O SYSTEM STRUCTURE		RACK & MODULE	RACK & MODULE	RACK & MODULE

CONTROL NETWORK

	PLC-3	PLC-2/20	PLC-2/30
TYPE	PEER - PEER	N/A	N/A
NUMBER OF PC'S 60 CHANNELS/PC	8/CHANNEL	N/A	N/A
MAXIMUM LENGTH	10,000	N/A	N/A
TYPE OF CABLE	TWIN AX	N/A	N/A
OPERATION W/PROCESSOR	YES	N/A	N/A
DISCRETE SIGNALS TRANSFERRED	YES	N/A	N/A
DATA SIGNALS TRANSFERRED	YES	N/A	N/A
COMMUNICATION SPEED	115 K BAUD	N/A	N/A

ALLEN-BRADLEY		PLC-3	PLC-2/20	PLC-2/30

DISCRETE I/O MODULES

AC VOLTAGES/CURRENT		24,120,220	24,120,220	24,120,220
DC VOLTAGES/CURRENT		5-30,5,12,24,48	5-30,5,12,24,48	5-30,5,12,24,48
ISOLATED AC VOLTAGES/CURRENT		120,220	120,220	120,220
ISOLATED DC VOLTAGES/CURRENT		CONTACT	CONTACT	CONTACT
CONTINUOUS OUTPUT RATING		1-5A @ 6 TOTAL	1-5A @ 6 TOTAL	1-5A @ 6 TOTAL
COMMON POINTS/MODULE		8	8	8
OUTPUT FUSING CONFIGURATION		8/8	8/8	8/8
INDICATORS	I/O POINTS	YES	YES	YES
	FUSING	YES	YES	YES
ISOLATION		1500V OPTICAL	1500V OPTICAL	1500V OPTICAL

ANALOG I/O MODULES

	PLC-3	PLC-2/20	PLC-2/30
VOLTAGES: SINGLE	1-5V,0-10V, 0-5V,-10-+10V	1-5V,0-10V, 0-5V,-10-+10V	1-5V,0-10V, 0-5V,-10-+10V
VOLTAGES: DIFFERENTIAL	1-5V,0-10V, 0-5V,-10-+10V	1-5V,0-10V, 0-5V,-10-+10V	1-5V,0-10V, 0-5V,-10-+10V
CURRENT: SINGLE	0-20MA,4-20MA, -20-+20MA	0-20MA,4-20MA, -20-+20MA	0-20MA,4-20MA, -20-+20MA
CURRENT: DIFFERENTIAL	0-20MA,4-20MA, -20-+20MA	0-20MA,4-20MA, -20-+20MA	0-20MA,4-20MA, -20-+20MA
SPECIAL PURPOSE INPUT	THERMOCOUPLE	THERMOCOUPLE	THERMOCOUPLE
SPECIAL PURPOSE OUTPUT	SERVO & PID	SERVO & PID	SERVO & PID
INDICATORS:	YES	YES	YES

SPECIAL I/O MODULES

	PLC-3	PLC-2/20	PLC-2/30
P.I.D. LOOP CONTROL	YES	YES	YES
ASCII	YES	YES	YES
AXIS POSITIONING - SERVO & STEPPER	YES	YES	YES
ENCODER, PROXIMITY SWITCH	YES	YES	YES

I/O NODE DATA

		PLC-3	PLC-2/20	PLC-2/30
TOTAL NODES/CHANNEL		16	14	14
TOTAL DISCRETE POINTS PER RACK		128	128	128
MAXIMUM POINTS ASSIGNABLE	INPUT	128	128	128
	OUTPUT	128	128	128
MAXIMUM ANALOG ASSIGNABLE	INPUT	64	64	64
	OUTPUT	32	32	32
DISTANCE FROM DROP		50 FT.	50 FT.	50 FT.

ALLEN-BRADLEY	PLC-3	PLC-2/20	PLC-2/30

PROGRAMMING TERMINAL

SCREEN SIZE (DIAGONAL)	9"	9"	9"
WEIGHT	35 LBS.	35 LBS.	35 LBS.
SIZE (W X H X D)	14.5"X10"X23"	14.5"X10"X23"	14.5"X10"X23"
POWER REQUIREMENTS	120 OR 220	120 OR 220	120 OR 220
DISPLAY SIZE (CONTACT • ROWS)	11X17	11X17	11X17
GRAPHICS	YES	YES	YES
VIDEO OUT	YES	YES	YES
LADDER LISTER	YES	YES	YES
OFF-LINE PROGRAM	NO	NO	NO
REMOTE OPERATION	5K'	NO	NO
SCREEN MNEMONICS	YES	NO	NO
BUILT-IN TAPE	NO	NO	NO
KEYBOARD	YES	YES	YES
HANDHELD	NO	YES	YES
DOUBLES AS A COMPUTER TERMINAL	YES	YES	YES

AMICON		800 SERIES	900 SERIES	1000 SERIES

PROCESSOR

YEAR INTRODUCED		1980	1981	1982
MEMORY SIZE TOTAL		48K	2K	2K
MEMORY INCREMENTS		1K	-	-
MEMORY TYPES		EPROM RAM	EPROM RAM	EPROM RAM
VOLTAGE		120/240	120/240	120/240
FREQUENCY		50/60	50/60	50/60
HUMIDITY	MIN	-	-	-
	MAX	-	-	-
TEMPERATURE	MIN	-	-	-
	MAX	-	-	-
USER MEMORY	MIN	2K	2K	1K
	MAX	32K	2K	1K
DATA MEMORY	MIN	-	-	-
	MAX	-	-	-
USER CONFIGURABLE		-	-	-
WORD SIZE		PER APPLICATION	PER APPLICATION	8 BIT
WORDS/CONTACT - COIL		PER APPLICATION	PER APPLICATION	1 BIT
WORDS/TIMER - COUNTER		PER APPLICATION	PER APPLICATION	2 BIT
SCAN TIME - USER LOGIC		PER APPLICATION	PER APPLICATION	VARIABLE
I/O SCAN TIME		PER APPLICATION	PER APPLICATION	VARIABLE

SUPPORT

	800 SERIES	900 SERIES	1000 SERIES
FACTORY APPLICATIONS ENGINEERS	YES	YES	YES
DOCUMENTATION SERVICE	YES	YES	YES
SERVICE CENTER (24 HRS.)	-	-	-
MAINTENANCE TRAINING	YES	YES	YES
APPLICATION TRAINING	YES	YES	YES
WARRANTY TERMS	12 MONTHS	12 MONTHS	12 MONTHS
DISTRIBUTION	SALES REPS	SALES REPS	SALES REPS
MAINTENANCE CENTERS	SALES REPS	SALES REPS	SALES REPS

AMICON	800 SERIES	900 SERIES	1000 SERIES

INSTRUCTION SET

CONTACTS	PER APPLICATION	PER APPLICATION	PER APPLICATION
TRANSITIONAL	PER APPLICATION	PER APPLICATION	-
LATCHES	PER APPLICATION	PER APPLICATION	-
RETENTIVE COILS	PER APPLICATION	PER APPLICATION	YES
SKIP/JUMP	PER APPLICATION	PER APPLICATION	-
SUBROUTINE	PER APPLICATION	PER APPLICATION	-
TIMERS	PER APPLICATION	PER APPLICATION	YES
UP COUNTER	PER APPLICATION	PER APPLICATION	-
DOWN COUNTER	PER APPLICATION	PER APPLICATION	-
ADD/SUBTRACT	PER APPLICATION	PER APPLICATION	-
MULTIPLY/DIVIDE	PER APPLICATION	PER APPLICATION	-
>,=,<	PER APPLICATION	PER APPLICATION	-
FILE TRANSFER	PER APPLICATION	PER APPLICATION	-
LOGICAL	PER APPLICATION	PER APPLICATION	-
BIT ROTATE/SHIFT	PER APPLICATION	PER APPLICATION	-
FIFO/LIFO	PER APPLICATION	PER APPLICATION	-
SEARCH	PER APPLICATION	PER APPLICATION	-
DIAGNOSTIC	PER APPLICATION	PER APPLICATION	-
SEQUENCER	PER APPLICATION	PER APPLICATION	-
PID	PER APPLICATION	PER APPLICATION	-
ASCII	PER APPLICATION	PER APPLICATION	-
AXIS POSITIONING	PER APPLICATION	NO	NO
MACHINE LANGUAGE	PER APPLICATION	PER APPLICATION	PER APPLICATION
BOOLEAN ALGEBRA	-	-	-
HIGH LEVEL LANGUAGE	-	-	-

SPECIAL I/O MODULES

SIMULATION	-	-	-
ASCII	YES	HALF DUPLEX	-
AXIS POSITIONING	YES	NO	NO

AMICON		800 SERIES	900 SERIES	1000 SERIES

I/O STATISTICS

TOTAL DISCRETE POINTS AVAILABLE		720	104	32
MAXIMUM POINTS ASSIGNABLE	INPUT	0-720	8-96	8-24
	OUTPUT	0-720	8-96	8-24
NUMBER OF CHANNELS		-	-	-
NUMBER OF NODES/CHANNEL		-	-	-
MODULE I/O POINT DENSITY		-	-	-
ISOLATED MODULE I/O DENSITY		-	-	-
TOTAL ANALOG CHANNELS AVAILABLE		120	10	0
MAXIMUM CHANNELS ASSIGNABLE	INPUT	120	8	0
	OUTPUT	60	2	0
I/O MAPPING		-	-	-
INTERRUPTS		-	-	-
I/O LOGIC UPDATING	ASYNC	-	-	-
	SYNC	-	-	-
ERROR CHECKING METHOD		PER APPLICATION	PER APPLICATION	-
MODULES PER CHANNEL		-	-	-
APPLICABLE PROCESSORS		8085	8085	8085
YEAR OF INTRODUCTION		1980	1981	1982
I/O SYSTEM STRUCTURE		RACK & BASE	RACK & BASE	RACK & BASE

CONTROL NETWORK

	800 SERIES	900 SERIES	1000 SERIES
TYPE	PER APPLICATION	PER APPLICATION	-
NUMBER OF PC'S	PER APPLICATION	PER APPLICATION	-
MAXIMUM LENGTH	PER APPLICATION	PER APPLICATION	-
TYPE OF CABLE	PER APPLICATION	PER APPLICATION	-
OPERATION W/PROCESSOR	PER APPLICATION	PER APPLICATION	-
DISCRETE SIGNALS TRANSFERRED	PER APPLICATION	PER APPLICATION	-
DATA SIGNALS TRANSFERRED	PER APPLICATION	PER APPLICATION	-
COMMUNICATION SPEED	PER APPLICATION	PER APPLICATION	-

AMICON	800 SERIES	900 SERIES	1000 SERIES

DISCRETE I/O MODULES

	800 SERIES	900 SERIES	1000 SERIES
AC VOLTAGES/CURRENT	12,24,120/240	12,24,120/240	12,24,120/240
DC VOLTAGES/CURRENT	12,24,48,60	12,24,48,60	12,24,48,60
ISOLATED AC VOLTAGES/CURRENT	12,24,120,240	12,24,120,240	12,24,120,240
ISOLATED DC VOLTAGES/CURRENT	12,24,48,60	12,24,48,60	12,24,48,60
CONTINUOUS OUTPUT RATING	-	-	-
COMMON POINTS/MODULE	PLUG-IN	PLUG-IN	PLUG-IN
OUTPUT FUSING CONFIGURATION	-	-	-
INDICATORS — I/O POINTS	-	-	-
INDICATORS — FUSING	YES	YES	YES
ISOLATION	15000 OPTIONAL	15000 OPTIONAL	15000 OPTIONAL

PROGRAMMING TERMINAL

	800 SERIES	900 SERIES	1000 SERIES
SCREEN SIZE (DIAGONAL)	PRE-PROGRAMMED	PRE-PROGRAMMED	-
WEIGHT	PRE-PROGRAMMED	PRE-PROGRAMMED	3•
SIZE (W X H X D)	PRE-PROGRAMMED	PRE-PROGRAMMED	7' X 3' X 10'
POWER REQUIREMENTS	PRE-PROGRAMMED	PRE-PROGRAMMED	120 OR BATTERY
DISPLAY SIZE (CONTACT • ROWS)	PRE-PROGRAMMED	PRE-PROGRAMMED	1 CONTACT
GRAPHICS	PRE-PROGRAMMED	PRE-PROGRAMMED	-
VIDEO OUT	PRE-PROGRAMMED	PRE-PROGRAMMED	-
LADDER LISTER	PRE-PROGRAMMED	PRE-PROGRAMMED	-
OFF-LINE PROGRAM	PRE-PROGRAMMED	PRE-PROGRAMMED	-
REMOTE OPERATION	PRE-PROGRAMMED	PRE-PROGRAMMED	-
SCREEN MNEMONICS	PRE-PROGRAMMED	PRE-PROGRAMMED	-
BUILT-IN TAPE	PRE-PROGRAMMED	PRE-PROGRAMMED	-
KEYBOARD	PRE-PROGRAMMED	PRE-PROGRAMMED	HEX
HANDHELD	PRE-PROGRAMMED	PRE-PROGRAMMED	YES
DOUBLES AS A COMPUTER TERMINAL	PRE-PROGRAMMED	PRE-PROGRAMMED	-

AMICON		800 SERIES	900 SERIES	1000 SERIES

I/O CHANNEL DATA

TOTAL DISCRETE POINTS		0-720	0-104	32
MAXIMUM POINTS ASSIGNABLE	INPUT	0-720	8-96	8-24
	OUTPUT	0-720	8-96	8-24
MAXIMUM ANALOG ASSIGNABLE	INPUT	120	8	0
	OUTPUT	60	2	0
MAXIMUM LOCAL CHANNELS		ALL	ALL	ALL
MAXIMUM REMOTE CHANNELS		ALL	ALL	ALL
LOCAL CHANNEL DISTANCE		-	-	-
REMOTE CHANNEL DISTANCE		7000'	7000'	-
TYPE LOCAL CHANNEL CABLE		PARALLEL	PARALLEL	PARALLEL
TYPE REMOTE CHANNEL CABLE		2 WIRE	2 WIRE	NONE
LOCAL CHANNEL COMMUNICATION		PARALLEL	PARALLEL	-
REMOTE CHANNEL COMMUNICATION		SERIAL	SERIAL	-
LOCAL CHANNEL DATA RATE		-	-	-
REMOTE CHANNEL DATA RATE		-	-	-
FAULT MONITORING		PER APPLICATION	PER APPLICATION	-

DATA NETWORK

	800 SERIES	900 SERIES	1000 SERIES
TYPE	PER APPLICATION	PER APPLICATION	-
NUMBER OF NODES	22	3	-
MAXIMUM LENGTH	7000'	7000'	-
TYPE OF CABLE	TWISTED PAIR	TWISTED PAIR	-
COMMUNICATION SPEED	1 MSEC	1 MSEC	-
COMPUTER INTERFACES	YES	YES	-
SPECIAL INTERFACES	PER APPLICATION	PER APPLICATION	-

ANALOG I/O MODULES

	800 SERIES	900 SERIES	1000 SERIES
VOLTAGES: SINGLE	0-100	0-100	-
VOLTAGES: DIFFERENTIAL	-	-	-
CURRENT: SINGLE	4-20 MA	4-20 MA	-
CURRENT: DIFFERENTIAL	-	-	-
SPECIAL PURPOSE INPUT	THERMOCOUPLE	THERMOCOUPLE	-
SPECIAL PURPOSE OUTPUT	-	-	-
INDICATORS:	PER APPLICATION	PER APPLICATION	-

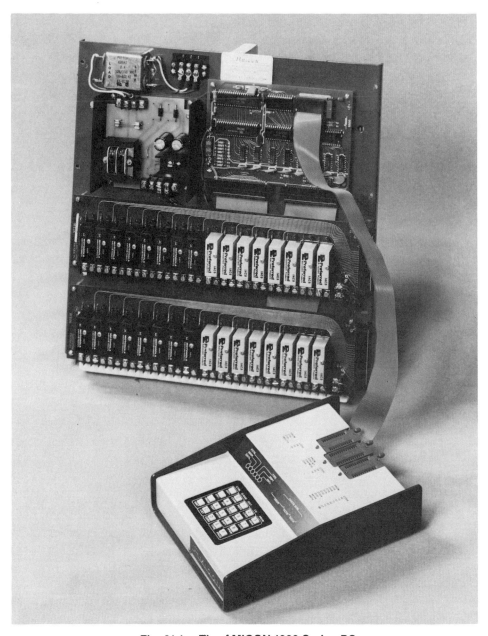

Fig. 21.1.   The AMICON 1000 Series PC

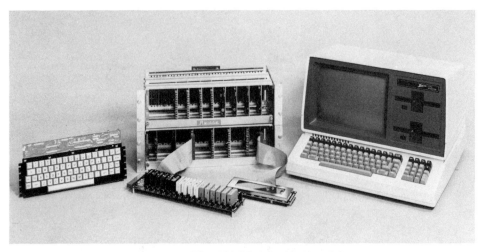

Fig. 21.2. The AMICON 800 Series PC

DIVELBISS		ICM-320A	ICM-BB

PROCESSOR

YEAR INTRODUCED		1978	1982
MEMORY SIZE TOTAL		4K	4K
MEMORY INCREMENTS		1,2,4K	2,4K
MEMORY TYPES		EPROM	EPROM
VOLTAGE (±)		120	12/24 VDC 24/120/240VAC
FREQUENCY (±)		60HZ	50/60HZ
HUMIDITY	MIN	10%	10%
	MAX	90%	90%
TEMPERATURE	MIN	0°C	0°C
	MAX	50°C	50°C
USER MEMORY	MIN	1K	2K
	MAX	4K	4K
DATA MEMORY	MIN	N/A	N/A
	MAX	N/A	N/A
USER CONFIGURABLE		YES (SEE NOTE 1)	YES (SEE NOTE 1)
WORD SIZE		SINGLE	SINGLE
WORDS/CONTACT - COIL		1	1
WORDS/TIMER - COUNTER		3	3
SCAN TIME - USER LOGIC		10USEC/BIT	5USEC/BIT
I/O SCAN TIME		20MSEC (SEE NOTE 2)	10MSEC (SEE NOTE 2)

SUPPORT

FACTORY APPLICATIONS ENGINEERS	YES	YES
DOCUMENTATION SERVICE	YES	YES
SERVICE CENTER (24 HRS.)	N/A	N/A
MAINTENANCE TRAINING	YES	YES
APPLICATION TRAINING	YES	YES
WARRANTY TERMS	1 YEAR	1 YEAR
DISTRIBUTION	DISTRIBUTORS	DISTRIBUTORS
MAINTENANCE CENTERS	DISTRIBUTORS	DISTRIBUTORS

NOTES :

1) USER CAN SELECT MANY PLUG-IN OPTIONS
2) ASSUME 2K OF MEMORY

DIVELBISS	ICM-320A	ICM-BB

INSTRUCTION SET

CONTACTS	YES	YES
TRANSITIONAL	ON/OFF	ON/OFF
LATCHES	YES	YES
RETENTIVE COILS (SEE NOTE 1)	YES	YES
SKIP/JUMP	YES	YES
SUBROUTINE	NO	NO
TIMERS	YES	YES
UP COUNTER	YES	YES
DOWN COUNTER	NO	NO
ADD/SUBTRACT	NO	NO
MULTIPLY/DIVIDE	NO	NO
>,=,<	YES	YES
FILE TRANSFER	NO	NO
LOGICAL	YES	YES
BIT ROTATE/SHIFT	NO	NO
FIFO/LIFO	NO	NO
SEARCH	YES	YES
DIAGNOSTIC	NO	NO
SEQUENCER (SEE NOTE 2)	YES	YES
PID	NO	NO
ASCII	NO	NO
AXIS POSITIONING (SEE NOTE 3)	YES	YES
MACHINE LANGUAGE	NO	NO
BOOLEAN ALGEBRA	NO	NO
HIGH LEVEL LANGUAGE	NO	NO

NOTES :

1) WITH BATTERY BACK-UP
2) USER PROGRAMMABLE
3) PLUG-IN CARDS

DIVELBISS	ICM-320A	ICM-BB

I/O STATISTICS

		ICM-320A	ICM-BB
TOTAL DISCRETE POINTS AVAILABLE		252	248
MAXIMUM POINTS ASSIGNABLE	INPUT	126	124
	OUTPUT	126	124
NUMBER OF CHANNELS		8	8
NUMBER OF NODES/CHANNEL		4	8
MODULE I/O POINT DENSITY		1	N/A
ISOLATED MODULE I/O DENSITY		1	8
TOTAL ANALOG CHANNELS AVAILABLE		48	25
MAXIMUM CHANNELS ASSIGNABLE	INPUT	48	25
	OUTPUT	N/A	N/A
I/O MAPPING		N/A	N/A
INTERRUPTS		N/A	N/A
I/O LOGIC UPDATING	ASYNC	N/A	N/A
	SYNC	N/A	N/A
ERROR CHECKING METHOD		N/A	N/A
MODULES PER CHANNEL		16	N/A
APPLICABLE PROCESSORS		ALL	ALL
YEAR OF INTRODUCTION		1978	1982
I/O SYSTEM STRUCTURE		PLUG-IN BOARDS BUILT-IN	PLUG-IN BOARDS BUILT-IN

DISCRETE I/O MODULES

		ICM-320A	ICM-BB
AC VOLTAGES/CURRENT		24,120,240	24,120,240
DC VOLTAGES/CURRENT		12,24,48,125	12,24
ISOLATED AC VOLTAGES/CURRENT		24,120,240	NEUTRAL IS COMMON
ISOLATED DC VOLTAGES/CURRENT		12,24,48,125	PLUS OR MINUS IS COMMON
CONTINUOUS OUTPUT RATING		2A	2A
COMMON POINTS/MODULE		NONE	4
OUTPUT FUSING CONFIGURATION		ALL	BY USER
INDICATORS	I/O POINTS	ALL	ALL
	FUSING	BY USER	BY USER
ISOLATION		1500V OPTICAL	1500V OPTICAL

DIVELBISS		ICM-320A	ICM-BB

I/O CHANNEL DATA

TOTAL DISCRETE POINTS		252	248
MAXIMUM POINTS ASSIGNABLE	INPUT	126	124
	OUTPUT	126	124
MAXIMUM ANALOG ASSIGNABLE	INPUT	N/A	N/A
	OUTPUT	N/A	N/A
MAXIMUM LOCAL CHANNELS		N/A	N/A
MAXIMUM REMOTE CHANNELS		N/A	N/A
LOCAL CHANNEL DISTANCE		N/A	N/A
REMOTE CHANNEL DISTANCE		N/A	N/A
TYPE LOCAL CHANNEL CABLE		N/A	N/A
TYPE REMOTE CHANNEL CABLE		N/A	N/A
LOCAL CHANNEL COMMUNICATION		N/A	N/A
REMOTE CHANNEL COMMUNICATION		N/A	N/A
LOCAL CHANNEL DATA RATE		N/A	N/A
REMOTE CHANNEL DATA RATE		N/A	N/A
FAULT MONITORING		N/A	N/A

ANALOG I/O MODULES

VOLTAGES: SINGLE	N/A	N/A
VOLTAGES: DIFFERENTIAL	0-5VDC	0-5 & 0-10VDC
CURRENT: SINGLE	NA/	N/A
CURRENT: DIFFERENTIAL	N/A	4-20MADC
SPECIAL PURPOSE INPUT	N/A	N/A
SPECIAL PURPOSE OUTPUT	N/A	N/A
INDICATORS:	N/A	N/A

SPECIAL I/O MODULES

SIMULATION	NO	NO
ASCII	NO	NO
AXIS POSITIONING	NO	NO

DIVELBISS	ICM-320A	ICM-BB

PROGRAMMING TERMINAL

	ICM-320A	ICM-BB
SCREEN SIZE (DIAGONAL)	12'	12'
WEIGHT	7 . 25	7 . 25
SIZE (W X H X D)	19X12X21	19X12X21
POWER REQUIREMENTS	120VAC	120VAC
DISPLAY SIZE (CONTACT ■ ROWS)	7 X 5	7 X 5
GRAPHICS	NO	NO
VIDEO OUT	NO	NO
LADDER LISTER	YES	YES
OFF-LINE PROGRAM	YES	YES
REMOTE OPERATION	NO	NO
SCREEN MNEMONICS	YES	YES
BUILT-IN TAPE	NO	NO
KEYBOARD	ASCII	ASCII
HANDHELD	NO	NO
DOUBLES AS A COMPUTER TERMINAL	IS A COMPUTER	IS A COMPUTER

EAGLE SIGNAL		EPTAK 210	EPTAK 220	EPTAK 240

PROCESSOR

YEAR INTRODUCED		1982	1982	1982
MEMORY SIZE TOTAL		2K	4K	8K
MEMORY INCREMENTS		NONE	4K	8K
MEMORY TYPES		PLUG-A-PROM (TM) EEPROM	PLUG-A-PROM (TM) EEPROM	PLUG-A-PROM (TM) EEPROM
VOLTAGE		102-132VAC 204-264VAC	102-132VAC 204-264VAC	102-132VAC 204-264VAC
FREQUENCY		47-63HZ	47-63HZ	47-63HZ
HUMIDITY	MIN	0	0	0
	MAX	95%	95%	5%
TEMPERATURE	MIN	0°C	0°C	0°C
	MAX	60°C	60°C	60°C
USER MEMORY	MIN	2K	4K	8K
	MAX	2K	4K	8K
DATA MEMORY	MIN	½K	½K	2K
	MAX	½K	½K	2K
USER CONFIGURABLE		N/A	N/A	N/A
WORD SIZE		8 BITS	8 BITS	8 BITS
WORDS/CONTACT - COIL		4	4	4
WORDS/TIMER - COUNTER		4	4	4
SCAN TIME - USER LOGIC		35 MICROSEC PER INSTRUCTION	35 MICROSEC PER INSTRUCTION	15 MICROSEC PER INSTRUCTION
I/O SCAN TIME		3MS TYP	3MS TYP	10M

SUPPORT

FACTORY APPLICATIONS ENGINEERS	YES	YES	YES
DOCUMENTATION SERVICE	YES	YES	YES
SERVICE CENTER (24 HRS.)	NO	NO	NO
MAINTENANCE TRAINING	YES	YES	YES
APPLICATION TRAINING	YES	YES	YES
WARRANTY TERMS	1 YEAR	1 YEAR	1 YEAR
DISTRIBUTION	DEALERS	DEALERS	DEALERS
MAINTENANCE CENTERS	DAVENPORT, IA	DAVENPORT, IA	DAVENPORT, IA

EAGLE SIGNAL	EPTAK 210	EPTAK 220	EPTAK 240

INSTRUCTION SET

	EPTAK 210	EPTAK 220	EPTAK 240
CONTACTS	YES	YES	YES
TRANSITIONAL	N/A	N/A	N/A
LATCHES	YES	YES	YES
RETENTIVE COILS	32	64	64
SKIP/JUMP	YES	YES	YES
SUBROUTINE	N/A	N/A	YES
TIMERS	16	32	32
UP COUNTER	16	32	32
DOWN COUNTER	16	32	32
ADD/SUBTRACT	N/A	YES	YES
MULTIPLY/DIVIDE	N/A	N/A	YES
>,=,<	N/A	YES	YES
FILE TRANSFER	N/A	N/A	NO
LOGICAL	YES	YES	YES
BIT ROTATE/SHIFT	YES	YES	YES
FIFO/LIFO	YES	YES	YES
SEARCH	YES	YES	YES
DIAGNOSTIC	YES	YES	YES
SEQUENCER	CAM TIMER	CAM TIMER	CAM TIMER
PID	N/A	N/A	16 LOOPS
ASCII	YES	YES	YES
AXIS POSITIONING	N/A	N/A	NO
MACHINE LANGUAGE	N/A	N/A	NO
BOOLEAN ALGEBRA	YES	YES	YES
HIGH LEVEL LANGUAGE	N/A	N/A	N/A

EAGLE SIGNAL		EPTAK 210	EPTAK 220	EPTAK 240

I/O STATISTICS

		EPTAK 210	EPTAK 220	EPTAK 240
TOTAL DISCRETE POINTS AVAILABLE		32	128	128
MAXIMUM POINTS ASSIGNABLE	INPUT	32	128	128
	OUTPUT	32	128	128
NUMBER OF CHANNELS		1	1	1
NUMBER OF NODES/CHANNEL		32	128	128
MODULE I/O POINT DENSITY		2	2	2
ISOLATED MODULE I/O DENSITY		2	2	2
TOTAL ANALOG CHANNELS AVAILABLE		N/A	N/A	32
MAXIMUM CHANNELS ASSIGNABLE	INPUT	N/A	N/A	16
	OUTPUT	N/A	N/A	16
I/O MAPPING		FIXED	FIXED	FIXED
INTERRUPTS		2	2	3
I/O LOGIC UPDATING	ASYNC			
	SYNC	YES	YES	YES
ERROR CHECKING METHOD		CHECK SUM PARITY	CHECK SUM PARITY	CHECK SUM PARITY
MODULES PER CHANNEL		16	64	16
APPLICABLE PROCESSORS		8049	8049	8051
YEAR OF INTRODUCTION		1980	1980	1980
I/O SYSTEM STRUCTURE		VARIABLE SIZE I/O TRACKS	VARIABLE SIZE I/O TRACKS	VARIABLE SIZE I/O TRACKS

DISCRETE I/O MODULES

		EPTAK 210	EPTAK 220	EPTAK 240
AC VOLTAGES/CURRENT		120,240	120,240	120,240
DC VOLTAGES/CURRENT		5,10-55	5,10-55	5,10-55
ISOLATED AC VOLTAGES/CURRENT		120,240	120,240	120,240
ISOLATED DC VOLTAGES/CURRENT		10-55	10-55	10-55
CONTINUOUS OUTPUT RATING		2 AMP	2 AMP	2 AMP
COMMON POINTS/MODULE		2/MODULE	2/MODULE	2/MODULE
OUTPUT FUSING CONFIGURATION		1/OUTPUT	1/OUTPUT	1/OUTPUT
INDICATORS	I/O POINTS	1/OUTPUT	1/OUTPUT	1/OUTPUT
	FUSING	N/A	N/A	N/A
ISOLATION		3500V OPTICAL	3500V OPTICAL	3500V OPTICAL

EAGLE SIGNAL		*EPTAK* 210	*EPTAK* 220	*EPTAK* 240

I/O CHANNEL DATA

TOTAL DISCRETE POINTS		32	128	128
MAXIMUM POINTS ASSIGNABLE	INPUT	32	128	128
	OUTPUT	32	128	128
MAXIMUM ANALOG ASSIGNABLE	INPUT	N/A	N/A	16
	OUTPUT	N/A	N/A	16
MAXIMUM LOCAL CHANNELS		1	1	1
MAXIMUM REMOTE CHANNELS		N/A	N/A	N/A
LOCAL CHANNEL DISTANCE		N/A	N/A	N/A
REMOTE CHANNEL DISTANCE		N/A	N/A	N/A
TYPE LOCAL CHANNEL CABLE		PARALLEL	PARALLEL	PARALLEL
TYPE REMOTE CHANNEL CABLE		N/A	N/A	N/A
LOCAL CHANNEL COMMUNICATION		PARALLEL	PARALLEL	PARALLEL
REMOTE CHANNEL COMMUNICATION		N/A	N/A	N/A
LOCAL CHANNEL DATA RATE		50KHZ	50KHZ	50KHZ
REMOTE CHANNEL DATA RATE		N/A	N/A	N/A
FAULT MONITORING		YES	YES	YES

I/O NODE DATA

TOTAL NODES/CHANNEL		1	4	4
TOTAL DISCRETE POINTS		32	128	128
MAXIMUM POINTS ASSIGNABLE	INPUT	32	128	128
	OUTPUT	32	128	128
MAXIMUM ANALOG ASSIGNABLE	INPUT	N/A	N/A	16
	OUTPUT	N/A	N/A	16
DISTANCE FROM DROP		LOCAL I/O	LOCAL I/O	LOCAL I/O

SPECIAL I/O MODULES

SIMULATION	I/O SIMULATOR MODULES	I/O SIMULATOR MODULES	I/O SIMULATOR MODULES
ASCII	N/A	N/A	N/A
AXIS POSITIONING	N/A	N/A	N/A

EAGLE SIGNAL	EPTAK 210	EPTAK 220	EPTAK 240
ANALOG I/O MODULES			
VOLTAGES: SINGLE	N/A	N/A	1-5V
VOLTAGES: DIFFERENTIAL	N/A	N/A	J THERMOCOUPLES
CURRENT: SINGLE	N/A	N/A	4-20MA 10-50MA
CURRENT: DIFFERENTIAL	N/A	N/A	N/A
SPECIAL PURPOSE INPUT	N/A	N/A	J/K THERMOCOUPLE
SPECIAL PURPOSE OUTPUT	N/A	N/A	N/A
INDICATORS:	N/A	N/A	N/A
PROGRAMMING TERMINAL			
SCREEN SIZE (DIAGONAL)	NON-CRT	NON-CRT	NON-CRT
WEIGHT	3.5 LBS	3.5 LBS	3.5 LBS
SIZE (W X H X D)	8.37X3X8.18	8.37X3X8.18	8.37X3X8.18
POWER REQUIREMENTS	102-132VAC 204-264VAC	102-132VAC 204-264VAC	102-132VAC 204-264VAC
DISPLAY SIZE (CONTACT × ROWS)	N/A	N/A	N/A
GRAPHICS	N/A	N/A	N/A
VIDEO OUT	N/A	N/A	N/A
LADDER LISTER	YES	YES	YES
OFF-LINE PROGRAM	YES	YES	YES
REMOTE OPERATION	1000 FT	1000 FT	1000 FT
SCREEN MNEMONICS	N/A	N/A	N/A
BUILT-IN TAPE	N/A	N/A	N/A
KEYBOARD	54-KEY: FULL TRAVEL	54-KEY: FULL TRAVEL	54-KEY: FULL TRAVEL
HANDHELD	YES	YES	YES
DOUBLES AS A COMPUTER TERMINAL	N/A	N/A	N/A

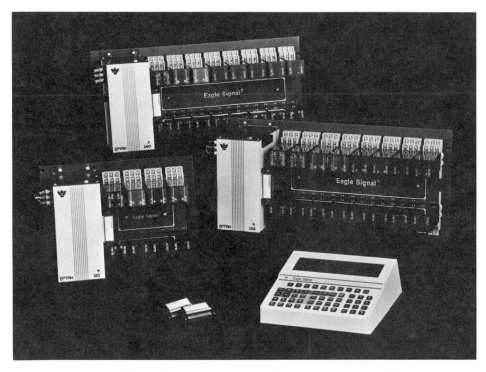

Fig. 21.3. The Eagle Signal EPTAK 210, 220, and 240 PC Families

EAGLE SIGNAL	EPTAK 700

PROCESSOR

YEAR INTRODUCED		1976
MEMORY SIZE TOTAL		48K
MEMORY INCREMENTS		16K
MEMORY TYPES		RAM,UV PROM
VOLTAGE		102-132VAC 204-264VAC
FREQUENCY		47-63HZ
HUMIDITY	MIN	0
	MAX	95%
TEMPERATURE	MIN	0°C
	MAX	60°C
USER MEMORY	MIN	16K
	MAX	48K
DATA MEMORY	MIN	16K
	MAX	48K
USER CONFIGURABLE		YES
WORD SIZE		8 BITS
WORDS/CONTACT - COIL		2
WORDS/TIMER - COUNTER		2
SCAN TIME - USER LOGIC		7MS PER 1000 INSTRUCTIONS
I/O SCAN TIME		2MS

SUPPORT

FACTORY APPLICATIONS ENGINEERS	YES
DOCUMENTATION SERVICE	YES
SERVICE CENTER (24 HRS.)	N/A
MAINTENANCE TRAINING	ES
APPLICATION TRAINING	YES
WARRANTY TERMS	1 YEAR
DISTRIBUTION	DEALERS
MAINTENANCE CENTERS	DAVENPORT, IA

EAGLE SIGNAL	EPTAK 700

INSTRUCTION SET

CONTACTS	YES
TRANSITIONAL	NO
LATCHES	YES
RETENTIVE COILS	YES
SKIP/JUMP	YES
SUBROUTINE	YES
TIMERS	255
UP COUNTER	255
DOWN COUNTER	255
ADD/SUBTRACT	YES
MULTIPLY/DIVIDE	YES
>,=,<	YES
FILE TRANSFER	YES
LOGICAL	YES
BIT ROTATE/SHIFT	YES
FIFO/LIFO	YES
SEARCH	YES
DIAGNOSTIC	YES
SEQUENCER	N/A
PID	255
ASCII	16 DEVICES
AXIS POSITIONING	N/A
MACHINE LANGUAGE	YES
BOOLEAN ALGEBRA	YES
HIGH LEVEL LANGUAGE	YES

SPECIAL I/O MODULES

SIMULATION	NO
ASCII	YES
AXIS POSITIONING	NO

EAGLE SIGNAL	EPTAK 700

I/O STATISTICS

TOTAL DISCRETE POINTS AVAILABLE		2048
MAXIMUM POINTS ASSIGNABLE	INPUT	2048
	OUTPUT	2048
NUMBER OF CHANNELS		8
NUMBER OF NODES/CHANNEL		256
MODULE I/O POINT DENSITY		1,8,32
ISOLATED MODULE I/O DENSITY		1
TOTAL ANALOG CHANNELS AVAILABLE		510
MAXIMUM CHANNELS ASSIGNABLE	INPUT	255
	OUTPUT	255
I/O MAPPING		ES
INTERRUPTS		8 LEVELS
I/O LOGIC UPDATING	ASYNC	YES
	SYNC	N/A
ERROR CHECKING METHOD		CHECK SUM PARITY
MODULES PER CHANNEL		256
APPLICABLE PROCESSORS		8080A
YEAR OF INTRODUCTION		1976
I/O SYSTEM STRUCTURE		1 I/O DRIVER PER 16 TRACKS

CONTROL NETWORK

TYPE	PROPRIETARY
NUMBER OF PC'S	16
MAXIMUM LENGTH	10000 FT
TYPE OF CABLE	TWISTED PAIR
OPERATION W/PROCESSOR	YES
DISCRETE SIGNALS TRANSFERRED	YES
DATA SIGNALS TRANSFERRED	YES
COMMUNICATION SPEED	UP TO 9600 BAUD

EAGLE SIGNAL	EPTAK 700

I/O CHANNEL DATA

TOTAL DISCRETE POINTS		2048
MAXIMUM POINTS ASSIGNABLE	INPUT	2048
	OUTPUT	2048
MAXIMUM ANALOG ASSIGNABLE	INPUT	255
	OUTPUT	255
MAXIMUM LOCAL CHANNELS		2048 I/O
MAXIMUM REMOTE CHANNELS		2048 I/O
LOCAL CHANNEL DISTANCE		10 FT
REMOTE CHANNEL DISTANCE		500 FT
TYPE LOCAL CHANNEL CABLE		PARALLEL
TYPE REMOTE CHANNEL CABLE		PARALLEL
LOCAL CHANNEL COMMUNICATION		PARALLEL
REMOTE CHANNEL COMMUNICATION		PARALLEL
LOCAL CHANNEL DATA RATE		500 KHZ
REMOTE CHANNEL DATA RATE		125 KHZ
FAULT MONITORING		YES

I/O NODE DATA

TOTAL NODES/CHANNEL		256
TOTAL DISCRETE POINTS		2048
MAXIMUM POINTS ASSIGNABLE	INPUT	2048
	OUTPUT	2048
MAXIMUM ANALOG ASSIGNABLE	INPUT	255
	OUTPUT	255
DISTANCE FROM DROP		500 FT

ANALOG I/O MODULES

VOLTAGES: SINGLE	
VOLTAGES: DIFFERENTIAL	
CURRENT: SINGLE	
CURRENT: DIFFERENTIAL	N/A
SPECIAL PURPOSE INPUT	
SPECIAL PURPOSE OUTPUT	N/A
INDICATORS:	N/A

EAGLE SIGNAL	EPTAK 700

DISCRETE I/O MODULES

AC VOLTAGES/CURRENT	24,120,240
DC VOLTAGES/CURRENT	4-250V
ISOLATED AC VOLTAGES/CURRENT	120,240
ISOLATED DC VOLTAGES/CURRENT	4-250V
CONTINUOUS OUTPUT RATING	3A
COMMON POINTS/MODULE	1
OUTPUT FUSING CONFIGURATION	PROVIDED BY USER
INDICATORS — I/O POINTS	1/1 STATUS
INDICATORS — FUSING	N/A
ISOLATION	2000V OPTICAL

PROGRAMMING TERMINAL

SCREEN SIZE (DIAGONAL)	7" AND 12"
WEIGHT	47 LBS
SIZE (W X H X D)	18X8 1/2X19 1/4
POWER REQUIREMENTS	120,240VAC
DISPLAY SIZE (CONTACT • ROWS)	8 X 5
GRAPHICS	YES
VIDEO OUT	NO
LADDER LISTER	YES
OFF-LINE PROGRAM	YES
REMOTE OPERATION	YES
SCREEN MNEMONICS	YES
BUILT-IN TAPE	TAPE OR FLOPPY
KEYBOARD	FULL TRAVEL
HANDHELD	NO
DOUBLES AS A COMPUTER TERMINAL	YES

Fig. 21.4.   The Eagle Signal EPTAK 700 PC System

EATON CUTLER-HAMMER		D120	MPCI

PROCESSOR

YEAR INTRODUCED		1977	1983
MEMORY SIZE TOTAL		1K	2K
MEMORY INCREMENTS		500.1K	2K
MEMORY TYPES		RAM,EPROM	RAM,EPROM
VOLTAGE		120 ±10%	115/230 ±20%
FREQUENCY		50/60	50/60
HUMIDITY	MIN	0	0
	MAX	95%	95%
TEMPERATURE (OPERATING)	MIN	0°C	0°C
	MAX	55°C	60°C
USER MEMORY	MIN	500	2K
	MAX	1K	2K
DATA MEMORY	MIN	1800 BITS	672 BITS
	MAX	1800 BITS	672 BITS
USER CONFIGURABLE		NO	NO
WORD SIZE		36	8
WORDS/CONTACT - COIL		1/3-1/3	2-2
WORDS/TIMER - COUNTER		SPECIAL BD.	32 SEP. REG
SCAN TIME - USER LOGIC		10-20 MS	25 MS MAX
I/O SCAN TIME		0	15 MS MAX

SUPPORT

	D120	MPCI
FACTORY APPLICATIONS ENGINEERS	YES	YES
DOCUMENTATION SERVICE	NO	NO
SERVICE CENTER (24 HRS.)	NO	NO
MAINTENANCE TRAINING	YES	YES
APPLICATION TRAINING	YES	YES
WARRANTY TERMS	12 MOS.	18 MOS.
DISTRIBUTION	DIRECT & DIST.	DIRECT & DIST.
MAINTENANCE CENTERS	FACTORY	FACTORY

EATON CUTLER-HAMMER	DI2O	MPCI

INSTRUCTION SET

CONTACTS	YES	YES
TRANSITIONAL (ONE SHOTS)	NO	YES
LATCHES	SPECIAL BD.	YES
RETENTIVE COILS	NO	YES
SKIP/JUMP	NO	NO
SUBROUTINE	NO	NO
TIMERS	SPECIAL BD.	YES
UP COUNTER	SPECIAL BD.	YES
DOWN COUNTER	NO	YES
ADD/SUBTRACT	NO	YES
MULTIPLY/DIVIDE	NO	NO
>,=,<	NO	YES
FILE TRANSFER	NO	NO
LOGICAL	NO	NO
BIT ROTATE/SHIFT	SPECIAL BD.	YES
FIFO/LIFO	NO	NO
SEARCH	NO	NO
DIAGNOSTIC	NO	NO
SEQUENCER	SPECIAL BD.	NO
PID	NO	NO
ASCII	NO	NO
AXIS POSITIONING	NO	NO
MACHINE LANGUAGE	NO	NO
BOOLEAN ALGEBRA	NO	NO
HIGH LEVEL LANGUAGE	NO	NO

EATON CUTLER-HAMMER		D120	MPC1

I/O STATISTICS

TOTAL DISCRETE POINTS AVAILABLE		400	128
MAXIMUM POINTS ASSIGNABLE	INPUT	400	128
	OUTPUT	400	128
NUMBER OF CHANNELS (I/O RACKS)		4	4
NUMBER OF NODES/CHANNEL		1	1
MODULE I/O POINT DENSITY		10	4
ISOLATED MODULE I/O DENSITY		5	2
TOTAL ANALOG CHANNELS AVAILABLE		0	0
MAXIMUM CHANNELS ASSIGNABLE	INPUT	4	4
	OUTPUT	4	4
I/O MAPPING		FIXED	FIXED
INTERRUPTS		NONE	NONE
I/O LOGIC UPDATING	ASYNC		YES
	SYNC	YES	
ERROR CHECKING METHOD		NONE	PARITY
MODULES PER CHANNEL		10	8
APPLICABLE PROCESSORS		D120	MPC1
YEAR OF INTRODUCTION		N/A	N/A
I/O SYSTEM STRUCTURE		PANEL MOUNTED	PANEL MOUNTED

DISCRETE I/O MODULES

AC VOLTAGES/CURRENT		12-48,120	10-32,110,230
DC VOLTAGES/CURRENT		12-48,120	10-32,110,230
ISOLATED AC VOLTAGES/CURRENT		120	10-32,110,230
ISOLATED DC VOLTAGES/CURRENT		12-28V	10-32,110,230
CONTINUOUS OUTPUT RATING		2 AMP	1.25 AMP
COMMON POINTS/MODULE		1/10	2/4
OUTPUT FUSING CONFIGURATION		10/10	2/4
INDICATORS	I/O POINTS	YES	YES
	FUSING	NO	NO
ISOLATION		1500V/OPTICAL	4000V/OPTICAL

EATON CUTLER-HAMMER		D120	MPCI

I/O CHANNEL DATA

TOTAL DISCRETE POINTS / CHANNEL		100	32
MAXIMUM POINTS ASSIGNABLE	INPUT	100	32
	OUTPUT	100	32
MAXIMUM ANALOG ASSIGNABLE	INPUT	0	0
	OUTPUT	0	0
MAXIMUM LOCAL CHANNELS		4	4
MAXIMUM REMOTE CHANNELS		0	3
LOCAL CHANNEL DISTANCE		500 FT	500 FT
REMOTE CHANNEL DISTANCE		N/A	2000 FT
TYPE LOCAL CHANNEL CABLE		PARALLEL	TWISTED PAIR
TYPE REMOTE CHANNEL CABLE		N/A	TWISTED PAIR
LOCAL CHANNEL COMMUNICATION		PARALLEL	SERIAL
REMOTE CHANNEL COMMUNICATION		N/A	SERIAL
LOCAL CHANNEL DATA RATE		50 KHZ	12 KHZ
REMOTE CHANNEL DATA RATE		N/A	12 KHZ
FAULT MONITORING		NO	YES

I/O NODE DATA

TOTAL NODES/CHANNEL		1	1
TOTAL DISCRETE POINTS		100	32
MAXIMUM POINTS ASSIGNABLE	INPUT	100	32
	OUTPUT	100	32
MAXIMUM ANALOG ASSIGNABLE	INPUT	0	0
	OUTPUT	0	0
DISTANCE FROM DROP		N/A	N/A

SPECIAL I/O MODULES

SIMULATION	SIM. PANEL	SIM. MODULE
ASCII	NO	NO
AXIS POSITIONING	NO	NO

EATON CUTLER-HAMMER	D120	MPCI

PROGRAMMING TERMINAL

SCREEN SIZE (DIAGONAL)	N/A	N/A
WEIGHT	7.5 LBS	1.4 LBS
SIZE (W X H X D)	13.5X5.25X16.5	5.5X8.5X2
POWER REQUIREMENTS	120VAC	5VDC
DISPLAY SIZE (CONTACT · ROWS)	1	9
GRAPHICS	NO	NO
VIDEO OUT	NO	NO
LADDER LISTER	YES	YES (OPTIONAL)
OFF-LINE PROGRAM	YES	YES
REMOTE OPERATION	YES	YES
SCREEN MNEMONICS	NO	YES
BUILT-IN TAPE	NO	NO
KEYBOARD	YES	YES
HANDHELD	NO	YES
DOUBLES AS A COMPUTER TERMINAL	NO	NO
EPROM PROG. CAPABILITY	YES	YES

GENERAL ELECTRIC		SERIES SIX	SERIES THREE	SERIES ONE

PROCESSOR

YEAR INTRODUCED		1981	1983	1983
MEMORY SIZE TOTAL		32K	4K	1724
MEMORY INCREMENTS		2K	-	1K
MEMORY TYPES		CMOS	CMOS. PROM	CMOS. PROM
VOLTAGE		90-260VAC	115/230VAC ±15%	115/230VAC ±15%
FREQUENCY		47-63HZ	48-63HZ	48-63HZ
HUMIDITY	MIN	5%	5%	5%
	MAX	95%	95%	95%
TEMPERATURE	MIN	0°C (32°F)	0°C (32°F)	0°C (32°F)
	MAX	60°C (140°F)	60°C (140°F)	60°C (140°F)
USER MEMORY	MIN	2K	4K	700
	MAX	32K	4K	1724
DATA MEMORY	MIN	256	256	64
	MAX	8192	256	64
USER CONFIGURABLE		NO	NO	NO
WORD SIZE		16 BIT	16 BIT	16 BIT
WORDS/CONTACT - COIL		1	1	1
WORDS/TIMER - COUNTER		2	2	2
SCAN TIME - USER LOGIC		1.9M SEC/K	30 M SEC/K	50 M SEC/K
I/O SCAN TIME		6 M SEC	7 M SEC	2 M SEC
REDUNDANCY		YES	-	-
DATA PROCESSOR		128K MEMORY	-	-

SUPPORT

	SERIES SIX	SERIES THREE	SERIES ONE
FACTORY APPLICATIONS ENGINEERS	YES	YES	YES
DOCUMENTATION SERVICE	YES	NO	NO
SERVICE CENTER (24 HRS..)	(804) 978-5747	(804) 978-5747	(804) 978-5747
MAINTENANCE TRAINING	YES	YES	NO
APPLICATION TRAINING	YES	YES	NO
WARRANTY TERMS (START-UP/SHIPMENT)	12/18 MOS.	12/18 MOS.	12/18 MOS.
DISTRIBUTION	YES	YES	YES
MAINTENANCE CENTERS	FACTORY	FACTORY	FACTORY

GENERAL ELECTRIC	SERIES SIX	SERIES THREE	SERIES ONE

INSTRUCTION SET

	SERIES SIX	SERIES THREE	SERIES ONE
CONTACTS	YES	YES	YES
TRANSITIONAL	ONE-SHOT COILS	YES	NO
LATCHES	YES	64	28
RETENTIVE COILS	YES	YES	YES
SKIP/JUMP	SKIP & MCR	MCR	MCR
SUBROUTINE	16	1	-
TIMERS	1K, 3 TYPES	128	64
UP COUNTER	1K	-	64
DOWN COUNTER	1K	128	-
ADD/SUBTRACT	±2,147,483,647	9999	-
MULTIPLY/DIVIDE	±2,147,483,647	99,999,999	-
>,=,<	YES	YES	NO
FILE TRANSFER	YES	YES	NO
LOGICAL	YES	YES	NO
BIT ROTATE/SHIFT	YES	YES	NO
FIFO/LIFO	YES	NO	NO
SEARCH	NO	NO	NO
DIAGNOSTIC	YES	YES	NO
SEQUENCER	NO	NO	YES
PID	SOFTWARE	NO	NO
ASCII	YES	NO	NO
AXIS POSITIONING	YES	NO	NO
MACHINE LANGUAGE	NO	NO	NO
BOOLEAN ALGEBRA	YES	YES	YES
HIGH LEVEL LANGUAGE	OPTIONS	NO	NO
SHIFT REGISTER	NO	YES	YES

GENERAL ELECTRIC		SERIES SIX	SERIES THREE	SERIES ONE

I/O STATISTICS

TOTAL DISCRETE POINTS AVAILABLE		4000	400	112
MAXIMUM POINTS ASSIGNABLE	INPUT	2000	400	112
	OUTPUT	2000	400	112
NUMBER OF CHANNELS		UP TO 50	1	1
NUMBER OF NODES/CHANNEL		MAX. 10/CHAN	3	3
MODULE I/O POINT DENSITY		8 AND 32	16	8
ISOLATED MODULE I/O DENSITY		6	8	4
TOTAL ANALOG CHANNELS AVAILABLE		992	-	-
MAXIMUM CHANNELS ASSIGNABLE	INPUT	496	-	-
	OUTPUT	496	-	-
I/O MAPPING		USER SELECTABLE	FIXED	FIXED
INTERRUPTS		16	-	-
I/O LOGIC UPDATING	ASYNC		X (OUTPUTS)	
	SYNC	X	X (INPUTS)	X
ERROR CHECKING METHOD		PARITY	LOCAL-NOT REQ'D	LOCAL-NOT REQ'D
MODULES PER CHANNEL		MAX 100	14	24
APPLICABLE PROCESSORS		ALL	ALL	ALL
YEAR OF INTRODUCTION		1981	1983	1983
I/O SYSTEM STRUCTURE		19" RACK	BASE	RACK

CONTROL NETWORK

	SERIES SIX	SERIES THREE	SERIES ONE
TYPE	MASTER/SLAVE	-	-
NUMBER OF PC'S	8	-	-
MAXIMUM LENGTH	4000'	-	-
TYPE OF CABLE	TWISTED PAIR	-	-
OPERATION W/PROCESSOR	ALL	-	-
DISCRETE SIGNALS TRANSFERRED	USER SELECTABLE	-	-
DATA SIGNALS TRANSFERRED	USER SELECTABLE	-	-
COMMUNICATION SPEED	19.2 K	-	-

GENERAL ELECTRIC		SERIES SIX	SERIES THREE	SERIES ONE

DISCRETE I/O MODULES

AC VOLTAGES/CURRENT	IN	12,24,48,115,230	115,230	115
	OUT	115,230	115	115
DC VOLTAGES/CURRENT (SINK AND SOURCE)	IN 5VTTL	12,24,48,115,230	5VTTL,12,24	24
	OUT 5VTTL	12,24,48,120	5VTTL,12,24	24
ISOLATED AC VOLTAGES/CURRENT		115,230	115,230(OUT)	115
ISOLATED DC VOLTAGES/CURRENT		-	-	-
CONTINUOUS OUTPUT RATING (TYP)		2 AMP	2 AMP	1 AMP
COMMON POINTS/MODULE		2/8 OR 1/32	2/16 (TYP)	2/8
OUTPUT FUSING CONFIGURATION (TYP)		EACH	4/8, 4/16 (TYP)	2/8
INDICATORS	I/O POINTS	EACH,FIELD	EACH,FIELD	EACH
	FUSING	EACH	-	-
ISOLATION (TYP)		2500V OPTICAL	1500V OPTICAL	1500V OPTICAL
REED RELAY OUTPUTS		100VA NO/NC	5 AMP NO	4 AMP NO

ANALOG I/O MODULES (12 BIT CONVERSION)

	SERIES SIX	SERIES THREE	SERIES ONE
VOLTAGES: SINGLE	0-10VDC	-	-
VOLTAGES: DIFFERENTIAL	±10VDC	-	-
CURRENT: SINGLE	4-20MA	-	-
CURRENT: DIFFERENTIAL	-	-	-
SPECIAL PURPOSE INPUT	T/C TYPE J,K,S,T	-	-
SPECIAL PURPOSE OUTPUT	-	-	-
HIGH SPEED COUNTER	50KHZ	10KHZ	10KHZ

SPECIAL I/O MODULES

	SERIES SIX	SERIES THREE	SERIES ONE
SIMULATION	-	-	-
ASCII	(2) PORTS, 16K, BASIC	-	-
AXIS POSITIONING	SERVO-RESOLVERS	-	-

I/O NODE DATA

		SERIES SIX	SERIES THREE	SERIES ONE
TOTAL NODES/CHANNEL		MAX 10	3	3
TOTAL DISCRETE POINTS		80 PER NODE	128 PER NODE	40 PER NODE
MAXIMUM POINTS ASSIGNABLE	INPUT	320 PER NODE	128 PER NODE	40 PER NODE
	OUTPUT	320 PER NODE	128 PER NODE	40 PER NODE
MAXIMUM ANALOG ASSIGNABLE	INPUT	80 PER NODE	-	-
	OUTPUT	40 PER NODE	-	-
DISTANCE FROM DROP		500'	1'	1'

GENERAL ELECTRIC		SERIES SIX	SERIES THREE	SERIES ONE

I/O CHANNEL DATA

TOTAL DISCRETE POINTS		MAX 800	400	112
MAXIMUM POINTS ASSIGNABLE	INPUT	800	400	112
	OUTPUT	800	400	112
MAXIMUM ANALOG ASSIGNABLE	INPUT	248	-	-
	OUTPUT	248	-	-
MAXIMUM LOCAL CHANNELS		50	1	1
MAXIMUM REMOTE CHANNELS		62	-	-
LOCAL CHANNEL DISTANCE		2000'	5'	5'
REMOTE CHANNEL DISTANCE		10,000'	-	-
TYPE LOCAL CHANNEL CABLE		16 PAIR PARALLEL	25 PAIR PARALLEL	20 WIRE PARALLEL
TYPE REMOTE CHANNEL CABLE		NAT. 22P1SLCB1	-	-
LOCAL CHANNEL COMMUNICATION		PARALLEL	PARALLEL	PARALLEL
REMOTE CHANNEL COMMUNICATION		SERIAL	-	-
LOCAL CHANNEL DATA RATE		350K	57K	57K
REMOTE CHANNEL DATA RATE		57.6K	-	-
FAULT MONITORING		YES	YES	YES
RS-232 REMOTE I/O MODEMS		YES	-	-

DATA NETWORK

	SERIES SIX	SERIES THREE	SERIES ONE
TYPE	PEER-TO-PEER	MASTER/SLAVE	MASTER/SLAVE
NUMBER OF NODES	250	FUTURE	FUTURE
MAXIMUM LENGTH	15,000'	FUTURE	FUTURE
TYPE OF CABLE	CATV	FUTURE	FUTURE
COMMUNICATION SPEED	2MB	FUTURE	FUTURE
COMPUTER INTERFACES	PDP-11/VAX	FUTURE	FUTURE
SPECIAL INTERFACES	CONSULT FACTORY	FUTURE	FUTURE

GENERAL ELECTRIC	SERIES SIX	SERIES THREE	SERIES ONE

PROGRAMMING TERMINAL

SCREEN SIZE (DIAGONAL)	12"	MANUAL	MANUAL
WEIGHT	57 LBS.	BUILT-IN	1/2LB.
SIZE (W X H X D)	21½X12½X16½	-	4.33X4.65X0.95
POWER REQUIREMENTS	115V/230V ±15%	FROM CPU	FROM CPU
DISPLAY SIZE (CONTACT • ROWS)	9 X 8 • COIL	NO LIMIT	NO LIMIT
GRAPHICS	NO	NO	NO
VIDEO OUT	COMP. VIDEO	NO	NO
LADDER LISTER	YES	OPTIONAL	OPTIONAL
OFF-LINE PROGRAM	YES	NO	NO
REMOTE OPERATION	YES	NO	NO
SCREEN MNEMONICS	NO	NO	NO
BUILT-IN TAPE	YES	NO	NO
KEYBOARD	FULL-TRAVEL	PUSHBUTTONS	PUSHBUTTONS
HANDHELD	NO	YES	YES
DOUBLES AS A COMPUTER TERMINAL	NO	NO	NO

GOULD INC.		MODICON 584L	MODICON 584M

PROCESSOR

YEAR INTRODUCED		1983	1983
MEMORY SIZE TOTAL		32KW	16KW
MEMORY INCREMENTS		4KW,16KW	4K,4K,4K
MEMORY TYPES		CMOS	CMOS
VOLTAGE (± 15)		115	115
FREQUENCY HZ		47-63	47-63
HUMIDITY	MIN	0	0
	MAX	95%	95%
TEMPERATURE	MIN	0°C	0°C
	MAX	60°C	60°C
USER MEMORY OR DATA MEM.	MIN	12KW	4KW
	MAX	32KW	16KW
DATA MEMORY OR USER MEM.	MIN	12KW	4KW
	MAX	32KW	16KW
USER CONFIGURABLE		YES	YES
WORD SIZE		16 BIT & 24 BIT	16 BIT
WORDS/CONTACT - COIL		ONE	ONE
WORDS/TIMER - COUNTER		THREE	THREE
SCAN TIME - USER LOGIC OR I/O SCAN		10-60 MS	30-100 MS
I/O SCAN TIME OR USER LOGIC		10-60 MS	30-100 MS

SUPPORT

FACTORY APPLICATIONS ENGINEERS	YES	YES
DOCUMENTATION SERVICE	YES	YES
SERVICE CENTER (24 HRS.)	YES	YES
MAINTENANCE TRAINING	YES	YES
APPLICATION TRAINING	YES	YES
WARRANTY TERMS	1 YEAR	1 YEAR
DISTRIBUTION	SALES	SALES
MAINTENANCE CENTERS	YES	YES

GOULD INC.	MODICON 584L	MODICON 584M

INSTRUCTION SET

CONTACTS	YES	YES
TRANSITIONAL	YES	YES
LATCHES	YES	YES
RETENTIVE COILS	YES	YES
SKIP/JUMP	YES	YES
SUBROUTINE	NO	NO
TIMERS	YES	YES
UP COUNTER	YES	YES
DOWN COUNTER	YES	YES
ADD/SUBTRACT	YES	YES
MULTIPLY/DIVIDE	YES	YES
>,=,<	YES	YES
FILE TRANSFER	NO	NO
LOGICAL	YES	YES
BIT ROTATE/SHIFT	YES	YES
FIFO/LIFO	YES	YES
SEARCH	YES	YES
DIAGNOSTIC	YES	YES
SEQUENCER	YES	YES
PID	YES	YES
ASCII	YES	YES
AXIS POSITIONING	NO	NO
MACHINE LANGUAGE	NO	NO
BOOLEAN ALGEBRA	YES	YES
HIGH LEVEL LANGUAGE	YES	YES

GOULD INC.		MODICON 584L	MODICON 584M

I/O STATISTICS

TOTAL DISCRETE POINTS AVAILABLE		8092	2048
MAXIMUM POINTS ASSIGNABLE	INPUT	4096	1024
	OUTPUT	4096	1024
NUMBER OF CHANNELS		32	32
NUMBER OF NODES/CHANNEL		256	256
MODULE I/O POINT DENSITY		16	16
ISOLATED MODULE I/O DENSITY		8	8
TOTAL ANALOG CHANNELS AVAILABLE		500	128
MAXIMUM CHANNELS ASSIGNABLE	INPUT	250	64
	OUTPUT	250	64
I/O MAPPING		YES	YES
INTERRUPTS		NO	NO
I/O LOGIC UPDATING	ASYNC	N/A	N/A
	SYNC	YES	YES
ERROR CHECKING METHOD		CRC-16	CRC-16
MODULES PER CHANNEL		16	16
APPLICABLE PROCESSORS		2901	2901
YEAR OF INTRODUCTION		1980	1980
I/O SYSTEM STRUCTURE		MULTI DROP	MULTI DROP

DISCRETE I/O MODULES

AC VOLTAGES/CURRENT		120/220	120/220
DC VOLTAGES/CURRENT		10-60 VDC	10-60 VDC
ISOLATED AC VOLTAGES/CURRENT		120V	120V
ISOLATED DC VOLTAGES/CURRENT		24VDC	24VDC
CONTINUOUS OUTPUT RATING		2A	2A
COMMON POINTS/MODULE		4 POINTS	4 POINTS
OUTPUT FUSING CONFIGURATION		SEPARATE	SEPARATE
INDICATORS	I/O POINTS	ONE	ONE
	FUSING	ONE	ONE
ISOLATION		1500V	1500V

GOULD INC.		MODICON 584L	MODICON 584M

I/O CHANNEL DATA

TOTAL DISCRETE POINTS		256	256
MAXIMUM POINTS ASSIGNABLE	INPUT	128	128
	OUTPUT	128	128
MAXIMUM ANALOG ASSIGNABLE	INPUT	8	8
	OUTPUT	8	8
MAXIMUM LOCAL CHANNELS		4	4
MAXIMUM REMOTE CHANNELS		32	28
LOCAL CHANNEL DISTANCE		15 FT	15 FT
REMOTE CHANNEL DISTANCE		1500 FT	1500 FT
TYPE LOCAL CHANNEL CABLE		COAXIAL	COAXIAL
TYPE REMOTE CHANNEL CABLE		COAXIAL	COAXIAL
LOCAL CHANNEL COMMUNICATION		PROPRIETARY	PROPRIETARY
REMOTE CHANNEL COMMUNICATION		HDLC	HDLC
LOCAL CHANNEL DATA RATE		PROPRIETARY	PROPRIETARY
REMOTE CHANNEL DATA RATE		1554 MB	1554 MB
FAULT MONITORING		YES	YES

DATA NETWORK

	MODICON 584L	MODICON 584M
TYPE	MODBUS	MODBUS
NUMBER OF NODES	257	257
MAXIMUM LENGTH	15000 FT	15000 FT
TYPE OF CABLE	SHIELDED	SHIELDED
COMMUNICATION SPEED	19.2KB	19.2KB
COMPUTER INTERFACES	YES	YES
SPECIAL INTERFACES	NO	NO

I/O NODE DATA

		MODICON 584L	MODICON 584M
TOTAL NODES/CHANNEL		256	256
TOTAL DISCRETE POINTS		256	256
MAXIMUM POINTS ASSIGNABLE	INPUT	128	128
	OUTPUT	128	128
MAXIMUM ANALOG ASSIGNABLE	INPUT	8	8
	OUTPUT	8	8
DISTANCE FROM DROP		100 FT	100 FT

GOULD INC.	MODICON 584L	MODICON 584M

ANALOG I/O MODULES

VOLTAGES: SINGLE	0-10VDC	0-10VDC
VOLTAGES: DIFFERENTIAL	0-10VDC	0-10VDC
CURRENT: SINGLE	4-20MA	4-20MA
CURRENT: DIFFERENTIAL	4-20MA	4-20MA
SPECIAL PURPOSE INPUT	THERMOCOUPLE	THERMOCOUPLE
SPECIAL PURPOSE OUTPUT	0-10VDC	0-10VDC
INDICATORS:	INDIVIDUAL	INDIVIDUAL

SPECIAL I/O MODULES

SIMULATION	N/A	N/A
ASCII	YES	YES
AXIS POSITIONING	NO	NO

PROGRAMMING TERMINAL

SCREEN SIZE (DIAGONAL)	9"	9"
WEIGHT	30 LBS	30 LBS
SIZE (W X H X D)	17.5X11X24	17.5X11X24
POWER REQUIREMENTS	100W	100W
DISPLAY SIZE (CONTACT • ROWS)	11 X 7	11 X 7
GRAPHICS	NO	YES
VIDEO OUT	YES	YES
LADDER LISTER	YES	YES
OFF-LINE PROGRAM	NO	NO
REMOTE OPERATION	YES	YES
SCREEN MNEMONICS	NO	NO
BUILT-IN TAPE	YES	YES
KEYBOARD	YES	YES
HANDHELD	NO	NO
DOUBLES AS A COMPUTER TERMINAL	NO	NO

Fig. 21.5. Gould–Modicon 584L and 584M Processors

IDEC SYSTEMS & CONTROLS CORP		PLE-30R	PLE-30P	PLB-48R

PROCESSOR

YEAR INTRODUCED		1982	1983	1983
MEMORY SIZE TOTAL		300 WORD		488 WORD
MEMORY INCREMENTS		FIXED	FIXED	FIXED
MEMORY TYPES		EPROM,RAM	EPROM,RAM	EPROM,RAM
VOLTAGE		120/240	120/240	120/240
FREQUENCY		50/60HZ	50/60HZ	50/60HZ
HUMIDITY	MIN	45%	45%	45%
	MAX	85%	85%	55%
TEMPERATURE STORE/OPERATE	MIN	-20°C/0°C	-20°C/0°C	-20°C/0°C
	MAX	·70°C/·55°C	·70°C/55°C	·70°C/55°C
USER MEMORY	MIN			
	MAX	FIXED 300 WORDS	FIXED 50 OR 90	FIXED 488 WORDS
DATA MEMORY	MIN			
	MAX	FIXED	FIXED	FIXED
USER CONFIGURABLE		NO	NO	NO
WORD SIZE		8 BIT	8 BIT	8 BIT
WORDS/CONTACT - COIL		ONE	ONE	ONE
WORDS/TIMER - COUNTER		ONE	ONE	ONE
SCAN TIME - USER LOGIC		30MS/256 WORDS		28MS/488 WORDS
I/O SCAN TIME		INC. IN ABOVE	30MS	INC. IN ABOVE

SUPPORT

FACTORY APPLICATIONS ENGINEERS	YES	YES	YES
DOCUMENTATION SERVICE	YES	YES	YES
SERVICE CENTER (24 HRS.)	DISTRIBUTOR	NO	NO
MAINTENANCE TRAINING	YES	YES	YES
APPLICATION TRAINING	YES	YES	YES
WARRANTY TERMS	1 YEAR	1 YEAR	1 YEAR
DISTRIBUTION	YES	YES	YES
MAINTENANCE CENTERS	YES	YES	YES

IDEC SYSTEMS & CONTROLS CORP	PLE-30R	PLE-30P	PLB-48R

INSTRUCTION SET

	PLE-30R	PLE-30P	PLB-48R
CONTACTS	YES	YES	YES
TRANSITIONAL	NO	NO	NO
LATCHES	YES	NO	YES
RETENTIVE COILS	YES	NO	YES
SKIP/JUMP	NO	YES	NO
SUBROUTINE	NO	YES	NO
TIMERS	YES	YES	YES
UP COUNTER	YES	YES	YES
DOWN COUNTER	NO	NO	NO
ADD/SUBTRACT	NO	NO	NO
MULTIPLY/DIVIDE	NO	NO	NO
>,=,<	NO	NO	NO
FILE TRANSFER	NO	NO	NO
LOGICAL	NO	NO	NO
BIT ROTATE/SHIFT	NO	NO	YES
FIFO/LIFO	NO	NO	NO
SEARCH	YES	YES	YES
DIAGNOSTIC	YES	YES	YES
SEQUENCER	NO	YES	NO
PID	NO	NO	NO
ASCII	NO	NO	NO
AXIS POSITIONING	NO	NO	NO
MACHINE LANGUAGE	NO	NO	NO
BOOLEAN ALGEBRA	NO	NO	NO
HIGH LEVEL LANGUAGE	NO	NO	NO

IDEC SYSTEMS & CONTROLS CORP		PLE-30R	PLE-30P	PLB-48R

I/O STATISTICS

TOTAL DISCRETE POINTS AVAILABLE		30 FIXED	30 FIXED	64
MAXIMUM POINTS ASSIGNABLE	INPUT	18	18	32
	OUTPUT	12	12	32
NUMBER OF CHANNELS		ONE	ONE	ONE
NUMBER OF NODES/CHANNEL				
MODULE I/O POINT DENSITY				
ISOLATED MODULE I/O DENSITY				
TOTAL ANALOG CHANNELS AVAILABLE				
MAXIMUM CHANNELS ASSIGNABLE	INPUT			
	OUTPUT			
I/O MAPPING				
INTERRUPTS				
I/O LOGIC UPDATING	ASYNC			
	SYNC			
ERROR CHECKING METHOD		LED DISPLAY	LED DISPLAY	LED DISPLAY
MODULES PER CHANNEL		FIXED	FIXED	TWO
APPLICABLE PROCESSORS				
YEAR OF INTRODUCTION		1982	1983	1983
I/O SYSTEM STRUCTURE		SELF-CONTAINED	SELF-CONTAINED	SELF-CONTAINED

DISCRETE I/O MODULES

AC VOLTAGES/CURRENT		0	0	0
DC VOLTAGES/CURRENT		24VDC	24VDC	24VDC
ISOLATED AC VOLTAGES/CURRENT		0	0	0
ISOLATED DC VOLTAGES/CURRENT		24VDC	VDC	24VDC
CONTINUOUS OUTPUT RATING		10 AMPS	10 AMPS	2 AMPS
COMMON POINTS/MODULE		TWO	TWO	TEN
OUTPUT FUSING CONFIGURATION		N/A	N/A	N/A
INDICATORS	I/O POINTS	YES	YES	YES
	FUSING	N/A	N/A	N/A
ISOLATION		YES	YES	YES

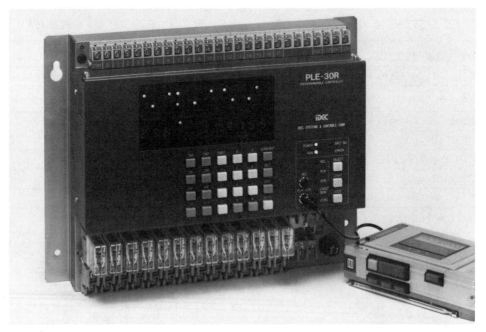

Fig. 21.6. The IDEC Systems' PLE-3OR PC System

INCON		C-64	C-80

PROCESSOR

YEAR INTRODUCED		1982	1983
MEMORY SIZE TOTAL		10K	10K
MEMORY INCREMENTS		2K	2K
MEMORY TYPES		EPROM/CMOSRAM	EPROM/CMOSRAM
VOLTAGE (±)		120/240VAC	120/240VAC
FREQUENCY (±)		50/60 HZ	50/60 HZ
HUMIDITY	MIN	5	5
	MAX	95	95
TEMPERATURE	MIN	0°C	0°C
	MAX	60°C	60°C
USER MEMORY	MIN	1K	1K
	MAX	10K	10K
DATA MEMORY	MIN		
	MAX		
USER CONFIGURABLE			
WORD SIZE		8 BITS	8 BITS
WORDS/CONTACT - COIL		1	1
WORDS/TIMER - COUNTER		3-2	3-2
SCAN TIME - USER LOGIC		35MS	35MS
I/O SCAN TIME			

SUPPORT

	C-64	C-80
FACTORY APPLICATIONS ENGINEERS	YES	YES
DOCUMENTATION SERVICE		
SERVICE CENTER (24 HRS.)		
MAINTENANCE TRAINING	YES	YES
APPLICATION TRAINING	YES	YES
WARRANTY TERMS	1 YEAR	1 YEAR
DISTRIBUTION	DISTRIBUTORS	DISTRIBUTORS
MAINTENANCE CENTERS	FACTORY	FACTORY

PROGRAMMING TERMINAL

EPROM PROGRAMMER WITH AN EMULATION FEATURE

INCON	C-64	C-80

INSTRUCTION SET

	C-64	C-80
CONTACTS	YES	YES
TRANSITIONAL		
LATCHES	YES	YES
RETENTIVE COILS	YES	YES
SKIP/JUMP	YES	YES
SUBROUTINE		
TIMERS	8	8
UP COUNTER	8	8
DOWN COUNTER		
ADD/SUBTRACT	YES	YES
MULTIPLY/DIVIDE		
>,=,<		
FILE TRANSFER		
LOGICAL		
BIT ROTATE/SHIFT	YES	YES
FIFO/LIFO		
SEARCH		
DIAGNOSTIC		
SEQUENCER	YES	YES
PID		
ASCII		YES
AXIS POSITIONING		
MACHINE LANGUAGE	ASSEMBLER	ASSEMBLER
BOOLEAN ALGEBRA		
HIGH LEVEL LANGUAGE		

INCON	C-64	C-80

I/O STATISTICS

TOTAL DISCRETE POINTS AVAILABLE		64	80
MAXIMUM POINTS ASSIGNABLE — INPUT		32	48
MAXIMUM POINTS ASSIGNABLE — OUTPUT		32	32
NUMBER OF CHANNELS			
NUMBER OF NODES/CHANNEL			
MODULE I/O POINT DENSITY		COMPLETE CONTROLLER	
ISOLATED MODULE I/O DENSITY		ON A SINGLE PRINTED CIRCUIBD	
TOTAL ANALOG CHANNELS AVAILABLE			
MAXIMUM CHANNELS ASSIGNABLE — INPUT			
MAXIMUM CHANNELS ASSIGNABLE — OUTPUT			
I/O MAPPING			
INTERRUPTS			
I/O LOGIC UPDA ING — ASYNC			
I/O LOGIC UPDA ING — SYNC			
ERROR CHECKING METHOD		CHECK SUM W/D TIMER	CHECK SUM W/D TIMER
MODULES PER CHANNEL			
APPLICABLE PROCESSORS			
YEAR OF INTRODUCTION			
I/O SYSTEM STRUCTURE			

DISCRETE I/O MODULES

AC VOLTAGES/CURRENT		24,120,240	N/A
DC VOLTAGES/CURRENT		5,24	5,24
ISOLATED AC VOLTAGES/CURRENT		120/240	N/A
ISOLATED DC VOLTAGES/CURRENT		24	24
CONTINUOUS OUTPUT RATING		1A	1A
COMMON POINTS/MODULE			
OUTPUT FUSING CONFIGURATION			
INDICATORS — I/O POINTS		STATUS	STATUS
INDICATORS — FUSING			
ISOLATION		OPTICAL	OPTICAL

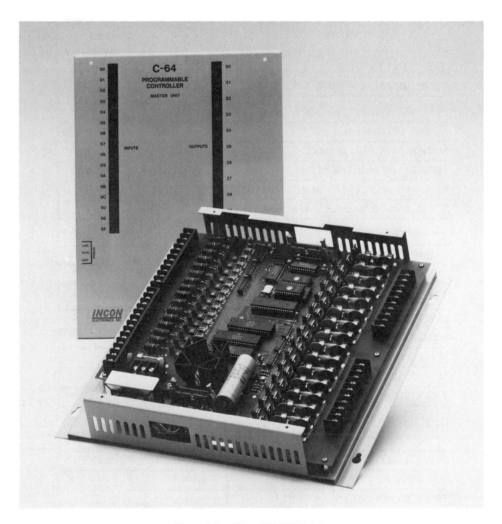

Fig. 21.7.   The INCON C-64

ISSC		620-10	620-15	620-20

PROCESSOR

YEAR INTRODUCED		1983	1983	1983
MEMORY SIZE TOTAL		2K	2K	8K
MEMORY INCREMENTS		1/2K,1K	1/2K,1K	2K,4K,8K
MEMORY TYPES		CMOS,EPROM	CMOS,EPROM	CMOS
VOLTAGE (±15)		115/230	115/230	114/230
FREQUENCY HZ		47-63	47-63	47-63
HUMIDITY	MIN	5	5	5
	MAX	95	95	95
TEMPERATURE °C	MIN	0	0	0
	MAX	60	60	60
USER MEMORY	MIN	1/2K	1/2K	2K
	MAX	2K	2K	8K
DATA MEMORY	MIN	256	256	2048
	MAX	256	256	2048
USER CONFIGURABLE		NO	NO	NO
WORD SIZE		24 BITS	24 BITS	24 BITS
WORDS/CONTACT - COIL		1	1	1
WORDS/TIMER - COUNTER		1	1	1
SCAN TIME - USER LOGIC		10 MSEC	10 MSEC	3.3 MSEC
I/O SCAN TIME		0.4 MSEC	0.4 MSEC	0.5 MSEC

SUPPORT

	620-10	620-15	620-20
FACTORY APPLICATIONS ENGINEERS	YES	YES	YES
DOCUMENTATION SERVICE	YES	YES	YES
SERVICE CENTER (24 HRS.)	YES	YES	YES
MAINTENANCE TRAINING	YES	YES	YES
APPLICATION TRAINING	YES	YES	YES
WARRANTY TERMS	1 YEAR	1 YEAR	1 YEAR
DISTRIBUTION	DISTRIBUTORS	DISTRIBUTORS	DISTRIBUTORS
MAINTENANCE CENTERS	DISTRIBUTORS	DISTRIBUTORS	DISTRIBUTORS

ISSC	620-10	620-15	620-20

INSTRUCTION SET

	620-10	620-15	620-20
CONTACTS	YES	YES	YES
TRANSITIONAL	ON & OFF	ON & OFF	ON & OFF
LATCHES	YES	YES	YES
RETENTIVE COILS	YES	YES	YES
SKIP/JUMP	YES	YES	YES
SUBROUTINE	NO	NO	NO
TIMERS	YES	YES	YES
UP COUNTER	YES	YES	YES
DOWN COUNTER	YES	YES	YES
ADD/SUBTRACT	NO	YES	YES
MULTIPLY/DIVIDE	NO	YES	YES
>,=,<	NO	YES	YES
FILE TRANSFER	NO	NO	NO
LOGICAL	NO	NO	NO
BIT ROTATE/SHIFT	NO	NO	NO
FIFO/LIFO	NO	NO	NO
SEARCH			
DIAGNOSTIC	YES	YES	YES
SEQUENCER	YES	YES	YES
PID	NO	YES	YES
ASCII	NO	NO	NO
AXIS POSITIONING	YES	YES	YES
MACHINE LANGUAGE	NO	NO	NO
BOOLEAN ALGEBRA	NO	NO	NO
HIGH LEVEL LANGUAGE	BASIC	BASIC	BASIC

ISSC		620-10	620-15	620-20

I/O STATISTICS

		620-10	620-15	620-20
TOTAL DISCRETE POINTS AVAILABLE		256	256	512
MAXIMUM POINTS ASSIGNABLE	INPUT	256	256	512
	OUTPUT	256	256	512
NUMBER OF CHANNELS		1	1	1
NUMBER OF NODES/CHANNEL		3	3	11
MODULE I/O POINT DENSITY		8/16	8/16	8/16
ISOLATED MODULE I/O DENSITY		6	6	6
TOTAL ANALOG CHANNELS AVAILABLE		0	256	512
MAXIMUM CHANNELS ASSIGNABLE	INPUT	1	1	1
	OUTPUT	1	1	1
I/O MAPPING		USER DEFINABLE	USER DEFINABLE	USER DEFINABLE
INTERRUPTS		0	0	0
I/O LOGIC UPDATING	ASYNC			
	SYNC	YES	YES	YES
ERROR CHECKING METHOD		REDUNDANCY	REDUNDANCY	REDUNDANCY
MODULES PER CHANNEL		32/16	32/16	64/32
APPLICABLE PROCESSORS		YES	YES	YES
YEAR OF INTRODUCTION		1983	1983	1983
I/O SYSTEM STRUCTURE				

CONTROL NETWORK

	620-10	620-15	620-20
TYPE	PEER TO PEER	PEER TO PEER	PEER TO PEER
NUMBER OF PC'S	8	8	8
MAXIMUM LENGTH	8000 FT	8000 FT	8000 FT
TYPE OF CABLE	TWISTED PAIR TWIN AXIAL	TWISTED PAIR TWIN AXIAL	TWISTED PAIR TWIN AXIAL
OPERATION W/PROCESSOR	ALL	ALL	ALL
DISCRETE SIGNALS TRANSFERRED	256	256	256
DATA SIGNALS TRANSFERRED	16	16	16
COMMUNICATION SPEED	47 KHZ	47 KHZ	47 KHZ

ISSC	*620-10*	*620-15*	*620-20*

DISCRETE I/O MODULES

AC VOLTAGES/CURRENT	24,115,230	24,115,230	24,115,230
DC VOLTAGES/CURRENT	5,12-24,48-66, 115,230	5,12-24,48-66, 115,230	5,12-24,48-66, 115,230
ISOLATED AC VOLTAGES/CURRENT	115,230	115,230	115,230
ISOLATED DC VOLTAGES/CURRENT	N/A	N/A	N/A
CONTINUOUS OUTPUT RATING	2 AMP	2 AMP	2 AMP
COMMON POINTS/MODULE	2/8	2/8	2/8
OUTPUT FUSING CONFIGURATION	8/8	8/8	8/8
INDICATORS I/O POINTS	POWER	POWER	POWER
FUSING	YES	YES	YES
ISOLATION	2500V OPTIONAL	2500V OPTIONAL	2500V OPTIONAL

ANALOG I/O MODULES

VOLTAGES: SINGLE	N/A	N/A	N/A
VOLTAGES: DIFFERENTIAL	0-10,-10--10, 0-5,-5-·5	0-10,-10--10, 0-5,-5-·5	0-10,-10--10, 0-5,-5-·5
CURRENT: SINGLE	N/A	N/A	N/A
CURRENT: DIFFERENTIAL	0-20,0-40	0-20,0-40	0-20,0-40
SPECIAL PURPOSE INPUT	N/A	N/A	N/A
SPECIAL PURPOSE OUTPUT	N/A	N/A	N/A
INDICATORS:	ACITVE	ACITVE	ACITVE

SPECIAL I/O MODULES

SIMULATION	YES	YES	YES
ASCII	NO	NO	NO
AXIS POSITIONING	YES	YES	YES

I/O NODE DATA

TOTAL NODES/CHANNEL	3	3	11
TOTAL DISCRETE POINTS	256	256	512
MAXIMUM POINTS ASSIGNABLE INPUT	256	256	512
OUTPUT	256	256	512
MAXIMUM ANALOG ASSIGNABLE INPUT	0	256	512
OUTPUT	0	128	256
DISTANCE FROM DROP	N/A	N/A	N/A

ISSC		620-10	620-15	620-20

I/O CHANNEL DATA

TOTAL DISCRETE POINTS			2048	
MAXIMUM POINTS ASSIGNABLE	INPUT		2048	
	OUTPUT		2048	
MAXIMUM ANALOG ASSIGNABLE	INPUT	0	256	512
	OUTPUT	0	128	256
MAXIMUM LOCAL CHANNELS		1	1	1
MAXIMUM REMOTE CHANNELS		0	0	4
LOCAL CHANNEL DISTANCE		50 FT	50 FT	100 FT
REMOTE CHANNEL DISTANCE		N/A	N/A	8000 FT
TYPE LOCAL CHANNEL CABLE		PARALLEL	PARALLEL	PARALLEL
TYPE REMOTE CHANNEL CABLE		TWISTED PAIR OR TWIN AXIAL	TWISTED PAIR OR TWIN AXIAL	TWISTED PAIR OR TWIN AXIAL
LOCAL CHANNEL COMMUNICATION		PARALLEL	PARALLEL	PARALLEL
REMOTE CHANNEL COMMUNICATION		SERIAL	SERIAL	SERIAL
LOCAL CHANNEL DATA RATE		1 MHZ	1 MHZ	1 MHZ
REMOTE CHANNEL DATA RATE		47 KHZ	47 KHZ	47 KHZ
FAULT MONITORING		YES	YES	YES

DATA NETWORK

	620-10	620-15	620-20
TYPE	COPNET MASTER-SLAVE	COPNET MASTER-SLAVE	COPNET MASTER-SLAVE
NUMBER OF NODES	254	254	254
MAXIMUM LENGTH	32000 FT	32000 FT	32000 FT
TYPE OF CABLE	TWISTED OR COAX	TWISTED OR COAX	TWISTED OR COAX
COMMUNICATION SPEED	115 MHZ	115 MHZ	115 MHZ
COMPUTER INTERFACES	YES	YES	YES
SPECIAL INTERFACES	YES	YES	YES

ISSC	620-10	620-15	620-20

PROGRAMMING TERMINAL

	620-10	620-15	620-20
SCREEN SIZE (DIAGONAL)	9"	9"	9"
WEIGHT	28 LBS.	28 LBS.	28 LBS.
SIZE (W X H X D)	17.5X8.5X15.5	17.5X8.5X15.5	17.5X8.5X15.5
POWER REQUIREMENTS	115 OR 230VAC	115 OR 230VAC	115 OR 230VAC
DISPLAY SIZE (CONTACT × ROWS)	9 X 5	9 X 5	9 X 5
GRAPHICS	YES	YES	YES
VIDEO OUT	YES	YES	YES
LADDER LISTER	YES	YES	YES
OFF-LINE PROGRAM	YES	YES	YES
REMOTE OPERATION	YES	YES	YES
SCREEN MNEMONICS	YES	YES	YES
BUILT-IN TAPE	YES	YES	YES
KEYBOARD	MEMBRANE OR MECHANICAL	MEMBRANE OR MECHANICAL	MEMBRANE OR MECHANICAL
HANDHELD	YES	YES	YES
DOUBLES AS A COMPUTER TERMINAL	YES	YES	YES

ISSC	*620-30*

PROCESSOR

YEAR INTRODUCED		1983
MEMORY SIZE TOTAL		24K
MEMORY INCREMENTS		2K,4K,8K
MEMORY TYPES		CMOS
VOLTAGE (± 15)		115/230
FREQUENCY (±)		47-63
HUMIDITY	MIN	5
	MAX	95
TEMPERATURE °C	MIN	0
	MAX	60
USER MEMORY	MIN	2K
	MAX	24K
DATA MEMORY	MIN	2048
	MAX	4096
USER CONFIGURABLE		NO
WORD SIZE		24 BITS
WORDS/CONTACT - COIL		1
WORDS/TIMER - COUNTER		1
SCAN TIME - USER LOGIC		3.3 MSEC
I/O SCAN TIME		2.0 MSEC

SUPPORT

FACTORY APPLICATIONS ENGINEERS	YES
DOCUMENTATION SERVICE	YES
SERVICE CENTER (24 HRS.)	YES
MAINTENANCE TRAINING	YES
APPLICATION TRAINING	YES
WARRANTY TERMS	1 YEAR
DISTRIBUTION	DISTRIBUTORS
MAINTENANCE CENTERS	DISTRIBUTORS

ISSC	*620-30*

INSTRUCTION SET

CONTACTS	YES
TRANSITIONAL	ON & OFF
LATCHES	YES
RETENTIVE COILS	YES
SKIP/JUMP	YES
SUBROUTINE	YES
TIMERS	YES
UP COUNTER	YES
DOWN COUNTER	YES
ADD/SUBTRACT	YES
MULTIPLY/DIVIDE	YES
>,=,<	YES
FILE TRANSFER	NO
LOGICAL	YES
BIT ROTATE/SHIFT	NO
FIFO/LIFO	NO
SEARCH	N/A
DIAGNOSTIC	YES
SEQUENCER	YES
PID	YES
ASCII	NO
AXIS POSITIONING	YES
MACHINE LANGUAGE	NO
BOOLEAN ALGEBRA	NO
HIGH LEVEL LANGUAGE	BASIC

ISSC	620-30

I/O STATISTICS

TOTAL DISCRETE POINTS AVAILABLE		2048
MAXIMUM POINTS ASSIGNABLE	INPUT	2048
	OUTPUT	2048
NUMBER OF CHANNELS		1
NUMBER OF NODES/CHANNEL		22
MODULE I/O POINT DENSITY		8/16
ISOLATED MODULE I/O DENSITY		6
TOTAL ANALOG CHANNELS AVAILABLE		2048
MAXIMUM CHANNELS ASSIGNABLE	INPUT	1
	OUTPUT	1
I/O MAPPING		USER DEFINABLE
INTERRUPTS		0
I/O LOGIC UPDATING	ASYNC	
	SYNC	YES
ERROR CHECKING METHOD		REDUNDANCY
MODULES PER CHANNEL		256/128
APPLICABLE PROCESSORS		YES
YEAR OF INTRODUCTION		1983
I/O SYSTEM STRUCTURE		N/A

CONTROL NETWORK

TYPE	PEER TO PEER
NUMBER OF PC'S	8
MAXIMUM LENGTH	8000 FT
TYPE OF CABLE	TWISTED PAIR OR TWIN AXIAL
OPERATION W/PROCESSOR	ALL
DISCRETE SIGNALS TRANSFERRED	256
DATA SIGNALS TRANSFERRED	16
COMMUNICATION SPEED	47 KHZ

ISSC	*620-30*

DISCRETE I/O MODULES

AC VOLTAGES/CURRENT	24,115,230
DC VOLTAGES/CURRENT	5,12-24,48-66, 115,230
ISOLATED AC VOLTAGES/CURRENT	115,230
ISOLATED DC VOLTAGES/CURRENT	N/A
CONTINUOUS OUTPUT RATING	2 AMP
COMMON POINTS/MODULE	2/8
OUTPUT FUSING CONFIGURATION	8/8
INDICATORS — I/O POINTS	POWER
INDICATORS — FUSING	YES
ISOLATION	2500V OPTIONAL

ANALOG I/O MODULES

VOLTAGES: SINGLE	N/A
VOLTAGES: DIFFERENTIAL	0-10,-10-+10, 0-5,-5-+5
CURRENT: SINGLE	N/A
CURRENT: DIFFERENTIAL	0-20,0-40
SPECIAL PURPOSE INPUT	N/A
SPECIAL PURPOSE OUTPUT	N/A
INDICATORS:	ACTIVE

SPECIAL I/O MODULES

SIMULATION	YES
ASCII	NO
AXIS POSITIONING	YES

I/O NODE DATA

TOTAL NODES/CHANNEL		22
TOTAL DISCRETE POINTS		2048
MAXIMUM POINTS ASSIGNABLE	INPUT	2048
	OUTPUT	2048
MAXIMUM ANALOG ASSIGNABLE	INPUT	2048
	OUTPUT	1024
DISTANCE FROM DROP		N/A

ISSC	620-30

I/O CHANNEL DATA

TOTAL DISCRETE POINTS		2048
MAXIMUM POINTS ASSIGNABLE	INPUT	2048
	OUTPUT	2048
MAXIMUM ANALOG ASSIGNABLE	INPUT	2048
	OUTPUT	1024
MAXIMUM LOCAL CHANNELS		1
MAXIMUM REMOTE CHANNELS		4
LOCAL CHANNEL DISTANCE		100 FT
REMOTE CHANNEL DISTANCE		8000 FT
TYPE LOCAL CHANNEL CABLE		PARALLEL
TYPE REMOTE CHANNEL CABLE		TWISTED PAIR OR TWIN AXIAL
LOCAL CHANNEL COMMUNICATION		PARALLEL
REMOTE CHANNEL COMMUNICATION		SERIAL
LOCAL CHANNEL DATA RATE		1 MHZ
REMOTE CHANNEL DATA RATE		47 MHZ
FAULT MONITORING		YES

DATA NETWORK

TYPE	COPNET MASTER/SLAVE
NUMBER OF NODES	254
MAXIMUM LENGTH	32000 FT
TYPE OF CABLE	TWISTED OR COAX
COMMUNICATION SPEED	115 MHZ
COMPUTER INTERFACES	YES
SPECIAL INTERFACES	YES

ISSC	620-30

PROGRAMMING TERMINAL

SCREEN SIZE (DIAGONAL)	9"
WEIGHT	28 LBS.
SIZE (W X H X D)	17.5X8.5X15.5
POWER REQUIREMENTS	115 OR 230VAC
DISPLAY SIZE (CONTACT ∗ ROWS)	9 X 5
GRAPHICS	YES
VIDEO OUT	YES
LADDER LISTER	YES
OFF-LINE PROGRAM	YES
REMOTE OPERATION	YES
SCREEN MNEMONICS	YES
BUILT-IN TAPE	YES
KEYBOARD	MEMBRANE OR MECHANICAL
HANDHELD	YES
DOUBLES AS A COMPUTER TERMINAL	YES

KLOCKNER-MOELLER		PS-24	PS-22	PS-21

PROCESSOR

YEAR INTRODUCED		1977	1980	1983
MEMORY SIZE TOTAL		4K	2K	4K
MEMORY INCREMENTS		1K	1K	4K
MEMORY TYPES		RAM/EPROM	EPROM	EPROM
VOLTAGE +15% -25%		+24 VDC	+24 VDC	110/220 VAC ±10%
FREQUENCY		N/A	N/A	50/60 HZ
HUMIDITY	MIN	10	10	10
	MAX	90	90	90
TEMPERATURE (TRUE AMBIENT)	MIN	0° C	0° C	0° C
	MAX	+55° C	+55° C	+55° C
USER MEMORY	MIN	1K	1K	4K
	MAX	4K	2K	4K
DATA MEMORY	MIN	0K	0K	0K
	MAX	4K	0K	0K
USER CONFIGURABLE		YES	YES	NO
WORD SIZE		16 BITS	16 BITS	16 BITS
WORDS/CONTACT - COIL		1	1	1
WORDS/TIMER - COUNTER		1	1	1
SCAN TIME - USER LOGIC (SEE NOTE 1)		1 MS/K	2.1 MS/K	2 MS/K
I/O SCAN TIME (SEE NOTE 1)		1 MS/K	2.1 MS/K	2 MS/K

SUPPORT

FACTORY APPLICATIONS ENGINEERS	YES	YES	YES
DOCUMENTATION SERVICE	YES	YES	YES
SERVICE CENTER (24 HRS.)	NO	NO	NO
MAINTENANCE TRAINING	YES	YES	YES
APPLICATION TRAINING	YES	YES	YES
WARRANTY TERMS	6 MOS.	6 MOS.	6 MOS.
DISTRIBUTION	K.M.BRANCHES	K.M.BRANCHES	K.M.BRANCHES
MAINTENANCE CENTERS	K.M.BRANCHES	K.M.BRANCHES	K.M.BRANCHES

NOTES :

    1) USER LOGIC AND I/O SCAN TIMES ARE NOT ADDITIVE.
    2) 12' PROGRAMMING TERMINAL IS A LEAR-SIEGLER ADM3A TELEVIDEO TVI920.
    3) 9' PROGRAMMING TERMINAL IS A KLOCKNER-MOELLER PRG-245.
    4) TERMINAL IS A KLOCKNER-MOELLER PRG-22.

KLOCKNER-MOELLER	PS-24	PS-22	PS-21

INSTRUCTION SET

	PS-24	PS-22	PS-21
CONTACTS	YES	YES	YES
TRANSITIONAL	NO	NO	NO
LATCHES	YES	YES	YES
RETENTIVE COILS	YES	YES	YES
SKIP/JUMP	CONDITIONAL & UNCONDITIONAL	CONDITIONAL & UNCONDITIONAL	CONDITIONAL & UNCONDITIONAL
SUBROUTINE	YES	YES	YES
TIMERS	PROGRAMMABLE & MANUAL	PROGRAMMABLE & MANUAL	PROGRAMMABLE & MANUAL
UP COUNTER	NO	NO	NO
DOWN COUNTER	NO	NO	NO
ADD/SUBTRACT	NO	NO	NO
MULTIPLY/DIVIDE	NO	NO	NO
>,=,<	NO	NO	NO
FILE TRANSFER	NO	NO	NO
LOGICAL	NO	NO	NO
BIT ROTATE/SHIFT	NO	NO	NO
FIFO/LIFO	NO	NO	NO
SEARCH	NO	NO	NO
DIAGNOSTIC	NO	NO	NO
SEQUENCER	NO	NO	NO
PID	NO	NO	NO
ASCII	NO	NO	NO
AXIS POSITIONING	NO	NO	NO
MACHINE LANGUAGE	NO	NO	NO
BOOLEAN ALGEBRA	YES	YES	YES
HIGH LEVEL LANGUAGE	LADDER & LOGIC SYMBOL	LADDER & LOGIC SYMBOL	LADDER & LOGIC SYMBOL

KLOCKNER-MOELLER		PS-24	PS-22	PS-21

I/O STATISTICS

TOTAL DISCRETE POINTS AVAILABLE		2048	1024	56
MAXIMUM POINTS ASSIGNABLE	INPUT	1024	512	32
	OUTPUT	1024	512	24
NUMBER OF CHANNELS		7	7	N/A
NUMBER OF NODES/CHANNEL		10 BASIC/19EXP.	4,8,12,OR 18 BASIC/19EXP.	N/A
MODULE I/O POINT DENSITY		16	16	N/A
ISOLATED MODULE I/O DENSITY		16	16	N/A
TOTAL ANALOG CHANNELS AVAILABLE		64	32	N/A
MAXIMUM CHANNELS ASSIGNABLE	INPUT	64	32	N/A
	OUTPUT	64	32	N/A
I/O MAPPING		USER-DEFINED	USER-DEFINED	FIXED
INTERRUPTS		N/A	N/A	N/A
I/O LOGIC UPDATING	ASYNC			
	SYNC	N/A	N/A	N/A
ERROR CHECKING METHOD		N/A	N/A	N/A
MODULES PER CHANNEL		N/A	N/A	N/A
APPLICABLE PROCESSORS		N/A	N/A	N/A
YEAR OF INTRODUCTION		1977	1980	1983
I/O SYSTEM STRUCTURE		RACK & BASE	RACK & BASE	INHERENT IN SYS.

DATA NETWORK

	PS-24	PS-22	PS-21
TYPE	MASTER-SLAVE	MASTER-SLAVE	N/A
NUMBER OF NODES	1	1	N/A
MAXIMUM LENGTH (W/20MA C/L)	1000 FT	1000 FT	N/A
TYPE OF CABLE	UNSHIELDED	UNSHIELDED	N/A
COMMUNICATION SPEED	9600	9600	N/A
COMPUTER INTERFACES	YES	YES	N/A
SPECIAL INTERFACES	RS-232	RS-232	N/A

KLOCKNER-MOELLER		*PS-24*	*PS-22*	*PS-21*

DISCRETE I/O MODULES

AC VOLTAGES/CURRENT		110 & 220	110 & 220	110 & 220
DC VOLTAGES/CURRENT		24	24	24
ISOLATED AC VOLTAGES/CURRENT		110 & 220	110 & 220	110 & 220
ISOLATED DC VOLTAGES/CURRENT		24	24	24
CONTINUOUS OUTPUT RATING		200MA,350MA,2.1A	200MA,350MA,2.1A	200MA,350MA,2.1A
COMMON POINTS/MODULE		4/4 OR 8/8	4/4 OR 8/8	4/16 OR 8/32
OUTPUT FUSING CONFIGURATION		4/4 OR 8/8	4/4 OR 8/8	N/A
INDICATORS	I/O POINTS	LOGIC & POWER	LOGIC & POWER	LOGIC & POWER
	FUSING	YES	YES	YES
ISOLATION		3000V	3000V	3000V

ANALOG I/O MODULES

VOLTAGES: SINGLE	0-10 V	0-10 V	N/A
VOLTAGES: DIFFERENTIAL	0-10 V	0-10 V	N/A
CURRENT: SINGLE	0-20 OR 4-20	0-20 OR 4-20	N/A
CURRENT: DIFFERENTIAL	0-20 OR 4-20	0-20 OR 4-20	N/A
SPECIAL PURPOSE INPUT	PT100 RTD	PT100 RTD	N/A
SPECIAL PURPOSE OUTPUT	NONE	NONE	N/A
INDICATORS:	NONE	NONE	N/A

SPECIAL I/O MODULES

SIMULATION	YES	YES	YES
ASCII	YES	YES	YES
AXIS POSITIONING	NO	NO	NO

I/O NODE DATA

TOTAL NODES/CHANNEL		10 BASIC/19EXP	4,8,12, OR 18 BASIC/19EXP	N/A
TOTAL DISCRETE POINTS		UP TO 304	UP TO 304	N/A
MAXIMUM POINTS ASSIGNABLE	INPUT	304	304	N/A
	OUTPUT	304	304	N/A
MAXIMUM ANALOG ASSIGNABLE	INPUT	19	19	N/A
	OUTPUT	19	19	N/A
DISTANCE FROM DROP		N/A	N/A	N/A

KLOCKNER-MOELLER		PS-24	PS-22	PS-21

I/O CHANNEL DATA

TOTAL DISCRETE POINTS		2048	1024	28 OR 56
MAXIMUM POINTS ASSIGNABLE	INPUT	1024	512	12 OR 24
	OUTPUT	1024	512	16 OR 32
MAXIMUM ANALOG ASSIGNABLE	INPUT	64	32	N/A
	OUTPUT	64	32	N/A
MAXIMUM LOCAL CHANNELS		64	32	N/A
MAXIMUM REMOTE CHANNELS		N/A	N/A	N/A
LOCAL CHANNEL DISTANCE		800'	800'	800'
REMOTE CHANNEL DISTANCE		N/A	N/A	N/A
TYPE LOCAL CHANNEL CABLE		UNSHIELDED	UNSHIELDED	UNSHIELDED
TYPE REMOTE CHANNEL CABLE		N/A	N/A	N/A
LOCAL CHANNEL COMMUNICATION		BIT SERIAL	BIT SERIAL	BIT SERIAL
REMOTE CHANNEL COMMUNICATION		N/A	N/A	N/A
LOCAL CHANNEL DATA RATE		1 MHZ	500 KHZ	500 KHZ
REMOTE CHANNEL DATA RATE		N/A	N/A	N/A
FAULT MONITORING		N/A	N/A	N/A

PROGRAMMING TERMINAL

	(SEE NOTE 2)	(SEE NOTE 3)	(SEE NOTE 4)
SCREEN SIZE (DIAGONAL)	12"	9"	N/A
WEIGHT	32 LBS	42 LBS	28 LBS
SIZE (W X H X D)	15.6"X20.2"X13.5		
POWER REQUIREMENTS	110VAC 60W	110VAC	110VAC
DISPLAY SIZE (CONTACT • ROWS)	6X10	6X10	1 LINE
GRAPHICS	NO	NO	NO
VIDEO OUT	NO	YES	NO
LADDER LISTER	YES	YES	NO-MNEMONIC
OFF-LINE PROGRAM	NO	YES	YES
REMOTE OPERATION	YES	YES	YES
SCREEN MNEMONICS	NO	YES	NO
BUILT-IN TAPE	NO	YES	NO
KEYBOARD	YES	YES	YES
HANDHELD	NO	NO	NO
DOUBLES AS A COMPUTER TERMINAL	YES	NO	NO

H.KUHNKE CO.	KUAX653	KUAX654

PROCESSOR

YEAR INTRODUCED		1979	1981
MEMORY SIZE TOTAL		4K	4K
MEMORY INCREMENTS		.5	.5
MEMORY TYPES		RAM-EPROM	RAM-EPROM
VOLTAGE		120/240	120/240
FREQUENCY		50/60	50/60
HUMIDITY	MIN	10%	10%
	MAX	90%	90%
TEMPERATURE	MIN	0°C	0°C
	MAX	55°C	55°C
USER MEMORY	MIN	1K	1K
	MAX	4K	4K
DATA MEMORY	MIN		
	MAX		
USER CONFIGURABLE		LADDER/DIA	LADDER/DIA
WORD SIZE		16 BITS	16 BITS
WORDS/CONTACT - COIL		1	1
WORDS/TIMER - COUNTER		1	1
SCAN TIME - USER LOGIC		5 MS/1K	5 MS/1K
I/O SCAN TIME			

SUPPORT

FACTORY APPLICATIONS ENGINEERS	YES	YES
DOCUMENTATION SERVICE	YES	YES
SERVICE CENTER (24 HRS.)	NO	NO
MAINTENANCE TRAINING	YES	YES
APPLICATION TRAINING	YES	YES
WARRANTY TERMS	1 YEAR	1 YEAR
DISTRIBUTION		
MAINTENANCE CENTERS		

H.KUHNKE CO.	KUAX653	KUAX654

INSTRUCTION SET

	KUAX653	KUAX654
CONTACTS	YES	YES
TRANSITIONAL	ON/OFF	ON/OFF
LATCHES	YES	NO
RETENTIVE COILS		
SKIP/JUMP	JUMP	JUMP
SUBROUTINE	YES	YES
TIMERS	256	256
UP COUNTER	256	256
DOWN COUNTER	NO	NO
ADD/SUBTRACT	NO	NO
MULTIPLY/DIVIDE	NO	NO
>,=,<	NO	NO
FILE TRANSFER	NO	NO
LOGICAL	NO	NO
BIT ROTATE/SHIFT		
FIFO/LIFO		
SEARCH	YES	YES
DIAGNOSTIC		
SEQUENCER		
PID		
ASCII		
AXIS POSITIONING		
MACHINE LANGUAGE	NO	NO
BOOLEAN ALGEBRA	YES	YES
HIGH LEVEL LANGUAGE	NO	NO

SPECIAL I/O MODULES

	KUAX653	KUAX654
SIMULATION	YES	
ASCII	NO	NO
AXIS POSITIONING	NO	NO

H.KUHNKE CO.		KUAX653	KUAX654

I/O STATISTICS

TOTAL DISCRETE POINTS AVAILABLE			
MAXIMUM POINTS ASSIGNABLE	INPUT	256	32
	OUTPUT	256	32
NUMBER OF CHANNELS		256	32
NUMBER OF NODES/CHANNEL		1	1
MODULE I/O POINT DENSITY		16	16
ISOLATED MODULE I/O DENSITY			
TOTAL ANALOG CHANNELS AVAILABLE			
MAXIMUM CHANNELS ASSIGNABLE	INPUT	256	32
	OUTPUT	256	32
I/O MAPPING			
INTERRUPTS			
I/O LOGIC UPDATING	ASYNC	YES	YES
	SYNC	N/A	N/A
ERROR CHECKING METHOD			
MODULES PER CHANNEL			
APPLICABLE PROCESSORS			
YEAR OF INTRODUCTION		1979	1981
I/O SYSTEM STRUCTURE		RACK-WALL	RACK-WALL

DISCRETE I/O MODULES

AC VOLTAGES/CURRENT			
DC VOLTAGES/CURRENT			
ISOLATED AC VOLTAGES/CURRENT		120-240	120-240
ISOLATED DC VOLTAGES/CURRENT		24	24
CONTINUOUS OUTPUT RATING		300 MA	300 MA
COMMON POINTS/MODULE		4-16	8-16
OUTPUT FUSING CONFIGURATION		SOLID STATE	SOLID STATE
INDICATORS	I/O POINTS	YES	YES
	FUSING	N/A	N/A
ISOLATION		1500V	1500V

H.KUHNKE CO.		KUAX653	KUAX654

I/O CHANNEL DATA

		KUAX653	KUAX654
TOTAL DISCRETE POINTS		512	64
MAXIMUM POINTS ASSIGNABLE	INPUT	256	32
	OUTPUT	256	32
MAXIMUM ANALOG ASSIGNABLE	INPUT		
	OUTPUT		
MAXIMUM LOCAL CHANNELS		256	32
MAXIMUM REMOTE CHANNELS			
LOCAL CHANNEL DISTANCE		16 FT	16 FT
REMOTE CHANNEL DISTANCE			
TYPE LOCAL CHANNEL CABLE		PLUG	PLUG
TYPE REMOTE CHANNEL CABLE			
LOCAL CHANNEL COMMUNICATION			
REMOTE CHANNEL COMMUNICATION			
LOCAL CHANNEL DATA RATE			
REMOTE CHANNEL DATA RATE			
FAULT MONITORING		YES	YES

PROGRAMMING TERMINAL

	KUAX653	KUAX654
SCREEN SIZE (DIAGONAL)		
WEIGHT		
SIZE (W X H X D)		
POWER REQUIREMENTS		
DISPLAY SIZE (CONTACT • ROWS)	1 ROW LED	1 ROW LED
GRAPHICS	NO	NO
VIDEO OUT	NO	NO
LADDER LISTER	NO	NO
OFF-LINE PROGRAM	NO	NO
REMOTE OPERATION	NO	NO
SCREEN MNEMONICS	NO	NO
BUILT-IN TAPE	NO	NO
KEYBOARD	NO	NO
HANDHELD	YES	YES
DOUBLES AS A COMPUTER TERMINAL	NO	NO

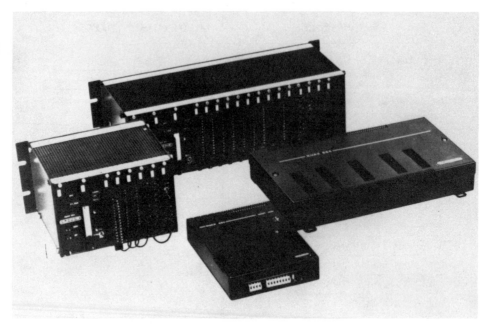

Fig. 21.8.   The H. Kuhnke Company PC System

MITSUBISHI ELECTRIC		F-20M	F-40M

PROCESSOR

YEAR INTRODUCED		1981	1981
MEMORY SIZE TOTAL		320 WORDS	890 WORDS
MEMORY INCREMENTS		N/A	N/A
MEMORY TYPES		RAM,EPROM	RAM,EPROM
VOLTAGE (±10)		120	120
FREQUENCY (±3)		45-63	45-63
HUMIDITY	MIN	0	0
	MAX	90	90
TEMPERATURE °C	MIN	0	0
	MAX	55	55
USER MEMORY	MIN		N/A
	MAX	320	890
DATA MEMORY	MIN	N/A	N/A
	MAX	N/A	N/A
USER CONFIGURABLE		NO	NO
WORD SIZE		8 BIT	8 BIT
WORDS/CONTACT - COIL		1	1
WORDS/TIMER - COUNTER		3	3
SCAN TIME - USER LOGIC		15 MSEC 320 WORD	40 MSEC 890 WORD
I/O SCAN TIME		TIME SHARE	TIME SHARE

SUPPORT

FACTORY APPLICATIONS ENGINEERS	YES	YES
DOCUMENTATION SERVICE	UPON REQUEST	UPON REQUEST
SERVICE CENTER (24 HRS.)	NO	NO
MAINTENANCE TRAINING	YES	YES
APPLICATION TRAINING	YES	YES
WARRANTY TERMS	YES	YES
DISTRIBUTION	YES	YES
MAINTENANCE CENTERS	YES	YES

MITSUBISHI ELECTRIC	F-20M	F-40M

INSTRUCTION SET

	F-20M	F-40M
CONTACTS	YES	YES
TRANSITIONAL	N/A	N/A
LATCHES	YES	YES
RETENTIVE COILS	YES	YES
SKIP/JUMP	NO	YES
SUBROUTINE	NO	NO
TIMERS	8	16
UP COUNTER	NONE	NONE
DOWN COUNTER	8	16
ADD/SUBTRACT	NO	NO
MULTIPLY/DIVIDE	NO	NO
>,=,<	NO	NO
FILE TRANSFER	NO	NO
LOGICAL	N/A	N/A
BIT ROTATE/SHIFT	NO	NO
FIFO/LIFO	N/A	N/A
SEARCH	YES	YES
DIAGNOSTIC	YES	YES
SEQUENCER	POSSIBLE	POSSIBLE
PID	NO	NO
ASCII	NO	NO
AXIS POSITIONING	NO	NO
MACHINE LANGUAGE	NO	NO
BOOLEAN ALGEBRA	YES	YES
HIGH LEVEL LANGUAGE	NO	NO

MITSUBISHI ELECTRIC		F-20M	F-40M

I/O STATISTICS

TOTAL DISCRETE POINTS AVAILABLE		40	80
MAXIMUM POINTS ASSIGNABLE	INPUT	24	48
	OUTPUT	16	32
NUMBER OF CHANNELS		ALL I/O BUILT-IN	ALL I/O BUILT-IN
NUMBER OF NODES/CHANNEL		ALL I/O BUILT-IN	ALL I/O BUILT-IN
MODULE I/O POINT DENSITY		ALL I/O BUILT-IN	ALL I/O BUILT-IN
ISOLATED MODULE I/O DENSITY		N/A	N/A
TOTAL ANALOG CHANNELS AVAILABLE		NONE	NONE
MAXIMUM CHANNELS ASSIGNABLE	INPUT	24	24
	OUTPUT	16	16
I/O MAPPING		DEFINED BY PC	DEFINED BY PC
INTERRUPTS		NONE	NONE
I/O LOGIC UPDATING	ASYNC		
	SYNC	YES	YES
ERROR CHECKING METHOD		SUMCHECK	SUMCHECK
MODULES PER CHANNEL		N/A	N/A
APPLICABLE PROCESSORS		8049	8039
YEAR OF INTRODUCTION		1981	1981
I/O SYSTEM STRUCTURE		BUILT-IN	BUILT-IN

DISCRETE I/O MODULES

AC VOLTAGES/CURRENT		NONE	NONE
DC VOLTAGES/CURRENT		24V CONTACT CLOSURE ONLY	24V CONTACT CLOSURE ONLY
ISOLATED AC VOLTAGES/CURRENT		NONE	NONE
ISOLATED DC VOLTAGES/CURRENT		24	48
CONTINUOUS OUTPUT RATING		2 AMP	2 AMP
COMMON POINTS/MODULE		3	3
OUTPUT FUSING CONFIGURATION		USER SUPPLIED	USER SUPPLIED
INDICATORS	I/O POINTS	LOGIC	LOGIC
	FUSING	NO	NO
ISOLATION		1500V OPTICAL	1500V OPTICAL

MITSUBISHI ELECTRIC		F-20M	F-40M

I/O CHANNEL DATA

TOTAL DISCRETE POINTS		40	80
MAXIMUM POINTS ASSIGNABLE	INPUT	24	48
	OUTPUT	16	32
MAXIMUM ANALOG ASSIGNABLE	INPUT	N/A	N/A
	OUTPUT	N/A	N/A
MAXIMUM LOCAL CHANNELS		40	80
MAXIMUM REMOTE CHANNELS		N/A	N/A
LOCAL CHANNEL DISTANCE		N/A	N/A
REMOTE CHANNEL DISTANCE		N/A	N/A
TYPE LOCAL CHANNEL CABLE		PARALLEL	PARALLEL
TYPE REMOTE CHANNEL CABLE		N/A	N/A
LOCAL CHANNEL COMMUNICATION		PARALLEL	PARALLEL
REMOTE CHANNEL COMMUNICATION		N/A	N/A
LOCAL CHANNEL DATA RATE		N/A	N/A
REMOTE CHANNEL DATA RATE		N/A	N/A
FAULT MONITORING		NO	NO

I/O NODE DATA

TOTAL NODES/CHANNEL		ALL I/O BUILT-IN	ALL I/O BUILT-IN
TOTAL DISCRETE POINTS		ALL I/O BUILT-IN	ALL I/O BUILT-IN
MAXIMUM POINTS ASSIGNABLE	INPUT	ALL I/O BUILT-IN	ALL I/O BUILT-IN
	OUTPUT	ALL I/O BUILT-IN	ALL I/O BUILT-IN
MAXIMUM ANALOG ASSIGNABLE	INPUT	N/A	N/A
	OUTPUT	N/A	N/A
DISTANCE FROM DROP		N/A	N/A

MITSUBISHI ELECTRIC	F-20M	F-40M

PROGRAMMING TERMINAL

	F-20M	F-40M
SCREEN SIZE (DIAGONAL)	LCD	LCD
WEIGHT	7 LBS	7 LBS
SIZE (W X H X D)		
POWER REQUIREMENTS	120 VAC	120 VAC
DISPLAY SIZE (CONTACT ▪ ROWS)	11 X 7	11 X 7
GRAPHICS	NO	NO
VIDEO OUT	NO	NO
LADDER LISTER	YES	YES
OFF-LINE PROGRAM	YES	YES
REMOTE OPERATION	NO	NO
SCREEN MNEMONICS	NO	NO
BUILT-IN TAPE	NO	NO
KEYBOARD	MEMBRANE	MEMBRANE
HANDHELD	YES	YES
DOUBLES AS A COMPUTER TERMINAL	NO	NO

MTS SYSTEMS CORP.	INCOL/470

PROCESSOR

YEAR INTRODUCED		1982
MEMORY SIZE TOTAL		64K
MEMORY INCREMENTS		4K
MEMORY TYPES		RA/EPROM/EAROM
VOLTAGE		110V
FREQUENCY		50-60
HUMIDITY	MIN	N/A
	MAX	95%
TEMPERATURE	MIN	0°C
	MAX	50°C
USER MEMORY	MIN	4K
	MAX	12K
DATA MEMORY	MIN	.5K
	MAX	4K
USER CONFIGURABLE		YES
WORD SIZE		8 BITS
WORDS/CONTACT - COIL		3
WORDS/TIMER - COUNTER		2/3
SCAN TIME - USER LOGIC		PROGRAM FLOW
I/O SCAN TIME		N/A

SUPPORT

FACTORY APPLICATIONS ENGINEERS	YES
DOCUMENTATION SERVICE	YES
SERVICE CENTER (24 HRS.)	YES
MAINTENANCE TRAINING	YES
APPLICATION TRAINING	YES
WARRANTY TERMS	1 YEAR
DISTRIBUTION	DISTRIBUTORS
MAINTENANCE CENTERS	FACTORY

MTS SYSTEMS CORP.	INCOL/470

INSTRUCTION SET

CONTACTS	INP/OUT
TRANSITIONAL	ON/OFF
LATCHES	YES
RETENTIVE COILS	NO
SKIP/JUMP	DO/CALL/START
SUBROUTINE	YES
TIMERS	10
UP COUNTER	10
DOWN COUNTER	10
ADD/SUBTRACT	999999
MULTIPLY/DIVIDE	999999/999999
>,=,<	YES
FILE TRANSFER	YES
LOGICAL	YES
BIT ROTATE/SHIFT	NO
FIFO/LIFO	NO
SEARCH	NO
DIAGNOSTIC	YES
SEQUENCER	YES
PID	PI
ASCII	YES
AXIS POSITIONING	YES
MACHINE LANGUAGE	ASSEMBLER
BOOLEAN ALGEBRA	YES
HIGH LEVEL LANGUAGE	INCOL

MTS SYSTEMS CORP.	INCOL/470

I/O STATISTICS

TOTAL DISCRETE POINTS AVAILABLE		160
MAXIMUM POINTS ASSIGNABLE	INPUT	160
	OUTPUT	160
NUMBER OF CHANNELS		1
NUMBER OF NODES/CHANNEL		32
MODULE I/O POINT DENSITY		32
ISOLATED MODULE I/O DENSITY		32
TOTAL ANALOG CHANNELS AVAILABLE		0
MAXIMUM CHANNELS ASSIGNABLE	INPUT	1
	OUTPUT	1
I/O MAPPING		FACTORY ASSIGNABLE
INTERRUPTS		
I/O LOGIC UPDATING	ASYNC	YES
	SYNC	
ERROR CHECKING METHOD		
MODULES PER CHANNEL		8
APPLICABLE PROCESSORS		
YEAR OF INTRODUCTION		1982
I/O SYSTEM STRUCTURE		RACK & BASE

DISCRETE I/O MODULES

AC VOLTAGES/CURRENT		OPTO 22
DC VOLTAGES/CURRENT		12,24
ISOLATED -AC VOLTAGES/CURRENT		N/A
ISOLATED DC VOLTAGES/CURRENT		12,24
CONTINUOUS OUTPUT RATING		160 MA
COMMON POINTS/MODULE		2
OUTPUT FUSING CONFIGURATION		OPTO 22
INDICATORS	I/O POINTS	YES
	FUSING	
ISOLATION		1500V OPTICAL

MTS SYSTEMS CORP.	INCOL/470

I/O CHANNEL DATA

TOTAL DISCRETE POINTS		255/MP
MAXIMUM POINTS ASSIGNABLE	INPUT	255/MP
	OUTPUT	255/MP
MAXIMUM ANALOG ASSIGNABLE	INPUT	
	OUTPUT	
MAXIMUM LOCAL CHANNELS		1
MAXIMUM REMOTE CHANNELS		0
LOCAL CHANNEL DISTANCE		N/A
REMOTE CHANNEL DISTANCE		N/A
TYPE LOCAL CHANNEL CABLE		N/A
TYPE REMOTE CHANNEL CABLE		N/A
LOCAL CHANNEL COMMUNICATION		PARALLEL
REMOTE CHANNEL COMMUNICATION		N/A
LOCAL CHANNEL DATA RATE		N/A
REMOTE CHANNEL DATA RATE		N/A
FAULT MONITORING		YES

DATA NETWORK

TYPE	N/A
NUMBER OF NODES	N/A
MAXIMUM LENGTH	N/A
TYPE OF CABLE	RG59U/GPIB
COMMUNICATION SPEED	9600 BAUD/450K
COMPUTER INTERFACES	YES
SPECIAL INTERFACES	RS232/GPIB 448

I/O NODE DATA

TOTAL NODES/CHANNEL		1
TOTAL DISCRETE POINTS		255
MAXIMUM POINTS ASSIGNABLE	INPUT	255
	OUTPUT	255
MAXIMUM ANALOG ASSIGNABLE	INPUT	N/A
	OUTPUT	N/A
DISTANCE FROM DROP		N/A

MTS SYSTEMS CORP.	INCOL/470

ANALOG I/O MODULES

VOLTAGES: SINGLE	
VOLTAGES: DIFFERENTIAL	
CURRENT: SINGLE	
CURRENT: DIFFERENTIAL	
SPECIAL PURPOSE INPUT	ENCODER
SPECIAL PURPOSE OUTPUT	SLIPPER,SERVO, TEMPORONIC
INDICATORS:	

SPECIAL I/O MODULES

SIMULATION	
ASCII	YES
AXIS POSITIONING	UP TO 32 AXES

PROGRAMMING TERMINAL

SCREEN SIZE (DIAGONAL)	9"
WEIGHT	20 LBS
SIZE (W X H X D)	18X18X18
POWER REQUIREMENTS	110V/60
DISPLAY SIZE (CONTACT ▪ ROWS)	80 X 24
GRAPHICS	NO
VIDEO OUT	YES
LADDER LISTER	NO
OFF-LINE PROGRAM	YES
REMOTE OPERATION	NO
SCREEN MNEMONICS	YES
BUILT-IN TAPE	NO
KEYBOARD	YES
HANDHELD	NO
DOUBLES AS A COMPUTER TERMINAL	YES

RELIANCE		A/5	A35	UDAC

PROCESSOR

YEAR INTRODUCED		1982	1978	1976
MEMORY SIZE TOTAL		4K X 8	24K X 8	40K X 16
MEMORY INCREMENTS		NONE	4K X 8	8K X 16
MEMORY TYPES		EPROM/NVRAM	CMOS/EPROM	CORE/MOS
VOLTAGE (±10)		120	120	120
FREQUENCY		20-80	20-80	20-80
HUMIDITY	MIN	0%	0%	0%
	MAX	95%	95%	95%
TEMPERATURE	MIN	0°C	0°C	0°C
	MAX	60°C	60°C	60°C
USER MEMORY	MIN	4K	4K	4K
	MAX	4K	24K	28K
DATA MEMORY	MIN	N/A	0	0
	MAX	N/A	24K	28K
USER CONFIGURABLE		NO	YES	YES
WORD SIZE		8 BIT	8 BIT	16 BIT
WORDS/CONTACT - COIL		4	2	2
WORDS/TIMER - COUNTER		8	12	12
SCAN TIME - USER LOGIC		1 µSEC/BYTE	4 µSEC/BYTE	10 µSEC
I/O SCAN TIME		1 µSEC/POINT	1 µSEC/POINT	1 SEC

SUPPORT

	A/5	A35	UDAC
FACTORY APPLICATIONS ENGINEERS	YES	YES	YES
DOCUMENTATION SERVICE	FUTURE	FUTURE	FUTURE
SERVICE CENTER (24 HRS.)	800-241-2886	800-241-2886	800-241-2886
MAINTENANCE TRAINING	YES	YES	YES
APPLICATION TRAINING	YES	YES	YES
WARRANTY TERMS	1 YEAR	1 YEAR	1 YEAR
DISTRIBUTION	DISTRIBUTORS & DIRECT	DISTRIBUTORS & DIRECT	DISTRIBUTORS & DIRECT
MAINTENANCE CENTERS	DISTRIBUTORS & FACTORY	DISTRIBUTORS & FACTORY	FACTORY

RELIANCE	A/5	A35	UDAC

INSTRUCTION SET

	A/5	A35	UDAC
CONTACTS	YES	YES	"SPEAK EASY" HIGH LEVEL REALTIME BASIC TYPE LANGUAGE
TRANSITIONAL	ON	NO	
LATCHES	YES	YES	
RETENTIVE COILS	YES	YES	YES
SKIP/JUMP	MCR	JUMP	YES
SUBROUTINE	NO	YES	YES
TIMERS	YES	YES	YES
UP COUNTER	YES	YES	YES
DOWN COUNTER	YES	YES	YES
ADD/SUBTRACT	NO	9999.9999	$\pm 10^{38}$
MULTIPLY/DIVIDE	NO	9999.9999	$\pm 10^{38}$
>,=,<	NO	YES	YES
FILE TRANSFER	NO	YES	YES
LOGICAL	NO	YES	YES
BIT ROTATE/SHIFT	NO	YES	YES
FIFO/LIFO	NO	NO	YES
SEARCH	NO	NO	YES
DIAGNOSTIC	NO	NO	YES
SEQUENCER	NO	YES	YES
PID	NO	YES	YES
ASCII	NO	YES	YES
AXIS POSITIONING	NO	NO	NO
MACHINE LANGUAGE	NO	YES	NO
BOOLEAN ALGEBRA	NO	YES	YES
HIGH LEVEL LANGUAGE	NO	YES	YES

RELIANCE		A5	A35	UDAC

I/O STATISTICS

TOTAL DISCRETE POINTS AVAILABLE		64	1920	640
MAXIMUM POINTS ASSIGNABLE	INPUT	64	1290	320
	OUTPUT	64	1920	320
NUMBER OF CHANNELS		1	30	30
NUMBER OF NODES/CHANNEL		4	4	16
MODULE I/O POINT DENSITY		2 PT/MOD	2 PT/MOD	32
ISOLATED MODULE I/O DENSITY		2 PT/MOD	2 PT/MOD	2 PT/MOD
TOTAL ANALOG CHANNELS AVAILABLE		0	64	320
MAXIMUM CHANNELS ASSIGNABLE	INPUT	1	30	160
	OUTPUT	1	30	160
I/O MAPPING			USER DEFINED	USER DEFINED
INTERRUPTS		NO	32	NO
I/O LOGIC UPDATING	ASYNC	N/A	N/A	N/A
	SYNC	YES	YES	YES
ERROR CHECKING METHOD		CHECK BITS	CHECK BITS	N/A
MODULES PER CHANNEL		32	32	16
APPLICABLE PROCESSORS		ALL	ALL	ALL
YEAR OF INTRODUCTION		1982	1982	1976
I/O SYSTEM STRUCTURE		BASE & MODULES	BASE & MODULES	BASE & PACK

CONTROL NETWORK

	A5	A35	UDAC
TYPE	TOKEN PASSING	TOKEN PASSING	N/A
NUMBER OF PC'S	256	256	N/A
MAXIMUM LENGTH	12000 FT	12000 FT	N/A
TYPE OF CABLE	RG11AU	RG11AU	N/A
OPERATION W/PROCESSOR	ALL	ALL	N/A
DISCRETE SIGNALS TRANSFERRED	PROGRAMMABLE	PROGRAMMABLE	N/A
DATA SIGNALS TRANSFERRED	PROGRAMMABLE	PROGRAMMABLE	N/A
COMMUNICATION SPEED	800K BAUD	800K BAUD	N/A

RELIANCE		A/5	A35	UDAC

I/O CHANNEL DATA

TOTAL DISCRETE POINTS		64	64	640
MAXIMUM POINTS ASSIGNABLE	INPUT	64	64	320
	OUTPUT	64	64	320
MAXIMUM ANALOG ASSIGNABLE	INPUT	N/A	32	160
	OUTPUT	N/A	32	160
MAXIMUM LOCAL CHANNELS		1	2	320
MAXIMUM REMOTE CHANNELS		0	28	0
LOCAL CHANNEL DISTANCE		4 FT	4 FT	1 FT
REMOTE CHANNEL DISTANCE		N/A	9000 FT	0
TYPE LOCAL CHANNEL CABLE		TWISTED PAIR	TWISTED PAIR	TWISTED PAIR
TYPE REMOTE CHANNEL CABLE		TWISTED PAIR	TWISTED PAIR	N/A
LOCAL CHANNEL COMMUNICATION		SERIAL	SERIAL	PARALLEL
REMOTE CHANNEL COMMUNICATION		N/A	SERIAL	N/A
LOCAL CHANNEL DATA RATE		250K BAUD	250K BAUD	
REMOTE CHANNEL DATA RATE		N/A	250K BAUD	N/A
FAULT MONITORING		YES	YES	NO

DATA NETWORK

	A/5	A35	UDAC
TYPE	TOKEN PASSING	TOKEN PASSING	MASTER/SLAVE
NUMBER OF NODES	256	256	9
MAXIMUM LENGTH	12000 FT	1200 FT	1000 FT
TYPE OF CABLE	RG11AU	RG11AU	TWISTED PAIR
COMMUNICATION SPEED	800K BAUD	800K BAUD	9600 BAUD
COMPUTER INTERFACES	YES	YES	RS232/20MA
SPECIAL INTERFACES	RS232	RS232	RS232/20MA

RELIANCE	A15	A35	UDAC
PROGRAMMING TERMINAL			
SCREEN SIZE (DIAGONAL)	9'	9'	12'
WEIGHT			35 LBS
SIZE (W X H X D)			15X13X20
POWER REQUIREMENTS	115/230VAC	115/230VAC	115/230VAC
DISPLAY SIZE (CONTACT ▪ ROWS)	6 X 10 · COIL	6 X 10 · COIL	N/A
GRAPHICS	YES	YES	NO
VIDEO OUT	NO	NO	NO
LADDER LISTER	YES	YES	NO
OFF-LINE PROGRAM	YES	YES	YES
REMOTE OPERATION	YES	YES	YES
SCREEN MNEMONICS	FUTURE	YES	YES
BUILT-IN TAPE	YES	YES	NO
KEYBOARD	TYPEWRITER	TYPEWRITER	TYPEWRITER
HANDHELD	YES	NO	NO
DOUBLES AS A COMPUTER TERMINAL	YES	YES	YES

Fig. 21.9. Reliance Electric Automate 15 Processor

Fig. 21.10. Reliance Electric Automate 35 PC

SIEMENS		105R	150K	150S

PROCESSOR

		105R	150K	150S
YEAR INTRODUCED		1983	1980	1981
MEMORY SIZE TOTAL		628 ELEMENTS	24K	48K·64K
MEMORY INCREMENTS		N/A	2,4,8,16,24K	4,8,16,24,32K
MEMORY TYPES		RAM,EPROM	RAM,EPROM	RAM,EPROM
VOLTAGE ·10-15%		110/115, 220/240,24	110/115, 220/240,24	110/115, 220/240,24
FREQUENCY		48-63HZ	47-63HZ	47-63HZ
HUMIDITY (RELATIVE)	MIN	0	0	0
	MAX	95	95	95
TEMPERATURE ·C	MIN	0	0	0
	MAX	60	55	55
USER MEMORY	MIN	628 ELEMENTS	2K	4K
	MAX	628 ELEMENTS	24K	48K
DATA MEMORY	MIN	52 REGISTERS	2K	4K
	MAX	52 REGISTERS	24K	112K
USER CONFIGURABLE		N/A	YES	YES
WORD SIZE		16 BITS	16 BITS	16 BITS
WORDS/CONTACT - COIL		1	1	1
WORDS/TIMER - COUNTER		3	1	1
SCAN TIME - USER LOGIC		2.5 MS/1/2K	4 MS/K	2 MS/K
I/O SCAN TIME		.3MS	1MS	1MS

SUPPORT

	105R	150K	150S
FACTORY APPLICATIONS ENGINEERS	YES	YES	YES
DOCUMENTATION SERVICE	NO	NO	NO
SERVICE CENTER (24 HRS.)	800/433-0954	800/433-0954	800/433-0954
MAINTENANCE TRAINING	YES	YES	YES
APPLICATION TRAINING	YES	YES	YES
WARRANTY TERMS	1 YEAR	1 YEAR	1 YEAR
DISTRIBUTION	DISTRIBUTORS	DISTRIBUTORS	DISTRIBUTORS
MAINTENANCE CENTERS	DISTRIBUTORS	DISTRIBUTORS	DISTRIBUTORS

SIEMENS	105R	150K	150S

INSTRUCTION SET

CONTACTS	YES	YES	YES
TRANSITIONAL	ON,OFF	ON,OFF	ON,OFF
LATCHES	YES	YES	YES
RETENTIVE COILS	YES	YES	YES
SKIP/JUMP	JUMP	YES	YES
SUBROUTINE	N/A	YES	YES
TIMERS	24	128	256
UP COUNTER	8	128	256
DOWN COUNTER	8	128	256
ADD/SUBTRACT	N/A	2^{31}	.1701412E · 39 TO
MULTIPLY/DIVIDE	N/A	2^{31}	.1469368E · 38
>,=,<	N/A	YES	YES
FILE TRANSFER	N/A	YES	YES
LOGICAL	N/A	YES	YES
BIT ROTATE/SHIFT	N/A	YES	YES
FIFO/LIFO	N/A	YES	YES
SEARCH	YES	YES	YES
DIAGNOSTIC	YES	YES	YES
SEQUENCER	YES	YES	YES
PID	N/A	SOFTWARE	SOFTWARE
ASCII	N/A	YES	YES
AXIS POSITIONING	N/A	YES	YES
MACHINE LANGUAGE	MC5	MC5	MC5
BOOLEAN ALGEBRA	N/A	YES	YES
HIGH LEVEL LANGUAGE	N/A	YES	YES

SIEMENS	*105R*	*150K*	*150S*

I/O STATISTICS

		105R	150K	150S
TOTAL DISCRETE POINTS AVAILABLE		32	2048	37888
MAXIMUM POINTS ASSIGNABLE	INPUT	20	1024	18944
	OUTPUT	12	1024	18944
NUMBER OF CHANNELS		I/O IN MAIN CHASSIS	I/O BUS	I/O BUS
NUMBER OF NODES/CHANNEL		I/O IN MAIN CHASSIS	I/O BUS	I/O BUS
MODULE I/O POINT DENSITY		8 (5/3)	6,16,32	8,16,32
ISOLATED MODULE I/O DENSITY		8 (5/3)	8,16,32	8,16,32
TOTAL ANALOG CHANNELS AVAILABLE		N/A	128	384
MAXIMUM CHANNELS ASSIGNABLE	INPUT	N/A	64	192
	OUTPUT	N/A	64	192
I/O MAPPING		FIXED	FIXED AND/OR PROGRAMMABLE	FIXED AND/OR PROGRAMMABLE
INTERRUPTS		N/A	64	64
I/O LOGIC UPDATING	ASYNC			
	SYNC	YES	YES	YES
ERROR CHECKING METHOD		N/A	PARITY	PARITY
MODULES PER CHANNEL		N/A	64	64
APPLICABLE PROCESSORS		N/A	ALL	ALL
YEAR OF INTRODUCTION		1983	1979	1981
I/O SYSTEM STRUCTURE		LOCAL	LOCAL & REMOTE	LOCAL & REMOTE

CONTROL NETWORK

	105R	150K	150S
TYPE	N/A	POINT TO POINT	POINT TO POINT
NUMBER OF PC'S	N/A	8	16
MAXIMUM LENGTH	N/A	3300 FT RADIUS	3300 FT RADIUS
TYPE OF CABLE	N/A	TWISTED PAIR	TWISTED PAIR
OPERATION W/PROCESSOR	N/A	ALL	ALL
DISCRETE SIGNALS TRANSFERRED	N/A	YES	YES
DATA SIGNALS TRANSFERRED	N/A	YES	YES
COMMUNICATION SPEED	N/A	9600 BAUD	9600 BAUD

SIEMENS		105R	150K	150S

DISCRETE I/O MODULES

AC VOLTAGES/CURRENT		110/115,220/240	110/115,220/240	110/115,220/240
DC VOLTAGES/CURRENT		24	5,24,48	5,24,48
ISOLATED AC VOLTAGES/CURRENT		110/115,220/240	110/115,220/240	110/115,220/240
ISOLATED DC VOLTAGES/CURRENT		24	5,24,48	5,24,48
CONTINUOUS OUTPUT RATING		2A	2A	2A
COMMON POINTS/MODULE		(5/5,3/3)	1/8,4/16,16/32	1/8,4/16,16/32
OUTPUT FUSING CONFIGURATION		3/8	8/8,16/16,32/32	8/8,16/16,32/32
INDICATORS	I/O POINTS	LOGIC & POWER	POWER	POWER
	FUSING	NO	YES	YES
ISOLATION		YES	YES	YES

ANALOG I/O MODULES

VOLTAGES: SINGLE		N/A	50,500MV 1,10V	50,500MV 1,10V
VOLTAGES: DIFFERENTIAL		N/A	50,500MV 1,10V	50,500MV 1,10V
CURRENT: SINGLE		N/A	20MA,4-20MA	20MA,4-20MA
CURRENT: DIFFERENTIAL		N/A	20MA,4-20MA	20MA,4-20MA
SPECIAL PURPOSE INPUT		N/A	THERMOCOUPLE	THERMOCOUPLE
SPECIAL PURPOSE OUTPUT		N/A	N/A	N/A
INDICATORS:		N/A	YES	YES

SPECIAL I/O MODULES

SIMULATION		N/A	YES	YES
ASCII		N/A	YES	YES
AXIS POSITIONING		/A	YES	YES

I/O NODE DATA

TOTAL NODES/CHANNEL		N/A	4	4
TOTAL DISCRETE POINTS		N/A	512	512
MAXIMUM POINTS ASSIGNABLE	INPUT	N/A	512	512
	OUTPUT	N/A	512	512
MAXIMUM ANALOG ASSIGNABLE	INPUT	N/A	64	192
	OUTPUT	N/A	64	192
DISTANCE FROM DROP		N/A	1 FT	1 FT

SIEMENS		105R	150K	150S

I/O CHANNEL DATA

		105R	150K	150S
TOTAL DISCRETE POINTS		N/A	2048	2048
MAXIMUM POINTS ASSIGNABLE	INPUT	N/A	1024	1024
	OUTPUT	N/A	1024	1024
MAXIMUM ANALOG ASSIGNABLE	INPUT	N/A	64	192
	OUTPUT	N/A	64	192
MAXIMUM LOCAL CHANNELS		N/A	ALL	ALL
MAXIMUM REMOTE CHANNELS		N/A	ALL	ALL
LOCAL CHANNEL DISTANCE		N/A	600 FT	600 FT
REMOTE CHANNEL DISTANCE		N/A	3300 FT	3300 FT
TYPE LOCAL CHANNEL CABLE		N/A	PARALLEL	PARALLEL
TYPE REMOTE CHANNEL CABLE		N/A	4 WIRES	4 WIRES
LOCAL CHANNEL COMMUNICATION		N/A	PARALLEL	PARALLEL
REMOTE CHANNEL COMMUNICATION		N/A	SERIAL	SERIAL
LOCAL CHANNEL DATA RATE		N/A	250 KHZ	250 KHZ
REMOTE CHANNEL DATA RATE		N/A	9600 BAUD	9600 BAUD
FAULT MONITORING		N/A	YES	YES

DATA NETWORK

	105R	150K	150S
TYPE	N/A	POINT TO POINT	POINT TO POINT
NUMBER OF NODES	N/A	N/A	N/A
MAXIMUM LENGTH	N/A	3300 FT RADIUS	3300 FT RADIUS
TYPE OF CABLE	N/A	TWISTED PAIR	TWISTED PAIR
COMMUNICATION SPEED	N/A	9600 BAUD	9600 BAUD
COMPUTER INTERFACES	N/A	YES	YES
SPECIAL INTERFACES	N/A	RS232 PROTOCOL	RS232 PROTOCOL

SIEMENS	105R	150K	150S
PROGRAMMING TERMINAL	605R, 655	675	
SCREEN SIZE (DIAGONAL)	2.5' LCD	9" CRT	9" CRT
WEIGHT	2 LBS	28 LBS	33 LBS
SIZE (W X H X D)	4.75X1.75X8.0	16X8X12	17X8.5X14
POWER REQUIREMENTS	5 VDC FROM PC	110/115,220/240	110/115,220/240
DISPLAY SIZE (CONTACT ▪ ROWS)	8 X 1	8 X 16	8 X 16
GRAPHICS	N/A	N/A	N/A
VIDEO OUT	N/A	YES	YES
LADDER LISTER	N/A	YES	YES
OFF-LINE PROGRAM	N/A	YES	YES
REMOTE OPERATION	N/A	N/A	YES
SCREEN MNEMONICS	N/A	N/A	YES
BUILT-IN TAPE	N/A	N/A	2 DISCS
KEYBOARD	PUSHBUTTON	PUSHBUTTON	PUSHBUTTON
HANDHELD	YES	N/A	N/A
DOUBLES AS A COMPUTER TERMINAL	N/A	N/A	YES

NOTES :

  1) 605R AND 655 PROGRAMMING TERMINALS ARE FOR USE WITH THE
     SIEMENS 105R PC.

  2) 675 PROGRAMMING TERMINAL IS FOR USE WITH THE 150K AND 150S PCS.

STRUTHERS-DUNN		DIRECTOR 4001	DIRECTOR 4002

PROCESSOR

YEAR INTRODUCED		1978	1983
MEMORY SIZE TOTAL		6K	6K
MEMORY INCREMENTS		2,4,6K	2,4,6K
MEMORY TYPES		CMOS/RAM EAROM	CMOS/RAM EAROM/LEROM
VOLTAGE		120/240	120/240
FREQUENCY		47-63HZ	47-63HZ
HUMIDITY	MIN	10	10
	MAX	95	95
TEMPERATURE	MIN	0°C	0°C
	MAX	55°C	60°C
USER MEMORY	MIN	2K	2K
	MAX	6K	6K
DATA MEMORY	MIN	2K	2K
	MAX	2K	2K
USER CONFIGURABLE		NO	NO
WORD SIZE		8 BIT	8 BIT
WORDS/CONTACT - COIL		1	1
WORDS/TIMER - COUNTER		3	3
SCAN TIME - USER LOGIC		VARIABLE	VARIABLE
I/O SCAN TIME		10MSEC MAX	5 MSEC MAX

SUPPORT

FACTORY APPLICATIONS ENGINEERS	YES	YES
DOCUMENTATION SERVICE	YES	YES
SERVICE CENTER (24 HRS.)	NO	NO
MAINTENANCE TRAINING	YES	YES
APPLICATION TRAINING	YES	YES
WARRANTY TERMS	1 YEAR	1 YEAR
DISTRIBUTION	DISTRIBUTORS	DISTRIBUTORS
MAINTENANCE CENTERS	DISTRIBUTORS FACTORY	DISTRIBUTORS FACTORY

STRUTHERS-DUNN	DIRECTOR 4001	DIRECTOR 4002

INSTRUCTION SET

CONTACTS	YES	YES
TRANSITIONAL	ON/OFF	ON/OFF
LATCHES	YES	YES
RETENTIVE COILS	YES	YES
SKIP/JUMP	JUMP	JUMP
SUBROUTINE	NO	NO
TIMERS	64	64
UP COUNTER	64	64
DOWN COUNTER	64	64
ADD/SUBTRACT	9999	9999
MULTIPLY/DIVIDE	9999.9999	9999.9999
>,=,<	YES	YES
FILE TRANSFER	YES	YES
LOGICAL	YES	YES
BIT ROTATE/SHIFT	NO	NO
FIFO/LIFO	FIFO	FIFO
SEARCH	YES	YES
DIAGNOSTIC	YES	YES
SEQUENCER	YES	YES
PID	8 LOOPS	8 LOOPS
ASCII	NO	NO
AXIS POSITIONING	NO	NO
MACHINE LANGUAGE	NO	NO
BOOLEAN ALGEBRA	NO	NO
HIGH LEVEL LANGUAGE	NO	NO

SPECIAL I/O MODULES

SIMULATION	INPUT SIMULATION	INPUT SIMULATION
ASCII	NO	NO
AXIS POSITIONING	NO	NO

STRUTHERS-DUNN		DIRECTOR 4001	DIRECTOR 4002

I/O STATISTICS

TOTAL DISCRETE POINTS AVAILABLE		384	64
MAXIMUM POINTS ASSIGNABLE	INPUT	384	64
	OUTPUT	384	64
NUMBER OF CHANNELS		12	1
NUMBER OF NODES/CHANNEL		32	64
MODULE I/O POINT DENSITY		4 OR 8	8
ISOLATED MODULE I/O DENSITY		4 OR 8	6
TOTAL ANALOG CHANNELS AVAILABLE		128	64
MAXIMUM CHANNELS ASSIGNABLE	INPUT	128	64
	OUTPUT	32	16
I/O MAPPING		USER DEFINED	USER DEFINED
INTERRUPTS		NO	NO
I/O LOGIC UPDATING	ASYNC	YES	YES
	SYNC	N/A	N/A
ERROR CHECKING METHOD		CHECK SUM	CHECK SUM
MODULES PER CHANNEL		8	8
APPLICABLE PROCESSORS		ALL	ALL
YEAR OF INTRODUCTION		1978	1982
I/O SYSTEM STRUCTURE		TRACK · MODULES	TRACK · MODULES

DISCRETE I/O MODULES

AC VOLTAGES/CURRENT		12,24,48,120,240	12,24,48,120,240
DC VOLTAGES/CURRENT		5,12,24,48,120	5,12,24,48,120
ISOLATED AC VOLTAGES/CURRENT		5,12,24,48,120	5,12,24,48,120
ISOLATED DC VOLTAGES/CURRENT		5,12,24,48,120	5,12,24,48,120
CONTINUOUS OUTPUT RATING		2 AMPS	2 AMPS
COMMON POINTS/MODULE		4-8	4-8
OUTPUT FUSING CONFIGURATION		BLOWN FUSE INDICATION	BLOWN FUSE INDICATION
INDICATORS	I/O POINTS	LOGIC/POWER	LOGIC/POWER
	FUSING	YES	YES
ISOLATION		1500V OPTIC	1500V OPTIC

STRUTHERS-DUNN		DIRECTOR 4001	DIRECTOR 4002

I/O CHANNEL DATA

TOTAL DISCRETE POINTS		64	64
MAXIMUM POINTS ASSIGNABLE	INPUT	64	64
	OUTPUT	64	64
MAXIMUM ANALOG ASSIGNABLE	INPUT	32	64
	OUTPUT	8	16
MAXIMUM LOCAL CHANNELS		16	1
MAXIMUM REMOTE CHANNELS		16	NONE
LOCAL CHANNEL DISTANCE		100 FT	N/A
REMOTE CHANNEL DISTANCE		4000 FT	N/A
TYPE LOCAL CHANNEL CABLE		IEEE 488	NONE
TYPE REMOTE CHANNEL CABLE		RS-422 FIBER OPTIC	NONE
LOCAL CHANNEL COMMUNICATION		PARALLEL	N/A
REMOTE CHANNEL COMMUNICATION		SERIAL	N/A
LOCAL CHANNEL DATA RATE			
REMOTE CHANNEL DATA RATE			
FAULT MONITORING		YES	YES

DATA NETWORK

	DIRECTOR 4001	DIRECTOR 4002
TYPE	MASTER/SLAVE	MASTER/SLAVE
NUMBER OF NODES	32	32
MAXIMUM LENGTH	4000 FT MIN	4000 FT MIN
TYPE OF CABLE	TWISTED PAIR	TWISTED PAIR
COMMUNICATION SPEED	56K BAUD	56K BAUD
COMPUTER INTERFACES	YES	YES
SPECIAL INTERFACES	RS-422	RS-422

ANALOG I/O MODULES

	DIRECTOR 4001	DIRECTOR 4002
VOLTAGES: SINGLE	0-1,0-10VDC	0-1,0-10VDC
VOLTAGES: DIFFERENTIAL	NO	NO
CURRENT: SINGLE	4-20MA	4-20MA
CURRENT: DIFFERENTIAL	NO	NO
SPECIAL PURPOSE INPUT	THERMOCOUPLE HIGH SPEED COUNT	
SPECIAL PURPOSE OUTPUT	STEPPER MOTOR	HIGH SPEED COUNT
INDICATORS:	NO	NO

STRUTHERS-DUNN	DIRECTOR 4001	DIRECTOR 4002

PROGRAMMING TERMINAL

	DIRECTOR 4001	DIRECTOR 4002
SCREEN SIZE (DIAGONAL)	9"	9"
WEIGHT	25 LBS	25 LBS
SIZE (W X H X D)	15X7X17	15X7X17
POWER REQUIREMENTS	120/240 50/60HZ	120/240 50/60HZ
DISPLAY SIZE (CONTACT ▪ ROWS)	4 X 8	4 X 8
GRAPHICS	OPTIONAL	OPTIONAL
VIDEO OUT	YES	YES
LADDER LISTER	YES	YES
OFF-LINE PROGRAM	YES	YES
REMOTE OPERATION	YES	YES
SCREEN MNEMONICS	YES	YES
BUILT-IN TAPE	NO	NO
KEYBOARD	MEMBRANE	MEMBRANE
HANDHELD	NO	NO
DOUBLES AS A COMPUTER TERMINAL	YES	YES

SQUARE D		SY/MAX MODEL 100	SY/MAX MODEL 300	SY/MAX MODEL 500

PROCESSOR

YEAR INTRODUCED - COMMERCIALLY		1982	1982	1983
MEMORY SIZE TOTAL		420 WORDS	2K	8K
MEMORY INCREMENTS		420 WORDS	1/2,1,2K	2,4,8K
MEMORY TYPES		RAM,UV-PROM	RAM,UV-PROM, COMBINATION	RAM,EEPROM
VOLTAGE		120	120/240VAC	120/240VAC
FREQUENCY		47/63HZ	50/60HZ	50/60HZ
HUMIDITY NON-CONDENSING	MIN	0%	0%	0%
	MAX	95%	95%	95%
TEMPERATURE °F	MIN	32°	32°	32°
	MAX	140°	140°	140°
USER MEMORY	MIN	420 WORDS	1/2K	2K
	MAX	420 WORDS	2K	8K
DATA MEMORY	MIN			
	MAX	38 REGISTERS	112 REGISTERS	460 REGISTERS
USER CONFIGURABLE		NO	YES	YES
WORD SIZE		16 BIT	16 BIT	16 BIT
WORDS/CONTACT - COIL		1	1	1
WORDS/TIMER - COUNTER		5/5	4/5	5/5
SCAN TIME - USER LOGIC		40 MS	30 MS/K	6 MS/K
I/O SCAN TIME		12 MS	1 MS	1 MS

SUPPORT

FACTORY APPLICATIONS ENGINEERS	YES	YES	YES
DOCUMENTATION SERVICE	NO (FACTORY)	NO (FACTORY)	NO (FACTORY)
SERVICE CENTER (24 HRS.)	YES	YES	YES
MAINTENANCE TRAINING	YES	YES	YES
APPLICATION TRAINING	YES	YES	YES
WARRANTY TERMS	1 YEAR	1 YEAR	1 YEAR
DISTRIBUTION	DISTRIBUTORS	DISTRIBUTORS	DISTRIBUTORS
MAINTENANCE CENTERS	YES	YES	YES

SQUARE D	SY/MAX MODEL 100	SY/MAX MODEL 300	SY/MAX MODEL 500

INSTRUCTION SET

	SY/MAX MODEL 100	SY/MAX MODEL 300	SY/MAX MODEL 500
CONTACTS	YES	YES	YES
TRANSITIONAL	ON/OFF	ON/OFF	ON/OFF
LATCHES	YES	YES	YES
RETENTIVE COILS	NO	YES	YES
SKIP/JUMP	NO	NO	YES
SUBROUTINE	NO	NO	YES
TIMERS	YES - 38	96	>1000
UP COUNTER	YES - 38	96	>1000
DOWN COUNTER	YES - 38	96	>1000
ADD/SUBTRACT	NO	±32767	±32767
MULTIPLY/DIVIDE	NO	±32767	±32767
>,=,<	YES	YES	YES
FILE TRANSFER	NO	YES	YES
LOGICAL	YES	YES	YES
BIT ROTATE/SHIFT	YES	YES	YES
FIFO/LIFO	SYNCHRONOUS	BOTH	BOTH
SEARCH	YES	YES	YES
DIAGNOSTIC	YES	YES	YES
SEQUENCER	USER SOFTWARE	USER SOFTWARE	USER SOFTWARE
PID	NO	USER SOFTWARE	USER SOFTWARE
ASCII	NO	YES	YES
AXIS POSITIONING	NO	FUTURE	FUTURE
MACHINE LANGUAGE	NO	LADDER	LADDER
BOOLEAN ALGEBRA	NO	YES	YES
HIGH LEVEL LANGUAGE	NO	NO	NO

SQUARE D	SY/MAX MODEL 100	SY/MAX MODEL 300	SY/MAX MODEL 500

I/O STATISTICS

		SY/MAX MODEL 100	SY/MAX MODEL 300	SY/MAX MODEL 500
TOTAL DISCRETE POINTS AVAILABLE		40	1792	>1000
MAXIMUM POINTS ASSIGNABLE	INPUT	24	256 TYPICAL	>1000
	OUTPUT	16	256 TYPICAL	>1000
NUMBER OF CHANNELS		1	2	14
NUMBER OF NODES/CHANNEL		N/A	8	8
MODULE I/O POINT DENSITY		NO MODULES - ON BOARD I/O	4 OR 8	4 OR 8
ISOLATED MODULE I/O DENSITY		NO MODULES - ON BOARD I/O	4 OR 8	4 OR 8
TOTAL ANALOG CHANNELS AVAILABLE		N/A	64	128
MAXIMUM CHANNELS ASSIGNABLE	INPUT	N/A	2	14
	OUTPUT	N/A	2	14
I/O MAPPING		FIXED	USER DEFINABLE	USER DEFINABLE
INTERRUPTS			YES	YES
I/O LOGIC UPDATING	ASYNC			YES
	SYNC	YES	YES	YES
ERROR CHECKING METHOD		CHECK SUM, PARITY	CHECK SUM, PARITY	CHECK SUM, PARITY
MODULES PER CHANNEL		N/A	4,8,16,32	4,8,16,32
APPLICABLE PROCESSORS			MODELS 300,500 (MOD 700 FUT.)	MODELS 300,500 (MOD 700 FUT.)
YEAR OF INTRODUCTION		1982	1982	1983
I/O SYSTEM STRUCTURE		ON BOARD	RACK	RACK

CONTROL NETWORK

	SY/MAX MODEL 100	SY/MAX MODEL 300	SY/MAX MODEL 500
TYPE	SLAVE-TO-MASTER	PEER-TO-PEER	PEER-TO-PEER
NUMBER OF PC'S	199/NETWORK	200/NETWORK	200/NETWORK
MAXIMUM LENGTH	15000 FT	15000 FT	15000 FT
TYPE OF CABLE	TWIN AX	TWIN AX	TWIN AX
OPERATION W/PROCESSOR	ALL	ALL	ALL
DISCRETE SIGNALS TRANSFERRED	YES	YES	YES
DATA SIGNALS TRANSFERRED	YES	YES	YES
COMMUNICATION SPEED	UP TO 500K	UP TO 500K	UP TO 500K

SQUARE D	SY/MAX MODEL 100	SY/MAX MODEL 300	SY/MAX MODEL 500

DISCRETE I/O MODULES

AC VOLTAGES/CURRENT	120 (24 FUTURE)	9 TO 240VAC	9 TO 240VAC
DC VOLTAGES/CURRENT	(24 FUTURE)	TTL TO 240VDC	TTL TO 240VDC
ISOLATED AC VOLTAGES/CURRENT	120	120	120
ISOLATED DC VOLTAGES/CURRENT	N/A	120	120
CONTINUOUS OUTPUT RATING	2 AMPS	2 AMPS	2 AMPS
COMMON POINTS/MODULE	N/A	4	4
OUTPUT FUSING CONFIGURATION	EXTERNAL	ON MODULE	ON MODULE
INDICATORS I/O POINTS	LOGIC	LOGIC	LOGIC
INDICATORS FUSING	EXTERNAL	YES	YES
ISOLATION	N/A	OPTICAL/POINT	OPTICAL/POINT

ANALOG I/O MODULES

VOLTAGES: SINGLE	N/A	1-5V,0-10V, -10 TO +10V	1-5V,0-10V, -10 TO +10V
VOLTAGES: DIFFERENTIAL	N/A	1-5V,0-10V, -10 TO +10V	1-5V,0-10V, -10 TO +10V
CURRENT: SINGLE	N/A	4-20MA	4-20MA
CURRENT: DIFFERENTIAL	N/A	4-20MA	4-20MA
SPECIAL PURPOSE INPUT	N/A	ISOLATED	ISOLATED
SPECIAL PURPOSE OUTPUT	N/A	ISOLATED	ISOLATED
INDICATORS:	N/A	RUN	RUN

SPECIAL I/O MODULES

SIMULATION	NO	YES	YES
ASCII	NO	NO	NO
AXIS POSITIONING	NO	NO	NO

I/O NODE DATA

TOTAL NODES/CHANNEL	N/A	8	8
TOTAL DISCRETE POINTS	40	1792	2048
MAXIMUM POINTS ASSIGNABLE INPUT	24	1792	2048
MAXIMUM POINTS ASSIGNABLE OUTPUT	16	1792	2048
MAXIMUM ANALOG ASSIGNABLE INPUT	N/A	112	2048
MAXIMUM ANALOG ASSIGNABLE OUTPUT	N/A	112	2048
DISTANCE FROM DROP	N/A	15000 FT	15000 FT

SQUARE D		SY/MAX MODEL 100	SY/MAX MODEL 300	SY/MAX MODEL 500

I/O CHANNEL DATA

TOTAL DISCRETE POINTS		20 OR 40 (2 VERSIONS)	128	>1000
MAXIMUM POINTS ASSIGNABLE	INPUT	24	128	>1000
	OUTPUT	16	128	>1000
MAXIMUM ANALOG ASSIGNABLE	INPUT	N/A	112	2008
	OUTPUT	N/A	112	2008
MAXIMUM LOCAL CHANNELS		N/A	2	0
MAXIMUM REMOTE CHANNELS		N/A	2	14
LOCAL CHANNEL DISTANCE		N/A	IN RACK	N/A
REMOTE CHANNEL DISTANCE		N/A	15000 FT	15000 FT
TYPE LOCAL CHANNEL CABLE		N/A	N/A	N/A
TYPE REMOTE CHANNEL CABLE		N/A	TWISTED PAIR BELDEN 9463	TWISTED PAIR BELDEN 9463
LOCAL CHANNEL COMMUNICATION		PARALLEL	PARALLEL	PARALLEL
REMOTE CHANNEL COMMUNICATION		N/A	SERIAL	SERIAL
LOCAL CHANNEL DATA RATE		225K HZ	225K HZ	225K HZ
REMOTE CHANNEL DATA RATE		N/A	31.25 HZ	31.25 HZ
FAULT MONITORING		YES	YES	YES

DATA NETWORK

	SY/MAX MODEL 100	SY/MAX MODEL 300	SY/MAX MODEL 500
TYPE	SLAVE-TO-MASTER	PEER-TO-PEER	PEER-TO-PEER
NUMBER OF NODES	N/A	200/NETWORK	200/NETWORK
MAXIMUM LENGTH	15000 FT	15000 FT	15000 FT
TYPE OF CABLE	TWIN AX	TWIN AX	TWIN AX
COMMUNICATION SPEED	UP TO 500K BAUD	UP TO 500K BAUD	UP TO 500K BAUD
COMPUTER INTERFACES	NO	STANDARD	STANDARD
SPECIAL INTERFACES	NO	YES	YES

SQUARE D	SY/MAX MODEL 100	SY/MAX MODEL 300	SY/MAX MODEL 500
PROGRAMMING TERMINAL			
SCREEN SIZE (DIAGONAL)	9"	9"	9"
WEIGHT	33 LBS	33 LBS	33 LBS
SIZE (W X H X D)	17X9.8X19.8	17X9.8X19.8	17X9.8X19.8
POWER REQUIREMENTS	120/60	120/60	120/60
DISPLAY SIZE (CONTACT ∗ ROWS)	11 X 7	11 X 7	11 X 7
GRAPHICS	YES	YES	YES
VIDEO OUT	YES	YES	YES
LADDER LISTER	YES	YES	YES
OFF-LINE PROGRAM	YES	YES	YES
REMOTE OPERATION	10000 FT	10000 FT	10000 FT
SCREEN MNEMONICS	YES	YES	YES
BUILT-IN TAPE	YES	YES	YES
KEYBOARD	STANDARD	STANDARD	STANDARD
HANDHELD	YES	YES	YES
DOUBLES AS A COMPUTER TERMINAL	YES	YES	YES

SQUARE D		SY/MAX -20	CLASS 8881

PROCESSOR

YEAR INTRODUCED - COMMERCIALLY		1978	1974
MEMORY SIZE TOTAL		2K	32K
MEMORY INCREMENTS		1,2K	4,8,16,32K
MEMORY TYPES		EAROM	CORE
VOLTAGE		93-132VAC	120/240VAC
FREQUENCY		48-63HZ	50/60HZ
HUMIDITY NON-CONDENSING	MIN	0%	0%
	MAX	95%	95%
TEMPERATURE ˚F	MIN	32˚	32˚
	MAX	140˚	140˚
USER MEMORY	MIN	1K	4K
	MAX	2K	64K
DATA MEMORY	MIN		224
	MAX	80 REGISTERS	991
USER CONFIGURABLE		NO	YES
WORD SIZE		16 BIT	16 BIT
WORDS/CONTACT - COIL		1	1
WORDS/TIMER - COUNTER		3	4/4
SCAN TIME - USER LOGIC		10 MS/K	5 MS/K
I/O SCAN TIME		10 MS/K	

SUPPORT

	SY/MAX -20	CLASS 8881
FACTORY APPLICATIONS ENGINEERS	YES	YES
DOCUMENTATION SERVICE	NO (FACTORY)	NO (FACTORY)
SERVICE CENTER (24 HRS.)	YES	YES
MAINTENANCE TRAINING	YES	YES
APPLICATION TRAINING	YES	YES
WARRANTY TERMS	1 YEAR	1 YEAR
DISTRIBUTION	DISTRIBUTORS	DISTRIBUTORS
MAINTENANCE CENTERS	YES	YES

SQUARE D	SY/MAX -20	CLASS 8881
INSTRUCTION SET		
CONTACTS	YES	YES
TRANSITIONAL	YES	YES
LATCHES	YES	YES
RETENTIVE COILS	YES	YES
SKIP/JUMP	NO	NO
SUBROUTINE	NO	NO
TIMERS	YES - 80	990
UP COUNTER	YES - 80	990
DOWN COUNTER	YES - 80	990
ADD/SUBTRACT	YES	999
MULTIPLY/DIVIDE	YES	SOFTWARE
>,=,<	YES	YES
FILE TRANSFER	YES	NO
LOGICAL	YES	YES
BIT ROTATE/SHIFT	YES	YES
FIFO/LIFO	FIFO	BOTH
SEARCH	YES	YES
DIAGNOSTIC	YES	YES
SEQUENCER	N/A	SOFTWARE
PID	YES	SOFTWARE
ASCII	N/A	YES
AXIS POSITIONING	N/A	FUTURE
MACHINE LANGUAGE	N/A	LADDER
BOOLEAN ALGEBRA	N/A	YES
HIGH LEVEL LANGUAGE	N/A	D-LOG

SQUARE D	SY/MAX -20	CLASS 8881

I/O STATISTICS

TOTAL DISCRETE POINTS AVAILABLE		511	2048
MAXIMUM POINTS ASSIGNABLE	INPUT	511	2048
	OUTPUT	511	2048
NUMBER OF CHANNELS		4	16 REMOTE CHANNELS
NUMBER OF NODES/CHANNEL		8	2
MODULE I/O POINT DENSITY		2	16
ISOLATED MODULE I/O DENSITY		2	8
TOTAL ANALOG CHANNELS AVAILABLE		16	16
MAXIMUM CHANNELS ASSIGNABLE	INPUT		16
	OUTPUT		16
I/O MAPPING		NO IMAGE TABLE REQUIRED	USER DEFINABLE
INTERRUPTS		NONE	NO
I/O LOGIC UPDATING	ASYNC		YES
	SYNC	YES	
ERROR CHECKING METHOD		CHECK SUM	CHECK SUM
MODULES PER CHANNEL		8	4 OR 8
APPLICABLE PROCESSORS		ALL	CLASS 8881
YEAR OF INTRODUCTION		1978	1974
I/O SYSTEM STRUCTURE		RACK	RACK

CONTROL NETWORK

TYPE	N/A	MASTER-SLAVE
NUMBER OF PC'S	N/A	200/NETWORK
MAXIMUM LENGTH	N/A	15000 FT
TYPE OF CABLE	N/A	TWIN AX
OPERATION W/PROCESSOR	N/A	ALL
DISCRETE SIGNALS TRANSFERRED	N/A	YES
DATA SIGNALS TRANSFERRED	N/A	YES
COMMUNICATION SPEED	N/A	UP TO 500K

SQUARE D		SY/MAX -20	CLASS 8881

DISCRETE I/O MODULES

AC VOLTAGES/CURRENT		6-120VAC/240VAC	0-240VAC
DC VOLTAGES/CURRENT		6-120VDC/ 240VDC/TTL	0-150VDC
ISOLATED AC VOLTAGES/CURRENT		6-120VAC/240VAC	120
ISOLATED DC VOLTAGES/CURRENT		6-120VDC/ 240VDC/TTL	120
CONTINUOUS OUTPUT RATING		1.1 AMPS/2 AMPS	2 AMP @ 25°C
COMMON POINTS/MODULE		1	16
OUTPUT FUSING CONFIGURATION		1/1	ON BOARD
INDICATORS	I/O POINTS	LOGIC/POWER	LOGIC/POWER
	FUSING	DIRECT FUSE	PULLER ON DELUXE
ISOLATION		OPTICAL	REED OR OPTICAL

ANALOG I/O MODULES

VOLTAGES: SINGLE	0-5/1-5/0-10	N/A
VOLTAGES: DIFFERENTIAL	-10 TO +10	1 TO 5,0 TO 10V
CURRENT: SINGLE	0-20,4-20 MA	N/A
CURRENT: DIFFERENTIAL	N/A	4-20,10-5, 0-10,0-20 MA
SPECIAL PURPOSE INPUT	N/A	N/A
SPECIAL PURPOSE OUTPUT	HIGH TEMP	N/A
INDICATORS:	N/A	N/A

SPECIAL I/O MODULES

SIMULATION	YES	YES
ASCII	N/A	NO
AXIS POSITIONING	N/A	NO

I/O NODE DATA

TOTAL NODES/CHANNEL		128	2
TOTAL DISCRETE POINTS		511	2048
MAXIMUM POINTS ASSIGNABLE	INPUT	511	2048
	OUTPUT	511	2048
MAXIMUM ANALOG ASSIGNABLE	INPUT	16	1024
	OUTPUT	16	512
DISTANCE FROM DROP		36 IN.	5000 FT

SQUARE D	SY/MAX -20	CLASS 8881

I/O CHANNEL DATA

		SY/MAX -20	CLASS 8881
TOTAL DISCRETE POINTS		511	2048
MAXIMUM POINTS ASSIGNABLE	INPUT	511	2048
	OUTPUT	511	2048
MAXIMUM ANALOG ASSIGNABLE	INPUT	16	1024
	OUTPUT	16	512
MAXIMUM LOCAL CHANNELS		4	16
MAXIMUM REMOTE CHANNELS		N/A	16
LOCAL CHANNEL DISTANCE		12 FT	135 FT
REMOTE CHANNEL DISTANCE		N/A	5000 FT
TYPE LOCAL CHANNEL CABLE		RIBBON	BELDON
TYPE REMOTE CHANNEL CABLE		N/A	BELDON TWISTED PAIR
LOCAL CHANNEL COMMUNICATION		PARALLEL	PARALLEL
REMOTE CHANNEL COMMUNICATION		N/A	SERIAL
LOCAL CHANNEL DATA RATE		MEMORY SIZE DEPENDENT	200K HZ
REMOTE CHANNEL DATA RATE		N/A	28.5 HZ
FAULT MONITORING		YES	YES

DATA NETWORK

	SY/MAX -20	CLASS 8881
TYPE	STAR	MASTER-SLAVE
NUMBER OF NODES	COMPUTER DEPENDENT	200/NETWORK
MAXIMUM LENGTH	2000 FT	15000 FT
TYPE OF CABLE	BELDEN 8723	TWIN AX
COMMUNICATION SPEED	19.2K	UP TO 500K BAUD
COMPUTER INTERFACES	YES	YES
SPECIAL INTERFACES	N/A	YES

SQUARE D	SY/MAX -20	CLASS 8881
PROGRAMMING TERMINAL		
SCREEN SIZE (DIAGONAL)	9"	9"
WEIGHT	33 LBS	33 LBS
SIZE (W X H X D)	17X9.8X19.8	17X9.8X19.8
POWER REQUIREMENTS	120/60	120/60
DISPLAY SIZE (CONTACT × ROWS)	10 X 6	11 X 7
GRAPHICS	YES	YES
VIDEO OUT	YES	YES
LADDER LISTER	YES	YES
OFF-LINE PROGRAM	YES	YES
REMOTE OPERATION	2000 FT	2000 FT
SCREEN MNEMONICS	YES	YES
BUILT-IN TAPE	YES	YES
KEYBOARD	STANDARD	STANDARD
HANDHELD	YES	NO
DOUBLES AS A COMPUTER TERMINAL	YES	YES
SPECIAL INTERFACE	CASSETTE	

Fig. 21.11. The Square D PC Family

TEXAS INSTRUMENTS		TI MODEL 530	TI MODEL 520	5TI

PROCESSOR

YEAR INTRODUCED		1982	1982	1974
MEMORY SIZE TOTAL		24K BYTES	4K BYTES	8K BYTES
MEMORY INCREMENTS		10,14,24K	4K	2,4,8K
MEMORY TYPES		RAM,EPROM	RAM,EPROM	RAM,EPROM
VOLTAGE		110,220VAC	110,220VAC	110,220VAC
FREQUENCY		50/60	50/60	50/60
HUMIDITY	MIN	0%	0%	%
	MAX	95% NC	95% NC	95% NC
TEMPERATURE	MIN	0°C	0°C	0°C
	MAX	60°C	60°C	60°C
USER MEMORY	MIN	4K	2K	2K
	MAX	16K	2K	8K
DATA MEMORY	MIN	6K	2K	N/A
	MAX	8K	2K	N/A
USER CONFIGURABLE		NO	NO	NO
WORD SIZE		16 BITS	16 BITS	16 BITS
WORDS/CONTACT - COIL		1 WORD/COIL	1 WORD/COIL	1 WORD/COIL
WORDS/TIMER - COUNTER		3 WORDS/TK	3 WORDS/TK	3 WORDS/TK
SCAN TIME - USER LOGIC		2 MILLISEC/K	2 MILLISEC/K	8.3 MS/K
I/O SCAN TIME		25 MICROSEC/IO	25 MICROSEC/IO	INCLUDED IN ABOVE

SUPPORT

	TI MODEL 530	TI MODEL 520	5TI
FACTORY APPLICATIONS ENGINEERS	YES	YES	YES
DOCUMENTATION SERVICE	NO	NO	NO
SERVICE CENTER (24 HRS.)	615-929-1141	615-929-1141	615-929-1141
MAINTENANCE TRAINING	YES	YES	YES
APPLICATION TRAINING	YES	YES	YES
WARRANTY TERMS	1 YEAR	1 YEAR	1 YEAR
DISTRIBUTION	DISTRIBUTORS	DISTRIBUTORS	DISTRIBUTORS
MAINTENANCE CENTERS	DISTRIBUTORS	DISTRIBUTORS	DISTRIBUTORS

TEXAS INSTRUMENTS	TI MODEL 530	TI MODEL 520	5TI

INSTRUCTION SET

	TI MODEL 530	TI MODEL 520	5TI
CONTACTS	YES	YES	YES
TRANSITIONAL	ON/OFF	ON/OFF	ON/OFF
LATCHES	YES	YES	YES
RETENTIVE COILS	YES	YES	YES
SKIP/JUMP	NO	NO	NO
SUBROUTINE	0	NO	NO
TIMERS	255	60	200
UP COUNTER	255	60	200
DOWN COUNTER	255	60	200
ADD/SUBTRACT	4×10^9	4×10^9	NO
MULTIPLY/DIVIDE	4×10^9	4×10^9	NO
>,=,<	YES	YES	NO
FILE TRANSFER	NO	NO	NO
LOGICAL			
BIT ROTATE/SHIFT	YES	YES	NO
FIFO/LIFO	FIFO	FIFO	NO
SEARCH	YES	YES	NO
DIAGNOSTIC	YES	YES	NO
SEQUENCER	YES	YES	YES
PID	SOFTWARE	SOFTWARE	
ASCII	YES	YES	NO
AXIS POSITIONING	YES	YES	NO
MACHINE LANGUAGE	NO	NO	YES
BOOLEAN ALGEBRA	YES	YES	YES
HIGH LEVEL LANGUAGE	BASIC	BASIC	NO

TEXAS INSTRUMENTS		TI MODEL 530	TI MODEL 520	5TI

I/O STATISTICS

		TI MODEL 530	TI MODEL 520	5TI
TOTAL DISCRETE POINTS AVAILABLE		1023	128	512
MAXIMUM POINTS ASSIGNABLE	INPUT	1023	128	512
	OUTPUT	1023	128	512
NUMBER OF CHANNELS		1	1	1
NUMBER OF NODES/CHANNEL		16	1	16
MODULE I/O POINT DENSITY		8	8	4
ISOLATED MODULE I/O DENSITY		6	6	N/A
TOTAL ANALOG CHANNELS AVAILABLE		512	64	A
MAXIMUM CHANNELS ASSIGNABLE	INPUT	12	64	/A
	OUTPUT	256	32	N/A
I/O MAPPING		USER DEFINABLE	USER DEFINABLE	USER DEFINABLE
INTERRUPTS		NONE	NONE	NONE
I/O LOGIC UPDATING	ASYNC			
	SYNC	YES	YES	YES
ERROR CHECKING METHOD		CRC-16	CRC-16	N/A
MODULES PER CHANNEL		128	16	128
APPLICABLE PROCESSORS		ALL	ALL	
YEAR OF INTRODUCTION		1982	1982	1974
I/O SYSTEM STRUCTURE		MODULE & BASE	MODULE & BASE	MODULE & BASE

CONTROL NETWORK

	TI MODEL 530	TI MODEL 520	5TI
TYPE	PEER-TO-PEER	PEER-TO-PEER	PEER-TO-PEER
NUMBER OF PC'S	65000	65000	65000
MAXIMUM LENGTH	50 MILES CATV	50 MILES CATV	50 MILES CATV
TYPE OF CABLE	COAX FIBER,CATV TWISTED PAIR	COAX FIBER,CATV TWISTED PAIR	COAX FIBER,CATV TWISTED PAIR
OPERATION W/PROCESSOR	YES	YES	YES
DISCRETE SIGNALS TRANSFERRED			
DATA SIGNALS TRANSFERRED			
COMMUNICATION SPEED	4M BAUD	4M BAUD	4M BAUD

TEXAS INSTRUMENTS		TI MODEL 530	TI 520	5TI

DISCRETE I/O MODULES

AC VOLTAGES/CURRENT		24,48,110,220	24,48,110,220	24,110,220
DC VOLTAGES/CURRENT		TTL 24,48	TTL 24,48	TTL 12,24,48
ISOLATED AC VOLTAGES/CURRENT		110	110	N/A
ISOLATED DC VOLTAGES/CURRENT		24	24	N/A
CONTINUOUS OUTPUT RATING		2 AMPS	2 AMPS	2 AMPS
COMMON POINTS/MODULE		4/8	4/8	4
OUTPUT FUSING CONFIGURATION		8/8	8/8	N/A
INDICATORS	I/O POINTS	LOGIC AND LOAD	LOGIC AND LOAD	LOGIC
	FUSING	YES	YES	NO
ISOLATION		2500V OPTICAL	2500V OPTICAL	3800V OPTICAL

ANALOG I/O MODULES

VOLTAGES: SINGLE			N/A
VOLTAGES: DIFFERENTIAL	0-5,0-10,±5,±10	0-5,0-10, 5, 10	N/A
CURRENT: SINGLE		-	N/A
CURRENT: DIFFERENTIAL	0-20MA	0-20MA	N/A
SPECIAL PURPOSE INPUT	N/A	N/A	N/A
SPECIAL PURPOSE OUTPUT	N/A	N/A	N/A
INDICATORS:	MODULE GOOD	MODULE GOOD	N/A

SPECIAL I/O MODULES

SIMULATION	YES	YES	YES
ASCII	YES	YES	NO
AXIS POSITIONING	YES	YES	NO

I/O NODE DATA

TOTAL NODES/CHANNEL		16	1	16
TOTAL DISCRETE POINTS		64	128	32
MAXIMUM POINTS ASSIGNABLE	INPUT	64	128	32
	OUTPUT	64	128	32
MAXIMUM ANALOG ASSIGNABLE	INPUT	32	64	N/A
	OUTPUT	16	32	N/A
DISTANCE FROM DROP		0	N/A	0

TEXAS INSTRUMENTS		TI MODEL 530	TI MODEL 520	5TI

I/O CHANNEL DATA

		TI MODEL 530	TI MODEL 520	5TI
TOTAL DISCRETE POINTS		1023	128	512
MAXIMUM POINTS ASSIGNABLE	INPUT	1023	128	512
	OUTPUT	1023	128	512
MAXIMUM ANALOG ASSIGNABLE	INPUT	512	64	N/A
	OUTPUT	256	32	N/A
MAXIMUM LOCAL CHANNELS		1	1	1
MAXIMUM REMOTE CHANNELS		1	0	0
LOCAL CHANNEL DISTANCE		1000 FT	0	64 FT
REMOTE CHANNEL DISTANCE		1000 FT	N/A	N/A
TYPE LOCAL CHANNEL CABLE		BELDEN 9271	N/A	N/A
TYPE REMOTE CHANNEL CABLE		BELDEN 9860	N/A	N/A
LOCAL CHANNEL COMMUNICATION		PARALLEL	PARALLEL	PARALLEL
REMOTE CHANNEL COMMUNICATION		SERIAL	N/A	N/A
LOCAL CHANNEL DATA RATE		2 MHZ	2 MHZ	N/A
REMOTE CHANNEL DATA RATE		2 MHZ	N/A	N/A
FAULT MONITORING		YES	YES	NO

DATA NETWORK

	TI MODEL 530	TI MODEL 520	5TI
TYPE	MASTER-SLAVE	MASTER-SLAVE	MASTER-SLAVE
NUMBER OF NODES	254	254	254
MAXIMUM LENGTH	25000 PT	25000 PT	25000 PT
TYPE OF CABLE	TWISTED PAIR	TWISTED PAIR	TWISTED PAIR
COMMUNICATION SPEED	112K BAUD	112K BAUD	112K BAUD
COMPUTER INTERFACES	YES	YES	YES
SPECIAL INTERFACES	NO	NO	NO

TEXAS INSTRUMENTS	TI MODEL 530	TI MODEL 520	5TI

PROGRAMMING TERMINAL

	TI MODEL 530	TI MODEL 520	5TI
SCREEN SIZE (DIAGONAL)	8"	8"	8"
WEIGHT	40 LBS	40 LBS	40 LBS
SIZE (W X H X D)	11.6X17.2X28.4	11.6X17.2X28.4	11.6X17.2X28.4
POWER REQUIREMENTS	110,220	110,220	110,220
DISPLAY SIZE (CONTACT · ROWS)	12 X 7	12 X 7	12 X 7
GRAPHICS	NO	NO	NO
VIDEO OUT	YES	YES	YES
LADDER LISTER	YES	YES	YES
OFF-LINE PROGRAM	YES	YES	YES
REMOTE OPERATION	YES	YES	YES
SCREEN MNEMONICS	YES	YES	YES
BUILT-IN TAPE	NO	NO	NO
KEYBOARD	MEMBRANE TYPEWRITER	MEMBRANE TYPEWRITER	MEMBRANE TYPEWRITER
HANDHELD	NO	NO	NO
DOUBLES AS A COMPUTER TERMINAL	YES	YES	YES

TEXAS INSTRUMENTS	*510*	*PM550*

PROCESSOR

YEAR INTRODUCED		1981	1979
MEMORY SIZE TOTAL		256 WORDS	28K WORDS
MEMORY INCREMENTS		N/A	1,2K
MEMORY TYPES		RAM,EPROM	RAM,EPROM,EEPROM
VOLTAGE		120/240	90-132VAC,24VDC, 180-268VAC
FREQUENCY		47-63	47-63HZ
HUMIDITY NON-CONDENSING	MIN	5%	5%
	MAX	95%	95%
TEMPERATURE OPERATING	MIN	0°C	0°C
	MAX	60°C	60°C
USER MEMORY	MIN	256 WORDS	2K WORDS
	MAX	256 WORDS	4K WORDS
DATA MEMORY	MIN	N/A	2K WORDS
	MAX	N/A	3K WORDS
USER CONFIGURABLE		NO	NO
WORD SIZE		1	16 BITS
WORDS/CONTACT - COIL		1	1
WORDS/TIMER - COUNTER		3	3
SCAN TIME - USER LOGIC		16.67 MSEC	40 MSEC
I/O SCAN TIME			1 MSEC

SUPPORT

FACTORY APPLICATIONS ENGINEERS	NO	YES
DOCUMENTATION SERVICE	NO	YES - LIMITED
SERVICE CENTER (24 HRS.)	YES	615-461-2501
MAINTENANCE TRAINING	NO	615-461-2500
APPLICATION TRAINING	YES	615-461-2500
WARRANTY TERMS	1 YEAR	1 YEAR
DISTRIBUTION	YES	WORLD WIDE
MAINTENANCE CENTERS	YES	WORLD WIDE

TEXAS INSTRUMENTS	510	PM550

INSTRUCTION SET

	510	PM550
CONTACTS	YES	YES
TRANSITIONAL	ON/OFF	YES
LATCHES	YES	USER BUILT
RETENTIVE COILS	YES	YES
SKIP/JUMP	JUMP	YES
SUBROUTINE	NO	NO
TIMERS	16	YES
UP COUNTER	16	YES
DOWN COUNTER	16	NO
ADD/SUBTRACT	N/A	YES
MULTIPLY/DIVIDE	N/A	YES
>,=,<	N/A	YES
FILE TRANSFER	N/A	YES
LOGICAL	S	YES
BIT ROTATE/SHIFT	N/A	NO
FIFO/LIFO	N/A	FIFO
SEARCH	YES	NO
DIAGNOSTIC	S	YES
SEQUENCER	YES	
PID	NO	YES
ASCII	NO	YES
AXIS POSITIONING	NO	O
MACHINE LANGUAGE	NO	NO
BOOLEAN ALGEBRA	YES	YES
HIGH LEVEL LANGUAGE	NO	NO

TEXAS INSTRUMENTS		*510*	*PM550*

I/O STATISTICS

TOTAL DISCRETE POINTS AVAILABLE		40	512
MAXIMUM POINTS ASSIGNABLE	INPUT	24	256
	OUTPUT	16	256
NUMBER OF CHANNELS		2	1
NUMBER OF NODES/CHANNEL		20	16
MODULE I/O POINT DENSITY		N/A	4
ISOLATED MODULE I/O DENSITY		N/A	128
TOTAL ANALOG CHANNELS AVAILABLE		/A	1
MAXIMUM CHANNELS ASSIGNABLE	INPUT	N/A	
	OUTPUT	N/A	
I/O MAPPING		NO	USER DEFINABLE
INTERRUPTS		N/A	N/A
I/O LOGIC UPDATING	ASYNC	N/A	WORDS MODULES
	SYNC	YES	DISCRETE I/O
ERROR CHECKING METHOD		N/A	N/A
MODULES PER CHANNEL		N/A	
APPLICABLE PROCESSORS		N/A	N/A
YEAR OF INTRODUCTION		1981	
I/O SYSTEM STRUCTURE		BUILT INTO CCU	

I/O NODE DATA

TOTAL NODES/CHANNEL		N/A	16
TOTAL DISCRETE POINTS		N/A	512
MAXIMUM POINTS ASSIGNABLE	INPUT	N/A	256
	OUTPUT	N/A	256
MAXIMUM ANALOG ASSIGNABLE	INPUT	N/A	USER MIX
	OUTPUT	N/A	USER MIX
DISTANCE FROM DROP		N/A	3 FT

SPECIAL I/O MODULES

SIMULATION	N/A	YES
ASCII	N/A	INTERNAL TO MACHINE
AXIS POSITIONING	N/A	N/A

TEXAS INSTRUMENTS		510	PM550

I/O CHANNEL DATA

TOTAL DISCRETE POINTS		N/A	512
MAXIMUM POINTS ASSIGNABLE	INPUT	N/A	256
	OUTPUT	N/A	256
MAXIMUM ANALOG ASSIGNABLE	INPUT	N/A	USER MIX
	OUTPUT	N/A	USER MIX
MAXIMUM LOCAL CHANNELS		N/A	1
MAXIMUM REMOTE CHANNELS		N/A	NONE
LOCAL CHANNEL DISTANCE		N/A	1,2,4 FT
REMOTE CHANNEL DISTANCE		N/A	N/A
TYPE LOCAL CHANNEL CABLE		N/A	STANDARD CABLE
TYPE REMOTE CHANNEL CABLE		N/A	N/A
LOCAL CHANNEL COMMUNICATION		N/A	SERIAL
REMOTE CHANNEL COMMUNICATION		N/A	N/A
LOCAL CHANNEL DATA RATE		N/A	1.25 MHZ
REMOTE CHANNEL DATA RATE		N/A	N/A
FAULT MONITORING		N/A	N/A

DATA NETWORK

TYPE	N/A	
NUMBER OF NODES	N/A	255
MAXIMUM LENGTH	N/A	10000 FT
TYPE OF CABLE	N/A	TWISTED PAIR
COMMUNICATION SPEED	N/A	115K BAUD
COMPUTER INTERFACES	N/A	YES
SPECIAL INTERFACES	N/A	

ANALOG I/O MODULES

VOLTAGES: SINGLE	N/A	N/A
VOLTAGES: DIFFERENTIAL	N/A	0-5,-5 TO +5, 0-10,-10-+10
CURRENT: SINGLE	N/A	0-20MA
CURRENT: DIFFERENTIAL	N/A	0-20MA
SPECIAL PURPOSE INPUT	N/A	THERMOCOUPLE RTD,PULSE
SPECIAL PURPOSE OUTPUT	N/A	
INDICATORS:	N/A	

TEXAS INSTRUMENTS	510	PM550

DISCRETE I/O MODULES

	510	PM550
AC VOLTAGES/CURRENT	85-132	24,120,240V
DC VOLTAGES/CURRENT	20.5-31	5,28,40-160V
ISOLATED AC VOLTAGES/CURRENT	N/A	3RD PARTY
ISOLATED DC VOLTAGES/CURRENT	N/A	3RD PARTY
CONTINUOUS OUTPUT RATING	2 AMPS	2 AMPS
COMMON POINTS/MODULE	N/A	4
OUTPUT FUSING CONFIGURATION	N/A	N/A
INDICATORS I/O POINTS	YES	YES
INDICATORS FUSING	N/A	N/A
ISOLATION	2500 OPTICAL	3800V

PROGRAMMING TERMINAL

	510	PM550
SCREEN SIZE (DIAGONAL)	9"	9"
WEIGHT	30 LBS	47 LBS
SIZE (W X H X D)	9.4X17.2X21.7	17X8.5X19
POWER REQUIREMENTS	137 WATTS	120/240
DISPLAY SIZE (CONTACT × ROWS)	11 X 7	11 X 7
GRAPHICS	YES	NO
VIDEO OUT	YES	NO
LADDER LISTER	YES	YES
OFF-LINE PROGRAM	NO	YES
REMOTE OPERATION	YES	
SCREEN MNEMONICS	NO	YES
BUILT-IN TAPE	NO	NO
KEYBOARD	MEMBRANE TYPEWRITER	QUERY FULL TRAVEL
HANDHELD	YES	NO
DOUBLES AS A COMPUTER TERMINAL	NO	NO
DISK DRIVE	N/A	YES
STATUS MONITORING	N/A	YES
I/O CROSS REFERENCE	N/A	YES

WESTINGHOUSE		HPPC 700	HPPC 900	HPPC 1100
PROCESSOR				
YEAR INTRODUCED				
MEMORY SIZE TOTAL		8K	2 1/2K	3 1/2K
MEMORY INCREMENTS		2K,4K,8K	1K,1 1/2K,2 1/2K	1/2K,1 1/2K, 2 1/2K,3 1/2K
MEMORY TYPES		CMOS	CMOS	CMOS
VOLTAGE (±)		120/240	120/240	120/240
FREQUENCY (±)		50/60 HZ	50/60 HZ	50/60 HZ
HUMIDITY	MIN	0%	0%	0%
	MAX	95%	95%	95%
TEMPERATURE	MIN	32'F	32'F	32'F
	MAX	140'F	140'F	140'F
USER MEMORY	MIN	2K	1K	1/2K
	MAX	8K	2 1/2K	3 1/2K
DATA MEMORY	MIN	2K	1K	1/2K
	MAX	8K	2 1/2K	3 1/2K
USER CONFIGURABLE		YES	YES	YES
WORD SIZE		16 BIT	16 BIT	16 BIT
WORDS/CONTACT - COIL		1	1	1
WORDS/TIMER - COUNTER		3	3	3
SCAN TIME - USER LOGIC		8 MS/1K	20 MS/1K	7 MSEC/WORD
I/O SCAN TIME				
SUPPORT				
FACTORY APPLICATIONS ENGINEERS		YES	YES	YES
DOCUMENTATION SERVICE		YES	YES	YES
SERVICE CENTER (24 HRS.)		YES	YES	YES
MAINTENANCE TRAINING		YES	YES	YES
APPLICATION TRAINING		YES	YES	YES
WARRANTY TERMS		1 YEAR	1 YEAR	1 YEAR
DISTRIBUTION		YES	YES	YES
MAINTENANCE CENTERS		YES	YES	YES

WESTINGHOUSE	HPPC 700	HPPC 900	HPPC 1100

INSTRUCTION SET

	HPPC 700	HPPC 900	HPPC 1100
CONTACTS	YES	YES	YES
TRANSITIONAL	YES	YES	YES
LATCHES	YES	YES	YES
RETENTIVE COILS	YES	YES	YES
SKIP/JUMP	YES	YES	YES
SUBROUTINE		YES	YES
TIMERS	YES	YES	YES
UP COUNTER	YES	YES	YES
DOWN COUNTER	YES	YES	YES
ADD/SUBTRACT	YES	YES	YES
MULTIPLY/DIVIDE	YES	YES	YES
>,=,<	YES	YES	YES
FILE TRANSFER	YES	YES	YES
LOGICAL	YES	YES	YES
BIT ROTATE/SHIFT	YES	YES	YES
FIFO/LIFO	YES	YES	YES
SEARCH	YES	YES	YES
DIAGNOSTIC	YES	YES	YES
SEQUENCER	YES	YES	YES
PID	YES		YES
ASCII	YES	YES	YES
AXIS POSITIONING	YES	YES	
MACHINE LANGUAGE			
BOOLEAN ALGEBRA			
HIGH LEVEL LANGUAGE	YES	YES	YES

WESTINGHOUSE		HPPC 700	HPPC 900	HPPC 1100

I/O STATISTICS

TOTAL DISCRETE POINTS AVAILABLE		512	256	128
MAXIMUM POINTS ASSIGNABLE	INPUT	256	128	64
	OUTPUT	256	128	64
NUMBER OF CHANNELS		4	2	1
NUMBER OF NODES/CHANNEL		2	1	1
MODULE I/O POINT DENSITY		4,8,16 POINT	4,8,16 POINT	8 POINT
ISOLATED MODULE I/O DENSITY		4,8 POINT	4,8 POINT	
TOTAL ANALOG CHANNELS AVAILABLE		64	32	16
MAXIMUM CHANNELS ASSIGNABLE	INPUT	32	16	8
	OUTPUT	32	16	8
I/O MAPPING		YES	YES	YES
INTERRUPTS		NO	NO	NO
I/O LOGIC UPDATING	ASYNC			
	SYNC			
ERROR CHECKING METHOD		LADDER CHECKSUM	LADDER CHECKSUM	LADDER CHECKSUM
MODULES PER CHANNEL		ANY MIX	ANY MIX	ANY MIX
APPLICABLE PROCESSORS		8 X 300	8085	Z80
YEAR OF INTRODUCTION		1983	1983	1983
I/O SYSTEM STRUCTURE		ADDRESSABLE	ADDRESSABLE	ADDRESSABLE

DISCRETE I/O MODULES

AC VOLTAGES/CURRENT		12,24,48,120,240	12,24,48,120,240	12-48,120,240
DC VOLTAGES/CURRENT		5,12,24,48, 125,240	5,12,24,48, 125,240	12-48,120,240
ISOLATED AC VOLTAGES/CURRENT		120	120	—
ISOLATED DC VOLTAGES/CURRENT		5,120	5,120	—
CONTINUOUS OUTPUT RATING		2AMP/120V	2AMP/120V	2AMP/120V
COMMON POINTS/MODULE		4,8,16	4,8,16	4,8,16
OUTPUT FUSING CONFIGURATION		4	4	4
INDICATORS	I/O POINTS	YES	YES	YES
	FUSING	YES	YES	YES
ISOLATION		OPTICAL	OPTICAL	OPTICAL

WESTINGHOUSE		*HPPC* 700	*HPPC* 900	*HPPC* 1100

I/O CHANNEL DATA

TOTAL DISCRETE POINTS				
MAXIMUM POINTS ASSIGNABLE	INPUT			
	OUTPUT			
MAXIMUM ANALOG ASSIGNABLE	INPUT	32	16	8
	OUTPUT	32	16	8
MAXIMUM LOCAL CHANNELS				
MAXIMUM REMOTE CHANNELS				
LOCAL CHANNEL DISTANCE		3 FT	3 FT	3 FT
REMOTE CHANNEL DISTANCE		5000 FT	5000 FT	—
TYPE LOCAL CHANNEL CABLE		MULTI-CONDUCTOR	MULTI-CONDUCTOR	MULTI-CONDUCTOR
TYPE REMOTE CHANNEL CABLE		TWISTED PAIR	TWISTED PAIR	—
LOCAL CHANNEL COMMUNICATION				
REMOTE CHANNEL COMMUNICATION				
LOCAL CHANNEL DATA RATE				
REMOTE CHANNEL DATA RATE				
FAULT MONITORING		YES	YES	YES

ANALOG I/O MODULES

	HPPC 700	*HPPC* 900	*HPPC* 1100
VOLTAGES: SINGLE	0-5,0-10,1-5, 0-1,-5-5,-10-10	0-5,0-10,1-5, 0-1,-5-5,-10-10	0-5,0-10
VOLTAGES: DIFFERENTIAL	0-5,0-10,1-5, 0-1,-5-5,-10-10	0-5,0-10,1-5, 0-1,-5-5,-10-10	0-5,0-10
CURRENT: SINGLE	0-20,1-5,10-50, 0-50,4-20	0-20,1-5,10-50, 0-50,4-20	4-20,0-20
CURRENT: DIFFERENTIAL	0-20,1-5,10-50, 0-50,4-20	0-20,1-5,10-50, 0-50,4-20	4-20,0-20
SPECIAL PURPOSE INPUT	TTL	TTL	TTL
SPECIAL PURPOSE OUTPUT	TTL	TTL	TTL
INDICATORS:	YES	YES	YES

SPECIAL I/O MODULES

	HPPC 700	*HPPC* 900	*HPPC* 1100
SIMULATION	YES	YES	YES
ASCII	NOT NEEDED	NOT NEEDED	NOT NEEDED
AXIS POSITIONING	YES	YES	NO

WESTINGHOUSE	HPPC 700	HPPC 900	HPPC 1100

PROGRAMMING TERMINAL

	HPPC 700	HPPC 900	HPPC 1100
SCREEN SIZE (DIAGONAL)	9"	9"	9"
WEIGHT	28 LBS	28 LBS	28 LBS
SIZE (W X H X D)	20X8.5X16	20X8.5X16	20X8.5X16
POWER REQUIREMENTS	120V,120 WATTS	120V,120 WATTS	120V,120 WATTS
DISPLAY SIZE (CONTACT × ROWS)	10 X 7	10 X 7	10 X 7
GRAPHICS	YES	YES	YES
VIDEO OUT	YES	YES	YES
LADDER LISTER	YES	YES	YES
OFF-LINE PROGRAM	YES	YES	YES
REMOTE OPERATION	YES	YES	YES
SCREEN MNEMONICS	YES	YES	YES
BUILT-IN TAPE	YES	YES	YES
KEYBOARD	ASCII	ASCII	ASCII
HANDHELD	AVAILABLE	AVAILABLE	AVAILABLE
DOUBLES AS A COMPUTER TERMINAL	YES	YES	YES

WESTINGHOUSE		HPPC 1500

PROCESSOR

YEAR INTRODUCED		1984
MEMORY SIZE TOTAL		128K
MEMORY INCREMENTS		8,16,32K
MEMORY TYPES		CMOS
VOLTAGE (±)		120/240
FREQUENCY (±)		50/60 HZ
HUMIDITY	MIN	0%
	MAX	95%
TEMPERATURE	MIN	32'F
	MAX	140'F
USER MEMORY	MIN	8K
	MAX	128K
DATA MEMORY	MIN	8K
	MAX	128K
USER CONFIGURABLE		YES
WORD SIZE		24 BIT
WORDS/CONTACT - COIL		1
WORDS/TIMER - COUNTER		3
SCAN TIME - USER LOGIC		
I/O SCAN TIME		

SUPPORT

FACTORY APPLICATIONS ENGINEERS	YES
DOCUMENTATION SERVICE	YES
SERVICE CENTER (24 HRS.)	YES
MAINTENANCE TRAINING	YES
APPLICATION TRAINING	YES
WARRANTY TERMS	1 YEAR
DISTRIBUTION	YES
MAINTENANCE CENTERS	YES

WESTINGHOUSE	HPPC 1500

INSTRUCTION SET

CONTACTS	YES
TRANSITIONAL	YES
LATCHES	YES
RETENTIVE COILS	YES
SKIP/JUMP	YES
SUBROUTINE	YES
TIMERS	YES
UP COUNTER	YES
DOWN COUNTER	YES
ADD/SUBTRACT	YES
MULTIPLY/DIVIDE	YES
>,=,<	YES
FILE TRANSFER	YES
LOGICAL	YES
BIT ROTATE/SHIFT	YES
FIFO/LIFO	YES
SEARCH	YES
DIAGNOSTIC	YES
SEQUENCER	YES
PID	YES
ASCII	YES
AXIS POSITIONING	YES
MACHINE LANGUAGE	
BOOLEAN ALGEBRA	
HIGH LEVEL LANGUAGE	YES

WESTINGHOUSE	HPPC 1500

I/O STATISTICS

TOTAL DISCRETE POINTS AVAILABLE		8192
MAXIMUM POINTS ASSIGNABLE	INPUT	4096
	OUTPUT	4096
NUMBER OF CHANNELS		32
NUMBER OF NODES/CHANNEL		1536
MODULE I/O POINT DENSITY		16
ISOLATED MODULE I/O DENSITY		8
TOTAL ANALOG CHANNELS AVAILABLE		
MAXIMUM CHANNELS ASSIGNABLE	INPUT	
	OUTPUT	
I/O MAPPING		YES
INTERRUPTS		NO
I/O LOGIC UPDATING	ASYNC	YES
	SYNC	N/A
ERROR CHECKING METHOD		CRC
MODULES PER CHANNEL		CONFIGURATION DEPENDENT
APPLICABLE PROCESSORS		8088
YEAR OF INTRODUCTION		1984
I/O SYSTEM STRUCTURE		DROP OR DAISY CHAIN

DISCRETE I/O MODULES

AC VOLTAGES/CURRENT		48-240
DC VOLTAGES/CURRENT		5-125
ISOLATED AC VOLTAGES/CURRENT		48-240
ISOLATED DC VOLTAGES/CURRENT		5-125
CONTINUOUS OUTPUT RATING		YES
COMMON POINTS/MODULE		4
OUTPUT FUSING CONFIGURATION		INDIVIDUALLY FUSED
INDICATORS	I/O POINTS	YES
	FUSING	YES
ISOLATION		OPTICAL

WESTINGHOUSE	HPPC 1500

I/O CHANNEL DATA

TOTAL DISCRETE POINTS		
MAXIMUM POINTS ASSIGNABLE	INPUT	
	OUTPUT	
MAXIMUM ANALOG ASSIGNABLE	INPUT	
	OUTPUT	
MAXIMUM LOCAL CHANNELS		32
MAXIMUM REMOTE CHANNELS		32
LOCAL CHANNEL DISTANCE		150 FT
REMOTE CHANNEL DISTANCE		10000FT
TYPE LOCAL CHANNEL CABLE		6 COND SHIELDED
TYPE REMOTE CHANNEL CABLE		CATV
LOCAL CHANNEL COMMUNICATION		SERIAL
REMOTE CHANNEL COMMUNICATION		SERIAL
LOCAL CHANNEL DATA RATE		826K
REMOTE CHANNEL DATA RATE		826K
FAULT MONITORING		YES

ANALOG I/O MODULES

VOLTAGES: SINGLE	5,10,I5,I10
VOLTAGES: DIFFERENTIAL	
CURRENT: SINGLE	4-20MA, 0-50MA,0-20MA
CURRENT: DIFFERENTIAL	
SPECIAL PURPOSE INPUT	HIGH SPEED PULSE INPUT
SPECIAL PURPOSE OUTPUT	
INDICATORS:	YES

SPECIAL I/O MODULES

SIMULATION	YES
ASCII	YES
AXIS POSITIONING	YES

WESTINGHOUSE	HPPC 1500

PROGRAMMING TERMINAL

SCREEN SIZE (DIAGONAL)	9"
WEIGHT	28 LBS
SIZE (W X H X D)	20X8.5X16
POWER REQUIREMENTS	120V/120 WATTS
DISPLAY SIZE (CONTACT × ROWS)	10 X 7
GRAPHICS	YES
VIDEO OUT	YES
LADDER LISTER	YES
OFF-LINE PROGRAM	YES
REMOTE OPERATION	YES
SCREEN MNEMONICS	YES
BUILT-IN TAPE	YES
KEYBOARD	ASCII
HANDHELD	
DOUBLES AS A COMPUTER TERMINAL	YES

Appendix

PC MANUFACTURERS

Adatek
Box 1339
Sandpoint, ID 83864
800/323-3343

Allen-Bradley
Systems Division
747 Alpha Drive
Highland Heights, OH 44143
216/449-6700

Amicon
245 West Roosevelt
West Chicago, IL 60185
312/293-0510

Apache Control Systems
1805 West County Road C
St. Paul, MN 55113
612/636-6530

Applied Systems
26401 Harper
St. Clair Shores, MI 48081
313/779-8700

August Systems
18277 Southwest Boones Ferry
Tigard, OR 97223
503/684-3550

Automation Systems
208 North 12th Street
Eldridge, IA 52748
319/285-8171

Barber-Coleman
Industrial Instruments Division
1354 Clifford
Box 2940
Loves Park, IL 61132
815/877-0241

Cincinnati Milacron
Electronic Systems Division
Lebanon, OH 45036
513/494-5275

Comptrol
9505 Midwest
Cleveland, OH 44125
216/587-5200

Cramer
Division of Conrac
Mill Rock
Old Saybrook, CT 06475
203/388-3574

Divelbiss
9776 Mount Gilead
Fredericktown, OH 43019
614/694-9015

Dynage
135 Prestige Park Circle
East Hartford, CT 06018
203/289-6831

Eagle Signal Controls
Division of Gulf & Western
736 Federal
Davenport, IA 52803
319/326-8100

Eaton Corporation
Industrial Automation Systems
9475 Center
Fenton, MI 48430
313/629-5361

EMICC
2871 Avondale Mill
Macon, GA 31206
912/784-5224

Encoder Products
1601B Dover
Sandpoint, ID 83864
208/263-8541

Entertron Industries
3857 Orangeport
Gasport, NY 14067
716/772-7216

General Electric
Box 8106
Charlottesville, VA 22906
804/978-5100

Giddings & Lewis Electronics
666 South Military
Fond du Lac, WI 54935
414/921-7100

Gould Incorporated
Modicon Division
Box 83 Shawsheen Village Station
Andover, MA 01810
617/475-4700

GTE/ECO
100 Endicott
Danvers, MA 01923
617/777-1900

GTE Sylvania
Electronic Control Operation
100 First
Waltham, MA 02254
617/890-9200

IDEC Systems and Controls
3050 Tasman
Santa Clara, CA 95050
408/988-7500

INCON
400 Matheson E
Unit 31
Mississauga, Ontario
Canada L4Z 1N8
416/273-4499

Industrial Solid State Controls
Box 934
435 West Philadelphia
York, PA 17405
717/848-1151

Klockner-Moeller
4 Strathmore
Natick, MA 01760
617/655-1910

Kuhnke
123 First
Atlantic Highlands, NJ 07716
201/291-3334

McGill Manufacturing
909 Lafayette
Valparaiso, IN 46383
219/465-2200

Minarik Electric
Box 54210
Los Angeles, CA 90054
213/624-8874

Mitsubishi Electric Sales America
3030 E. Victoria
Rancho Dominguez, CA 90221
213/537-7132

MTS Systems
Machine Controls Division
Box 24012
Minneapolis, MN 55424
612/937-4095

Omron Electronics
Control Components Division
650 Woodfield
Schaumburg, IL 60195
312/843-7900

Reliance Electric
4900 Lewis
Stone Mount, GA 30083
404/938-4888

Siemens-Allis
Central Freeway N
Wichita Falls, TX 76307
817/855-4980

Solid Controls
6925 Washington South
Minneapolis, MN 55435
612/941-6110

Square "D"
4041 North Richards, Box 472
Milwaukee, WI 53201
414/332-2000

Struthers-Dunn
Box 1327
Bettendorf, IA 52722
319/359-7501

Tenor
17020 West Rogers
New Berlin, WI 53151
414/782-3800

Texas Instruments
Drawer 1255
Johnson City, TN 37601
615/461-2500

TRIUS
2904 Corvin
Santa Clara, CA 95051
408/736-4141

Westinghouse Electric
Industry Electronics Division
Numa-Logic Department
1521 Avis
Madison Heights, MI 48071
313/588-1540

MAGAZINES

Control Engineering
875 Third Avenue
New York, NY 10022

Instruments and Control Systems
Chilton Company
Chilton Way
Radnor, PA 19089

Programmable Controls
InterTech Publications
25875 Jefferson Street
Saint Clair Shores, MI 48081

ORGANIZATIONS

Engineering Society of Detroit
100 Farnsworth
Detroit, MI 48202

Instrument Society of America
67 Alexander Drive
Post Office Box 12277
Research Triangle Park, NC 27709

Index